MODERN MANUFACTURING SYSTEMS

Trends and Developments

Frontiers of Mechanical and Industrial Engineering

MODERN MANUFACTURING SYSTEMS

Trends and Developments

Edited by

Rajiv Kumar Garg, PhD

Ravi Pratap Singh, PhD

Rajeev Trehan, PhD

Ramesh Singh, PhD

AAP | APPLE ACADEMIC PRESS

First edition published 2023

Apple Academic Press Inc.
1265 Goldenrod Circle, NE,
Palm Bay, FL 32905 USA

760 Laurentian Drive, Unit 19,
Burlington, ON L7N 0A4, CANADA

CRC Press
6000 Broken Sound Parkway NW,
Suite 300, Boca Raton, FL 33487-2742 USA

4 Park Square, Milton Park,
Abingdon, Oxon, OX14 4RN UK

Library and Archives Canada Cataloguing in Publication

Title: Modern manufacturing systems : trends and developments / edited by Rajiv Kumar Garg, PhD, Ravi Pratap Singh, PhD, Rajeev Trehan, PhD, Ramesh Singh, PhD.

Names: Garg, Rajiv Kumar, editor. | Singh, Ravi Pratap, editor. | Trehan, Rajeev, editor. | Singh, Ramesh (Lecturer in mechanical and materials engineering), editor.

Description: First edition. | Series statement: The frontiers of mechanical and industrial engineering book series | Includes bibliographical references and index.

Identifiers: Canadiana (print) 20220267936 | Canadiana (ebook) 20220267979 | ISBN 9781774910443 (hardcover) | ISBN 9781774910450 (softcover) | ISBN 9781003284024 (ebook)

Subjects: LCSH: Manufacturing processes—Case studies. | LCSH: Manufacturing processes—Technological innovations—Case studies. | LCSH: Machining—Case studies. | LCSH: Production engineering—Case studies. | LCGFT: Case studies.

Classification: LCC TS183 .M63 2023 | DDC 670—dc23

Library of Congress Cataloging-in-Publication Data

CIP data on file with US Library of Congress

ISBN: 978-1-77491-044-3 (hbk)
ISBN: 978-1-77491-045-0 (pbk)
ISBN: 978-1-00328-402-4 (ebk)

DEDICATION

This book is dedicated to the entire research society

ABOUT THE FRONTIERS OF MECHANICAL AND INDUSTRIAL ENGINEERING BOOK SERIES

EDITORS-IN-CHIEF

Dr. Prasenjit Chatterjee
Department of Mechanical Engineering, MCKV Institute of Engineering, West Bengal, India; E-Mail: prasenjit2007@gmail.com

Dr. Dilbagh Panchal
Department of Industrial and Production Engineering, Dr. B. R. Ambedkar National Institute of Technology (NIT) Jalandhar, Punjab, India E-Mail: panchald@nitj.ac.in

Description of the Series

Mechanical engineering applies the principles of mechanics and materials science for analysis, design, manufacturing, and maintenance of mechanical systems, whereas industrial engineering involves figuring out how to make or do things better with minimum resource consumption. Mechanical and industrial engineering utilize a deep foundational understanding of mathematics, physics, and analysis to develop machines and systems. The former specializes in thermodynamics, combustion, and electricity to make complex machines. The latter focuses on designing workflow and making production efficient.

This series aims to report the cutting-edge developments to present novel innovations in the field of mechanical and industrial engineering. Volumes published in this series aim to embrace all aspects, subfields, and new challenges of mechanical and industrial engineering and to be of interest and use to all those concerned with research in various related domains of mechanical and industrial engineering disciplines. The series aims to provide comprehensive value-added resources to the body of the literature to serve as essential references for years ahead. Features of the volumes under this series will explore recent trends, model extensions, developments, solution to real-time problems, and case studies.

The coverage of the series is committed to provide a diverse yet rigorous research experience that builds on a solid knowledge of science and engineering fundamentals and application-oriented knowledge of mechanical and industrial engineering practices for efficient functioning in today's global multidisciplinary rapidly changing work environment. It includes all new theoretical and experimental findings in mechanical, industrial engineering, and allied engineering fields.

Topics and Themes:

The primary endeavor of this series is to introduce and explore contemporary research developments in a variety of rapidly growing research areas. The volumes will deal with the following topics but not limited to:

- Automotive engineering
- Aerospace technology
- Big data in mechanical / manufacturing / industrial engineering
- Computational methods for optimizing manufacturing technology, robotics, mechatronics
- Decision science applications in mechanical, manufacturing, and industrial engineering
- Design and optimization of mechanical components
- Dynamical systems, control
- Engineering design
- Engineering thermodynamics, heat, and mass transfer
- Evolutionary computation
- Finite element method (FEM) modeling/simulation
- Fluid mechanics
- Fuzzy logic and neuro-fuzzy systems for relevant engineering applications
- Green and sustainable engineering techniques for modern manufacturing
- Innovative approaches for modeling relevant applications and real-life problems
- Life cycle engineering
- Machinery and machine elements
- Manufacturing
- Mathematical concepts and applications in mechanical and industrial engineering
- Materials engineering
- Mechanical structures and stress analysis
- MEMS
- Modeling in engineering applications
- Nanotechnology and microengineering

- Non-conventional machining in modern manufacturing systems
- Precision engineering, instrumentation and measurement
- Numerical simulations
- Soft computing techniques
- Sustainability in mechanical, manufacturing, and industrial engineering
- Theoretical and applied mechanics
- Tribology and surface technology

The series aims to serve as a valuable resource for undergraduate, postgraduate, doctoral students, researchers, academicians, and industry professionals in mechanical, manufacturing, industrial engineering, product design, management, and more.

BOOKS IN THE SERIES

Optimization Methods for Engineering Problems
Editors: Dilbagh Panchal, PhD, Prasenjit Chatterjee, PhD, Mohit Tyagi, PhD, and Ravi Pratap Singh, PhD

Modern Manufacturing Systems: Trends and Developments
Editors: Rajiv Kumar Garg, PhD, Ravi Pratap Singh, PhD, Rajeev Trehan, PhD, and Ramesh Singh, PhD

Multi-Criteria Decision-Making Methods in Manufacturing Environments: Models and Applications
Shankar Chakraborty, PhD, Prasenjit Chatterjee, PhD, and Partha Protim Das, PhD

ABOUT THE EDITORS

Rajiv Kumar Garg, PhD, is Professor (HAG) in the Industrial and Production Engineering Department at Dr. B. R. Ambedkar National Institute of Technology Jalandhar, India. His areas of interest are nontraditional machining, supply chain management, industrial management, and systems dynamics. He has over 28 years of industry and teaching experience. He has guided 47 MTech and 13 PhD candidates for pursuing their PhD degree. He has over 138 publications in international and national journals and conference proceedings. Dr. Garg is a reviewer for several international journals, associate editor of the *Global Journal of Flexible Systems Management,* and chief editor of the *International Journal of Machining and Machinability of Materials.* Dr. Garg has organized four international conferences at NIT Jalandhar as chairman and convener. He holds a BSc Engineering (Hons.) degree in Mechanical Engineering from the Regional Engineering College, Kurukshetra, and M. degree (Hons.) in Industrial Engineering, and a PhD from Thapar Institute of Engineering and Technology Patiala, India.

Ravi Pratap Singh, PhD, is Assistant Professor in the Department of Industrial and Production Engineering at Dr. B. R. Ambedkar National Institute of Technology, Jalandhar, Punjab, India. Prior to this, he has also served in the Department of Technical Education, Uttar Pradesh, India, for about two years. He is a life member of the Indian Institution of Industrial Engineering, Mumbai, India, and Science and Engineering Institute, Los Angeles, USA. Dr. Singh is also a senior member of the Indian Society of Mechanical Engineers, Chennai, India. He has published more than 75 research articles and papers at national and international conferences. His broad areas of research include conventional and advanced manufacturing processes, modeling and optimization methods, empirical modeling of operational systems, advanced engineering materials, and micro-structure analysis. He is also a guest editor of the *World Journal of Engineering* and the *International Journal of Six Sigma and Competitive Advantage*. He has received several awards for his work, including a Best Researcher Award from the International Scientist Awards on Engineering, Science, and Medicine, New Delhi, India. He is also serving as an ambassador of Bentham Open, Sharjah, UAE as well as a member of the Visor Academic Committee, Singapore. Dr. Singh has received his PhD degree in the area of Rotary Ultrasonic Machining of Advanced Engineering Materials and MTech in Mechanical Engineering (Industrial and Production Engineering) from the Department of Mechanical Engineering, National Institute of Technology, Kurukshetra, Haryana, India.

Rajeev Trehan, PhD, is Associate Professor and Head of the Department in Industrial and Production Engineering at Dr. B. R. Ambedkar National Institute of Technology Jalandhar, India. His areas of interest are quality management, supply chain management, industrial management, and systems dynamics. He has over 28 years of industry and teaching experience. He has published his work in international and national journals and proceedings of international conferences to his credit. Dr Trehan has organized several international conferences at NIT Jalandhar as chairman and convener. Dr. Trehan holds an MTech and PhD in Industrial Engineering from Dr. B. R. Ambedkar National Institute of Technology Jalandhar, India.

Ramesh Singh, PhD, is an outstanding scholar and a Fellow of the Academy of Sciences, Malaysia. He has contributed significantly to the development and enhancement in engineering, science, and technology in Malaysia, particularly in the field of engineering education and ceramics engineering. He is a Fellow of five prominent professional societies, that is, The Institution of Mechanical Engineers (United Kingdom), The Institution of Engineers (Australia), Institution of Engineers (Malaysia),

Institute of Materials (Malaysia), and Malaysian Scientific Association. He is on the international editorial board of *Ceramics International* (a Web of Science, Tier 1, Elsevier Publication), the editor-in-chief of the *International Journal of Mechanical and Materials Engineering* (an ESCI-listed Journal), and editorial board member of *Advances in Materials Research, Journal of The Institute of Materials Malaysia,* and *Journal of Industrial Technology.* Besides research excellence, Prof. Singh is also a distinguished professor in mechanical and materials engineering in Malaysia. He has made over 350 scientific publications, including nearly 240 peer-reviewed research papers. He has filed numerous patents for his inventions and received many accolades as a distinguished scholar. He has secured many research grants and completed the supervision of 74 postgraduate students (i.e., 31 PhD and 43 master, all by research mode). He was a recipient of the prestigious IEM Young Engineer Award and won more than 75 awards at international/national exhibitions for his research and inventions. He was recognized as one of the top 12 Malaysia Engineer Stars by the Institution of Engineers Malaysia in 2011. He was also a recipient of the prestigious Science & Technology Award 2014, which was awarded by the Malaysia Toray Science Foundation (MTSF) in recognition for his outstanding research achievement in Malaysia. Dr. Singh was named Fellow of the Academy of Sciences Malaysia for his excellent contribution in the field of engineering education and the advancement of science and technology in the country. Dr. Singh received his PhD from the University of Sunderland, the United Kingdom in 1997, with a specialization in advanced ceramics for mechanical application.

CONTENTS

CONTRIBUTORS

Naseem Ahamad
Department of Mechanical Engineering, Jamia Millia Islamia, New Delhi, India;
E-mail: naseem.ahamad9412@gmail.com

S. K. Ajmani
Research and Development, Tata Steel, Jamshedpur, Jharkhand, India

Pradyumn K. Arya
Discipline of Mechanical Engineering, Indian Institute of Technology Indore, Madhya Pradesh, India

M. K. Banerjee
Metallurgical and Materials Engineering, Jaipur, India

Rajdeep Banik
Department of Metallurgical and Materials Engineering, National Institute of Technology, Durgapur, India

Nikhil Bari
Department of Mechanical Engineering, Sardar Vallabhabhai National Institute of Technology Surat, India; E-mail: nikhilbari1665@gmail.com

R. M. Belokar
Production and Industrial Engineering Department, Punjab Engineering College
(Deemed To Be University), Chandigarh 160012, India

Shailender Singh Bhadauria
Department of Industrial and Production Engineering, Dr. B R Ambedkar National Institute of Technology, Jalandhar, Punjab 144011, India

Bhuvnesh Bhardwaj
Department of Mechanical Engineering, Jaipur Engineering College & Research Centre, Jaipur 302022, Rajasthan, India

Neeraj Kumar Bhoi
Department of Mechanical Engineering, PDPM Indian Institute of Information Technology Design and Manufacturing, Jabalpur 482005, Madhya Pradesh, India; E-mail: neerajbitd@gmail.com

Ravi Butola
Department of Mechanical, Production & Industrial and Automobile Engineering,
Delhi Technological University, Delhi 110042, India; E-mail: ravibutola33855@gmail.com

Bhavesh Chaudhary
Discipline of Mechanical Engineering, Indian Institute of Technology Indore, Madhya Pradesh, India

Shailendra Chauhan
Department of Industrial Production Engineering, Dr. B.R. Ambedkar National Institute of Technology, Jalandhar, Punjab, India

Ashok Kumar Dargar
Department of Mechanical-Mechatronics Engineering, The LNM Institute of Information Technology, Jaipur, 302031, India; E-mail: ashok.dargar@lnmiit.ac.in

Sudhansu Ranjan Das
Production Engineering Department, Veer Surendra Sai University of Technology, Burla, Sambalpur 768018, Odisha, India

Kishore Debnath
Department of Mechanical Engineering, National Institute of Technology Meghalaya, Shillong 793003, India

Debabrata Dhupal
Production Engineering Department, Veer Surendra Sai University of Technology, Burla, Sambalpur 768018, Odisha, India

Akshay Dvivedi
Department of Mechanical and Industrial Engineering, IIT Roorkee, Roorkee, India

Kuldeep Dwivedi
Department of Mechanical Engineering, Maulana Azad National Institute of Technology, Bhopal, India; E-mail: kshudwivedi@gmail.com

Ravi K. Dwivedi
Department of Mechanical Engineering, Maulana Azad National Institute of Technology, Bhopal, India

Kapil Kumar Goyal
Department of Industrial and Production Engineering, Dr B R Ambedkar National Institute of Technology, Jalandhar, India

Ashutosh Kumar Gupta
Department of Industrial & Production Engineering, Dr. B R Ambedkar National Institute of Technology Jalandhar, Punjab, India

Pallav Gupta
Department of Mechanical Engineering, A.S.E.T., Amity University Uttar Pradesh, Noida, India

Abid Haleem
Department of Mechanical Engineering, Jamia Millia Islamia, New Delhi, India

R. S. Jadoun
Department of Industrial and Production Engineering, College of Technology, Pantnagar, Uttarakhand, India

Neelesh Kumar Jain
Discipline of Mechanical Engineering, Indian Institute of Technology Indore, Madhya Pradesh, India

Prashant K. Jain
Department of Mechanical Engineering, PDPM Indian Institute of Information Technology Design and Manufacturing Jabalpur, India

Sunil Kumar Jatav
Metallurgical and Materials Engineering, Jaipur, India; E-mail: sunilkjatavnitjsr@gmail.com

Mohd Javaid
Department of Mechanical Engineering, Jamia Millia Islamia, New Delhi, India

M. Jayaprakash
Discipline of Metallurgy Engineering and Materials science, Indian Institute of Technology Indore, Madhya Pradesh, India

Neeraj Kamboj
Department of Mechanical Engineering, National Institute of Technology, Kurukshetra, Haryana, India

Luckshaya Kem
Department of Mechanical, Production & Industrial and Automobile Engineering, Delhi Technological University, Delhi 110042, India

Faim Khan
Department of Mechanical, Production & Industrial and Automobile Engineering, Delhi Technological University, Delhi 110042, India

Akshat Srivastava Kulshrestha
Department of Mechanical-Mechatronics Engineering, The LNM Institute of Information Technology, Jaipur 302031, India

Ankesh Kumar
Department of Civil Engineering, National Institute of Technology Kurukshetra, India

B. S. Ajay Kumar
Department of Mechanical Engineering, KS Institute of Technology, Bengaluru, India

K. S. Anil Kumar
Department of Mechanical Engineering, J.S.S. Academy of Technical Education, Bangalore 560060, India; E-mail: anilkumar_aks@rediffmail.com

Krishna Kumar
Metallurgical and Materials Engineering, Jaipur, India

Pawan Kumar
Department of Civil Engineering, National Institute of Technology Kurukshetra, India

Pradeep Kumar
Department of Mechanical and Industrial Engineering, IIT Roorkee, Roorkee, India

Rakesh Kumar
Department of Industrial and Production Engineering, Dr B R Ambedkar National Institute of Technology, Jalandhar, India

Rajan Kumar
Department of Mechanical Engineering, Maulana Azad National Institute of Technology, Bhopal, India

Shailendra Kumar
Department of Mechanical Engineering, Sardar Vallabhabhai National Institute of Technology Surat, India

Subodh Kumar
Department of Mechanical Engineering, DIT University, Dehradun 248009, India; E-mail: subodh.kumar@dituniversity.edu.in

M. Vijay Kumar
Department of Industrial Engineering and Management, JSS Academy of Technical Education, Bengaluru, India

Pranita Madalli
Department of Industrial Engineering and Management, JSS Academy of Technical Education, Bengaluru, India

H. C. Madhu
Department of Mechanical Engineering, Siddaganga Institute of Technology, Tumakuru 572103, India

D. Prem Madhur
Department of Metallurgical and Materials Engineering, National Institute of Technology, Durgapur, India

Sourav Majumdar
Department of Mechanical Engineering, PDPM Indian Institute of Information Technology Design and Manufacturing, Jabalpur 482005, Madhya Pradesh, India

Manab Mallik
Department of Metallurgical and Materials Engineering, National Institute of Technology, Durgapur, India

Anand Kumar Mandal
Department of Industrial and Production Engineering, Dr B R Ambedkar National Institute of Technology, Jalandhar, India
Department of Industrial and Production Engineering, College of Technology, Pantnagar, Uttarakhand, India; E-mail: anandkm.ip.19@nitj.ac.in

Pooja Maurya
Department of Metallurgical and Materials Engineering, National Institute of Technology, Durgapur, India

Richa Mishra
Department of Mechanical Engineering, PDPM Indian Institute of Information Technology Design and Manufacturing, Jabalpur 482005, Madhya Pradesh, India

Rahul Mitra
Department of Metallurgical and Materials Engineering, Indian Institute of Technology, Kharagpur, India

Aas Mohammad
Department of Mechanical Engineering, Jamia Millia Islamia, New Delhi, India

Manas Kumar Mondal
Department of Metallurgical and Materials Engineering, National Institute of Technology, Durgapur, India

Yatan Nagpal
Amity University Uttar Pradesh, Noida, India; E-mail: yatannagpal@gmail.com

Madhusudan Painuly
Department of Industrial and Production Engineering, Dr. B. R Ambedkar National Institute of Technology, Jalandhar, India

Deepak Kumar Patel
Department of Mechanical Engineering, PDPM Indian Institute of Information Technology Design and Manufacturing, Jabalpur 482005, Madhya Pradesh, India

Tanay Rudra Paul
Department of Metallurgical and Materials Engineering, National Institute of Technology, Durgapur, India

Subhadip Pradhan
Production Engineering Department, Veer Surendra Sai University of Technology, Burla, Sambalpur 768018, Odisha, India; E-mail: subha.mid94@gmail.com

Sourajit Pramanik
Department of Metallurgical and Materials Engineering, National Institute of Technology, Durgapur, India

S. Vandan Raj
Department of Metallurgical and Materials Engineering, National Institute of Technology, Durgapur, India

H. Rajneesh
Department of Mechanical Engineering, Indian Institute of Science, Bangalore 560012, India

Ramver
Department of Mechanical and Industrial Engineering, IIT Roorkee, Roorkee, India

Mainak Saha
Department of Metallurgical and Materials Engineering, National Institute of Technology, Durgapur, India; E-mail: mainaksaha1995@gmail.com

Sushreesudha Sahu
Production Engineering Department, Veer Surendra Sai University of Technology, Burla, Sambalpur 768018, Odisha, India

Babita Saini
Department of Civil Engineering, National Institute of Technology Kurukshetra, India

Sandip Sarkar
Research and Development, Tata Steel, Jamshedpur, Jharkhand, India

Pardeep Sharma
Department of Mechanical Engineering, Panipat Institute of Engineering and Technology, Panipat, India

Rohit Sharma
Department of Mechanical Engineering, Amity University Uttar Pradesh, Noida, India

Varun Sharma
Department of Industrial and Production Engineering, Dr. B R Ambedkar National Institute of Technology, Jalandhar 144011, India

Abhishek Singh
Department of Mechanical Engineering, National Institute of Technology Patna, Patna 800005, India

Harpreet Singh
Department of Mechanical Engineering, PDPM Indian Institute of Information Technology Design and Manufacturing, Jabalpur 482005, Madhya Pradesh, India; E-mail: hps.dme@gmail.com

Maninder Singh
Department of Civil Engineering, National Institute of Technology Kurukshetra, India

Nikhilesh Singh
Production and Industrial Engineering Department, Punjab Engineering College (Deemed to be University), Chandigarh 160012, India; E-mail: nikhileshsingh.phdprod@pec.edu.in

Ravi Pratap Singh
Department of Industrial Production Engineering, Dr. B.R. Ambedkar National Institute of Technology, Jalandhar, Punjab, India; E-mail: singhrp@nitj.ac.in

Satnam Singh
Department of Mechanical Engineering, National Institute of Technology, Kurukshetra 136119, India

Simrat Singh
Department of Metallurgical and Materials Engineering, National Institute of Technology, Durgapur, India

Vikas Sisodia
Department of Manufacturing Engineering & Industrial Management, College of Engineering, Pune 411005, Maharashtra, India

Rahul Soni
Department of Industrial and Production Engineering, Dr. B. R. Ambedkar National Institute of Technology Jalandhar, Punjab 144011, India

Rajiv Suman
Department of Industrial & Production Engineering, G. B. Pant University of Agriculture & Technology, Pantnagar, Uttarakhand, India

Ajaykumar Thakur
Sanjivani College of Engineering, SPPU Pune University, Kopargaon, Maharashtra 423603, India

Lalit Thakur
Department of Mechanical Engineering, National Institute of Technology, Kurukshetra, Haryana, India

Mahendra Singh Tomar
Department of Mechanical Engineering, National Institute of Technology, Kurukshetra 136119, India;
E-mail mahendra_31806209@nitkkr.ac.in

Rajeev Trehan
Department of Industrial Production Engineering, Dr. B.R. Ambedkar National Institute of Technology, Jalandhar, Punjab, India

Sourav Sanket Tripathy
Production Engineering Department, Veer Surendra Sai University of Technology, Burla, Sambalpur 768018, Odisha, India

Lakshay Tyagi
Department of Mechanical, Production & Industrial and Automobile Engineering, Delhi Technological University, Delhi 110042, India

Deepak Rajendra Unune
Department of Mechanical-Mechatronics Engineering, The LNM Institute of Information Technology, Jaipur, 302031, India; E-mail: ashok.dargar@lnmiit.ac.in
Department of Materials Science and Engineering, INSIGNEO Institute for in silico Medicine, University of Sheffield, Sheffield, S1 3JD, United Kingdom

Anuj Vats
Department of Mechanical and Industrial Engineering, IIT Roorkee, Roorkee, India;
E-mail: vatsanuj26@gmail.com

Kiran Wakchaure
Sanjivani College of Engineering, SPPU Pune University, Kopargaon, Maharashtra 423603, India;
E-mail: kiran.wakchaure@yahoo.com

R. S. Walia
Production and Industrial Engineering Department, Punjab Engineering College (Deemed To Be University), Chandigarh 160012, India

Harsh Yadav
Department of Civil Engineering, National Institute of Technology Kurukshetra, India

Santosh Kumar Yadav
Department of Mechanical Engineering, National Institute of Technology Patna, Patna 800005, India;
E-mail: santosh.me18@nitp.ac.in

N. Yuvaraj
Department of Mechanical, Production & Industrial and Automobile Engineering, Delhi Technological University, Delhi 110042, India

ABBREVIATIONS

AA	aluminium alloy
ABS	acrylonitrile butadiene styrene
ADOC	average diametral overcut
Al-MMC	aluminium metal matrix composites
ANOVA	analysis of variance
ANNs	artificial neural networks
APS	atmospheric plasma spraying
AS	advancing side
ASS	austenitic stainless steels
ASTM	American Society for Testing of Materials
AWJC	abrasive water jet cutting
AWJMM	abrasive water jet micromachining
AWSJM	abrasive water suspension jet machining
BBD	Box-Behken's design
BSE	backscattered electron
CA	coarse aggregates
CAD	computer-aided design
CAM	computer-aided manufacturing
CBN	cubic boron nitride
CCW	counter clockwise
CFD	computational fluid dynamics
CFSW	conventional in air fraction stir welding
CNC	computer numeric controlled
CRF	central research facility
CS	compressive strength
CTE	coefficient of linear thermal expansion
CW	clockwise
DC	direct current
DLEPR	double loop electrochemical potentiokinetic reactivation
DNS	direction numerical simulations
DOE	design of experiment
DOS	degree of sensitization
ECC	engineered cementitious composite
ECJM	electrochemical jet machining process

ECM	electrochemical machining
EDCG	electro discharge cut-off grinding
EDFG	electro discharge face grinding
EDM	electro discharge machining
EDS	energy dispersive x-ray analyzer
EDSG	electro discharge surface grinding
EMBR	electromagnetic braking
EMLA	electromagnetic level accelerator
EMLS	electromagnetic level stabilizer
EMRS	electromagnetic rotary stirring
ETL	exposed tool lengths
FA	fine aggregates
FDM	fused deposition method
FEM	finite element method
FRC	fiber reinforced concrete
FRCC	fiber reinforced cementitious composite
FRP	fiber-reinforced plastic
FSP	friction stir processing
FSW	friction stir welding
GA	generic algorithm
GRA	gray relational analysis
GRC	generation of grey relation coefficient
GRG	grey relation grade
HAJMM	hot abrasive jet micromachining
HAZ	heat-affected zone
HDS	hot die steel
HMMC	hybrid metal matrix composite
HPC	high performance concrete
HVOF	high velocity oxy-fuel
HVPS	high velocity plasma spraying
IEG	interelectrode gap
IoT	Internet of Things
ISF	incremental sheet forming
LES	large eddy simulation
MHD	magnetohydrodynamic
MHH	microwave hybrid heating
MMCs	metal matrix composites
MMNC	metal matrix nano composites
MMT	montmorillonite

MRR	material removal rate
MSPIF	multistage single point incremental forming
MWC	microwave casting
NC	normal concrete
NCM	non-conventional machining
NP	nanoparticle
NTM	non-traditional manufacturing
OM	optical microscopy
PC	Portland cement
PDMS	polydimethylsiloxane
PESM	pulse electrochemical sawing machining
PMEDM	powder mixed EDM
PSO	particle swarm optimization
RANS	Reynolds-average Navier–Stokes
RPC	reactive powder concrete
RS	retreating side
RSM	response surface methodology
RSW	resistance spot welding
RTM	resin transfer molding
RWR	relative electrode wear ratio
SA	sustainability assessment
SC	straight cylindrical
SCM	supplementary cementitious material
SE	secondary electron
SEM	scanning electron microscopy
SEM-EDS	scanning electron microscopy-energy dispersive spectroscopy
SEN	submerged entry nozzle
SOD	stand-off-distance
SPIF	single point incremental forming
SRB	solid rocket booster
SS	straight square
SS	stainless steels
STC	straight threaded cylindrical
STED	shaped tube electrolytic drilling
TC	truncated cone
TC	tapered cylindrical
TGRA	Taguchi-based grey relational analysis
TIG	tungsten inert gas

TMAZ	thermo-mechanically affected zone
TMEMM	through-mask electrochemical micromachining
TPD	tool plunge depth
TPS	thermal protection systems
TRS	tool rotation speed
TTC	taper threaded cylindrical
TWI	the welding Institute
TWR	tool wear rate
TWS	tool welding speed
UFSW	underwater friction stir welding
URAN	unsteady Reynolds-averaged Navier–Stokes
UTM	universal testing machine
UTS	ultimate tensile strength
VARTM	vacuum-assisted resin transfer molding
VR	virtual reality
WECMM	wire electrochemical micromachining
WN	weld nugget
WNZ	weld nugget zone
XRD	x-ray diffraction
μEDM	micro-electrical discharge machining

PREFACE

Observation and analysis of manufacturing systems and their various relevant advancements are becoming an issue of paramount importance in the trendy competitive culture of industrial applications. The studies related to the fabrication of typical engineering materials, favorable processing methods, effective overall process control, and optimization are some of the spheres that reflect the importance of learning these modern manufacturing aspects at a deeper level.

This book explores various research works carried out in production engineering and its allied technologies. It provides deep insights to readers about the current scenarios and future advancements in production engineering under the context of optimum value. Optimization and its applicability in various areas of production and manufacturing system engineering are discussed vividly. These include selection of designing parameters and decisions related to conditions of optimum process/operation parameters, behavior of response variables, 3D printing, advanced manufacturing processes, robotics and automation, hybrid machining, surface engineering, welding technology, casting processes, microstructure analysis, materials science, and metallurgical behavior analysis.

Nowadays, in the competitive structure of industrial advancements, the advanced machining processes are well proven and commercialized for fruitful processing of technically advanced materials as compared with the traditional machining solutions. However, these new machining processes still need to be optimized under the given cutting conditions and work material properties. The optimization of the machining responses further enables the operator to attain desired characteristics within the required ranges along with the accuracy, which ultimately helps in providing quality products to the end customers. These optimization methods are based on the evolutionary algorithms, and their performance is well proven around the globe.

The chapters are targeted and are compiled by including real-life case studies on advanced machining methods, traditional manufacturing technologies, advanced and composite materials, processing with hybrid manufacturing techniques, various joining processes and its applications, microstructure analysis, which further makes the book more meaningful and attractive for the young scholars and researchers from the sphere of manufacturing and its allied domains.

—Editors

PART I

Studies on Conventional Manufacturing Practices

CHAPTER 1

REMOVAL OF STEPPED FEATURE IN MULTISTAGE SINGLE POINT INCREMENTAL FORMING

NIKHIL BARI[1*], VIKAS SISODIA[2], SHAILENDRA KUMAR[1], and PRASHANT K. JAIN[3]

[1]*Department of Mechanical Engineering, Sardar Vallabhabhai National Institute of Technology Surat, Surat, India*

[2]*Department of Manufacturing Engineering & Industrial Management, College of Engineering Pune, Pune, Maharashtra 411005, India*

[3]*Department of Mechanical Engineering, PDPM Indian Institute of Information Technology Design and Manufacturing Jabalpur, Jabalpur, India*

[*]*Corresponding author. E-mail: nikhilbari1665@gmail.com*

ABSTRACT

Single point incremental forming (SPIF) is an advanced sheet metal forming process in which the requirement of a press tool is completely eliminated. Due to this, it is also called die-less forming process. Complex geometries can be easily formed with this process. But in SPIF process, steeper wall angles are difficult to achieve because of the deformation characteristics of the process. To overcome this, a specific number of intermediate stages are added before achieving final geometry with a steeper wall angle. But during the multistage SPIF process, some extra deformed residues of previous stages appear at the bottom of the formed part. These residues are known as stepped features.

Modern Manufacturing Systems: Trends and Developments. Rajiv Kumar Garg, Ravi Pratap Singh, Rajeev Trehan, & Ramesh Singh (Eds.)

These undesirable step features increase the overall forming depth of the formed part, which significantly contributes to the geometrical error, as the obtained forming depth (after forming operation) is more than the designed forming depth. This chapter describes an approach to reduce the stepped features in multistage SPIF process. It is done by placing a bolted support with backing plate at the bottom of the formed part. Bolt support can also be used to provide back support to different geometries. The backing plate restricts the rigid body translation of the formed part, thereby reducing steps in the intermediate stages. It is observed that the step features are removed to the most extent and the final forming depth in multistage SPIF process is effectively brought close to the desired forming depth. With the proposed methodology wall thickness distribution is also improved.

1.1 INTRODUCTION

Sheet metal parts are widely used in a variety of applications ranging from household appliances such as kitchen utensils to many high-end applications, such as parts used in automotive, aerospace, and biomedical applications. Traditionally, sheet metal parts are formed using a press tool that consists of a punch and a matching female die. Small batch production in conventional sheet metal forming is generally costly. It is because, for each part, geometry dedicated punch and die are required. Moreover, for a small change in part geometry, a whole new set of punch and die setup is required. Hence for low volume production, conventional sheet metal forming processes are expensive.[2] With the recent advancement in sheet metal forming, incremental sheet forming (ISF), also known as the die-less sheet metal forming, process is a novel sheet metal forming method for the fabrication of complex geometries. In this process, the part is directly formed from computer-aided design (CAD) model itself, thus eliminating the need for molds, fixture, and dedicated punch and die setup. The process can be executed on any three-axis conventional CNC milling machine without investing on new equipment and tools.[15] Tool paths can be simulated and optimized on simple CNC simulator software. It is one of the emerging sheet metal forming process with very less machine and tool cost.[7] Because of the incremental nature (i.e., layer by layer or loop by loop deformation characteristics) of the process, forming force is also less as compared to conventional sheet metal forming process.[3] It is applicable for small batch production. In this process, the requirement of dedicated press tools is completely eliminated. Also, the time and cost involved in design and manufacturing of such press tools are saved.

Due to its flexibility, simplicity, and die-less in nature, it is applicable to variety of industry-specific free-form shapes like vehicle headlight reflector, exhaust manifold, etc.[17] Also, classification of the ISF process is done based on different requirements.[12] For example, in some variant of ISF process, single sheet is deformed[11] while in some cases two sheets are simultaneously deformed.[25,28] In the ISF process, the final forming depth of the formed part is acquired incrementally by the precise control of the movement of forming tools.[19] For particular geometry, the CNC program can be extracted with the help of computer-aided manufacturing from a CAD model of the geometry. Figure 1.1 shows the schematic of the ISF process.

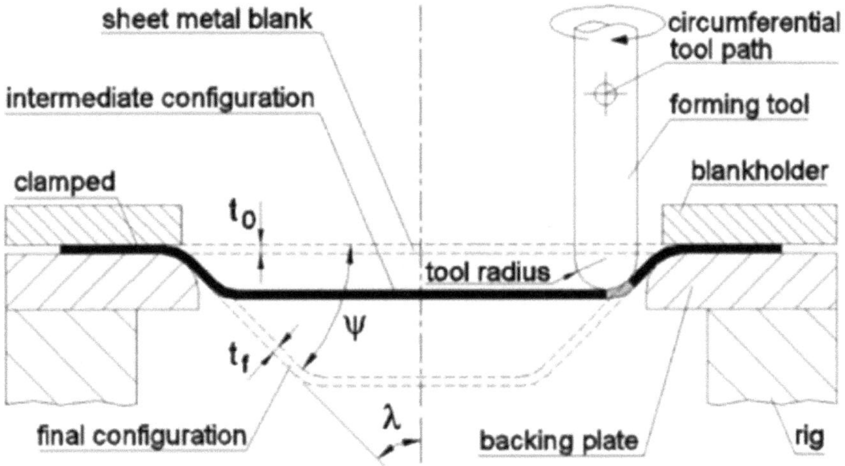

FIGURE 1.1 Schematic of the ISF process.

Source: Used with permission from Ref. [27]. © 2022 by Institution of Mechanical Engineers.

Single point incremental forming (SPIF) is one of the basic and simple variants of ISF process. The sheet is held along its edges while the rest of its portion is intentionally left unsupported. Whereas in TPIF (another variant of ISF process) sheet is supported by either full die[4] or partial die[23]. In another variant of TPIF process, two tools were used, one for deforming the sheet (master tool) and another for supporting the sheet which is termed as slave tool.[20] The axial movement and rotational motions can be given to both the tools.[5,8,18] One of the biggest challenges in SPIF process is poor geometrical accuracy and uneven wall thickness distribution. In SPIF process, final thickness of the formed part can be roughly approximated by the sine law of conventional sheet metal forming processes. It is given by eq 1.1.

$$t_f = t_0 \times \sin(90 - \psi) \tag{1.1}$$

From eq 1.1, the sine law can be used to determine the final thickness (t_f) of the part from the initial thickness of the sheet (t_0) and the part wall angle (ψ).[1] In SPIF process, there is a characteristic thinning band at a particular location along the forming depth where the wall thickness is minimum.[22,24] This characteristic thinning band occurs mainly due to severe stretching and localized necking, which results in minimum wall thickness at such location. Therefore there is a very high probability of fracture at this location. Significant geometric error in the formed part is another matter of concern in conventional SPIF process.[26] In SPIF process, uneven wall thickness distribution, spring back, and bending along the clamped contributes in geometric error.[6] In SPIF process, the sheet is deformed by stretching and shearing.[10] Due to this strain, hardening is induced in the sheet metal blank which changes the deformation capability of sheet. This strain hardening effect will be higher if the deformation is imposed on a relatively larger area. Therefore in SPIF process, forming steeper wall angles in single stage, the more strain hardening occurs.[19] Sheet blank experiences higher stress and may get fractured at a large strain hardening effect. In SPIF process, steeper wall angles are difficult to achieve because of deformation characteristics of the process. To overcome such limitations, a specific number of intermediate stages are added before achieving final geometry with steeper wall angle. When intermediate stages are introduced in SPIF process, strain hardening effect reduces.[14] By adding intermediate stages, steeper wall angles can be comfortable achieved and relatively uniform thickness distribution is obtained. This is so because the overall deformation gets distributed in intermediate stages. The process used to deform the sheet with intermediate stages is called multistage single point incremental forming (MSPIF). But during multistage SPIF process, some extra deformed residues of previous stages appear at the bottom of the formed part. These residues are known as stepped features, as shown in Figure 1.2. Skjodt et al.[27] termed it as the residual cone. These undesirable step features increase the overall forming depth of formed part, which significantly contributes to the geometrical error of formed part, as the obtained forming depth (after forming operation) is more than designed forming depth in the multistage SPIF process. The stepped features are formed due to increase in the depth of subsequent stage. In every stage, the tool path is restricted to a forming depth. Due to repeated rubbing action between tool and sheet material, unwanted flow of material takes place, and this flow of material occurs for every upcoming stage. This phenomenon is known as rigid body translation.[14] This chapter describes an approach to reduce the stepped features in multistage SPIF process. It is done by placing a bolted support with a backing plate at the bottom of the formed part. Bolt support can also be used to provide back support to various different geometries. The backing plate restricts the rigid body translation of the formed part, thereby reducing steps in the intermediate stages.

FIGURE 1.2 Stepped features in MSPIF.

Source: Used with permission from Ref. [13]. © 2013 Taylor & Francis.

The supports play a vital role in reducing the geometric error and achieving uniform wall thickness distribution of formed part in SPIF process. In SPIF process, geometric error arises mainly due to (1) bending of sheet along the clamped edges from where the plastic deformation initiates, (2) spring-back in formed part after tool/punch is raised after forming operation, and (3) pillow effect, as shown in Figure 1.3. If a simple truncated conical shape is to be deformed, these three different errors can be detected on it after forming operation.

Worldwide researchers have made efforts and suggested different supporting techniques for SPIF process. For example,[9] used water jet to provide counter support to the sheet during forming. It is observed that a jet of water exerts uniform pressure. Micari et. al.[16] made use of an elastic rubber tool under the sheet to reduce the geometric error. It is reported that the rubber tool does not produce dent on sheet blank. Sun et al.[29] reported that with the use of different types of elastic support made by rubber, polyurethane, and wood desired level of accuracy of formed can be achieved. Skjodt et al.[27] investigated the formability in MSPIF. Good formability was observed. Thickness distribution was more uniform as compared to single-stage SPIF process. Panjwani et al.[21] studied geometric inaccuracy arising due to the bending of sheet along the clamped edges. A supporting fixture was developed to reduce the bending of sheet along the edges. The fixture contains an array of bolts fabricated over a mild steel plate of 10 mm thickness. These bolts could rise up to a certain height and act as a partial die for formed part, thereby reducing the geometric error. The CAD model of the support structure is shown in Figure 1.4a. Figure 1.4b shows the fabricated support structure on the machine bed.

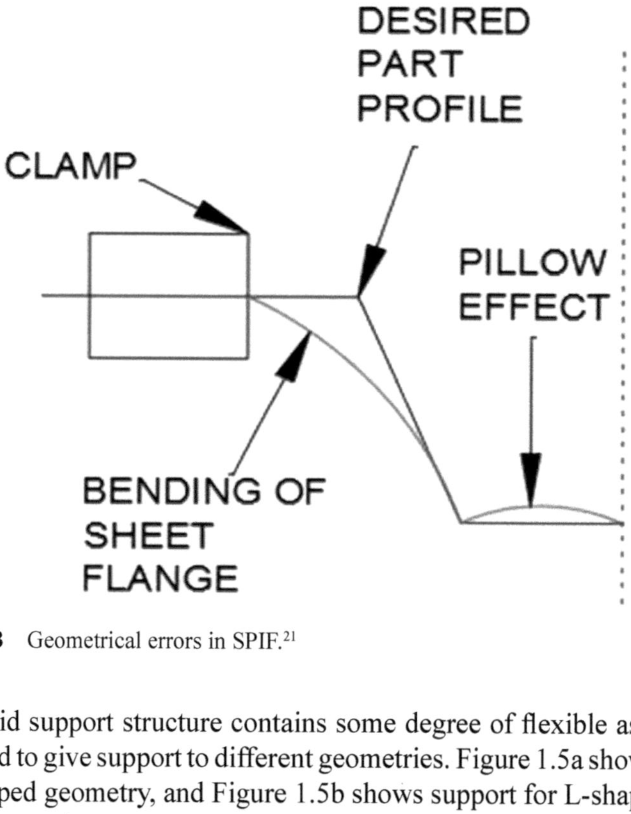

FIGURE 1.3 Geometrical errors in SPIF.[21]

The said support structure contains some degree of flexible as the same can be used to give support to different geometries. Figure 1.5a shows support for M-shaped geometry, and Figure 1.5b shows support for L-shaped geometry. However, the support structure was applicable to reduce the bending of sheet along the clamped edges only. But in the present study, its applicability is further extended to reduce stepped features also. In this study, stepped features, observed during MSPIF, are reduced by providing flexible support with backing plate at the base of forming the object.

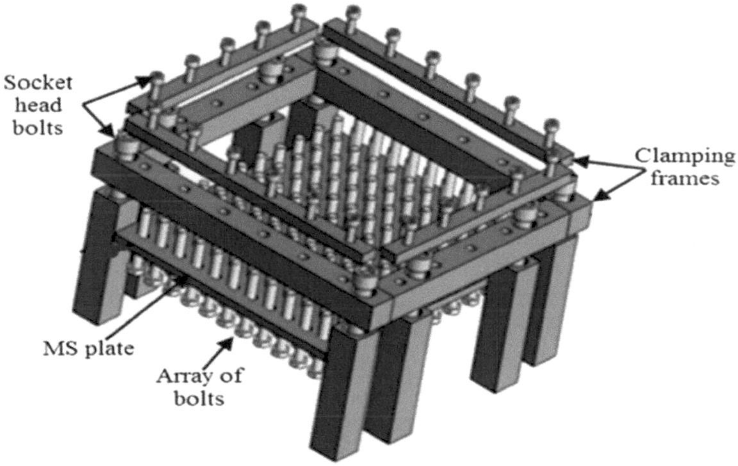

(a)

(b)

FIGURE 1.4 (a) CAD model of bolted support structure; (b) fabricated bolted support structure.[21]

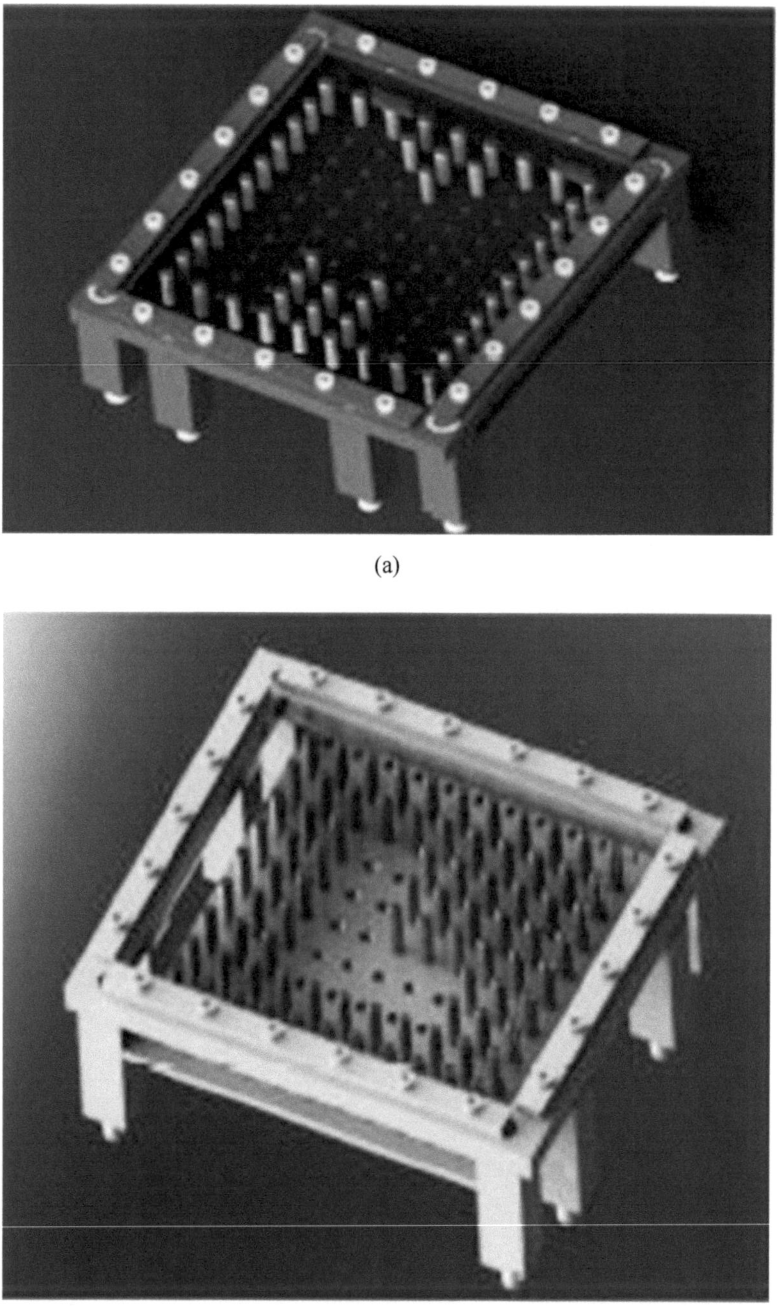

(a)

(b)

FIGURE 1.5 Bolted support for (a) M-shape geometry; (b) L-shape geometry.[21]

1.2 EXPERIMENTAL PLAN AND METHODOLOGY

In the present study, AA 1050 aluminum alloy with a thickness of 0.95 mm is used as sheet material. In the sheet metal forming process, the material deforms plastically and the stresses are beyond yield point and below the ultimate stress point. During this region, material experiences the strain hardening and stress–strain curve can be approximated by the Hollomon power law and is given by eq 1.2[9]

$$\sigma = 150 \times \varepsilon^{0.14} \tag{1.2}$$

where σ is true stress and ε is true strain.

The value of major process parameters considered is listed in Table 1.1. The experiments are performed on the OKUMA Center MB-46VAE-R (Fanuc controller) three-axis CNC vertical milling machine. The toolpath is extracted from CAD model for desired part geometry. First, the required CAD model was converted into ".STL" format thereafter slicing was done in CATIA modeling software using "digitized shape editor tool" to generate point cloud of CAD model. The point cloud obtained is then imported in MATLAB to convert step points cloud into a spiral point cloud.

TABLE 1.1 Input Parameters Used in Experiments.

Parameters	Value
Tool size	8 mm
Tool type	Hemispherical
Tool rotation	50 RPM
Tool path	Spiral (clockwise)
Forming depth	30 mm
Feed rate	1500 mm/min
Incremental depth	0.5 mm
Sheet blank thickness	0.95 mm
First stage wall angles	45°
Second stage wall angle	50°
Third stage wall angle	55°
Fourth stage wall angle	60°
Fifth stage wall angle	65°
Forming geometry	Square pyramid
Lubricant oil	Hydraulic oil (grade-68)

Further this spiral point cloud is converted into G codes & M codes by using MATLAB itself. Figure 1.6a shows the CAD model of fixture used for the present study and Figure 1.6b shows the fixture mounted on CNC machine bed. The gap between the machine bed and the bottom of the plate (which holds bolts) is around 285 mm. The gap was provided to accommodate the space for stepper motor assembly, which controls the height of bolts precisely. Bolts are fitted at the outer and inner sections of the plate. The bolts on the outer section are so arranged that they support the sheet along the edges according to part geometry.[21] Also, they restricts the bending action of a sheet during forming operation. Bolts at the inside section are used to restrict the undesired rigid body translation of the sheet by giving support to backing plate. The upward and downward movement of the backing plate can be controlled by controlling the movement of bolts placed at the inner section. Figure 1.7 depicts the schematics of the present fixture. The backing plate is placed on bolts placed at inner section. The stopper limits the movements of backing plate with respect to bolts. In the present work, the final forming depth is fixed to 30 mm and backing plate is placed at a height of 32 mm. A 2 mm clearance is required in order to accommodate height variation caused due to the bending effect.

1.2.1 BOLT HEIGHT CONTROLLING SET UP

The precise control of height of the backing plate is important because if the gap between the bottom of the formed component and backing plate is less then tool may collide with the backing plate due to this machine spindle stands the risk of damaged and if the gap is large then the rigid body translation will freely occur which results in the generation of stepped feature and there is no meaning of providing backing plate because the bottom of part would move in downward and the stepped features will generate. Therefore there should be optimum clearance between sheet and backing plate. For this purpose, a computerized controlled stepper motor with adequate precision is used. When counter-clockwise rotational movement is given to the bolts then linear upward movement of bolts occurs which pushes the backing plate upward and when the bolts are given rotational movement in a clockwise direction then liner downward movement of bolts occurs which takes the backing plate downward. For controlling of stepper motor, an Arduino board, DC motor driver, a DC 12-V power supply is required. The coding for Arduino board is done on Arduino software. The

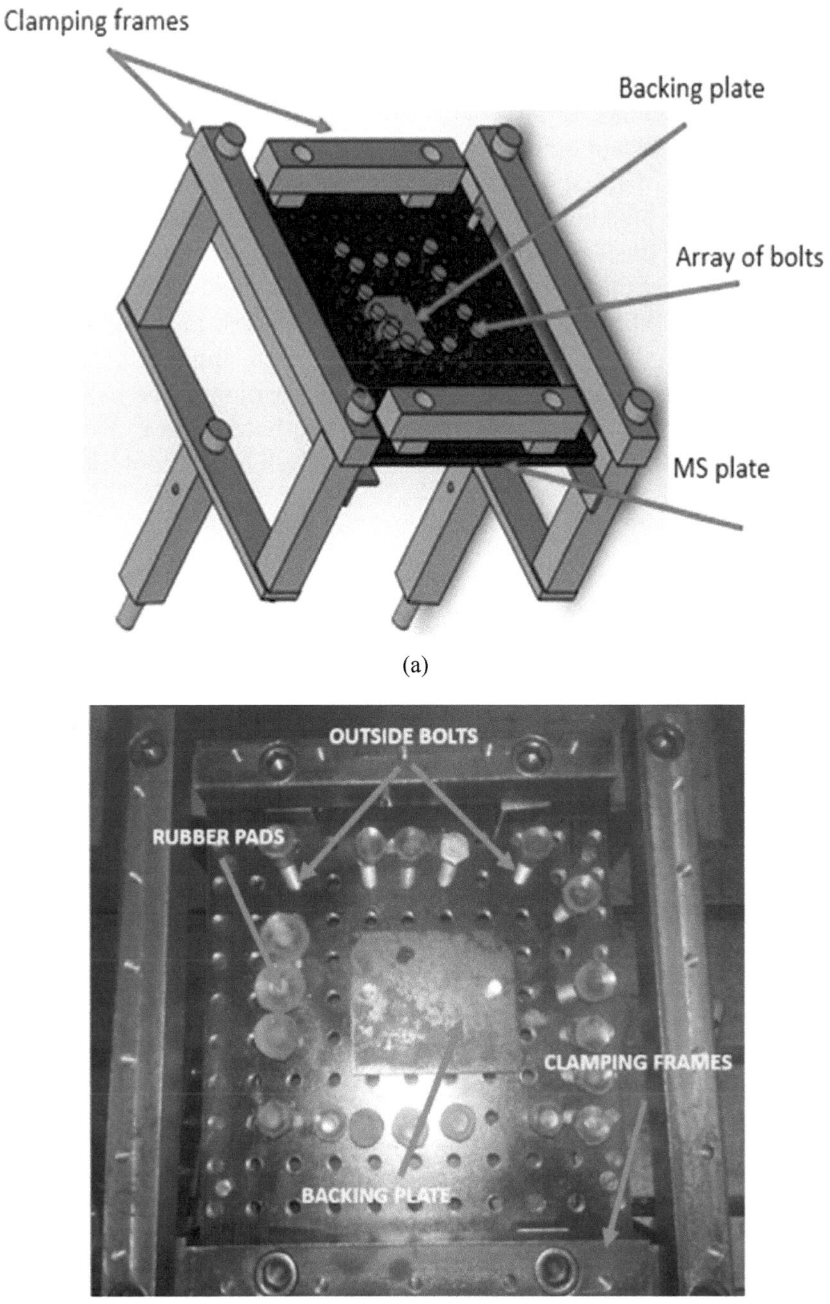

(a)

(b)

FIGURE 1.6 Modified flexible support, (a) CAD model, (b) fixture mounted on machine bed.

purpose of the code is to run the motor for a specific number of rotations in a clockwise and counter-clockwise direction. The height control setup of to run is depicted in Figure 1.8. The arrow indicated the direction of signals. For connecting the shaft of the motor with a hexagonal head of the bolt, an application-specific coupling is designed and fabricated using the 3D printer of fused deposition method type. The material of coupling is acrylonitrile butadiene styrene. It has an ultimate tensile strength of 40 MPa and can withstand the torsional moment, which is applied on the bolt. Figure 1.9 shows the height controlling set-up mounted on the machine bed. Since in MSPIF process, final part shape is achieved in a specific number of stages, so in the first stage, part is formed at an angle of 45°. Now just before the initiation of the second stage, the position of the backing plate is adjusted that it is exactly 2 mm below the bottom face of the formed part. This position is precisely achieved with bolt height controlling set-up using a stepper motor.

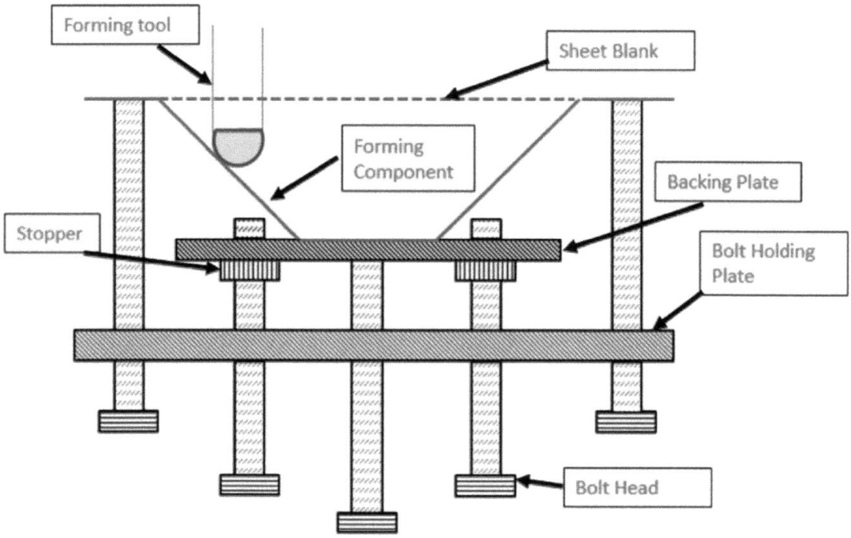

FIGURE 1.7 Schematics of the present fixture.

The backing plate is now rigidly held at this position, and it imposes the restriction on undesired rigid body translation of base of the forming geometry. Due to this, the reduction in stepped features is observed. The stepped features formed after the final stage of the MSPIF process with and without bolt height controlling setup are shown in Figure 1.10.

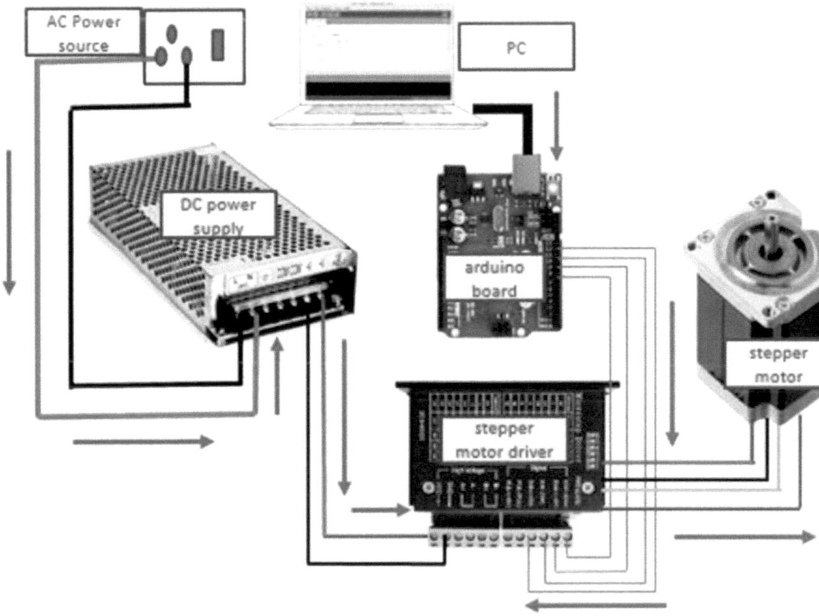

FIGURE 1.8 Height controlling set-up.

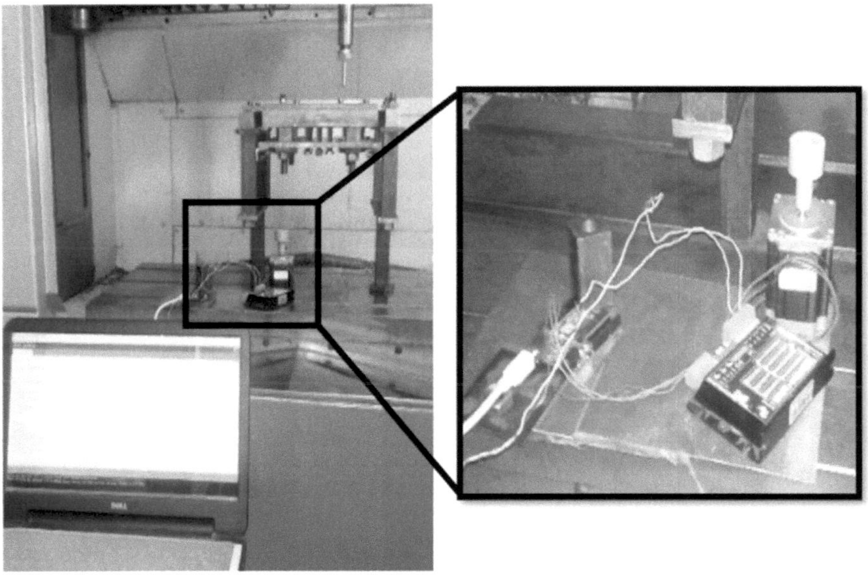

FIGURE 1.9 Height controlling set-up on CNC milling machine bed.

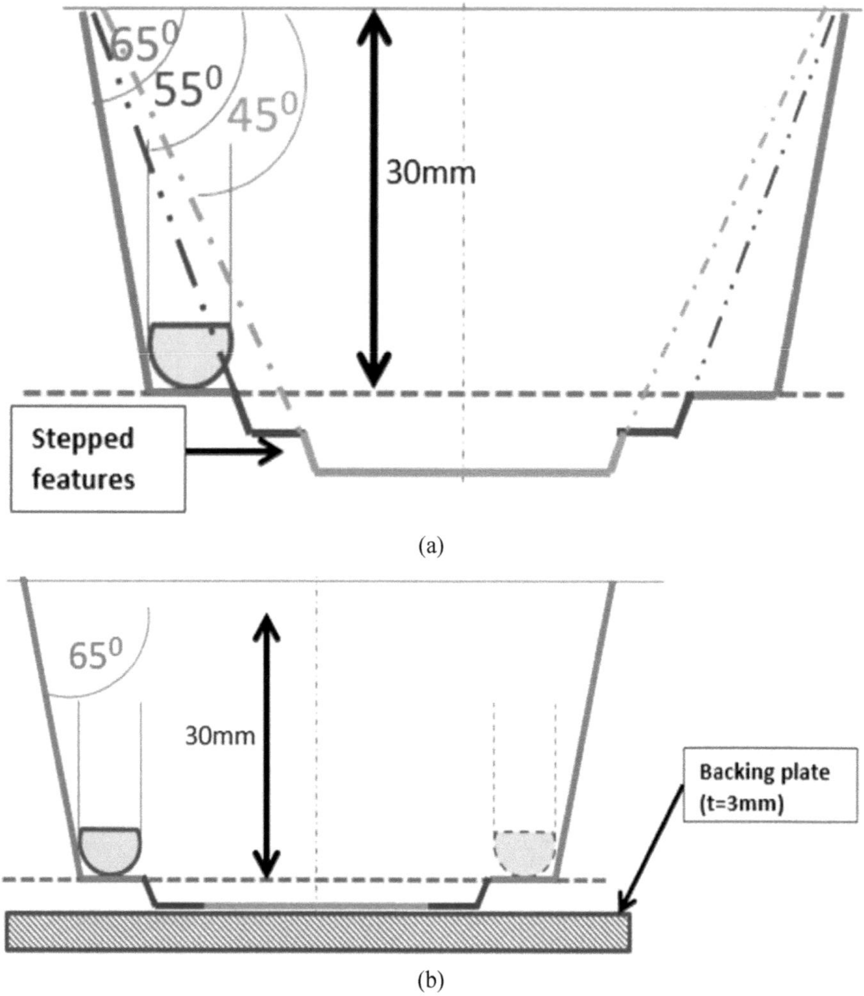

FIGURE 1.10 Forming (a) without backing plate (b) with backing plate.

1.3 RESULTS AND DISCUSSION

The major advantage of the MSPIF process is that steeper wall angles can be formed with a relatively uniform thickness distribution. But at the same time, due to the generation of stepped feature total forming depth increases, which contribute to the geometrical deviation of formed components. In the present work, efforts have been made to reduce undesired stepped features. The final wall angle is achieved in the specific number of stage (i.e., 45°–50°–55°–60°–65°). Experiments are performed with and without bolt height control support structure, in order to quantify its influence on part formed, as depicted in Figure 1.11. Reduction in stepped features is observed by using bolt height support structure. It is because of the absence of the least rigid body translation. It is achieved by the backing plate on bolt height support structure. The vertical positioning of the backing plate is set accurately with the help of a computerized controlled stepper motor. Thus due to the absence of the stepped feature, final forming depth is close to desired forming depth. This enhances the geometrical precision of formed part. Further wall thickness distribution in MSPIF process is also studied. For this, a 3D scanner makes GOM model ATOS is used. As a standard operating procedure of the scanner, some black markers are placed at different locations of formed part, as depicted in Figure 1.11. Now scanning process initiates by clicking photographs of formed part in every possible direction. The camera of the 3D scanner consists of two lenses. The images are captured with these lenses. The processor of the scanner identifies the position of each marker and accordingly generates the topological properties at each marker position. After recording all images, the scanner ultimately merges these images to CAD model. In the present study, GOM Inspect software is used to process the scanned CAD model of formed part. From the analysis of results, it is observed that the minimum thickness in the case of formed part with the stepped feature is 0.11 mm, whereas the minimum thickness in case of the part with the reduced stepped feature is around 0.20 mm. The minimum thickness was improved because the restrictions imposed on the flow of material in terms of rigid body translation. This is shown in Figure 1.12.

1.4 CONCLUSION

The major limitation of single stage SPIF process is that it is unable deform the component at steeper wall angle due to thinning. But this is not the case in MSPIF process. In MSPIF process due to the presence of undesired stepped features, the part quality is reduced. The present study is focused on developing a methodology to reduce stepped feature in MSPIF process. These undesirable step features increase the overall forming depth of the formed part, which significantly contributes in undesired geometrical error in formed part. With the proposed methodology, part with steeper wall angles and reduced stepped feature along with the improved wall thickness distribution is formed. This is due to restrictions imposed by bolts height support structure on rigid body translation of sheet at the bottom of the part. Further improvement in the minimum thickness of the formed part is also observed. The findings of the present work are useful in forming parts with less stepped features and improved thickness distribution by using MSPIF process.

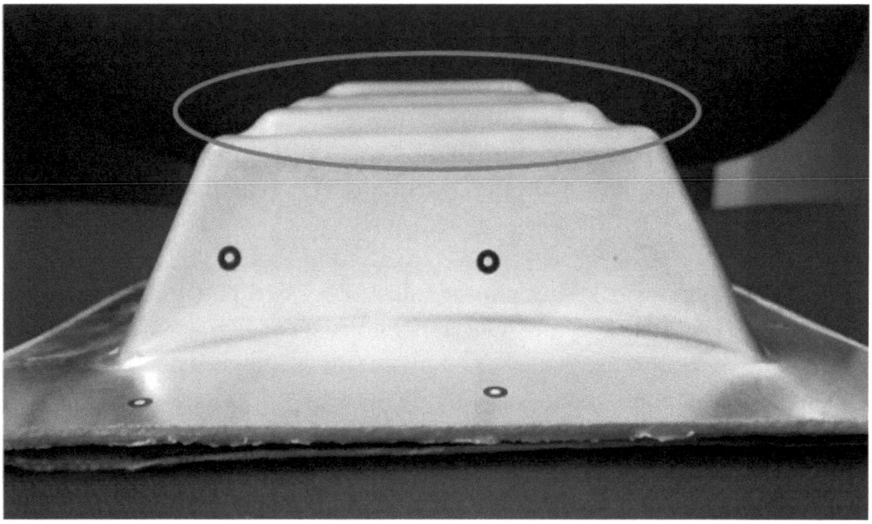

(a)

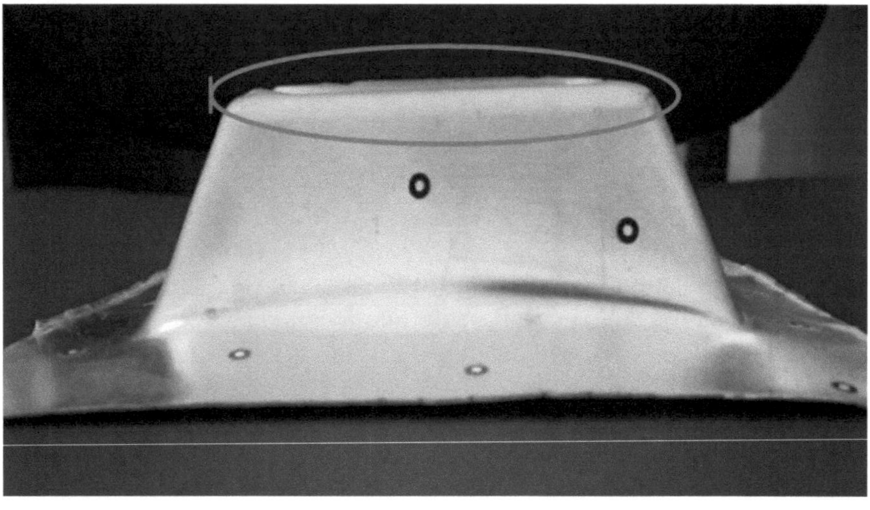

(b)

FIGURE 1.11 Components with (a) stepped features; (b) reduced stepped features.

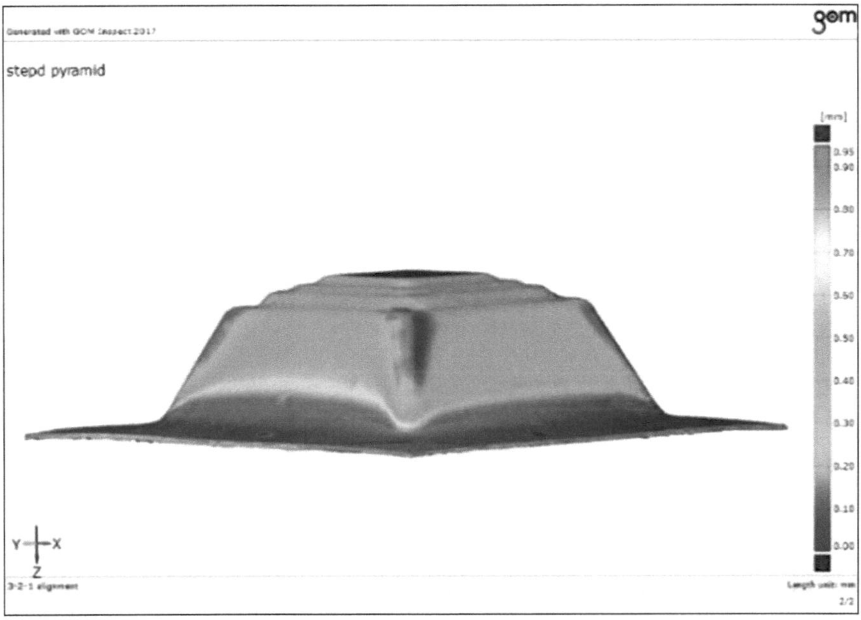

(a)

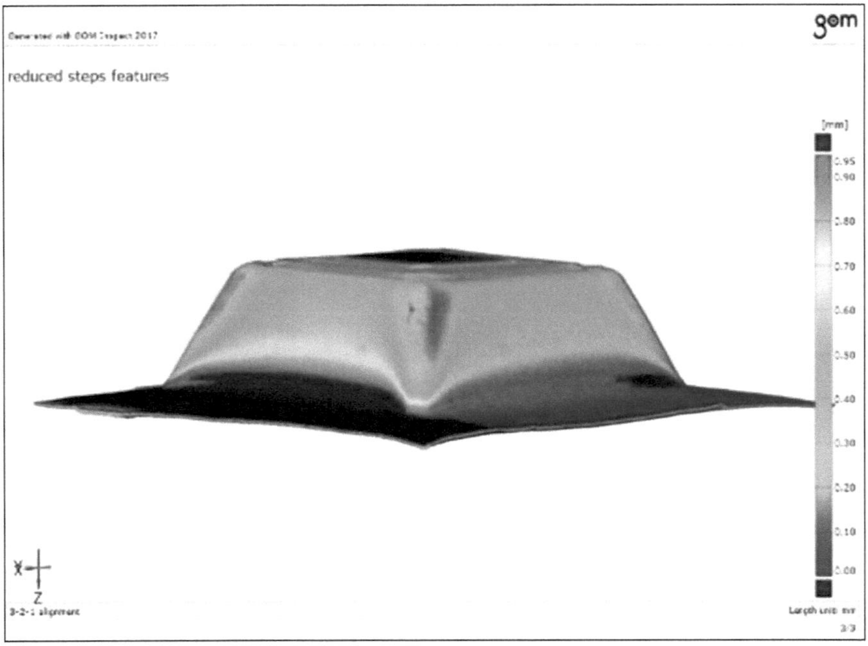

(b)

FIGURE 1.12 Comparison of wall thickness distribution in (a) pyramid with stepped feature; (b) pyramid with reduced stepped feature.

KEYWORDS

- **single point incremental forming**
- **multistage SPIF process**
- **bolt support**
- **wall angle**
- **stepped features**
- **wall thickness distribution**

REFERENCES

1. Allwood, J. M.; King, G. P. F.; Duflou, J. A structured Search for Applications of the Incremental Sheet-Forming Process by Product Segmentation. *Proc. Inst. Mech. Eng. B J. Eng. Manuf.* **2005,** *219* (2), 239–244.
2. Ambrogio, G.; Filice, L.; Gagliardi, F. Improving Industrial Suitability of Incremental Sheet Forming Process. *Int. J. Adv. Manuf. Technol.* **2012,** *58* (9–12), 941–947.
3. Baharudin, B. T. H. T.; Azpen, Q. M.; Sulaima, S.; Mustapha, F. Experimental Investigation of Forming Forces in Frictional Stir Incremental Forming of Aluminum Alloy AA6061-T6. *Metals* **2017,** *7* (11), 484.
4. Bambach, M.; Cannamela, M.; Azaouzi, M.; Hirt, G.; Batoz, J. L. Computer-Aided Tool Path Optimization for Single Point Incremental Sheet Forming. In *Advanced Methods in Material Forming*; Springer: Berlin, Heidelberg, 2007; pp 233–250.
5. Behera, A. K.; Lu, B.; Ou, H. Characterization of Shape and Dimensional Accuracy of Incrementally Formed Titanium Sheet Parts with Intermediate Curvatures between Two Feature Types. *Int. J. Adv. Manuf. Technol.* **2016,** *83* (5–8), 1099–1111.
6. Duflou, J.; Lauwers, B.; Verbert, J.; Tunckol, Y.; De Baerdemaeker, H. Achievable Accuracy in Single Point Incremental Forming: Case Studies. In *Proceedings of the 8th European Scientific Association for Material Forming Conference on Material Forming*, 2005; pp 675–678.
7. Franzen, V.; Kwiatkowski, L.; Martins, P. A. F.; and Tekkaya, A. E. (2009). Single Point Incremental Forming of PVC. *J. Mater. Process. Technol.* **2009,** *209* (1), 462–469.
8. Ilangovan, B.; Monfared, R. P.; Jackson, M. An Automated Solution for Fixtureless Sheet Metal Forming. *Int. J. Adv. Manuf. Technol.* **2016,** *82* (1–4), 315–326.
9. Iseki, H. Flexible and Incremental Bulging of Sheet Metal Using High-Speed Water Jet. *JSME Int. J. Ser. C Mech. Syst. Mach. Elements Manuf.* **2001,** *44* (2), 486–493.
10. Jackson, K.; Allwood, J. The Mechanics of Incremental Sheet Forming. *J. Mater. Process. Technol.* **2009,** *209* (3), 1158–1174.
11. Jeswiet, J.; Micari, F.; Hirt, G.; Bramley, A.; Duflou, J.; Allwood, J. Asymmetric Single Point Incremental Forming of Sheet Metal. *CIRP Ann. Manuf. Technol.* **2005,** *54* (2), 88–114.

12. Kumar, A.; Gulati, V.; Kumar, P.; Singh, H. Forming Force in Incremental Sheet Forming: A Comparative Analysis of the State of the Art. *J. Braz. Soc. Mech. Sci. Eng.* **2019,** *41* (6), 251.
13. Liu, Z.; Li, Y.; Meehan, P. A. Vertical Wall Formation and Material Flow Control for Incremental Sheet Forming by Revisiting Multistage Deformation Path Strategies. *Mater. Manuf. Process.* **2013,** *28* (5), 562–571.
14. Malhotra, R.; Bhattacharya, A.; Kumar, A.; Reddy, N. V.; Cao, J. A New Methodology for Multi-Pass Single Point Incremental Forming with Mixed Toolpaths. *CIRP Ann.* **2011,** *60* (1), 323–326.;
15. Matsubara, S. A Computer Numerically Controlled Dieless Incremental Forming of a Sheet Metal. *Proc. Inst. Mech. Eng. B J. Eng. Manuf.* **2001,** *215* (7), 959–966.
16. Micari, F. A. B. R. I. Z. I. O.; Ambrogio, G.; Filice, L. Shape and Dimensional Accuracy in Single Point Incremental Forming: State of the Art and Future Trends. *J. Mater. Process. Technol.* **2007,** *191* (1–3), 390–395.
17. More, K.R.; Sisodia, V.; Kumar, S. A Brief Review on Formability, Wall Thickness Distribution and Surface Roughness of Formed Part in Incremental Sheet Forming. In *Advances in Manufacturing Processes*; Dave, H. K., Nedelcu, D., Eds.; Lecture Notes in Mechanical Engineering; Springer: Singapore, 2021.
18. Moser, N.; Zhang, Z.; Ren, H.; Zhang, H.; Shi, Y.; Ndip-Agbor, E. E.; Lu, B.; Chen, J.; Ehmann, K. F.; Cao, J. Effective Forming Strategy for Double-Sided Incremental Forming Considering in-Plane Curvature and Tool Direction. *CIRP Ann.* **2016,** *65* (1), 265–268.
19. Nirala, H. K.; Jain, P. K.; Roy, J. J.; Samal, M. K.; Tandon, P. An Approach to Eliminate Stepped Features in Multistage Incremental Sheet Forming Process: Experimental and FEA Analysis. *J. Mech. Sci. Technol.* **2017,** *31* (2), 599–606.
20. Paniti, I.; Somló, J. Novel Incremental Sheet Forming System with Tool-Path Calculation Approach. *Acta Polytech. Hungarica* **2014,** *11* (7), 43–60.
21. Panjwani, D.; Priyadarshi, S.; Jain, P. K.; Samal, M. K.; Roy, J. J.; Roy, D.; Tandon, P. A Novel Approach Based on Flexible Supports for Forming Non-Axisymmetric Parts in SPISF. *Int. J. Adv. Manuf. Technol.* **2017,** *92* (5–8), 2463–2477.
22. Salem, E.; Shin, J.; Nath, M.; Banu, M.; Taub, A. I. Investigation of Thickness Variation in Single Point Incremental Forming. *Procedia Manuf.* **2016,** *5*, 828–837.
23. Silva, M. B.; Martins, P. A. F. Two-Point Incremental Forming with Partial Die: Theory and Experimentation. *J. Mater. Eng. Perform.* **2013,** *22* (4), 1018–1027.
24. Sisodia, V.; Kumar, S. Experimental Study Of Single Point Incremental Forming with Dummy Sheet. *Int. J. Mater. Eng. Innov.* **2019,** *10* (1), 60–82.
25. Sisodia, V.; Kumar S. Study of AA-1050 Sheet Metal Parts Processed by Single Point Incremental Forming with Dummy Sheet. *Int. J. Mater. Eng. Innov.* **2020,** *11* (2), 105–126.
26. Sisodia, V.; Kumar, S.; Jagtap, R.; More, K. Experimental Investigation on Geometric Error in Single-Point Incremental Forming with Dummy Sheet. In *Innovative Design, Analysis and Development Practices in Aerospace and Automotive Engineering*; Springer: Singapore, 2020; pp 81–92.
27. Skjødt, M.; Silva, M. B.; Martins, P. A. F.; Bay, N. (2010). Strategies and limits in multi-stage single-point incremental forming. The journal of strain analysis for engineering design, 45(1), 33-44.

28. Skjoedt, M.; Bay, N.; Ingarao, B. E. G.; Endelt, G. Single Point Incremental Forming Using a Dummy Plate-Multiplate Forming. In *2nd International Conference on New Forming Technology, Bremen, Conference Proceedings*; BIAS-Verlag, 2007; pp 267–276.
29. Sun, L.; Zhao, Q.; Xiang, J.; Shi, J.; Wang, L.; Hu, S.; Su, S. Adsorption of NO and NH_3 over $CuO/\gamma\text{-}Al_2O_3$ Catalyst by DRIFTS. *J. Chem. Ind. Eng. Soc. China* **2009,** *2009*, 444–449.

CHAPTER 2

MICROWAVE CASTING OF METALLIC MATERIALS: A REVIEW

DEEPAK KUMAR PATEL, NEERAJ KUMAR BHOI*, and HARPREET SINGH

Department of Mechanical Engineering, PDPM Indian Institute of Information Technology Design and Manufacturing, Jabalpur 482005, Madhya Pradesh, India

**Corresponding author. E-mail: neerajbitd@gmail.com*

ABSTRACT

The reduction of power consumption for making the process more efficient is the focal point of interest and research in today's manufacturing industry. The use of microwave energy has widely been used in the field of sintering, cladding, and joining processes. However, its use in the field of casting has not been explored for different kinds of metallic materials. Microwave casting has several advantages over the traditional casting method such as reduced processing time, uniform bulk volumetric heating, improved mechanical properties, and improved microstructures. This chapter illustrates the various developments made in the timeline for casting with microwave energy. The possibilities and challenges in the direction of microwave casting are given in brief to identify potential parameters for the casting of different metallic materials.

2.1 INTRODUCTION

Casting of material is the widely used process in every material processing industry. The conventional way of heating and casting has many limitations,

Modern Manufacturing Systems: Trends and Developments. Rajiv Kumar Garg, Ravi Pratap Singh, Rajeev Trehan, & Ramesh Singh (Eds.)

such as (1) time consuming, (2) excessive consumption of power, and (3) non-uniform heating of the material. These drawbacks of the conventional casting process can be minimized by using a new novel technique known as microwave casting (MWC). The use of microwave irradiation for casting is a significant development of research, which is gaining popularity. MWC of material has several advantages over conventional casting process, such as (1) time and energy saving, (2) volumetric heating, (3) reduction in processing time, (4) microstructure property enhancement, (5) better product performance, enhanced mechanical and tribological behavior.[1–7] The use of microwave energy is not just confined to casting but there are several other applications of microwave energies in the field of metals for (1) sintering, (2) cladding, (3) joining, (4) melting, and (5) drilling. Microwaves are a form of electromagnetic radiation with a wavelength in the range of 10–300 mm and frequency in the range of 0.3–300 GHz. However, the most commonly used microwave frequency and power for material processing is 2.45 GHz with different power range. In conventional heating of material, thermal energy mitigation is occurring through the surface of the metallic part to the core via conduction heating mode,[5–7] whereas, in microwave heating, the thermal mitigation occurs from the inner core of the material to the surface due to interaction of molecules with electromagnetic energy resulting in (1) effective heating, (2) selective heating, and (3) volumetric heating depending on different material processing operation.[7–11] Figure 2.1 explains the conventional and microwave heating process for a single molecule.[4] In recent year, the development of (1) in-situ casting and (2) ex-situ casting products developed through microwave is gaining popularity due to its numerous advantages. In context, various researchers were tried to cast lightweight material such AA-7039 (Al-Mg-Zn), Al-1050 with the help of different microwave power and with the application of susceptor material.[6,12–14] On a similar note, the development of cast component with the help of metallurgical milled powder sample were explored with conventional microwave oven.[7,8]

The present chapter aims to review the process fundamental lies in the MWC process. The heating mechanism and different (1) in-situ and (2) ex-situ method for MWC are described with the aim to identify the process variable and their effect on the material outcomes. The difference in the microstructure of the in-situ and ex-situ method is described to understand the input variable effect on the cast component. Finally, the developments made in the timeline of MWC are given to select and most influencing process parameters for the processing. The chapter will help the researchers to identify the input conditions and variables affecting the casting with microwave energy.

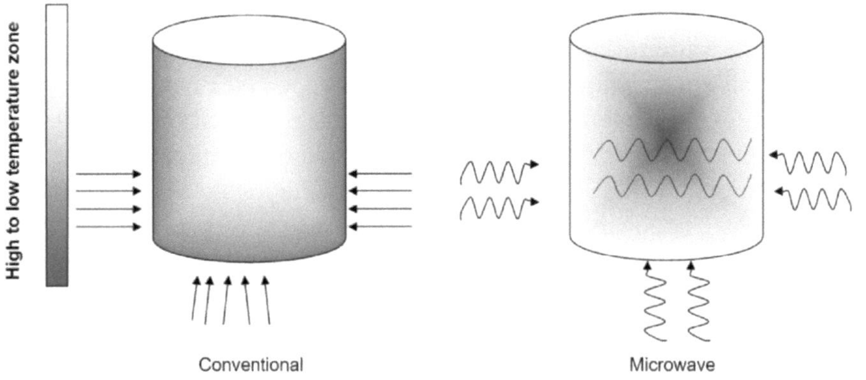

FIGURE 2.1 Schematic of heating pattern in conventional and microwave heating mode.
Source: Reprinted with permission from Ref. [4]. © 2019 Elsevier.

2.2 HISTORICAL BACKGROUND

The use of microwave energy started a way back in 1948 after the Second World War because at that time there was a need for processing techniques that require lesser time with improved product quality and reduced cycle time. The microwave was discovered in 1948 but its use was mainly confined to liquid heating and food processing only. During the 1960s, the use of microwave energy was confined to the curing of wood, rubber vulcanization, and its processing. In 1988, a successful experimentation on microwave heating was conducted by McGill et al., on selected chemicals and minerals.[15] Subsequently in the year 1990, the first patent was filed by Sheinberg et al., for heating of opaque and semiopaque substances that were considered unviable to heating through the microwave.[16] Later in 1999, the first successful experiment for microwave sintering of metals powders of Fe–Ni and Fe–Cu at 2.45 GHz frequency was conducted by Roy et al.[17] Their experimental results show enhanced mechanical properties through microwave sintering as compared to conventional sintering. Ever since, the use of microwaves has seen a rapid growth in the field of sintering, cladding, joining, melting, and drilling. In the last two decades, many metals powders and their alloy compositions including Fe, Cu, Al, Ni, Mo, Co, Ti, W, Sn, steel, and the alloys have been successfully sintered with better mechanical properties through microwave sintering.

In the case of casting process, the first and foremost step is to melt the charge and then allow this molten charge to flow through various gating elements to the mold cavity. The use of microwave energy for melting of materials was first reported in 2006 when a patent was filed by Moore et al., for the method and apparatus of melting metals using microwave energy.[18] Ever since, various literature on microwave melting of commercially available metals such aluminum, stainless steel, copper, etc., have been reported but none of them mentioned about the melting characteristics until 2010; Chandrasekaran et al. first reported the melting of aluminum, lead, copper, and tin through MHH.[19] In 2013, the first dissertation on MWC of lead was reported by Choudhary and Puri.[20] In 2015, a patent for a new in-situ MWC technique at 2.45 GHz was filed by Mishra et al.[21] Since then, lot of work have been reported on AA 7039, AA 1050, MMC's, Copper, and its alloys. Figure 2.2 shows the progressive development in microwave processing of materials.

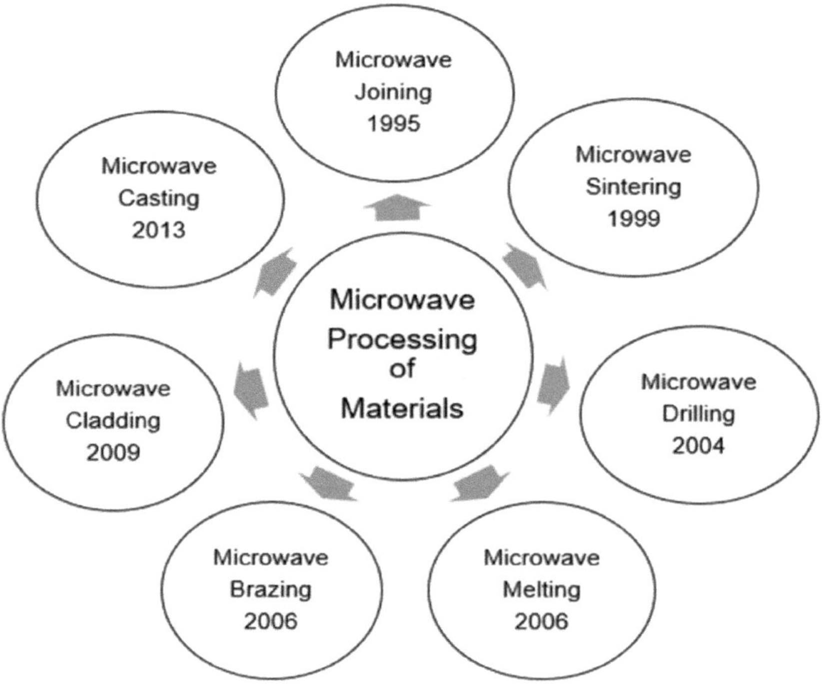

FIGURE 2.2 Schematic of the progressive development in microwave processing of materials.

2.3 WORKING MECHANISM AND PROCESS ADVANCEMENT

Over the last two decades, microwave energy has emerged as a novel way for metallic materials processing. However, this technique was highly efficient for powder metals as they can easily absorb microwave energy at room temperature and thus powder metals are easy to sinter and melt. When it comes to bulk metals due to their reflective nature at room temperature as bulk metallic material does not absorb microwave energy at room temperature. But when the temperature of bulk metals reaches a critical temperature, that is, 400°C–600°C then they start absorbing microwave energy. So, for enhancement of heating capabilities, microwave hybrid heating (MHH) is used. In MHH, susceptor material has been placed around bulk material.[2,4,5,9] The susceptor material absorbs microwave irradiation and further converts it to heat energy, which results into the increase in temperature of bulk metal. Once the temperature reaches a critical temperature, the bulk metal starts absorbing microwave energy, as shown in Figure 2.3. Melting of bulk metal using MHH can significantly save time and energy. This motivated the researchers to use MHH in casting and provide energy and time-efficient alternative. The energy required for MWC of bulk metal is one-third of the conventional casting.[12–14]

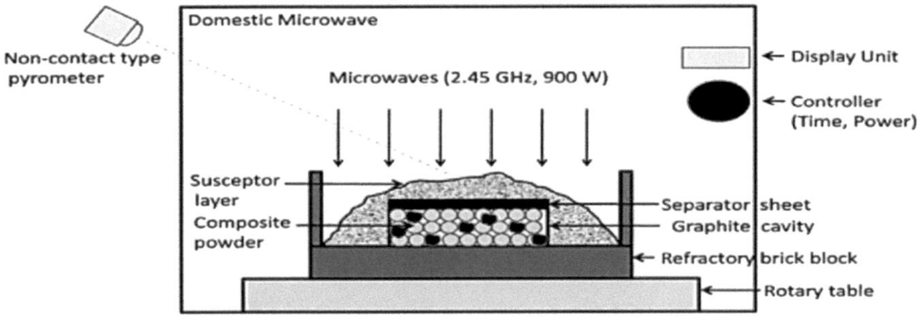

FIGURE 2.3 Schematic representation of microwave hybrid heating process for the casting of composite powder.

Source: Reprinted with permission from Ref. [8]. © 2016 Elsevier.

In MWC, a mold assembly and microwave cavity are used where the bulk metal is placed. The time required for complete casting depends on (1) melting time, (2) pouring time, and (3) solidification time of material. The main components of MWC are (1) pouring basin, (2) sprue, (3) split mold assembly (cope box and drag box), and (4) base. The most commonly used material used for pouring basin is silicon carbide as it can easily absorb microwave irradiation for volumetric heating and melting of charge. It can also resist high temperature thus maintaining its shape and size at high temperature. The melting of charge is done in pouring basin. Once the charge is melted, it flows into the mold cavity via sprue. The material used for sprue is alumina as it helps in maintaining liquid metal in molten state due to its high melting temperature and its behavior for microwave irradiation at room temperature is transparent. Thus, they can easily couple with microwave energy. The material used for mold assembly is graphite which is also a microwave irradiation absorbing material and it helps in eliminating the preheating of mold assembly during pouring.[11–14] The complete assembly is placed inside a refractory brick which is made of refractory material which can resist high temperature during heating. Figure 2.4 shows the setup of MWC.

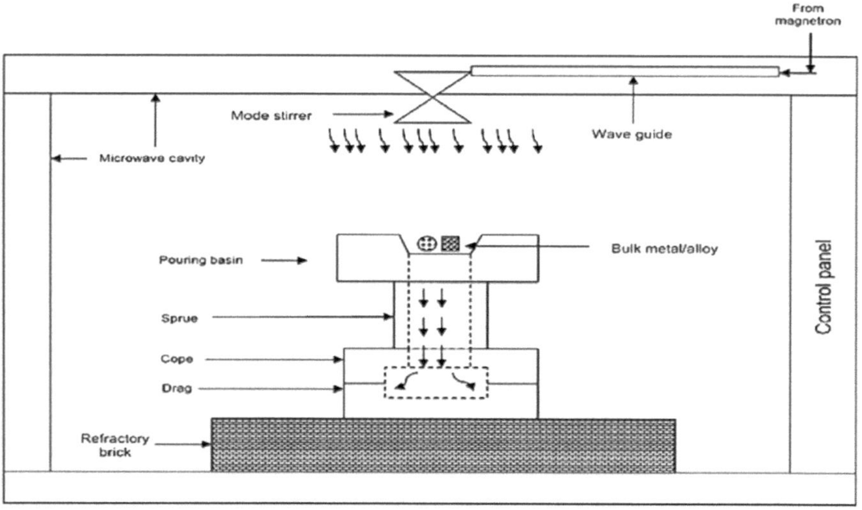

FIGURE 2.4 Basic schematic of microwave casting.

Source: Reprinted with permission from Ref. [4]. © 2019 Elsevier.

MWC can further be classified into (1) in-situ MWC and (2) ex-situ MWC. During in-situ MWC, the charge melting has been carried out inside the applicator cavity and then transferred into the mold assembly which is also placed inside applicator cavity. Thus melting, pouring, and solidification, all three steps are done inside the cavity itself. Similarly, in ex-situ MWC, the charge melting has been carried out inside applicator cavity and then transferred into the mold assembly placed outside applicator cavity. Thus, the microwave irradiation usage only takes place for charge melting but pouring and solidification take place outside the applicator cavity similar to conventional casting technique.[14,22–28] In order to determine the most favorable technique for MWC processes, several studies were carried out and it was observed that in-situ casting produces finer grains, dense structures, and better mechanical properties than ex-situ casting. Figure 2.5 shows the steps involved in (1) in-situ and (2) ex-situ MWC. A comparative assessment was made for the in-situ and ex-situ casting of the AA-7039 alloy material. Figure 2.6 depicts the BSE (backscattering images) for in-situ casts (a,b) and ex-situ casts (c,d) from the BSE images, black, white, grey, and dark grey phases can be observed.[22]

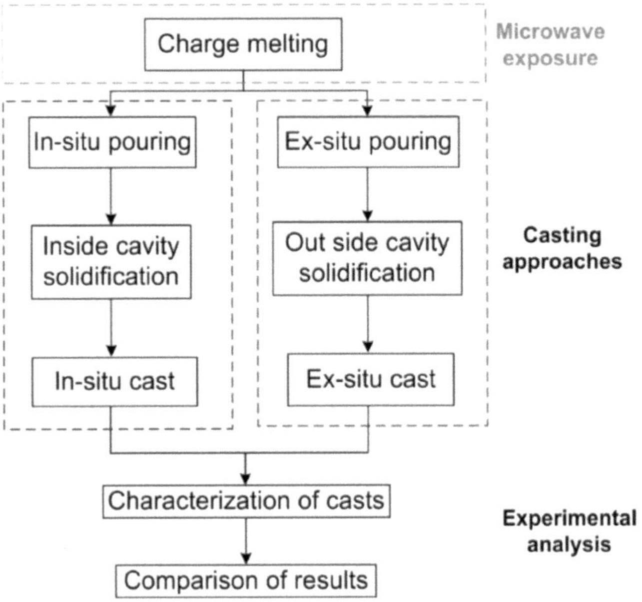

FIGURE 2.5 Steps involved in (i) in-situ and (ii) ex-situ casting.

Source: Reprinted with permission from Ref. [22]. © 2019 IOP Publishing.

2.4 PROCESS PARAMETERS AND OUTCOMES FROM THE MICROWAVE CASTING PROCESS

In this section, the process conditions and different outcomes reported by the researcher in the field of MWC are listed in Table 2.1.

2.5 CONCLUSION

The detailed archived reports on MWC have been discussed in this chapter. MWC of metals has been emerged as a research area due to its distinct advantages such as improved metallurgical and mechanical properties, energy and

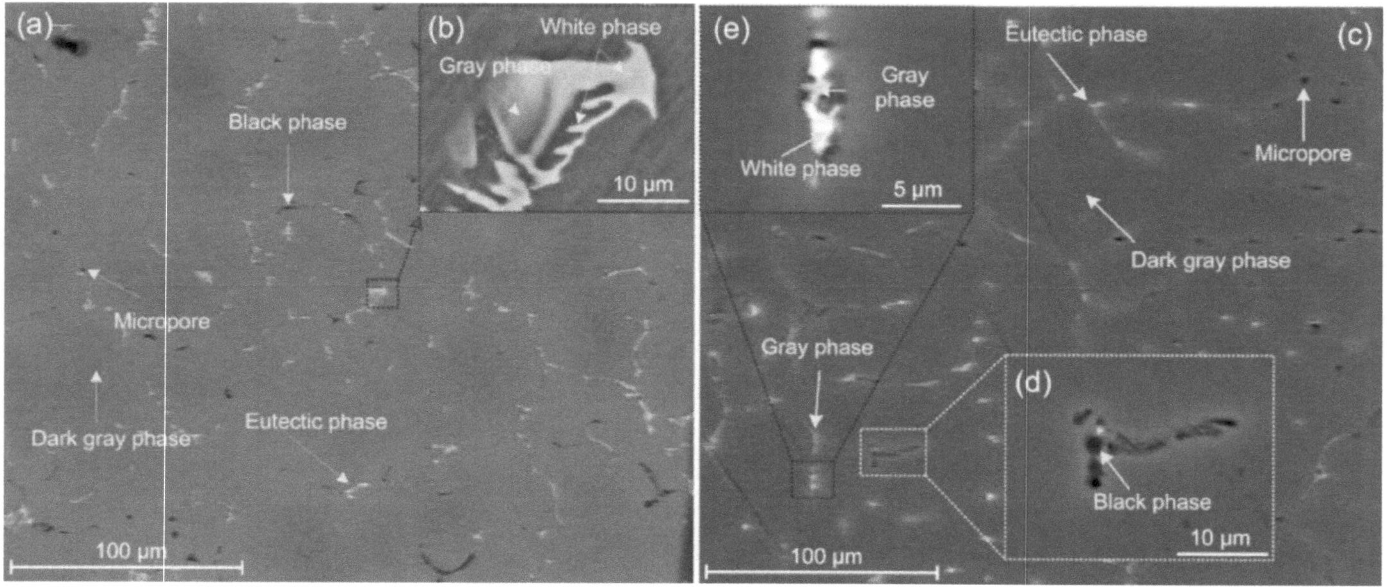

FIGURE 2.6 Typical backscattered images of the in-situ cast ((a) distribution of the eutectic phases and (b) magnified view of the gray eutectic phase) and ex situ cast ((c) distribution of the eutectic phases, and magnified views of (d) the gray eutectic phase and (e) the black eutectic phase)

Source: Reprinted with permission from Ref. [22]. © 2019 IOP Publishing.

TABLE 2.1 The Type of Material and Process Conditions with Their Outcomes Reported for Different Bulk Metallic Material.

Material and reported by	Process parameters					Characterizations	Remarks
	Casting type	Power & frequency	Susceptor	Exposure time (min)	Temp.		
Al 1050 (Lingappa et al., 2017)[6]	Ex-situ	900 W & 2.45 GHz	Silicon carbide	20 min	K type thermocouple	Microstructure, Microhardness, Microtensile test, SEM, EDS, XRD	Dense structure with equiaxed grains, Avg. tensile strength- 83.9 MPa, % Elongation—12.93%, Microhardness—121.1±10 HV
MMC (EWAC + 10% SiC) (Singh et al., 2016)[7]	In-situ	900 W & 2.45 GHz	Pulverized charcoal	18–24 min	–	Microstructure, Microhardness, SEM, XRD	Equiaxed grains with uniform reinforcement dispersion in matrix, Microhardness—950±208 HV
MMC (EWAC+ 5&10% SiC) (Singh et al., 2016)[8]	In-situ	900 W & 2.45 GHz	Pulverized charcoal	18–24 min	–	Microstructure, Microhardness, SEM, EDS, XRD	Equiaxed grains with uniform reinforcement dispersion in matrix phase, Microhardness—1085±210 HV
MMC (EWAC+ 10% SiC) (Singh et al., 2019)[10]	In-situ	900 W & 2.45 GHz	Pulverized charcoal	22±3 min	Noncontact type pyrometer	Microstructure, Microhardness, Microtensile test SEM, EDS, XRD	Equiaxed grains with matrix phase and uniform reinforcement dispersion, Microhardness—980±208 HV, Tensile strength—442±20 MPa, % Elongation—11±3%
MMC (EWAC+ 5&10% Ni-WC-8Co) (Singh et al., 2017)[11]	In-situ	900 W & 2.45 GHz	Pulverized charcoal	25 min	Noncontact type pyrometer	Microstructure, Microhardness, SEM, EDS, XRD	Equiaxed grains with uniform WC-particles dispersion in matrix phase, Microhardness—788±52 HV
Al-7039 (Al-Zn-Mg) (Mishra and Sharma 2016)[12]	In-situ	1400 W & 2.45 GHz	Silicon carbide	15.5 min	Build-In IR pyrometer	Microstructure, Microhardness, SEM, EDS, XRD, Temp–Time, Porosity	Microstructure study reveals uniform dense structure with porosity less than 2%, Microhardness—85.5±26.5 HV

TABLE 2.1 *(Continued)*

Material and reported by	Process parameters					Characterizations	Remarks
	Casting type	Power & frequency	Susceptor	Exposure time (min)	Temp.		
Al-7039 (Al-Zn-Mg) (Mishra and Sharma 2017)[13]	In-situ	1400 W & 2.45 GHz	Silicon carbide	8 min 25 s	Build-In IR pyrometer	Microstructure, Microhardness, Microtensile test SEM, EDS, XRD, Porosity, EDS Fractography	Dense structure with porosity less than 2%, Avg. Microhardness—132 HV near grain boundary, Avg. Microhardness—120 HV inside ∝ Al grain, UTS was 33% higher than as received alloy, Yield strength was 22% higher than as received alloy, % Elongation was 20% higher than as received alloy
Al-7039 (Al-Zn-Mg) (Mishra and Sharma 2017)[14]	In-situ	1400 W & 2.45 GHz	Silicon carbide (SiC) & charcoal (CC)	–	Build-In IR pyrometer	Microstructure, Microhardness, SEM, EDS, XRD, Temp–Time	Microstructure-Alumina mold with CC—coarse grain, Alumina mold with SiC—finer grain, Graphite mold with CC—Equiaxed grain, Graphite mold with SiC—Equiaxed grain, MICRO-HARDNESS-AMC-90±20 HV, AMSi-146±10 HV, GMC-107±20 HV, GMSi-121±15 HV
Al-7039 (Al-Zn-Mg) (Mishra and Sharma 2017)[22]	In-situ/ Ex-situ	1400 W & 2.45 GHz	Silicon carbide	–	Build-In IR pyrometer	Microstructure, Microhardness, SEM, EDS, XRD	Yield strength was observed to be 10% higher in in-situ cast than ex-situ cast, UTS was 16% higher in-situ cast, due to better cooling rates, grain size of in-situ cast was observed to be smaller than ex-situ cast

TABLE 2.1 *(Continued)*

Material and reported by	Process parameters					Characterizations	Remarks
	Casting type	Power & frequency	Susceptor	Exposure time (min)	Temp.		
Al-7039 (Al-Zn-Mg) (Mishra and Sharma 2019)[23]	In-situ	1400 W & 2.45 GHz	Silicon carbide	–	Build-In IR pyrometer	Microstructure, Microhardness, SEM, EDS, XRD	The cast made through water cooling cavity has smaller equiaxed grains with dense microstructure, Avg. Microhardness of cast made by conventional microwave casting was observed to be 191±32 HV
Copper (Cu) (Mishra and Sharma 2018)[24]	In-situ	1400 W & 2.45 GHz	Silicon carbide	14 min 40 s–15 min	Build-In IR pyrometer	Microstructure, Microhardness, SEM, EDS, XRD	Dense structure, porosity of in-situ cast was 2%–5%, avg. microhardness—93±20 HV
Naval Brass (Marahadige et al., 2018)[25]	–	3300 W & 2.45 GHz	Silicon carbide	17 min	K-type thermocouple	Microstructure, Microhardness, Microtensile test SEM, EDS, XRD, Porosity	Microstructure reveals formation of α & β phases with negligible porosity, yield strength—185.14 MPa, UTS—347.24 MPa, microhardness—145±3 HV

SEM, scanning electron microscopy; EDS, energy dispersive spectroscopy; XRD, X-ray diffraction analysis.

time saving, and so on, over the conventional processes. The archived reports or literature review reveals the research work in the area of aluminum alloys, metal matrix composites, copper, and naval brass. The successful completion of MWC of aluminum alloy, MMCs with better mechanical properties, and fine microstructures have opened the door for MWC of several other metallic materials. MWC has been a recently developed novel technique that is gaining popularity with the aim to fulfill the research areas, such as optimization of process parameters, mathematical modeling of MWC, tool designing, percentage increase in the utilization of microwave energy, in process estimation of properties, study of flow, and temperature on quality of MWC.

KEYWORDS

- **microwave material processing**
- **casting**
- **mechanical behavior**
- **microstructure**

REFERENCES

1. Singh, H.; Jain, P. K.; Bhoi, N.; Pratap, S. Experimental Study Pertaining to Microwave Sintering (MWS) of Al-Metal Matrix Composite—A Review. *Mater. Sci. Forum* **2018,** *928* (c), 150–155, DOI: 10.4028/www.scientific.net/MSF.928.150.
2. Bhoi, N. K.; Singh, H.; Pratap, S. Synthesis and Characterization of Zinc Oxide Reinforced Aluminium Metal Matrix Composite Produced by Microwave Sintering. *J. Compos. Mater.* **2020,** *54* (24), 3625–3636. DOI: 10.1177/0021998320918646.
3. Bhoi, N. K.; Singh, H.; Pratap, S. Developments in the Aluminum Metal Matrix Composites Reinforced by Micro / Nano Particles—A Review. *J. Compos. Mater.* **2020,** *54* (6), 813–833. DOI: 10.1177/0021998319865307.
4. Bhoi, N. K.; Singh, H.; Pratap, S.; Jain, P. K. Microwave Material Processing: A Clean, Green, and Sustainable Approach. In *Sustainable Engineering Products and Manufacturing Technologies*, 2019' pp 3–23. DOI: 10.1016/B978-0-12-816564-5.00001-3.
5. Samyal, R.; Bagha, A. K.; Bedi, R. The Casting of Materials Using Microwave Energy: A Review. *Mater. Today Proc.* **2020,** *26* (2), 1279–1283. DOI: 10.1016/j.matpr.2020.02.255.
6. Lingappa, M. S.; Srinath, M. S.; Amarendra, H J. Microstructural and Mechanical Investigation of Aluminium Alloy (Al 1050) Melted by Microwave Hybrid Heating. *Mater. Res. Express* **2017,** *4* (7). DOI: 10.1088/2053-1591/aa7aaf.

7. Singh, S.; Gupta, D.; Jain, V. Microwave Melting and Processing of Metal-Ceramic Composite Castings. *Proc. Inst. Mech. Eng. B J. Eng. Manuf.* **2016,** *232* (7), 1235–1243. DOI: 10.1177/0954405416666900.
8. Singh, S.; Gupta, D.; Jain, V. Novel Microwave Composite Casting Process: Theory, Feasibility, and Characterization. *Mater. Design* **2016,** *111*, 51–59. DOI: 10.1016/j.matdes.2016.08.071.
9. Bhoi, N. K.; Singh, H.; Pratap, S. A Study on Microwave Susceptor Material for Hybrid Heating. *J. Phys. Conf. Ser.* **2019,** *1240012097*, 1–6. DOI: 10.1088/1742-6596/1240/1/012097.
10. Singh, S.; Gupta, D.; Jain, V. Microwave Processing and Characterization of Nickel Powder-Based Metal Matrix Composite Castings. *Mater. Res. Express.* **2019,** *6*, 0865b1. DOI: 10.1088/2053-1591/ab2138.
11. Singh, S.; Gupta, D.; Jain, V. Processing of Ni-WC-8Co MMC Casting through Microwave Melting. *Mater. Manuf. Process.* **2017,** *33* (1), 26–34. DOI: 10.1080/10426914.2017.1291954.
12. Mishra, R. R.; Sharma, A. K. On the Mechanism of In-Situ Microwave Casting of Aluminium Alloy 7039 and Cast Microstructure. *Mater. Design* **2016,** *112*, 97–106. DOI: 10.1016/j.matdes.2016.09.041.
13. Mishra, R. R.; Sharma, A. K. Structure-Property Correlation in Al–Zn–Mg Alloy Cast Developed through In-Situ Microwave Casting. *Mater. Sci. Eng. A* **2017,** *688*, 532–544. DOI: 10.1016/j.msea.2017.02.021.
14. Mishra, R. R.; Sharma, A. K. Effect of Susceptor and Mold Material on the Microstructure of In-Situ Microwave Casts of Al-Zn-Mg Alloy. *Mater. Design* **2017,** *131*, 428–440. DOI: 10.1016/j.matdes.2017.06.038.
15. McGill, S. L.; Walkiewicz, J. W.; Smyres, G A. The Effect of Power Level on the Microwave Heating of Selected Chemicals and Minerals. *MRS Online Proc. Lib. (OPL) Symp. M—Microwave Process. Mater. I* **1988,** *124*, 247. DOI: https://doi.org/10.1557/PROC-124-247.
16. Sheinberg, H.; Meek, T.; Blake, R. D. Microwave of Normally Opaque and Semi Opaque Substances. *United States Patent* **1990,** *4*, 942, 278.
17. Roy, R.; Agrawal, D.; Cheng, J.; Gedevanlshvili, S. Full Sintering of Powdered-Metal Bodies in a Microwave Field. *Nature* **1999,** *399*, 668–670. https://doi.org/10.1038/21390.
18. Moore, A. F.; Schechter, D. E.; Morrow, M. S. Method and Apparatus for Melting Metals. *45*, 57–68. *United States Patent* **2006,** *7*, 011, 136 B2.
19. Chandrasekaran, S.; Basak, T.; Ramanathan, S. Experimental and Theoretical Investigation on Microwave Melting of Metals. *J. Mater. Process. Technol.* **2011,** *211* (3), 482–487. DOI: 10.1016/j.jmatprotec.2010.11.001.
20. Choudhury, M.; Puri, S. *Development of Microwave Casting, Dissertation, Roorkee (India)*; Indian Institute of Technology Roorkee: Roorkee, India, 2013.
21. Mishra, R. R.; Sharma, A. A. A New In-Situ Casting Technique Using Microwave Energy at 2.45 GHz. *Proc. India Int. Sci. Festiv.- Young Sci. Meet, DST, Govt. India, Design* **2015,** *58*, 1–7.
22. Mishra, R. R.; Sharma, A. K. Microstructural Characteristics and Tensile Properties of In-Situ and Ex-Situ Microwave Casts of Al-7039 Alloy. *Mater. Res. Express.* **2019,** *6*, 126591. DOI:10.1088/2053-1591/ab5ab1.
23. Mishra, R. R.; Sharma, A. K. Effect of Solidification Environment on Microstructure and Indentation Hardness of Al–Zn–Mg Alloy Casts Developed Using Microwave Heating. *Int. J. Metal Cast.* **2017,** *12* (2), 370–382. DOI: 10.1007/s40962-017-0176-1.

24. Mishra, R. R.; Sharma, A. K. Experimental Investigation on In-Situ Microwave Casting of Copper. *IOP Conf. Ser. Mater. Sci. Eng.* **2018,** *346*, 012052. DOI: 10.1088/1757-899x/346/1/012052.
25. Marahadige, S. L.; Sridharmurthy, S. M.; Jayraj, A. H.; Mahabaleshwar, U. S.; Lorenzini, G.; Lorenzini, E. Development of Copper Alloy by Microwave Hybrid Heating Technique and Its Characterization. *Int. J. Heat Technol.* **2018,** *36* (4), 1343–1349. DOI: 10.18280/ijht.360425.
26. Kumar, A.; Bhoi, N. K.; Singh, H. Corrosion Behavior of Microwave Clad Material Under Different Acidic Environment. In *Manufacturing Engineering*; Sharma, V., Dixit, U., Sørby, K., Bhardwaj, A., Trehan, R., Eds.; Lecture Notes on Multidisciplinary Industrial Engineering; Springer: Singapore, 2020; pp 603–612. https://doi.org/10.1007/978-981-15-4619-8_43.
27. Bhoi, N. K.; Singh, H.; Saurabh, P. Promise of Self-lubricating Aluminum-Based Composite Material. In *Functional and Smart Materials*; Prakash, C., Singh, S., Davin J. P., Eds.; CRC Press Taylor & Francis Group, 2020; pp 65–82. https://doi.org/10.1201/9780429298035.
28. Kumar, Y.; Singh, H. Multi-response Optimization in Dry Turning Process Using Taguchi's Approach and Utility Concept. *Procedia Mater. Sci.* **2014,** *5*, 2142–2151. https://doi.org/10.1016/j.mspro.2014.07.417.

CHAPTER 3

CUTTING FORCE AND CUTTING TEMPERATURE MODELING IN MILLING OF SUPERALLOYS AND COMPOSITES: A REVIEW

SHAILENDRA CHAUHAN, RAJEEV TREHAN, and RAVI PRATAP SINGH*

Department of Industrial Production Engineering, Dr. B.R. Ambedkar National Institute of Technology, Jalandhar, Punjab, India

**Corresponding author. E-mail: singhrp@nitj.ac.in*

ABSTRACT

Superalloys and composites are increasingly used in aerospace, biomedical applications, etc., and the complicated machinability of the material has drawn numerous research efforts. To prepare and optimize certain procedures, precise prediction of forces in the machining phase of superalloys is having great significance. There is no systematic analysis of cutting force and cutting temperature simulation of superalloys machining. The development of cutting forces, cutting temperature models created for the cutting process of superalloys and composites, is mainly reviewed in this paper. Documented models are explained, such as models of finite elements and models of neural networks, which are built either from the central theory of classical theoretical analysis or from computational simulations. Modeling rules and future developments in science are also presented and outlined in this chapter.

Modern Manufacturing Systems: Trends and Developments. Rajiv Kumar Garg, Ravi Pratap Singh, Rajeev Trehan, & Ramesh Singh (Eds.)

3.1 INTRODUCTION

Since superalloys have superior mechanical properties; such as high strength-to-weight ratio than conventional metallic materials.[1] A high ratio of rigidity to weight, good resistance to fatigue, good dimensional flexibility, and respectable corrosion confrontation.[2,3] In the aerospace, aeronautics, automotive, shipbuilding, chemical manufacturing equipment, and chemical processing sectors, they have been increasingly used.

In microinstruments for surgical treatment as well as in small size biomedical implantations such as miniature left ventricular assist devices, finger joint replacements and small bone implants, superalloys and composites are commonly used. Because of their thermomechanical properties, titanium alloys are known to be difficult for machining materials. Calculation of the workpiece and instrument temperature fields during the microparts.

A series of researchers have developed cutting temperature and force models for various superalloys and composite machining processes over the past few years. Table 3.1 lists the models of cutting temperature and force of various machining operations set up in the literature. The current review chapter is dedicated to the modeling of cutting temperature and forces models in milling of superalloys and composites, to best knowledge of the authors, and this study therefore offers a thorough analysis of the research findings on cutting force models for superalloy and composite machining.

TABLE 3.1 Models of Cutting Force and Cutting Temperature of Various Machining Operations Set Up in Literature.

Machining operation	Models	Authors and references
Milling	ANN	Sheikh-Ahmad et al.,[2] Li et al.,[4] Kalla et al.,[3] Abdul Shahed et al.,[23] Lopes Da Silva et al.[24]
	Finite element method	Rentsch et al.,[1] Dang et al.,[7] Matsumura and Usui,[8] Rao et al.,[6] Denkana et al.,[9] Özel and Altan,[5] Özel et al.,[10] Thepsonthi,[10] Özel,[11] Wu et al.,[12] Puls et al.,[13] Lazoglu and Bugdayci,[14] Zhongtao Fu et al.,[15] Davoudinejad et al.,[16] Grzesik, Wit et al.,[17] Denkena et al.,[18] Baohai et al.,[19] Mamedov and Lazoglu,[20] Ucun et al.,[21] Cui et al.[22]

3.2 MATERIAL AND METHODOLOGY

For hard-to-machine materials like aeronautical, die materials, and medicinal components made of metallic alloys, the milling process is widely used.

While parts made from superalloys are often manufactured to the net form, to fulfill tolerances, they often involve the removal of excess material. The most expedient machining process for such a purpose is milling. Milling force models include several models, such as FEMs, ANNs, and mechanical models; according to different modeling approaches. Superalloys are high-performance alloys that show magnificent creep confrontation and mechanical strength at elevated temperatures, superior surface stability, and oxidation and corrosion resistance. They have a base alloying element of nickel, nickel-iron, or cobalt with an austenitic FCC structure.[31] Power and aerospace industries have mainly contributed in the development of superalloys. Their development has ponderously relied on process and chemical innovations. Elevated temperature strength is created by strengthening superalloys by solid solution. Strengthening process is the amplification of precipitation, which occurs in the secondary step of carbides and precipitates such as gamma prime.[32,33] Elements such as chromium and aluminum offer resistance to corrosion and oxidation since these alloys are intended for applications at high temperatures and are of primary importance for oxidation and creep resistance. Ni (Nickel) based superalloys have appeared as one of the materials of preference for such applications.[34] Mainly attributed to their corrosion-resistant properties and their high strength-to-weight ratio, superalloys are used in various applications to sustain high temperature in nuclear reactors and stream turbines, in addition to being used in the aerospace industry.[35] The properties of these nickel-based superalloys can be made-to-order to a certain extent; cobalt, examples of alloying elements include iron, chromium, tungsten, molybdenum, titanium, tantalum, aluminum, niobium, zirconium, vanadium, yttrium, rhenium, boron or hafnium, and carbon elements used.[36–41]

Each modification has been selected to satisfy the specific tenacity of improving the properties for use at high temperatures. Thanks to their superior properties and excellent corrosion resistance at elevated temperatures among commercially available Ni-based alloys, nickel-based superalloys are also used in the manufacture of aerospace turbine blades.[42–49] While superalloys based on Ni are tremendous elevated temperature materials and have recognized to be very valuable, Cobalt-based superalloys may have superior wear resistance, oxidation, and hot corrosion compared to superalloys based on Ni. Therefore, attempts have been made over the past few years to create superalloys based on cobalt.[50–57] Instead, conventional superalloys based on cobalt have not been commonly used because their strength at high temperatures is lower than superalloys based on nickel. Essentially, it is possible to classify superalloys into three types: nickel base, cobalt base, and iron base. This can also be divided into different forms, as shown in Figure 3.1.[59]

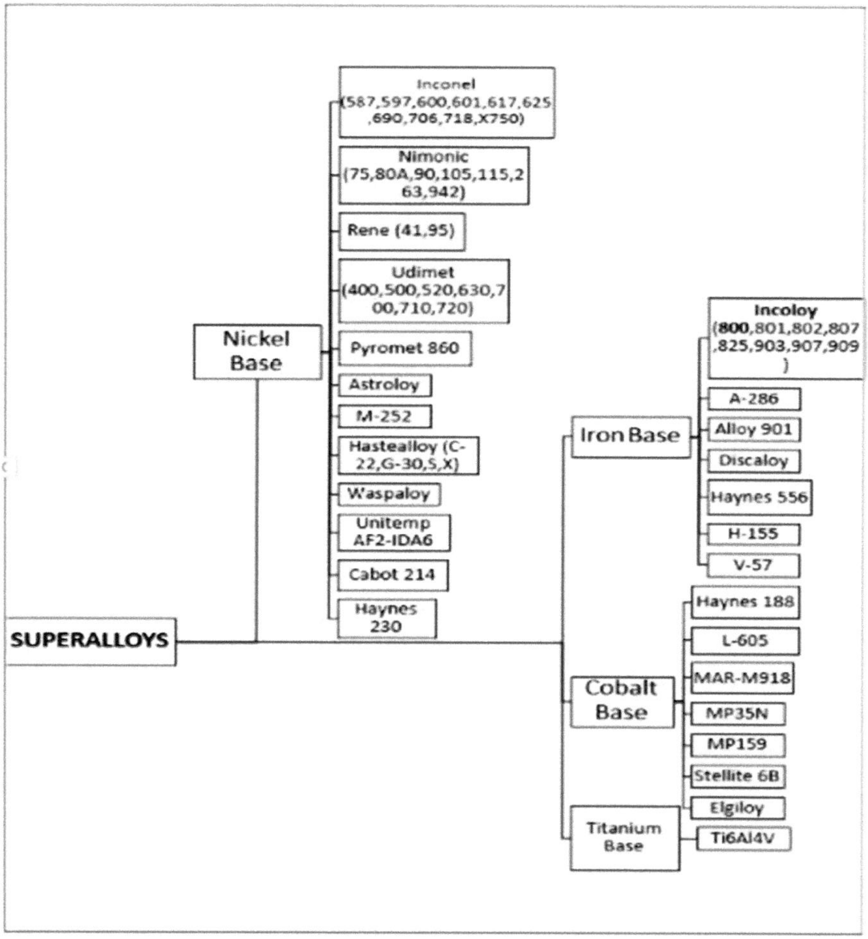

FIGURE 3.1 Classification of superalloys.[59]

Source: Adapted with permission from Ref. [59].

The milling of superalloys and composites has become a hot research area as a new approach to production technologies, and relevant literature is also reviewed in this chapter.

3.2.1 TI6AL4V

With its extensive use in the medical, shipbuilding, chemical, and aerospace industries, Ti6Al4V is by far the most common among the various titanium alloys. The main reason for the wide use of this titanium alloy is its superior

TABLE 3.2 Chemical Composition of Various Superalloys.

Alloy	Ni	Co	Cr	Mo	Mn	Cu	W	Nb	Fe	Al	V	Ti	C
Haste alloys	Base	–	21	09	–	–	–	–	18	–	–	–	0.05
Udimet 700	Base	15	15	05	–	–	–	–	3.5	4.3	–	3.5	0.05
Inconel 718	Base	0.04	19	03	–	–	–	5.1	18.5	0.5	–	0.9	–
Haynes No. 25	10	Base	20	–	–	–	14	–	03	–	–	–	0.05
Inconel 825	38–46	–	19.5–23.5	2.5–3.5	–	1.5–3	–	–	22	–	–	0.6–1.2	–
Incoloy 800	34.991	0.07	20.096	0.335	0.501	–	0.066	–	42.821	0.302	0.027	0.304	0.096
Inconel 601	61	–	23	–	<01	< 01	–	–	14	01	–	–	<0.10
Nimonic C-263	49	19–21	19–21	5.6–6.1	–	0.2 max	–	–	0.7 max	<0.6	–	1.9–2.4	0.04–0.08
Inconel 690	58	–	27–31	–	0.50 max	0.50 max	–	–	07–11	–	–	–	0.050 max
Inconel 625	Balance	01	20–23	08–10	0.5	–	–	3.5–1.5 (Nb+Ta)	05	0.4	–	0.4	–
Ti6Al4V	–	–	–	–	–	–	–	–	0.3	6.75	4.5	Balance	0.08

properties, which can be preserved at elevated temperatures and outstanding corrosion and fracture resistance. However, owing to its low thermal conductivity, which causes high tool temperatures and strong chemical affinity to most tool materials, Ti6Al4V is infamous for improper machinability, contributing to premature tool failure. In addition, it is inhomogeneous deformation by devastating shear that the cutting force fluctuates and the tool-wear and chatter worsen.

The detailed definition of the material properties of any FEM model is an important input. The workpiece is normally subjected to intense deformation under machining conditions involving high levels of stresses and rapid rise in temperature. Constitutive equations explaining stress-stress reactions, along with their dependency on strain rate, temperature, and hardening of the workpiece, are typically modeled on the workpiece material.[6]

3.2.2 INCONEL 718

Inconel alloys are superalloys based on Ni-Cr that cover a broad spectrum of composition and mechanical properties. Oxidation, corrosion, carburization, and other damage mechanisms operating at high temperatures are prevented by Ni and Cr. Inconel alloys have strong fatigue, improved cryogenic properties, and mechanical strength and reasonably good creep behavior at moderate temperatures.[39,40] Inconel alloys are typically extra-alloyed with Co, Nb, Ti, Al, W, and Cu to increase mechanical resistance and corrosion. Iron can be contained in concentrations ranging from 1% to 20% as well. For heat treatment receivers, aviation, engines, nuclear power plants, and so on,[41,42] these superalloys are sunken.

Unlike steel alloys, the nickel-base alloys have high yield and ultimate strength over a large range of temperature with lower thermal conductivity. Inconel alloys are corrosion resist materials that are ideally suited for use under pressure and heat in harsh conditions. Inconel forms a thick, robust, surface passivation oxide layer when hot, shielding the surface from further attack.

Owing to the accelerated hardening of the job, Inconel 718 is a challenging metal to shape and machine using traditional cold forming techniques. Age-hardened Inconel's, like Inconel 718, are machined with a hard instrument using a violent yet slow cut, which decreases the amount of passes needed. Inconel 718 is a high-strength nickel-chromium substance, corrosion-resistant, and used at −423°–1300°F. As compared to steel alloys,

Inconel 718 has comparatively low friction coefficients.[17–19] Inconel is also used in extreme conditions. Inconel 718 is primarily used in nuclear power plants, the aerospace, chemical, and pressure vessel industries.[13]

3.2.3 COMPOSITES

Composite material produces from two or more different materials with significantly different chemical and physical properties that, when combined, create a material with distinctive properties from the individual components. Inside the finished structure, the discrete elements remain distinct and separate, distinguishing composites from mixtures and solid solutions. Fiber-reinforced plastic (FRP) is a composite material composed of a fiber-reinforced polymer matrix. Although phenol-formaldehyde resins are still in usage, the material is usually an epoxy, vinyl ester, or polyester thermosetting component. FRPs are commonly used in the aviation, shipping, marine, and industrial industries.

3.2.4 ARTIFICIAL NEURAL NETWORKS MODELS

By learning from a varying value database to forecast cutting temperature and force in milling, artificial neural network (ANN) models have an implicit relationship. The ANN model is used by many researchers to construct prediction models. However, in order to train the ANN, the models also require a lot of experimental data and do not show the fundamental cutting process mechanism. Ahmad et al.[2] built mechanical strength models for straight edge milling of unidirectional and multidirectional fiber-reinforced polymers. The models were based independently on the Multiple Regression Models and Committee on Neural Networks of the basic functions of energy cutting. The input variables for both versions are chip width, and responses are average real cutting energies in the parallel (Kc) and regular (Kt) directions to the cutting speed. Kalla et al.[3] simulated and modeled cutting energy for milling process of the composite with the end mill cutter and then used this function to approximate the milling power of each fiber orientation. Planned forces associated with unidirectional laminate are more precise compared to laboratory results than those associated with multidirectional laminate.

Until an adaptive neuro-fuzzy inference method (ANFIS) is used to build two thermal prediction models, Abdulshahed et al.[23] have reviewed various approaches for modeling thermal error models. The Gray method theory is used to characterize the influence of all possible temperature sensors on the thermal reaction of the configuration of the device. Based on cutting power and AE signals, tool wear at the end of milling was calculated in each step. The wear conditions were then classified as good or bad, the signal parameters were derived and the neural probabilistic network was used. Figure 3.2 shows: (a) the Boxplots display a full signal behavior using cutting strength, skewness Acoustic Emission, and Root Mean Square Spectra Acoustic Emission parameters for all tool life tests, (b) normalized life skewness, and (c) fast Fourier acoustic emission transition in the tool life.[24]

3.2.5 FINITE ELEMENT MODELING

FEM uses finite element templates to simulate the milling process of superalloys and composites. There are few reports on milling operation modeling simulation along with orthogonal cutting. Rentsch et al.[1] used FEM with a disc milling cutter to study the slot milling mechanism of UD-CFRP and created two types of orthogonal cutting models, i.e., miniscule and macroscopical cutting models. A more visual explanation of the method of substance removal is possible in the miniscule model. The cutting force pattern expected by the microscopic model is closer to the evidence from the experiment. However, it is very difficult to use a miniscule model to simulate the milling process. The milling model has actually only been condensed into a two-dimensional model of orthogonal cutting. A novel cutting force model involving the bottom edge cutting effect was proposed by Dang et al.[7] to describe and thoroughly analyze the real cases of flat end milling. When the axial cutting depth is comparatively limited and the spindle speed determines the distribution of cutting power, they find that bottom edge cutting has a surprising impact on overall cutting force. Experiments also resulted in the cutting force being small as the spindle speed is raised. Matsumura and Usui[8] suggested a cutting force model to simulate the machining processes of complicated-shaped end mills and concluded that chip flow angle depends only on inclination of the local edge.

The experimentation and numerical analysis of face milling of Ti6Al4V with uncoated tungsten carbide in the presence of coolant was carried out by Rao et al.[6] To measure specific energy cutting, surface integrity, and tool quality, an experimental analysis was carried out to measure. The experimental study of thermo-mechanical effects on milling of thin-walled

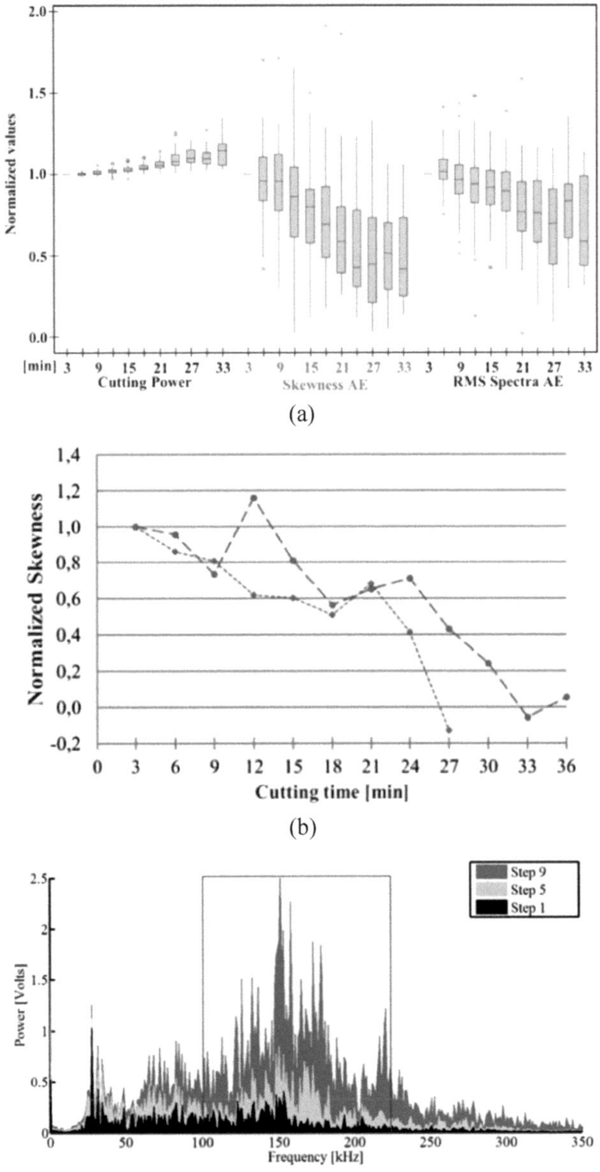

FIGURE 3.2 (a) Boxplots display a full signal behavior using cutting strength, skewness Acoustic Emission, and Root Mean Square Spectra Acoustic Emission parameters for all tool life tests, (b) normalized life skewness, and (c) fast Fourier acoustic emission transition in the tool life.

Source: Reprinted with permission from Ref. [24]. © 2016 Taylor & Francis.

structural components was described by Denkana et al.[9] A process model that can predict the generation of design deviations based on the calculated workpiece deflections is introduced. The results of the simulation are correlated to the measured surface and the surface is separated into major shape domains influenced by thermal and mechanical loads. Thepsonthi and Özel[5] have proposed a tool for modeling the cutting process using FEM in the flat end milling process and for forecasting chip flow, cutting strain, tool stress, and temperature. The comparison of the expected cutting forces with the calculated forces demonstrated fair agreement and shows that the stresses and temperature of the instrument are also predicted with sufficient precision. Özel et al.[10] examined simulations of finite elements on Ti6Al4V alloy micromilling with uncoated fine grain and CBN coated micromilling. Finite Element simulation is used for uncoated and CBN-coated microinstruments to measure forces, temperature, and wear speed. The advantages of CBN coatings reveal the expected instrument wear and temperature contours for micro edge-tool coated with uncoated CBN.[11] An attempt has been made to boost the performance of micro-end carbide mills by applying a layer of cubic boron nitride (CBN). FE simulations were used to analyze the process of chip creation in micromilling to identify the effects of CBN-coated micromilling with a higher edge radius in terms of cutting force and tool temperature.

A deduced cutting force forecast model for the circular end milling process was suggested by Wu et al.[12] Based on the standard linear milling force model, they proposed an improved cutting force model for the circular end milling process. An experimental test to analyze the friction mechanisms inside the tool-chip interface in metal cutting was suggested by Puls et al.[13] A temperature-dependent friction model is proposed and validated by a finite element model to account for the observed phenomenon. Investigations were carried out in conjunction with uncoated WC-6Co cemented carbide tools for Inconel 718.[14] For rapid and precise temperature estimation in end milling processes, novel thermal modeling is implemented. For different cutting environments, a theoretical modeling method and experimental validations are provided. In a small cutting field, combined with intense shearing, plastic deformation, and friction, milling temperatures can have a very critical impact on parts and machinery, such as dimensional defects, residual stress, and tool wear. Temperature in the milling process is a nonlinear complex simulation challenge. Zhongtao Fu et al.[15] proposed an empirical method focused on the predictive machining theory for the simulation of milling forces in the helical end milling process, which considers the properties of the workpiece material, instrument geometry, cutting conditions, and milling forms as input data. Using this predictive oblique cutting model with the effect of cutting-edge radius, the three-dimensional cutting force components applied on each element are analytically predicted. A finite element simulation was developed by Davoudinejad et al.[16] to study the simulation of the dry and cryogenic orthogonal cutting of ti-6al-4v using carbide-coated inserts. The numerical prediction of the effect of superior cryogenic cooling on the shape, cutting, and thrust forces of the chips was investigated. Comparison with experimental results validated the simulation. Denkena et al.[18] suggested a model for estimating temperature-mediated differences in dry milling based on simulation. Cutting forces and heat flux are calculated using a semiempirical process model based on the computed contact field.

A new analytical model approach to forecast the temperature of cutting tool in final milling has been proposed by Baohai et al.[19] The temperature forecast model considered the real friction state between the chip and the instrument, is suggested for the temperature increase stage, and the heat flux and the interaction length of the tool chip are then obtained by simulating the finite elements. A model of temperature reduction based on one-dimensional plate heat convection is proposed in the method of temperature reduction. Lazoglu and Mamedov[20] give a discussion using finite element simulations and experimental validity for the micromilling of titanium alloy. In the micromilling process of Ti6Al4V under various cutting conditions, a finite element model was proposed to simulate tool and workpiece temperature fields. Temperature simulations were verified by thermocouple measurements in the micromilling of Ti6Al4V. For micromilling of Ti6Al4V alloy, Ucun et al.[21] performed three-dimensional (3D) finite element analysis. By taking into account the influence of strain, strain rate, and temperature on material properties, the Johnson-Cook material model equation was used. DEFORM was used to simulate the milling process of Ti-6Al-4V based on the proven 3D finite element modeling technique.

Figure 3.3a shows the finite element model of micromilling, and (b) an illustration of the temperature fields at various immersion angles.

The temperature of the tool in the final milling process was examined by Cui et al.[22] by the theoretical and experimental method, taking into account the influence of flank wear. Taking into account the impact of flank wear, a temperature prediction model of the milling tool is presented, which was seldom taken into account in the previous modeling of the temperature during milling. Figure 3.4 shows the end milling diagram and approximation heating model.

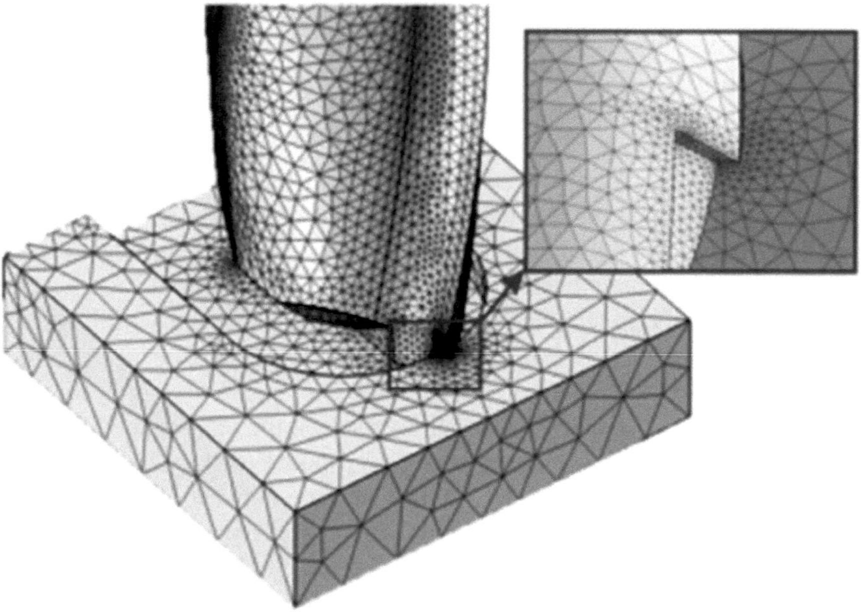

(a)

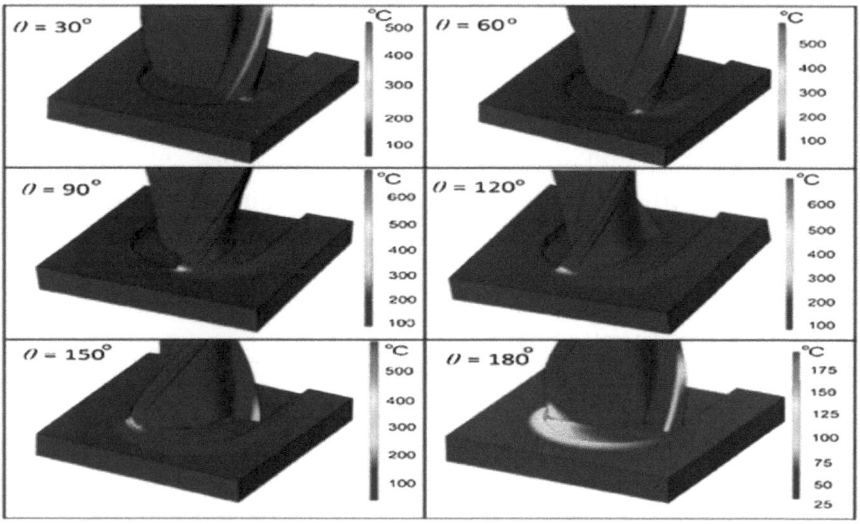

(b)

FIGURE 3.3 (a) Finite element model of micromilling. (b) An illustration of the temperature fields at various immersion angles.

Source: Reprinted with permission from Ref. [20]. © 2016 Elsevier.

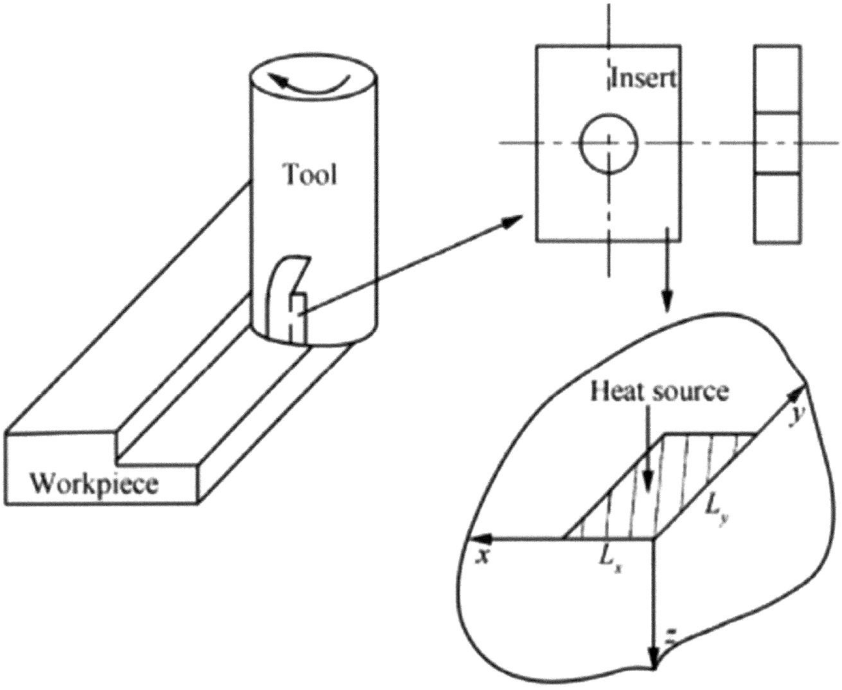

FIGURE 3.4 End milling diagram and approximation heating model.
Source: Reprinted from [19]. © 2016 Chinese Society of Aeronautics and Astronautics.

Different important points from the previous paper are summarized; process variants, w/p and tool materials, machining characteristic, and summary of various research papers are summarised in Table 3.3.

3.3 CONCLUSIONS

Superalloys and composites are commonly used in industries with many excellent mechanical properties. The paper addresses cutting temperature and forces simulation approaches in the milling of superalloys and composites. Out of the investigations carried out in last 10–15 years, most of the models have been revealed to be conducted in the domain of milling process and the same has been summarised and detailed out in proposed article methods of modeling are provided, starting from simple numerical simulations to fine-tuned analytical analyses, aimed at divulging the physical attributes of milling. Any findings and analysis directions are outlined from the articles above as follows.

TABLE 3.3 Literature Summary.

Sr. No.	Process variants	Workpiece material	Tool material	Machining characteristics investigated	Summary	Author and references
1	Slot milling	Composites	–	Cutting forces	A macro cutting force model is shown and this model was correlated with experimental findings based on inhomogeneous material properties with implicitly defined fiber orientation and simulation performance.	Rentsch et al.[1]
2	Flat end milling	Ti6Al4V	–	Cutting forces, and Cutting Temperature	Built a technique to simulate a flat end milling process and used a finite element analysis to model chip flow, forces, and pressures in cutting, stresses and temperatures on instrument and temperatures (FEA). Comparisons of the cutting forces (projected) with the cutting forces (measured) have shown indicated that the forces and temperatures of the instrument are also estimated with appropriate precision. Maximum tool temperatures, regardless of the cutting conditions, were required at the principal edge of cutting of the insert in flat end mill.	Özel and Altan[5]
3	Face milling	Ti6Al4V	Tungsten Carbide	Cutting force	An experimental investigation and numerical analysis in face milling of the Ti6Al4V were carried out. A 3-D Finite element modeling simulation was developed and the temperature of the tool chip interface was expected.	Rao et al.[6]
4	Flat end milling	–	–	Cutting forces	A model for new mechanical cutting force in flat end milling was developed. Principal advantage of the developed model was that both the edges of the flank and the bottom cutting edge take into account the overall cutting forces at the same time. The measured flank cutting force coefficients are calibrated instantaneously by the force variable obtained by deducting the flanking force factor from the total calculated force for the experimental validation demonstration.	Dang et al.[7]

TABLE 3.3 *(Continued)*

Sr. No.	Process variants	Workpiece material	Tool material	Machining characteristics investigated	Summary	Author and references
5	Ball end milling	–	–	Cutting force	Taking into consideration the slope of the pre-assembled surface, the model for chip flow was created using various machining data. The chip flow path was governed in order to mitigate the cutting energy i.e., the addition of the shear energy on shear plane and the friction energy on the rake surface. The CF is then estimated to be the specific cutting energy for the chip flow model.	Matsumura and Usui[8]
6	End milling	–	–	Cutting temperature	In order to discern the results of thermal deviations from process-force induced deflections, an experimental analysis of process-induced temperature and finite element simulations was performed.	Denkana et al.[9]
7	Micromilling	Ti6Al4V	Tungsten and CBN coated carbide	Cutting forces and cutting temperature	Finite element simulations of Ti6Al4V micromilling with noncoated and CBN coated micromilling were introduced. These calculations were carried out for modeling the CFs and temperatures and showed the advantages of CBN coatings.	Özel et al.[10]
8	Micromilling	Ti6Al4V	Tungsten and CBN coated carbide	Cutting forces and cutting temperature	Experimentation and simulations based on the FEM were used to investigate the effects of the CBN coated tool in micromilling of Ti6Al4V. In order to examine the mechanism of chip formation in micromilling, FE simulations were used to analyze the results of CBN coated micromilling with enhanced radius of edge in terms of generated CF and tool temperature.	Thepsonthi and Özel[11]
9	Circular end milling	Ti6Al4V	Tungsten Carbide	Cutting forces	Centered on the traditional linear milling strength model, an enhanced cutting strength model for the circular end milling process was presented. The proposed model predicted rapid CFs during the described machining process.	Wu et al.[12]

TABLE 3.3 *(Continued)*

Sr. No.	Process variants	Workpiece material	Tool material	Machining characteristics investigated	Summary	Author and references
10	Orthogonal cutting	INCONEL 718	Uncoated WC-6Co cemented carbide	Cutting temperature	An experimental analysis of the friction phenomena in machining operations within the tool-chip interface was presented. Investigations were performed in addition with noncoated WC-6Co cemented carbide tool inserts for Inconel 718. A friction model depending on temperature was developed and validated by a FEM to compensate the thermal softening of the contact interface.	Puls et al.[13]
11	End milling	–	–	Thermal modeling	For the sake of predicting temperatures rapidly and accurately in the milling process, a new thermal simulation was carried out. A theoretic simulation tactic and experimental evaluation are given for the different machining conditions.	Lazoglu and Bugdayci[14]
12	Helical end milling	Ti6Al4V	Tungsten Carbide	Cutting forces	An empirical approach to model the CFs in the helical end milling process was proposed on the basis of predictive machining that uses the predictive oblique cutting model to analyze the work material properties, geometry of the tool, conditions machining, and types of milling as input data for 3-D CF components implemented to each part and the suggested model was evaluated by simulation values using the advantages of FEM software.	Zhongtao Fu et al.[15]
13	Orthogonal cutting	Ti6Al4V	TiAlN coated carbide inserts	Cutting forces	The FEM model was developed for simulating the dry and cryogenic cutting of Ti6Al4V with the help of TiAlN coated carbide inserts and correlated the simulations to the experimental validation findings.	Davoudinejad et al.[16]

TABLE 3.3 *(Continued)*

Sr. No.	Process variants	Workpiece material	Tool material	Machining characteristics investigated	Summary	Author and references
14	Face milling	INCONEL 718	Tungsten Carbide	Cutting temperature	Three-dimensional FEM simulation findings obtained using the Johnson-Cook material modeling approach and varying cutting environments for milling operations with nickel-chromium alloy (Inconel 718) were presented. The mean and maximum values for the cutting temperature are used in FEM simulations.	Grzesik, Wit, et al.[17]
15	Face milling	–	–	Cutting force, Thermal error	A simulation-based predictive model for dry milling of temperature-induced shape deviations was introduced. CFs and heat flux are calculated using a semiempirical process model on the basis of computed contact field. The simulation results are compared to the calculated data, which points out that the workpiece temperatures, the workpiece deformation, and differences in shape are accurately expected in different workpiece regions.	Denkena et al.[18]
16	End milling	INCONEL 718	–	Cutting tool temperature	A new computational model-based methodology to determining cutting tool temperature in final milling has been introduced. A temperature forecast model is presented for the temperature increase process, taking into account the real friction condition between the chip and the tool, heat flux and the interaction period of the tool chip are then determined by finite element simulation. A temperature-reduction model based on a 1-D plate. Convection of heat is presented in the temperature-reduction method and the comparison revealed a high degree of agreement among the machining experiments and the suggested temperature-prediction framework of the cutting instrument.	Baohai et al.[19]

TABLE 3.3 *(Continued)*

Sr. No.	Process variants	Workpiece material	Tool material	Machining characteristics investigated	Summary	Author and references
17	Micromilling	Ti6Al4V	–	Cutting temperature	A FEM model was evolved to approximate the temperature of the tool and the workpiece area under various cutting conditions in the micromilling of Ti6Al4V. Thermocouple measurements in the Ti6Al4V micromilling validate temperature simulations.	Mamedov and Lazoglu[20]
18	Micromilling	Ti6Al4V	–	Cutting forces, Cutting temperature	The 3D FEM was conducted for the micromilling of the Ti6Al4V alloy. Centered on the current 3D FEM, the simulations of the Ti6Al4V milling process were done using DEFORM. The chip formation cutting force and hip shape were determined in numerical experiments according to the experimental findings.	Ucun et al.[21]
19	End milling	–	–	Cutting temperature	Taking into account the tool wear effect by hypothetical and experimental methodology, the instrument temperature was investigated in end milling. The theoretical model was developed to predict the temperature of the final milling instrument. Experimental tests have been performed to validate the proposed model and have shown that the proposed model can accurately estimate the sharp and worn, temperature of the tool during the final milling process.	Cui et al.[22]
20	Orthogonal cutting	Composites	–	Cutting force	The mechanistic simulation approach was used to estimate cutting forces and simulate the fiber-reinforced polymer (FRP) milling process with a straight cutting point. Neural network estimation and multiple regression analysis of milling strength data and cutting model were formed and were able to approximate cutting strengths in unidirectional and multidirectional composite milling.	Sheikh-Ahmad et al.[2]

TABLE 3.3 *(Continued)*

Sr. No.	Process variants	Workpiece material	Tool material	Machining characteristics investigated	Summary	Author and references
21	Helical end mill	Composites	–	Cutting force	The forces in helical end milling are modeled on the basis of a predictive machining theory in which the general features of the workpiece sections, the geometry of the machinery, and the cutting conditions are predicted from the input data for the characteristic machining factors. The cutting forces are predicted using the theory of statistical machining in oblique cutting processes.	Li et al.[4]
22	Helical end mill	Composites	–	Cutting force	A mechanical simulation technique was used to simulate the cutting of CFRP with an end mill. The technique for forecasting cutting forces is developed by converting specific cutting energies from orthogonal cutting to oblique cutting. This evolved method is capable of predicting the cutting powers of unidirectional and multidirectional composite material in helical end milling.	Kalla et al.[3]
23	Face milling	–	–	Thermal error	Until the use of the Adaptive Neuro Fuzzy Inference Scheme for the construction of two thermal prediction models was decided, different methods for modeling thermal error models were tested. A limited CNC milling machine research is used to provide training data for the proposed models, and the study results suggest that the ANFIS-FCM model is superior to the precision of its predictive capabilities, benefiting from fewer rules.	Abdul-shahed et al.[23]

TABLE 3.3 *(Continued)*

Sr. No.	Process variants	Workpiece material	Tool material	Machining characteristics investigated	Summary	Author and references
24	End milling	–	–	Tool wear	A novel approach based on AE signal energy was tested to improve the response. The skewness of the cutting power, root mean square, and mean of the spectral acoustic emission density showed sensitivity with the aid of the neural network and was applied with approximately 91% accuracy. There is no question that the combination of cutting power and AE with the signal energy parameter would be applied in the tool control unit.	Lopes Da Silva et al.[24]
25	Milling	Composites	–	Cutting force	The modeling of the cutting force of the composite polymer matrix machining was tested. Models of artificial neural networks and finite elements are described, based either on the fundamental principle of classical theoretical analyses, or on numerical simulations.	Wan et al.[30]

- The cutting of superalloys and composites is more complex than the cutting of metal due to the anisotropic and nonuniform properties of superalloys and composites. FEM and other various modeling techniques are reliable only under certain stated situations; stated cutting conditions and special cutting tools.
- The series of researches on CF projections in milling is comparatively limited, and recent research focuses primarily on the creation of mechanical and mathematical models. Some experiments were performed using regression analysis and ANN, while the basis for others were conventional milling operations theory and at the same time incorporated superalloy and composite coefficients.
- Mathematical orthogonal cutting models have progressed from 2- to 3-dimensional, which can help in simulating the more complex machining technologies. Numerical simulation is seen as a promising way not only to optimize the manufacturing conditions and configuration of the cutting tools but also to explore the physical structure of the workpiece-tool interactions. Many machining parameters such as tool deformation, thermal forces, friction in different conditions, cutting heat, and tool wear are needed in the future to take into account comprehensive numerical simulation.

KEYWORDS

- **superalloys**
- **ANN**
- **milling**
- **finite element modeling**
- **cutting force model**
- **cutting temperature model**

REFERENCES

1. Rentsch, R.; Pecat, O.; Brinksmeier, E. Macro and Micro Process Modeling of the Cutting of Carbon Fiber Reinforced Plastics Using FEM. *Procedia Eng.* **2011,** *10,* 1823–1828.

2. Sheikh-Ahmad, J.; Twomey, J.; Kalla, D.; Lodhia, P. Multiple Regression and Committee Neural Network Force Prediction Models in Milling FRP. *Mach. Sci. Technol.* **2007,** *11* (3), 391–412.
3. Kalla, D.; Sheikh-Ahmad, J.; Twomey, J. Prediction of Cutting Forces in Helical End Milling Fiber Reinforced Polymers. *Int. J. Mach. Tools Manuf.* **2010,** *50* (10), 882–891.
4. Li, H. Z.; Zhang, W. B.; Li, X. P. Modelling of Cutting Forces in Helical End Milling Using a Predictive Machining Theory. *Int. J. Mech. Sci.* **2001,** *43* (8), 1711–1730.
5. Özel, T.; Altan, T. Process Simulation Using Finite Element Method—Prediction of Cutting Forces, Tool Stresses and Temperatures in High-Speed Flat End Milling. *Int. J. Mach. Tools Manuf.* **2000,** *40* (5), 713–738.
6. Rao, B.; Dandekar, C. R.; Shin, Y. C. An Experimental and Numerical Study on the Face Milling of Ti–6Al–4V Alloy: Tool Performance and Surface Integrity. *J. Mater. Process. Technol.* **2011,** *211* (2), 294–304.
7. Dang, J. W.; Zhang, W. H.; Yang, Y.; Wan, M. Cutting Force Modelling for Flat End Milling Including Bottom Edge Cutting Effect. *Int. J. Mach. Tools Manuf.* **2010,** *50* (11), 986–997.
8. Matsumura, T.; Usui, E. Predictive Cutting Force Model in Complex-Shaped End Milling Based on Minimum Cutting Energy. *Int. J. Mach. Tools Manuf.* **2010,** *50* (5), 458–466.
9. Denkena, B.; Schmidt, C.; Krüger, M. Experimental Investigation and Modeling of Thermal and Mechanical Influences on Shape Deviations in Machining Structural Parts. *Int. J. Mach. Tools Manuf.* **2010,** *50* (11), 1015–1021.
10. Özel, T.; Thepsonthi, T.; Ulutan, D.; Kaftanoğlu, B. Experiments and Finite Element Simulations on Micro-Milling of Ti–6Al–4V Alloy with Uncoated and cBN Coated Micro-Tools. *CIRP Ann.* **2011,** *60* (1), 85–88.
11. Thepsonthi, T.; Özel, T. Experimental and Finite Element Simulation-Based Investigations on Micro-Milling Ti-6Al-4V Titanium Alloy: Effects of cBN Coating on Tool Wear. *J. Mater. Process. Technol.* **2013,** *213* (4), 532–542.
12. Wu, B.; Yan, X.; Luo, M.; Gao, G. Cutting Force Prediction for Circular End Milling Process. *Chinese J. Aeronaut.* **2013,** *26* (4), 1057–1063.
13. Puls, H.; Klocke, F.; Lung, D. Experimental Investigation on Friction under Metal Cutting Conditions. *Wear* **2014,** *310* (1–2), 63–71.
14. Lazoglu, I.; Bugdayci, B. Thermal Modelling of End Milling. *CIRP Ann.* **2014,** *63* (1), 113–116.
15. Fu, Z.; Yang, W.; Wang, X.; Leopold, J. Analytical Modelling of Milling Forces for Helical End Milling Based on a Predictive Machining Theory. *Procedia Cirp* **2015,** *31*, 258–263.
16. Davoudinejad, A.; Chiappini, E.; Tirelli, S.; Annoni, M.; Strano, M. Finite Element Simulation and Validation of Chip Formation and Cutting Forces in Dry and Cryogenic Cutting of Ti–6Al–4V. *Procedia Manuf.* **2015,** *1*, 728–739.
17. Grzesik, W.; Nieslony, P.; Habrat, W.; Laskowski, P. Influence of Cutting Conditions on Temperature Distribution in Face Milling of Inconel 718 Nickel-Chromium Alloy. *J. Mach. Eng.* **2015,** *15*.
18. Denkena, B.; Schmidt, A.; Maaß, P.; Niederwestberg, D.; Niebuhr, C.; Vehmeyer, J. Prediction of Temperature Induced Shape Deviations in Dry Milling. *Procedia CIRP* **2015,** *31*, 340–345.
19. Baohai, W.; Di, C.; Xiaodong, H.; Dinghua, Z.; Kai, T. Cutting Tool Temperature Prediction Method Using Analytical Model for End Milling. *Chinese J. Aeronaut.* **2016,** *29* (6), 1788–1794.

20. Mamedov, A.; Lazoglu, I. Thermal Analysis of Micro Milling Titanium Alloy Ti–6Al–4V. *J. Mater. Process. Technol.* **2016,** *229*, 659–667.
21. Ucun, İ.; Aslantas, K.; Özkaya, E.; Cicek, A. 3D Numerical Modelling of Micro-Milling Process of Ti6Al4V Alloy and Experimental Validation. *Adv. Mater. Process. Technol.* **2017,** *3* (3), 250–260.
22. Cui, D.; Zhang, D.; Wu, B.; Luo, M. An Investigation of Tool Temperature in End Milling Considering the Flank Wear Effect. *Int. J.Mech. Sci.* **2017,** *131*, 613–624.
23. Abdulshahed, A. M.; Longstaff, A. P.; Fletcher, S. The Application of ANFIS Prediction Models for Thermal Error Compensation on CNC Machine Tools. *Appl. Soft Comput.* **2015,** *27*, 158–168.
24. da Silva, R. H. L.; da Silva, M. B.; Hassui, A. A Probabilistic Neural Network Applied in Monitoring Tool Wear in the End Milling Operation via Acoustic Emission and Cutting Power Signals. *Mach. Sci. Technol.* **2016,** *20* (3), 386–405.
25. Arrazola, P. J.; Özel, T.; Umbrello, D.; Davies, M.; Jawahir, I. S. Recent Advances in Modelling of Metal Machining Processes. *CIRP Ann.* **2013,** *62* (2), 695–718.
26. Rech, J.; Arrazola, P. J.; Claudin, C.; Courbon, C.; Pusavec, F.; Kopac, J. Characterisation of Friction and Heat Partition Coefficients at the Tool-Work Material Interface in Cutting. *CIRP Ann.* **2013,** *62* (1), 79–82.
27. Putz, M.; Schmidt, G.; Semmler, U.; Dix, M.; Bräunig, M.; Brockmann, M.; Gierlings, S. Heat Flux in Cutting: Importance, Simulation and Validation. *Procedia CIRP* **2015,** *31*, 334–339.
28. Sima, M.; Özel, T. Modified Material Constitutive Models for Serrated Chip Formation Simulations and Experimental Validation in Machining of Titanium Alloy Ti–6Al–4V. *Int. J. Mach. Tools Manuf.* **2010,** *50* (11), 943–960.
29. Basturk, S.; Senbabaoglu, F.; Islam, C.; Erten, M.; Lazoglu, I.; Gulmez, T. Titanium Machining with New Plasma Boronized Cutting Tools. *CIRP Ann.* **2010,** *59* (1), 101–104.
30. Wan, M.; Li, S. E.; Yuan, H.; Zhang, W. H. Cutting Force Modelling in Machining of Fiber-Reinforced Polymer Matrix Composites (PMCs): A Review. *Compos. A Appl. Sci. Manuf.* **2019,** *117*, 34–55.
31. Akhtar, A.; Hegde, S.; Reed, R. C. The Oxidation of Single-Crystal Nickel-Based Superalloys. *Jom* **2006,** *58* (1), 37–42.
32. Durand-Charre, M. *The Microstructure of Superalloys*; Routledge: London, 2017.
33. Pollock, T. M.; Tin, S. Nickel-Based Superalloys for Advanced Turbine Engines: Chemistry, Microstructure and Properties. *J. Propulsion Power* **2006,** *22* (2), 361–374.
34. Baker, I. Nickel-Based Superalloys. In *Fifty Materials That Make the World*; Springer: Cham, 2018; pp 131–136.
35. Thomas, A.; El-Wahabi, M.; Cabrera, J. M.; Prado, J. M. High Temperature Deformation of Inconel 718. *J. Mater. Process. Technol.* **2006,** *177* (1–3), 469–472.
36. Furrer, D.; Fecht, H. Ni-Based Superalloys for Turbine Discs. *Jom* **1999,** *51* (1), 14–17.
37. Reed, R. C.; Rae, C. M. F. Physical Metallurgy of the Nickel-Based Superalloys. In *Physical Metallurgy*; Elsevier, 2014; pp 2215–2290.
38. Singh, R.P.; Singhal, S. Rotary Ultrasonic Machining: A Review. *Mater. Manuf. Process.* **2016,** *31*, 1795–1824
39. Singh, R.P.; Singhal, S. Investigation of Machining Characteristics in Rotary Ultrasonic Machining of Alumina Ceramic. *Mater. Manuf. Process.* **2017,** *32*, 309–326.
40. Singh, R.P.; Tyagi, M.; Kataria, R. Selection of the Optimum Hole Quality Conditions in Manufacturing Environment Using MCDM Approach: A Case Study. In *Operations Management and Systems Engineering*; Springer: Singapore, 2019; pp 133–152.

41. Singh, R.P.; Singhal, S. Rotary Ultrasonic Machining of Macor Ceramic: An Experimental Investigation and Microstructure Analysis. *Mater. Manuf. Process.* **2017,** *32*, 927–939.
42. Kim, Y. K.; Kim, D.; Kim, H. K.; Oh, C. S.; Lee, B. J. An Intermediate Temperature Creep Model for Ni-Based Superalloys. *Int. J. Plasticity* **2016,** *79*, 153–175.
43. Coutsouradis, D.; Davin, A.; Lamberigts, M. Cobalt-Based Superalloys for Applications in Gas Turbines. *Mater. Sci. Eng.* **1987,** *88*, 11–19.
44. Singh, R.P.; Singhal, S. Experimental Investigation of Machining Characteristics in Rotary Ultrasonic Machining of Quartz Ceramic. *J. Mater. Des. Appl.* **2018,** *232,* 870–889.
45. Singh, R.P.; Kataria, R.; Kumar, J.; Verma, J. Multi-Response Optimization of Machining Characteristics in Ultrasonic Machining of WC-Co Composite through Taguchi Method and Grey-Fuzzy Logic. *AIMS Mater. Sci.* **2018,** *5*, 75–92.
46. Singh, R.P.; Kumar, J.; Kataria, R.; Singhal, S. Investigation of the Machinability of Commercially Pure Titanium in Ultrasonic Machining Using Graph Theory and Matrix Method. *J. Eng. Res.* **2015,** *3*, 75–94.
47. Singh, R.P.; Singhal, S. An Experimental Study on Rotary Ultrasonic Machining of Macor Ceramic. *J. Eng. Manuf.* **2018,** *232,* 1221–1234.
48. Singh, R.P.; Singhal, S. Rotary Ultrasonic Machining of Alumina Ceramic: Experimental Study and Optimization of Machining Responses. *J. Eng. Res.* **2018,** *6*, 01–24.
49. Tyagi, M.; Panchal, D.; Singh, R. P.; Sachdeva, A. Modeling and Analysis of Critical Success Factors for Implementing the IT-Based Supply-Chain Performance System. In *Operations Management and Systems Engineering*; Springer: Singapore, 2019; pp 51–67.
50. Stewart, C. A.; Rhein, R. K.; Suzuki, A.; Pollock, T. M.; Levi, C. G. Oxide Scale Formation in Novel γ-γ′ Cobalt-Based Alloys. In *Superalloys 2016: Proceedings of the 13th Intenational Symposium of Superalloys*; John Wiley & Sons, Inc.: Hoboken, NJ, USA, Aug 2016; pp. 991–999.
51. Ezugwu, E. O. Key Improvements in the Machining of Difficult-to-Cut Aerospace Superalloys. *Int. J. Mach. Tools Manuf.* **2005,** *45* (12–13), 1353–1367.
52. Decker, R. F. The Evolution of Wrought Age-Hardenable Superalloys. *Jom* **2006,** *58* (9), 32–36.
53. Donachie, M. J.; Donachie, S. J. *Superalloys: A Technical Guide*; ASM international, 2002.
54. Singh, R. P.; Kataria, R.; Singhal, S. Decision-Making in Real-Life Industrial Environment through Graph Theory Approach. In *Computer Architecture in Industrial, Biomechanical and Biomedical Engineering*; IntechOpen, 2019. DOI: 10.5772/intechopen.82011
55. Singh, R.; Singh, R. P.; Trehan, R. State of the Art in Processing of Shape Memory Alloys with Electrical Discharge Machining: A Review. *Proc. Inst. Mech. Eng. B J. Eng. Manuf.* **2020,** 0954405420958771.
56. Singh, R.; Singh, R. P.; Tyagi, M.; Kataria, R. Investigation of Dimensional Deviation in Wire EDM of M42 HSS Using Cryogenically Treated Brass Wire. *Mater. Today Proc* **2019**.
57. Geddes, B.; Leon, H.; Huang, X. *Superalloys: Alloying and Performance*; Asm International, 2010.
58. Thellaputta, G. R.; Chandra, P. S.; Rao, C. S. P. Machinability of Nickel-Based Superalloys: A Review. *Mater. Today Proc.* **2017,** *4* (2), 3712–3721.
59. Choudhury, I. A., and El-Baradie, M. A. Machinability of nickel-base super alloys: a general review, Journal of Materials Processing Technology, Volume 77, Issues 1–3, 1998, 278–284, https://doi.org/10.1016/S0924-0136(97)00429-9.

CHAPTER 4

TO IMPROVE WEAR RESISTANCE OF CAST IRON DEPOSITED BY DETONATION SPRAYED NI–CR COATING

YATAN NAGPAL[1*], PARDEEP SHARMA[2], and ROHIT SHARMA[3]

[1]*Amity University Uttar Pradesh, Noida, India*

[2]*Department of Mechanical Engineering, Panipat Institute of Engineering and Technology, Panipat, India*

[3]*Department of Mechanical Engineering, Amity University Uttar Pradesh, Noida, India*

[*]*Corresponding author. E-mail: yatannagpal@gmail.com*

ABSTRACT

Cast iron is widely used as a cast material in different engineering fields such as manufacturing of internal combustion engines, pump housings, lathe bed, machine tools, etc. In the present work, an experimental study was conducted using thermal spray coating methods to study and enhance the wear resistance of cast iron. Gray cast iron was used as a specimen and nickel–chromium (Ni–Cr) was selected as a coating material in powdered form using detonation-gun (D-gun) as a thermal spray coating method. Wear behavior was investigated on both coated and uncoated cast iron specimens using pin-on disk apparatus at 30, 40, and 50 N, and a comparison analysis was conducted using scanning electron micrographs. Results showed that

Modern Manufacturing Systems: Trends and Developments. Rajiv Kumar Garg, Ravi Pratap Singh, Rajeev Trehan, & Ramesh Singh (Eds.)

wear rate increases on increasing the load and wear resistance on coated cast iron specimen has been enhanced.

4.1 INTRODUCTION

With rapid increase in technologies, there has been enormous and stiff competition among several industries including manufacturing and automation sector.[1] Significant economic losses can happen owing to rapid wear of different parts in order to satisfy the requirement within the most manufacturing sections. Materials such as cast iron that are used as castings in various manufacturing sectors, in manufacturing automotive parts, in machining parts, and tools like lathe bed, etc. are subjected to more wear and corrosion. Thus, minimizing and reducing the impact of component wear and tear and increasing the lifetime of product or material thermal spray coating alternatives are initiated in production and are being developed to meet even more challenging wear applications. A proven industrial method for resurfacing metal components is the application of coatings using thermal spray technique. Several form of coatings can be used either in the form of powder or wire of certain inorganic elements like composites, ceramics, metals, or even alloys. The process is characterized by the simultaneous melting of these powder coatings and transporting them to the required substrate or base metal. Thermal spraying techniques that seems to deposit the coatings on substrates are D-gun, high velocity oxy-fuel (HVOF), flame spraying, electric arc wire spraying, plasma spraying, etc.[2] Among all the above thermal spray coating methods, the use of D-gun coating method is most popular.[3]

4.1.1 DETONATION GUN SPRAYING

It is a form of combustion method that generated low porosity and very excellent adhesive strength along with compressive residual stresses. Gas inlet is shielded by nitrogen gas cover to prevent back firing. The combustion chamber is packed with predetermined quantity of coating powder at the same moment. Spark plug ignites the combustion chamber, which ignites the gas combination into the chamber. The high-pressure shock wave is produced by gas mixture combustion and begins to propagate through the gas stream in the chamber. Initially at a very high speed, these gases travel down into the barrel and the powder metal begins to melt and accelerate within the barrel at certain speeds. These fast-moving particles after appearing strike on to the surface thus making a thick coating. The density of the coating created on the base surface per layer depends on a certain parameters such as particles size, frequency, distance between the barrel and base components, rate of flow of carrier gas, etc. The coating produced by the D-gun depends on the number of shots produced per second and the type of coating material required.

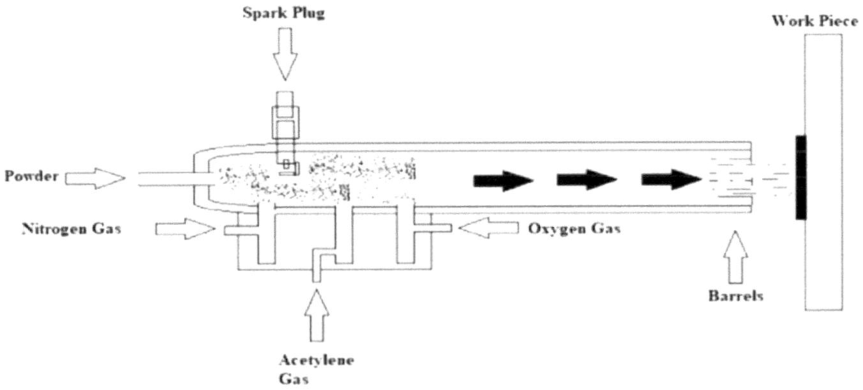

FIGURE 4.1 D-gun spraying.

4.2 RESEARCH BACKGROUND

Several studies have undergone related to thermal spray coating techniques for improving the wear and corrosion resistance which are as follows: The microstructure and mechanical characteristics were examined by Kamal et al.[4] using NiCrAlY+CeO_2 alloy coatings sprayed on superalloys using D-gun spraying technique. The effect of cold spray and gas detonation coating was studied by Sova et al.[5] for multimaterial coatings. The effect of grinding on the erosion behavior of a WC-Co-Cr coated layer was studied and investigated by Murthy et al.[6] Also, as a part of his experiment, comparison between two coatings by high velocity oxy fuel and D-gun was analyzed. Wang et al.[7] estimated the residual stresses and studied its effects on mechanical properties of WC-Co sprayed by D-gun method. The tribological results of 200 μm dense TiMo(CN)-28Co and TiMo(CN)-36NiCo coatings were studied by Sundararajan et al.[8] using D-gun method. Senderowsk et al.[9] performed microstructural characterization on Fe–Al intermetallic coatings using D-gun method. Zeng et al.[10] examined the formation and corrosive behavior of Fe-based amorphous metallic coatings by using D-gun spraying method. The microstructure and sliding wear resistance of aluminum coatings deposited on cast iron substrates were examined by Psyllaki et al.[11] using atmospheric plasma spraying (APS) and D-gun spraying method. As per their experiment, D-gun produced dense coatings and showed better wear resistance when compared with APS method. The rolling contact fatigue failure process in alumina coatings sprayed on multiple steel substrates using D-gun, HVOF, and HVPS (high velocity plasma spraying) was studied and discussed by Ahmed and Hadfield.[12] Singla et al.[13] studied and stated that when two surfaces are in contact with each other, a condition called wear occurs on both the surfaces so in order to prevent it, a technique called surface coating is introduced and is the best way to improve surface properties such as improving steel wear and corrosion resistance. Singh et al.[14] stated that materials are very precious and valuable resources so in order to safeguard and protect them from being degraded different coating techniques are used. Xiong et al.[15] examined that on increasing the speed, load and distance, the resulting temperature increases due to friction which leads to more wear rate and later on discussed that the applied load is a most important parameter that depends on pressure and resulting contact stress which affects the rate of wear and wear mechanism. Odabas et al. and Abdou et al.[16,17] discussed and concluded that on varying time, wear rate varies linearly and generally increases on increasing time. Abdou et al. and Terheci et al.[17,18] stated that increase in speed results in the increase in wear which mainly cause vibrations. The objective of this present work was to study the wear behavior of gray cast iron using pin-on disc apparatus and to increase its wear resistance using D-gun as thermal spray coating technique. In this work, Ni–Cr was used as coating material whose properties are also presented and micro-analysis was done for comparison between coated and uncoated specimen. D-gun method has been selected as thermal spray coating process as it has remarkably good adhesive strength.[3]

4.3 MATERIALS AND EXPERIMENTATION DETAILS

4.3.1 BASE MATERIAL

Base metal selected for this present work was −25 graded gray cast iron; chemical composition is shown in Table 4.1.

Properties of grade 25 gray cast iron are as follows:

- Melting temperature: 1380°C
- Hardness (Brinell): 170

- Density: 7.5 g/cm^3
- Coefficient of thermal expansion: 11 μm/m K

TABLE 4.1 Elements Present in 25 Graded Gray Cast Iron (wt.%) Composition.

Grade	Elements	C	Si	Mn	P	S	Fe	Hardness (HB)
ASTM 25 gray cast iron	Min	3.1	2.5	0.5	–	–	92.1	170
	Max	3.4	2.8	0.7	0.9	0.15	93.9	

4.3.2 COATING MATERIAL

The coating material selected for this research work is NiCr because of its good wear and corrosion resistant properties.

Properties of NiCr are as follows:

- Melting temperature: 1400°C
- Corrosion and wear resistant
- Coefficient of thermal expansion: $14 \times 10^{-6} K^{-1}$

4.3.3 SPECIMEN PREPARATION AND WEAR TEST (PIN-ON-DISC)

To perform the wear test; two cylindrical-shaped specimens were prepared with diameter 10 mm and length 30 mm from 25 grade gray cast iron without coating. All these specimens before performing wear test were polished and grinded with well-standard emery papers of various grades like 220, 400, 600, 800, etc., according to standard metallographic procedure. In the wear test, the initial and final weight of both the specimens before and after coating were taken and measured.

Wear test was conducted on both coated and uncoated specimens at various loads of 30, 40, and 50 N using pin-on disc apparatus conforming to ASTM G99 standards. During experimentation work several parameters such as load, time, and rotational speed were selected as shown in Table 4.2 and based on literature review, rotational speed was taken as a constant parameter. After wear test, results were analyzed to evaluate wear resistance of specimen and microanalysis was done for comparison of uncoated and coated specimens using SEM.

TABLE 4.2 Parameters of Wear Test.

Load (N)	Sliding speed (m/s)	Track radius (mm)	Disc speed (rpm)	Total sliding distance(m)	Total cycle (min)
30, 40, 50	1	80	239	5400	6 (5, 5, 10, 10, 20, 40)

4.4 RESULTS AND DISCUSSION

4.4.1 WEAR TEST

Wear test was performed on pin-on-disc apparatus and their results were shown in Tables 4.3 and 4.4 that show the wear rate depends upon load and time. It has been clearly observed that rate of wear was increased on varying the load and time and wear rate was observed more in case of uncoated specimen as compared to coated specimen at varying loads of 30, 40, and 50 N, respectively. Moreover, it shows that wear resistance has been enhanced by coating the specimen with NiCr by D-gun method.

TABLE 4.3 Wear Test of Uncoated Cast Iron Specimen at 30, 40 and 50 N Respectively.

Sample no.	Load (N)	Time (min)	Initial wt (g)	Final wt (g)	Weight loss (g)	Cumulative wt loss (g)
		5	17.198	17.1971	0.0009	0.0009
		5	17.1971	17.1967	0.0004	0.0013
		10	17.1967	17.1955	0.0012	0.0025
1	30	10	17.1955	17.1941	0.0014	0.0039
		20	17.1941	17.1924	0.0017	0.0056
		40	17.1924	17.1911	0.0013	0.0069
		5	17.3111	17.31	0.0011	0.0011
		5	17.31	17.3087	0.0013	0.0024
		10	17.3087	17.3059	0.0014	0.0038
2	40	10	17.3059	17.3036	0.0028	0.0066
		20	17.3036	17.3013	0.0023	0.0089
		40	17.3031	17.2999	0.0033	0.0112
		5	17.231	17.2295	0.0015	0.0015
		5	17.2295	17.2281	0.0014	0.0029
		10	17.2281	17.2251	0.003	0.0059
3	50	10	17.2251	17.2218	0.0033	0.0092
		20	17.2218	17.2174	0.0044	0.0136
		40	17.2174	17.2109	0.0065	0.0201

TABLE 4.4 Wear Test of Ni–Cr Coated Cast Iron Specimen at 30, 40 and 50 N Respectively.

Sample no.	Load (N)	Time (min)	Initial wt. (g)	Final wt. (g)	Weight loss (g)	Cumulative wt. loss (g)
		5	17.126	17.126	0	0
		5	17.126	17.1259	0.0001	0.0001
		10	17.1259	17.1258	0.0001	0.0002
1	30	10	17.1258	17.1256	0.0002	0.0004
		20	17.1256	17.1254	0.0002	0.0006
		40	17.1254	17.1252	0.0002	0.0008
		5	17.1216	17.1213	0.0003	0.0003
		5	17.1213	17.1211	0.0002	0.0005
		10	17.1211	17.1209	0.0002	0.0007
2	40	10	17.1209	17.1207	0.0002	0.0009
		20	17.1207	17.1204	0.0003	0.0012
		40	17.1204	17.1202	0.0002	0.0014
		5	17.2189	17.2187	0.0002	0.0002
		5	17.2187	17.2186	0.0001	0.0003
		10	17.2186	17.2184	0.0002	0.0005
3	50	10	17.2184	17.2184	0	0.0005
		20	17.2184	17.2179	0.0005	0.001
		40	17.2179	17.2173	0.0006	0.0016

Table 4.5 shows the comparative results of cumulative weight loss of NiCr coated cast iron specimen and it can be observed that by increasing the load, wear rate increases in both the work but cumulative weight loss of NiCr coated cast iron in the present work is significantly less than the cumulative weight loss of 80%Cr3C2+20%NiCr.[19]

TABLE 4.5 Comparative Results of Cumulative wt. Loss (in Grams) of Coated Cast Iron Specimen at Different Loads.

Current work (NiCr)		Priyan et al.19 (80%Cr3C2+20%NiCr)	
Load (N)	wt. Loss (g)	Load (N)	wt. Loss (g)
30	0.0008	10	0.009
40	0.0014	20	0.019
50	0.0016	30	0.026

4.4.2 *MICROSTRUCTURE ANALYSIS*

Figure 4.2 presents the SEM images of 25 graded gray cast iron without coating (Figure 4.2a) and with NiCr coating (Figure 4.2b). From Figure 4.2a, it has been revealed that uncoated specimen of cast iron suffered more material loss; hence wear rate was more rapid and certain cracks were also observed in it whereas on comparison with Figure 4.2b, when specimen was coated with NiCr by D-gun spraying technique strong bonding between specimen and coated particles was observed and hence wear rate has been decreased.

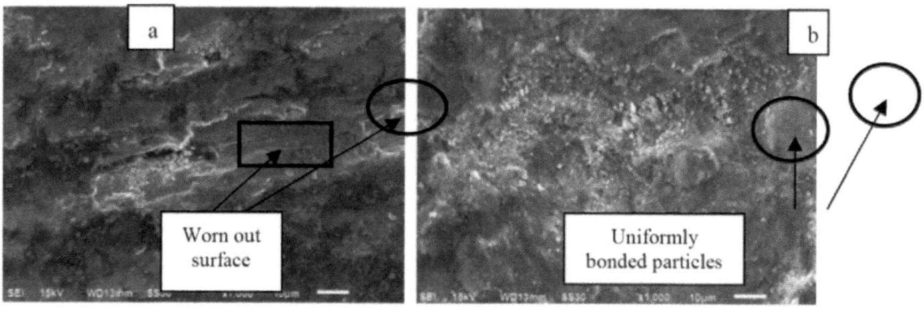

FIGURE 4.2 SEM micrograph of cast iron specimen (a) uncoated; (b) NiCr coated.

4.5 CONCLUSION

NiCr coatings were successfully deposited on gray cast iron specimen using detonation gun spray coated technique. Certain conclusions were being made as follows:

1. Rate of wear varies along with varying load and time and found to be increasing on increasing the load.
2. Cumulative weight loss of 25 graded gray cast iron specimen without coating at 30, 40, and 50 N load was found to be 0.0069, 0.0112, and 0.0201 g, respectively, which clearly shows rate of wear increases on increasing the load.
3. Cumulative weight loss of 25 graded gray cast iron specimen with NiCr coating at 30, 40, and 50 N load was found to be 0.0008, 0.0014, and 0.0016 g, respectively, which clearly shows wear rate increases on increasing the load but wear rate was observed comparatively less than that of uncoated gray cast iron specimen.

4. Micrographs of both the uncoated and coated specimens were observed using SEM and certain cracks were found in uncoated specimen that clearly indicates more material loss and hence more wear while NiCr coated particles shows uniform bonding with the specimen along with comparative less wear rate and hence wear resistance of gray cast iron specimen has been enhanced.

KEYWORDS

- **coating**
- **wear**
- **cast iron**
- **D-gun**
- **SEM**

REFERENCES

1. Singh, B. Recent Advancements in Thermal Spray Coatings. *Int. J. Res. Mech. Eng. Technol.* **2013,** *3* (1), 55.
2. Parshad, J.; Grover, N. K. Improving Wear Resistance of Grey Cast Iron Using Detonation Gun Sprayed Coatings: A Review. *Int. J. Eng. Manage. Res. (IJEMR)* **2015,** *5* (1), 303–307.
3. Rajasekaran, B.; Raman, S. G. S.; Joshi, S. V.; Sundararajan, G. Influence of Detonation Gun Sprayed Alumina Coating on AA 6063 Samples under Cyclic Loading with and Without Fretting. *Tribol. Int.* **2008,** *41* (4), 315–322. doi:10.1016/j.triboint.2007.08.007
4. Kamal, S.; Jayaganthan, R.; Prakash, S. Mechanical and Microstructural Characteristics of Detonation Gun Sprayed NiCrAlY+ 0.4 wt% CeO2 Coatings on Superalloys. *Mater. Chem. Phys.* **2010,** *122* (1), 262–268.
5. Sova, A.; Pervushin, D.; Smurov, I. Development of Multimaterial Coatings by Cold Spray and Gas Detonation Spraying. *Surf. Coatings Technol.* **2010,** *205* (4), 1108–1114. DOI: 10.1016/j.surfcoat.2010.07.092
6. Murthy, J. K. N.; Rao, D. S.; Venkataraman, B. Effect of Grinding on the Erosion Behaviour of a WC–Co–Cr Coating Deposited by HVOF and Detonation Gun Spray Processes. *Wear* **2001,** *249* (7), 592–600.
7. Wang, T. G.; Zhao, S. S.; Hua, W. G.; Li, J. B.; Gong, J.; Sun, C. Estimation of Residual Stress and Its Effects on the Mechanical Properties of Detonation Gun Sprayed WC–Co Coatings. *Mater. Sci. Eng. A* **2010,** *527* (3), 454–461.
8. Sundararajan, G.; Sivakumar, G.; Sen, D.; Rao, D. S.; Ravichandra, G. The Tribological Behaviour of Detonation Sprayed TiMo (CN) Based Cermet Coatings. *Int. J. Refractory Metals Hard Mater.* **2010,** *28* (1), 71–81. DOI:/10.1016/j.ijrmhm.2009.07.007

9. Senderowski, C.; Bojar, Z.; Wołczyński, W.; Pawłowski, A. Microstructure Characterization of D-Gun Sprayed Fe–Al Intermetallic Coatings. *Intermetallics* **2010,** *18* (7), 1405–1409.
10. Zheng, Z. H. O. U.; Lu, W. A. N. G.; Wang, F. C.; Liu, Y. B. Formation and Corrosion Behavior of Fe-Based Amorphous Metallic Coatings Prepared by Detonation Gun Spraying. *Trans. Nonferrous Metals Soc. China* **2009,** *19*, s634–s638.
11. Psyllaki, P. P.; Jeandin, M.; Pantelis, D. I. Microstructure and Wear Mechanisms of Thermal-Sprayed Alumina Coatings. *Mater. Lett.* **2001,** *47* (1–2), 77–82. DOI:/10.1016/S0167-577X(00)00215-9
12. Ahmed, R.; Hadfield, M. Rolling Contact Fatigue Performance of Detonation Gun Coated Elements. *Tribol. Int.* **1997,** *30* (2), 129–137. doi:/10.1016/0301-679X(96)00036-9
13. Singla, J. P.; Grewal, J. S.; Chawla, V. A Survey of the Detonation Gun Sprayed Wear Resistant Coatings. In *National Conference on Advancements and Futuristic Trends in Mechanical and Materials Engineering*, Talwandi Sabo, 2010; pp 355–359.
14. Singh, L.; Chawla, V.; Grewal, J. S. A Review on Detonation Gun Sprayed Coatings. *J. Miner. Mater. Charact. Eng.* **2012,** *11* (03), 243. DOI:/10.4236/jmmce.2012.113019
15. Xiong, F.; Manory, R. R. The Effect of Test Parameters on Alumina Wear under Low Contact Stress. *Wear* **1999,** *236* (1–2), 240–245.
16. Odabas, D. Effects of Load and Speed on Wear Rate of Abrasive Wear for 2014 Al Alloy. *IOP Conf. Ser. Mater. Sci. Eng.* 2018, *295* (1), 012008. DOI:10.1088/1757-899X/295/1/012008
17. Abdou, S.; Elkaseer, A.; Kouta, H.; Abu Qudeiri, J. 2018. Wear Behaviour of Grey Cast Iron with the Presence of Copper Addition. *Adv. Mech. Eng.* **2018,** *10* (10), 1687814018804741. DOI: 10.1177/1687814018804741
18. Terheci, M.; Manory, R. R.; Hensler, J. H. The Friction and Wear of Automotive Grey Cast Iron under Dry Sliding Conditions Part 1-Relationships between Wear Loss and Testing Parameters. *Wear* **1995,** *180* (1–2), 73–78.
19. Priyan, M. S.; Azad, A.; Araffath, S. Y. Influence of HVOF Parameters on the Wear Resistance of Cr3C2-NiCr Coating. *Carbon* **2016,** *3* (3.4), 3–40.

CHAPTER 5

EFFECT OF MAGNETIC FIELD MEDIA ON THE FLUID FLOW OF CONTINUOUS CASTING MOLD WITH APPLICATION OF MAGNETIC FIELD IN *X*-, *Y*-, *Z*-DIRECTIONS

SUNIL KUMAR JATAV[1*], KRISHNA KUMAR[1], SANDIP SARKAR[2], S. K. AJMANI[2], and M. K. BANERJEE[1]

[1]*Metallurgical and Materials Engineering, Jaipur, India*

[2]*Research and Development, Tata Steel, Jamshedpur, Jharkhand, India*

[*]*Corresponding author. E-mail: sunilkjatavnitjsr@gmail.com*

ABSTRACT

The turbulent flow in continuous casting mold leads to various kinds of defects including surfaces defects, internal defects, and inclusion-related defects in the final cast products. Applications of magnetic field are the advanced technology that was used to improve final cast products. The magnetic field stabilizes and minimizes the turbulent effect in the continuous casting mold. In this research work, a computational model is used to analyze the effect of magnetic field media (conducting and non-conducting) on the fluid flow in the continuous casting mold. The magnetic field of 0.2 tesla was applied in *x*-direction (along mold thickness), *y*-direction (along mold width), and *z*-direction (along mold length). The effect of magnetic field media (conducting and non-conducting) was studied in *x*-, *y*-, and *z*-directions in continuous casting mold. A three-dimentional realiable k-ε turbulence model was used

Modern Manufacturing Systems: Trends and Developments. Rajiv Kumar Garg, Ravi Pratap Singh, Rajeev Trehan, & Ramesh Singh (Eds.)

with magnetohydrodynamic (MHD)-induced Maxwell equation. The effect of magnetic field media along with the direction of magnetic field was studied in this research work. The magnetic field was favored to suppress the turbulent flow in the continuous casting mold; so, it leads to improving the quality of final cast products.

5.1 INTRODUCTION

The turbulent flow is responsible for different types of defects like surface defects, inclusions, and internal defects in the final cast product. The application of magnetic field is used to minimize the turbulent flow in the mold of continuous casting process that leads to improving the quality of final cast product.[1–4] The mold powder is used on the top surface of a mold to keep it hot during the melt on meniscus. Due to slow flow, the melt solidifies on the meniscus. The high surface fluctuations lead to formation of vertex at either side of the submerged entry nozzle in the continuous casting mold. The solidified melt on meniscus entraps in the cast products and made the product dirty. Therefore, it is important to control the flow in continuous casting mold for improving the quality of final cast product.[4,5] For controlling the fluid flow in the mold, it is necessary to optimize the casting parameters like submerged entry nozzle (SEN), casting speed, submergence depth, ports shape, mold size, inert gas injection, and application of magnetic field.[6] The application of magnetic field plays an important role to control the fluid flow in the continuous casting mold. One factor must be important to adjust with the other factors to provide optimize flow pattern.[7,8] The application of magnetic field can be used to stirrer, accelerate, or stabilize the fluid flow in mold. The electromagnetic level stabilizer (EMLS) slows the fluid flow, electromagnetic level accelerator (EMLA) accelerates the velocity of steel, and electromagnetic rotary stirring (EMRS) rotates the fluid.[9–12] MM-EMS favors double-roll flow pattern and minimize slivers and cracks in the ending cast product.[13] The effect of varying SEN depth and strength of electromagnetic braking (EMBR) field on fluid flow in a mold was studied by B. G. Thomas and Kevin Cukiershi et at.[14] Rajneesh et al. have found the influence of magnetic field on turbulent nature of flow in continuous casting mold. To perform the numerical study on fluid flow, the following transient and steady state computation models are useful:

a. Direction numerical simulations (DNS),
b. Large eddy simulation (LES),

c. Filtered unsteady Reynolds-averaged Navier-Stokes (URANS)
d. Steady Reynolds-average Navier-Stokes (RANS) with and without using magnetic field

Xincjeng Miao et al. also performed the numerical and experimental study on the fluid flow of continuous casting mold to understand the influence of magnetic field. They used a software package CFX with an implemented RANS and SST turbulence method for numerical calculation.[16] The LES used finds the effect of double-ruler electromagnetic field on flow pattern in mold that was investigated by Singh.[17] Hsueh-Min Chen et al. investigated the effect of magnetic field on electrically conducting fluid flow of continuous casting mold. This study was performed using commercial software packages Ansys fluent.[18] Sandip et al. have used a three-dimensional computation modal to study the influence of superimposed magnetic field on the fluid flow of continuous casting mold. They also investigated the fluid flow with single and two phases.[19]

In this research work, a computational model is adopted to study the influence of magnetic field media (conducting and non-conducting) and orientation of magnetic field on the fluid flow of the continuous casting mold. The magnetic field of 0.2 tesla was applied in the different direction like *x*-direction (along mold thicness), *y*-direction (along mold width), and *z*-direction (along mold length). The magnetic field media (conducting and non-conducting field media) and orientation of magnetic affect the fluid flow in continuous casting mold and control the trubulent flow. A three-dimentional RANS equation was used to simulate the fluid flow with realiable k-ε turbulence model and MHD-induced Maxwell equation.

5.2 METHODOLOGY

5.2.1 PROBLEMS GEOMETRY AND SIMULATION CONDITIONS

The geometry of SEN and mold made of around 9,56,937 cells of hexahedron mesh. The full physical modal of continuous casting mold and SEN with name of different part is shown in the Figure 5.1. The geometry was designed and meshed with help of ANSYS and used in the FLUENT for the experiments. The nozzle has two 15° downward angle ports with oval shape. The dimensions of mold and common casting conditions have been shown in Table 5.1 for all cases. The boundary conditions are also appended in Table 5.1. The modal was validated with help of experimental results.[19]

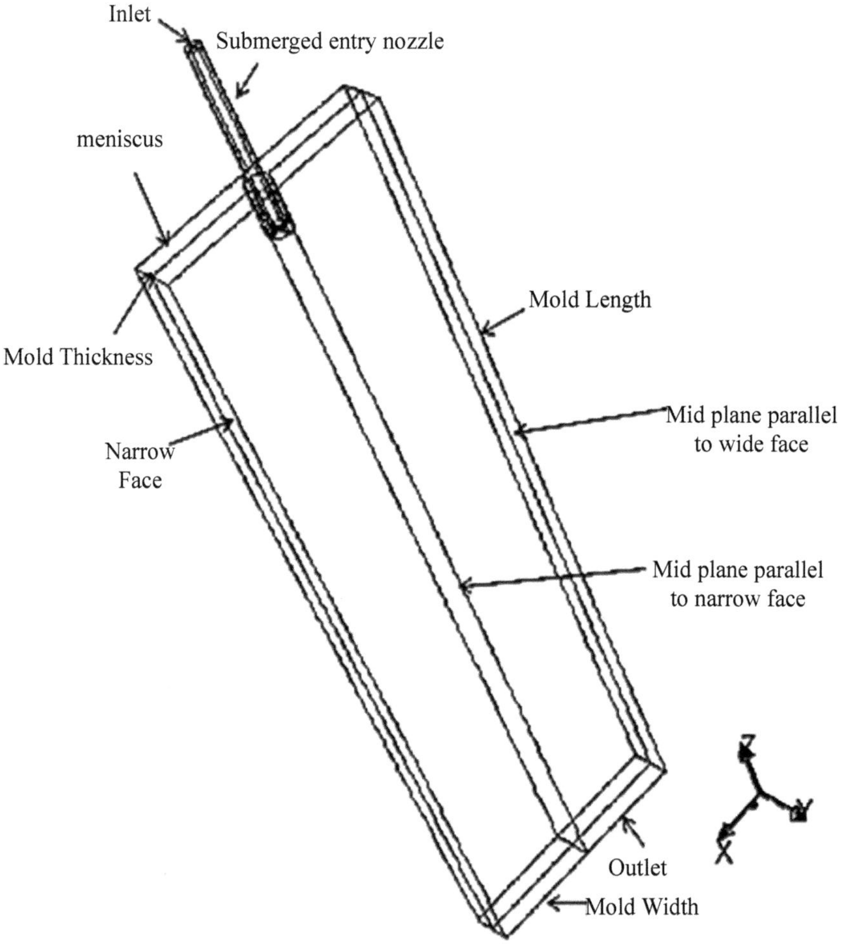

FIGURE 5.1 Physical modal of continuous casting mold and submerged entry nozzle (SEN).

A three-dimentional RANS with realiable *k*-ε turbulence model was used with MHD-induced Maxwell equation. The electrical potential method was used in this research work for the MHD simulantions by commerially available software, Ansys Fluent. All the simulations were divided in to seven cases that were shown in Table 5.2. One simulation was performed without using magnetic field and other six simulations were performed with using magnetic field of 0.2 tesla in different orientations with conducting and non-conducting magnetic field media.

TABLE 5.1 Simulation Parameters for All Cases.

Mold width	1300 mm
Mold thickness	218 mm
Mold length	5000 mm
Casting speed	1.3 m/min
SEN submergence depth	160 mm
Density of steel (ρ_{steel})	7200 kg/m^3
Viscosity of steel (μ_{steel})	0.6200 kg/m-s
Conductivity of steel (σ_{steel})	7,14,000 1/Ω m
Gravity	9.81 m/s^2
Magnetic field region along mold width	−0.65 to 0.65 m
Magnetic field region along mold thickness	−0.109 to 0.109 m
Magnetic field region along mold length	Up to 1 m from meniscus
Inlet velocity	1.39 m/s
Outlet gauge pressure	0 Pascal
Wall motion of meniscus, narrow face, wide face and SEN	Stationary wall
Shear conditions of meniscus, narrow face wide face and SEN	No slip

TABLE 5.2 Variation in Experimental Parameters by Cases.

Cases	**Magnetic field (in Tesla)**	**Magnetic field media**	**Direction of magnetic field**
Case 1	Without	–	–
Case 2(a)	0.2	Conducting	*x*-direction
Case 3(a)	0.2	Non-conducting	*x*-direction
Case 2(b)	0.2	Conducting	*y*-direction
Case 3(b)	0.2	Non-conducting	*y*-direction
Case 2(c)	0.2	Conducting	*z*-direction
Case 3(c)	0.2	Not-conducting	*z*-direction

5.2.2 GOVERNING EQUATIONS

The three-dimensional Naiver-Stokes equations are solved for transient, incompressible fluid in continuous casting mold domain. According to continuity equation, the rate of accumulation of mass in fluid element is equal to the net flow mass rate in to the fluid element.[20] The continuity equation (mass balance equation) is written as:

$$\rho(\nabla.\vec{v}) = 0 \tag{5.1}$$

where, ρ is density of molten steel and $\vec{v}$ is velocity,

The momentum conservation equation is given as follows:

$$\rho(\nabla.(\vec{v}\vec{v})) = -\nabla p + \nabla(\mu_{eff}(\nabla.\vec{v})) + \rho\vec{g} + \vec{F} \tag{5.2}$$

where, p = pressure,

$\vec{F}$ = momentum source/sink term used for electromagnetic force,

μ_{eff} = effective viscosity

K-ε Model for turbulence

This work is used reliable *k-ε* model of turbulent model.

$$\mu_{eff} = \mu_0 + \mu_t \tag{5.3}$$

where, μ_0 = molecular viscosity or laminar viscosity and μ_t = turbulent viscosity

$$\mu_t = \rho C_\mu \frac{K^2}{\varepsilon} \tag{5.4}$$

To calculate the turbulent kinetic energy K (m^2/s^2) and dissipation rate of Kinetic turbulent energy, ε (m^2/s^3).[5] The *k-ε* model needs to solve two other equations for the transport flow.

For kinetic turbulent energy, K

$$\rho(\nabla(K\vec{v})) = \nabla\cdot\left(\left(\mu_0 + \frac{\mu_t}{\sigma_K}\right)\nabla K\right) + G_K - \rho\varepsilon \tag{5.5}$$

$$\rho(\nabla(\varepsilon\vec{v})) = \nabla\cdot\left(\left(\mu_0 + \frac{\mu_t}{\sigma_\varepsilon}\right)\nabla\varepsilon\right) + C_1\frac{\varepsilon}{K}G_K - C_2\rho\frac{\varepsilon^2}{K} \tag{5.6}$$

$G_K = -\rho\overline{u_i'u_j'}\frac{\partial u_j}{\partial x_i}$ = Generation of kinetic energy

where σ_K, σ_ε, C_1, and C_2 are empirical constants which values are 1.0, 1.3, 1.44, and 1.92, respectively, and i and j are coordinate direction indices.

MHD Equations

The electric potential method of MHD theory was used for solving the current density using ohm's law. As compared to applied external magnetic field ($\vec{B_0}$) in Tesla, the induced magnetic field ($\vec{b}$) is insignificant.[1,2]

$$J = \sigma\left(-\nabla\varphi + \left(\vec{v}\times\vec{B_0}\right)\right) \tag{5.7}$$

$$\nabla^2\varphi = \nabla.\left(\vec{v} \times \overrightarrow{B_0}\right) \tag{5.8}$$

For the momentum equation, the additional source term is the Lorentz force (unit is N/m^3 in SI system)

$$\vec{F} = \vec{J} \times \overrightarrow{B_0} \tag{5.9}$$

where, J = current density,

φ = electric potential field

The electric potential method uses only one transport equation. One user-defined scalar is solved for the electric potential field (φ) and has unit of Volt. In this method, there are two options: patch or impart, used impart that has required of a mag data file. Imparted the file from the system and run the simulation with using both magnetic field data media (conducting media or non-conducting media).

5.2.3 EFFECT OF MAGNETIC FIELD MEDIA AND ITS ORIENTATION ON ELECTRIC POTENTIAL

The magnetic field of 0.2 tesla was applied along with mold length from meniscus up to 1 m in casting direction (in the *z*-direction). The contours of applied constant magnetic in mid-plane parallel to wide face and narrow face are shown in Figure 5.2a and b, respectively. The graphical representation of magnetic field with vertical distance from the meniscus is shown in Figure 5.2c. Figure 5.3 shows the contours of electric potential for different cases. The effect of magnetic field media on contour of electric potential is shown in these figures. The magnetic field has a great impact on the electric potential in mold domain when it is applied in the *y*-direction. The magnetic field media slightly affect the contour of electric potential, but orientation of magnetic field is greatly affecting the contour of electric potential. Therefore, the fluid flow of the continuous casting will also affect due to the orientation of magnetic field. The influence of orientation on the magnetic field with magnetic field media on the contour of electric potential is shown in the Figure 5.4.

5.3 RESULTS AND DISCUSSION

As discussed earlier, seven cases have been simulated to study the fluid flow of continuous casting mold in the direction to understand the influence of different magnetic field with different magnetic field media. One of the cases is simulated without using the magnetic field and other six cases were simulated with using the application of magnetic field. The magnetic field of 0.2 Tesla was applied along the different direction along mold thickness, mold width, and mold length. The contours of velocity with vector when magnetic field applied in x-direction are shown in Figure 5.4. Here, effect of magnetic field media (conducting and non-conducting) on fluid flow in mold of continuous casting was observed when magnetic field is applied along the x-direction (along the narrow face means thickness of continuous casting mold). The higher meniscus velocity with single roll flow in the both side of SEN was observed in the Case 1 in Figure 5.4. The arrows in the figure are direction of flow, which are called velocity vector. The turbulent flow in continuous casting mold at meniscus was decreased due to the magnetic field in x-direction but conducting magnetic field media has highly impact on the fluid flow at meniscus. The magnetic field along the x-direction suppresses the meniscus velocity with symmetry to lower level that will be helpful to improve the cast quality. The conducting magnetic field media has higher impact on velocity contour than non-conducting field media. In the real continuous casting practice, the non-conducting magnetic field is generally used. If the conducting field media is used in the real continuous casting process, the cleaner cast product can be produced.

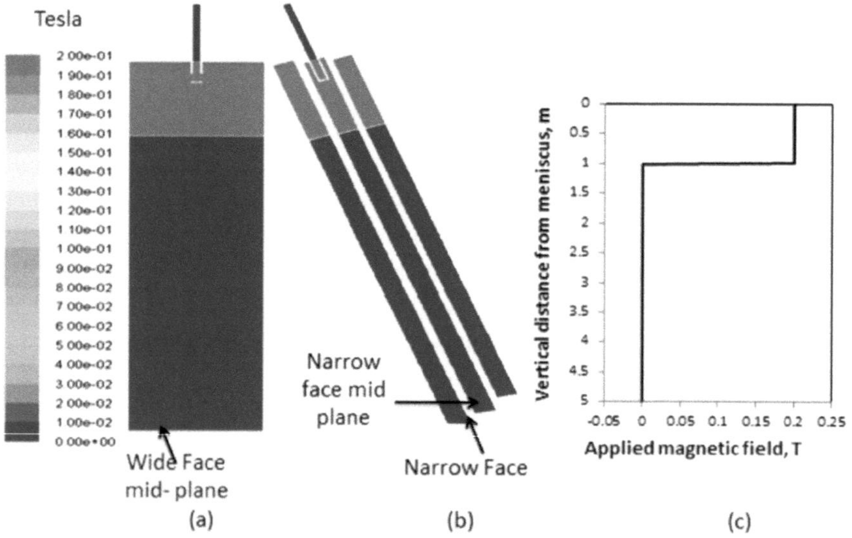

FIGURE 5.2 Contour of applied mangetic field at (a) wide face; (b) narrow face; and (c) variations with vertical distance from meniscus.

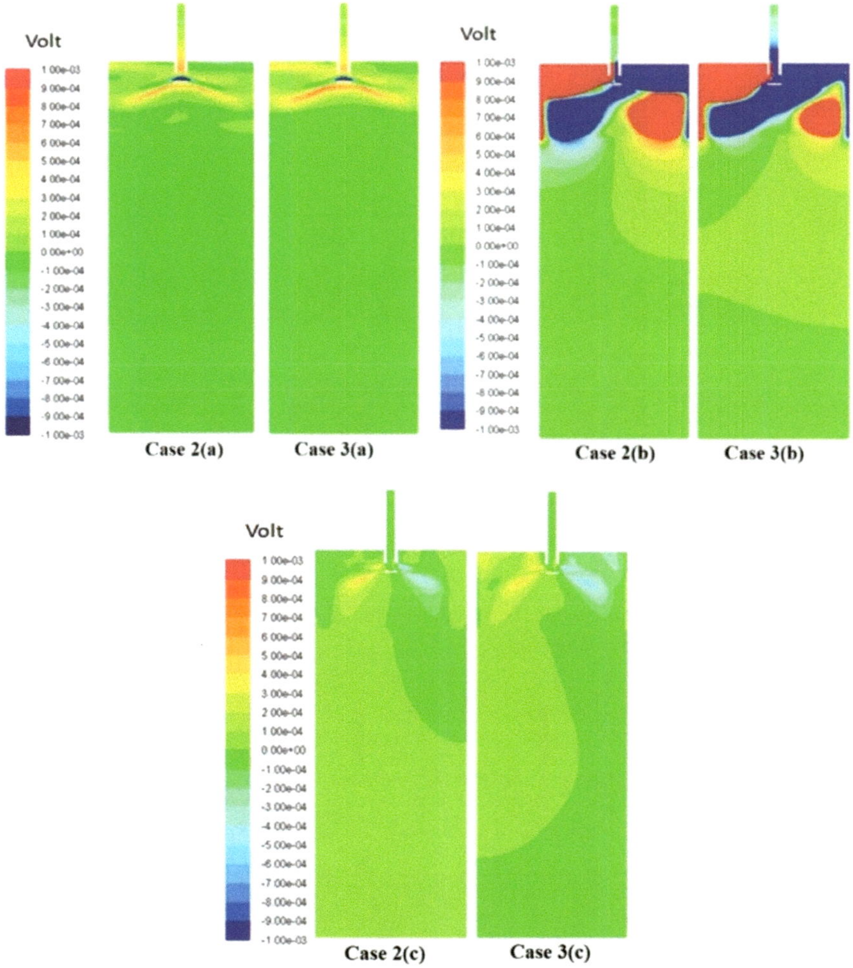

FIGURE 5.3 Effect of magnetic field media on contours of electric potential.

The flow in continuous casting was altered when magnetic field was worked in *y*-direction (along the mold width). The magnetic field increases the meniscus velocity in the casting mold when magnetic field is applied in *y*-direction as shown in the Figure 5.5. The conducting field media produced more symmetric flow in the mold of continuous casting as compared to non-conducting magnetic field media. The velocity contour with velocity vector for conducting magnetic media and non-conducting field media when the magnetic field was operated in *y*-direction as shown in Figure 5.5. The asymmetric flow with high velocity is very dangerous for final cast product

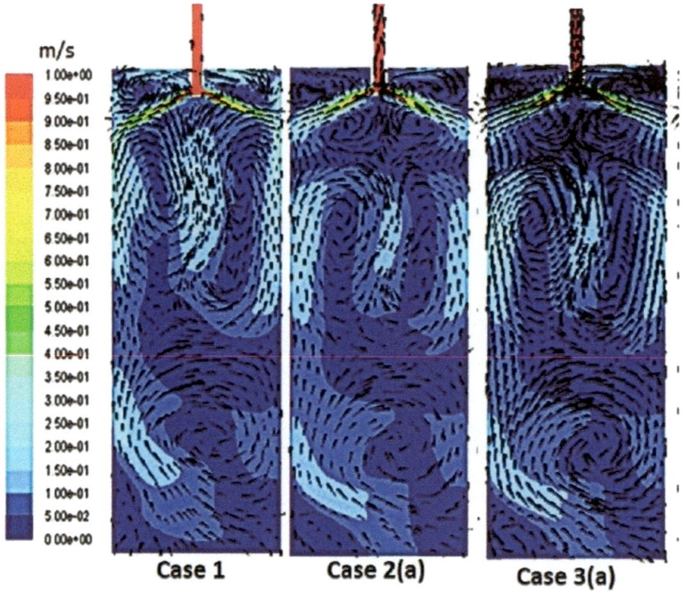

FIGURE 5.4 Effect of magnetic field media on fluid flow of continuous casting mold (when magnetic field applied in *x*-direction).

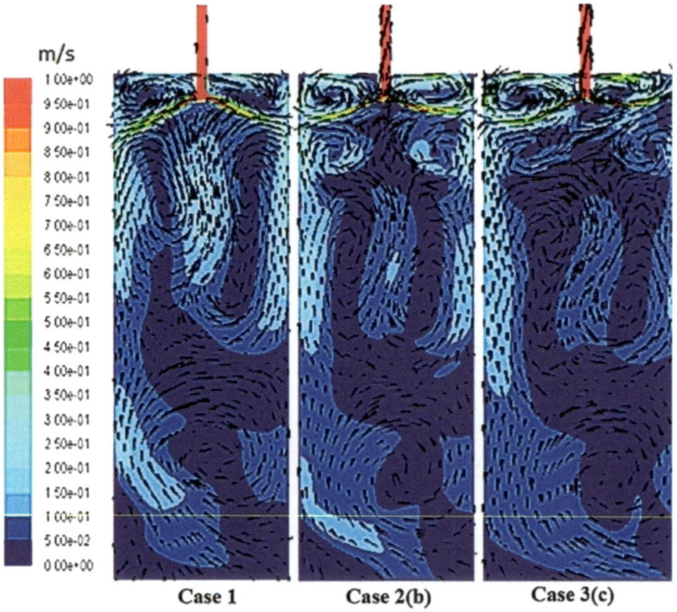

FIGURE 5.5 Effect of magnetic field media on fluid flow of continuous casting mold (when magnetic field applied in *y*-direction).

because it leads to formation of vertex, which is responsible for the ingression of inclusion inside the cast product. The symmetric flow is important for the quality of final cast product. When magnetic field is used in *y*-direction, it increased the meniscus velocity, which is useful to increase the meniscus velocity where meniscus velocity is very slow. When magnetic field was applied in the *y*-direction, it produced the double roll flow (means circulation of fluid above and below of submerged entry nozzle) that may be harmful for the quality of steel product. The conducting field media has also produced more symmetric flow than non-conducting field media for the application of magnetic field in the *y*-direction.

The more symmetric flow with lower velocity was observed when magnetic field was employed in *z*-direction as shown in Figure 5.6. The circulation of flow above and below the submerged entry nozzle in the continuous casting mold was removed and leads to the molten steel in the direction mold. It can be seen clearly with the help of velocity vector as shown in the Figure 5.6. Here, it is also noticed that the conducting field media produced extra symmetric flow as compared to non-conducting field media in the mold of continuous casting process in both sides of submerge entry nozzle. If it is required to increase velocity at the meniscus, the magnetic may be applied in *y*-direction and, for the slow down, the meniscus velocity, the magnetic field can be used in the *z*-direction. The optimum flow was observed in the magnetic field in the *x*-direction in the continuous casting mold. From Figures 5.4–5.6, the conducting field media produced highly symmetric flow field and orientation of magnetic field greatly affected the flow in the mold.

The effect of magnetic field media and orientation of magnetic field was discussed earlier on the fluid flow in the mold of continuous casting with using a plane that was cut parallel to the wide face of the mold. Now, the flow on top surface of mold can be discussed using a different plane parallel to top surface of mold. The influence of magnetic field media and orientation of magnetic field on meniscus velocity are shown in Figure 5.7.

The velocity contour with velocity vector at the meniscus of continuous casting mold for conducting and non-conducting magnetic field media when magnetic field was worked in different direction along mold thickness, mold width, and mold length as shown Figure 5.7. The fluid flow was altered with changing the orientation of applied magnetic field. The fluid flow was also affected due to magnetic field data media. The higher velocity was observed when the magnetic field was operated in the *y*-direction (along the width of mold) with higher fluctuations. The conducting magnetic field media produced more symmetry than non-conducting media as shown in

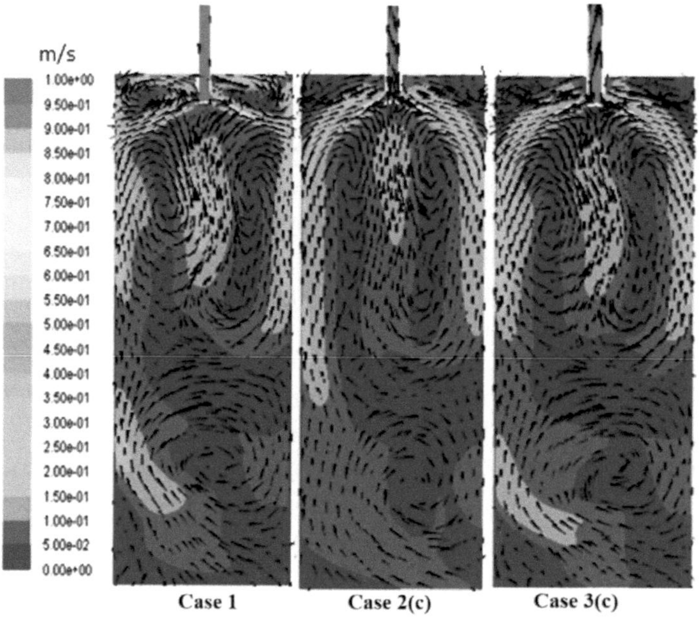

FIGURE 5.6 Effect of magnetic field media on fluid flow of continuous casting mold (when magnetic field applied in z-direction).

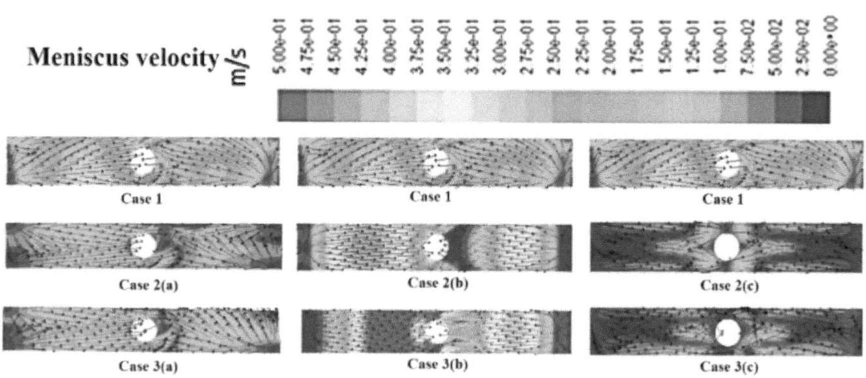

FIGURE 5.7 Effect of magnetic field media on meniscus velocity of continuous casting mold (when magnetic field applied in x, y, and z-direction).

the Figure 5.7. The magnetic field slows down the steel velocity at meniscus when the magnetic was applied in z-direction (along the length of mold). The conducting magnetic field data media produced symmetry in fluid flow of continuous casting mold. The fluctuations in meniscus with magnetic field media (conducting and non-conducting) was altered when magnetic field was employed in *x*, *y*, and *z*-direction as shown in Figure 5.8. The highest meniscus velocity is 0.3 m/s for Case 1 (without magnetic field), which is reduced to 0.15 and 0.13 m/s when magnetic was applied in *x*-direction with non-conducting and conducting field media, respectively, as shown in Figure 5.8b. The meniscus velocity was increased up to 0.56 m/s in one side of SEN due to the application of magnetic field in the y-direction, but the meniscus velocity is greatly decreased up to 0.12 m/s for small range along mold width with stable flow due to the application of magnetic field in *z*-direction as shown in Figure 5.8d. When magnetic field was applied in the *x*- and *z*-direction, the velocity at the meniscus was decreased as compared to the velocity of steel without magnetic field, but it is increased when the magnetic

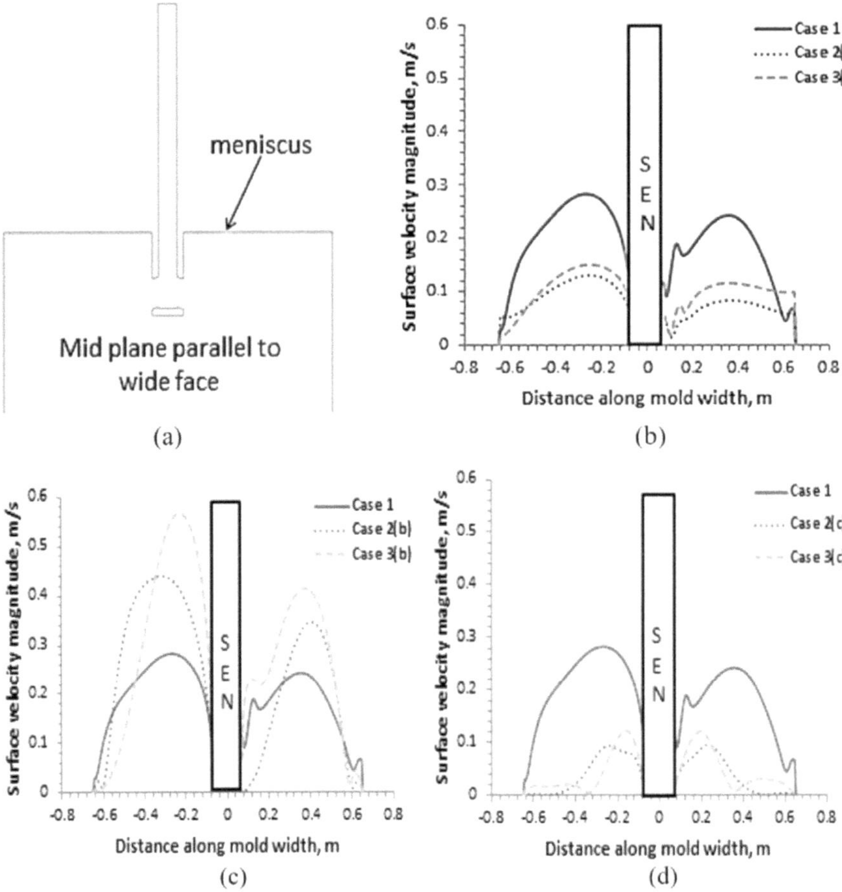

FIGURE 5.8 Effect of magnetic field media on meniscus velocity of continuous casting mold when magnetic field applied in (a) *x*-direction (b) *y*-direction, and (c) *z*-direction.

field was applied in the *y*-direction. The symmetric flow is produced with lower meniscus velocity when magnetic field was operated in *z*-direction as compared to *y*- and *x*-directions. It is clearly seen from Figure 5.8 that the conducting magnetic field media produced more symmetric flow than non-conducting field media. So, the conducting field media may produce more clean steel than the non-conducting field media. The turbulent flow can be controlled with the help of the application of magnetic field and it is altered with changing the direction of magnetic field. It is also affected due to the magnetic field media.

5.4 CONCLUSION

The influences of mangetic field media (conducting and non-conducting) and the orientation of magnetic field on the liquid steel flow in the continuous casting mold have been discussed with the help of a three-dimensional computational model in this research work. For this study, the continuity equation and momentum equation were solved with the help of Ansys Fluent. The modal was validated with help of experimental results. Here, it observed that the conducting media has higher impact on the fluid flow of continuous casting mold than non-conducting field media. The electromagnetic field highly suppressed meniscus velocity symmetrically with conducting magnetic field media. The fluid flow of continuous casting mold was altered with altering the orientation of magnetic field. The meniscus velocity of steel (0.3 m/s) was decreased to 0.15 and 0.13 m/s when the magnetic field is employed in *x*- and *z*-directions, respectively, but the velocity was more decreased due to the application of magnetic field in the *z*-direction. The meniscus velocity was accelerated up to 0.56 m/s due to the application of magnetic field in *y*-direction and flow was observed highly turbulent with asymmetry. The magnetic field has higher impact when it is applied when conducting the magnetic field media than the non-conducting field media. The conducting field media produced more asymetric flow with lower mescus velocity than the non-conducting field media with the effect of orientation of magnetic field. From this study, it was observed that the magnetic field media and orientation of magnetic field altered the fluid flow of continuous casting mold. So, the magnetic field media and orientation play an important role to control the turbulent flow in the continuous casting mold to produce clean steel product after casting.

ACKNOWLEDGMENT

The author is thankful to the Research and Development, Tata Steel, Jamshedpur, for providing the opportunity to do this work.

KEYWORDS

- **continuous casting mold**
- **fluid flow**
- **magntohydrodynamics**
- **conducting or non-conducting media**
- **reliable k-ε model**
- **casting defects**
- **computational model**

REFERENCES

1. Thomas, B. G. Continuous Casting of Steel. *Chapter* **2001,** *15*, 499–540.
2. Versteeg, H. K.; Malalasekera, W. *An Introduction to Computational Fluid Dynamics: The Finite Volume Method*; Pearson Education, 2007.
3. Davidson, P. A. An Introduction to Magnetohydrodynamics, 2002.
4. Chaudhary, R.; Thomas, B. G.; Vanka, S. P. Effect of Electromagnetic Ruler Braking (EMBr) on Transient Turbulent Flow in Continuous Slab Casting Using Large Eddy Simulations. *Metallurg. Mater. Trans. B* **2012,** *43* (3), 532–553.
5. Cukierski, K.; Thomas, B. G. Flow Control with Local Electromagnetic Braking in Continuous Casting of Steel Slabs. *Metallurg. Mater. Trans. B* **2008,** *39* (1), 94–107.
6. Singh, R.; Thomas, B. G.; Vanka, S. P. Effects of a Magnetic Field on Turbulent Flow in the Mold Region of a Steel Caster. *Metallurg. Mater. Trans. B* **2013,** *44* (5), 1201–1221.
7. Kunstreich, S.; Dauby, P. H. Effect of Liquid Steel Flow Pattern on Slab Quality and the Need for Dynamic Electromagnetic Control in the Mould. *Ironmak. Steelmak.* **2005,** *32* (1), 80–86.
8. Šarler, B.; Vertnik, R.; Mramor, K. A Numerical Benchmark Test for Continuous Casting of Steel. *IOP Conf. Ser. Mater. Sci. Eng.* 2012, *33* (1), 012012.
9. Thomas, B. G.; Chaudhary, R. State of the Art in Electromagnetic Flow Control in Continuous Casting of Steel Slabs: Modeling and Plant Validation. In *6th International Conference on Electromagnetic Processing of Materials EPM* (Vol. 2009), Oct 2009.
10. Hackl, H. R.; Lehman, A. F.; Eriksson, J. E. A.; Kollberg, S. G. *AIS Tech Preliminary Program*; AIST: Warrendale, PA, 2006.

11. Dauby, P. H. Continuous Casting: Make Better Steel and More of It! *Revue de Métallurgie–Int. J. Metallurg.* **2012,** *109* (2), 113–136.
12. Ishii, T.; Sazhin, S. S.; Makhlouf, M. Numerical Prediction of Magnetohydrodynamic Flow in Continuous Casting Process. *Ironmak. Steelmak.* **1996,** *23* (3), 267–272.
13. Kunstreich, S.; Dauby, P. H. Effect of Liquid Steel Flow Pattern on Slab Quality and the Need for Dynamic Electromagnetic Control in the Mould. *Ironmak. Steelmak.* **2005,** *32* (1), 80–86.
14. Singh, R.; Thomas, B. G.; Vanka, S. P. Effect of Electromagnetic Braking (EMBR) on Turbulent Flow in Continuous Casting. In *AIS Tech 2013 Iron and Steel Technology Conference*, Aug 2013; pp 1323–1336.
15. Chaudhary, R. *Studies of Turbulent Flows in Continuous Casting of Steel with and without Magnetic Field*; Doctoral dissertation, University of Illinois at Urbana-Champaign, 2011.
16. Miao, X., Timmel, K., Lucas, D., Ren, Z., Eckert, S.; Gerbeth, G. Effect of an Electromagnetic Brake on the Turbulent Melt Flow in a Continuous-Casting Mold. *Metallurg. Mater. Trans. B* **2012,** *43* (4), 954–972.
17. Singh, R.; Thomas, B. G.; Vanka, S. P. Large Eddy Simulations of Double-Ruler Electromagnetic Field Effect on Transient Flow during Continuous Casting. *Metallurg. Mater. Trans. B* **2014,** *45* (3), 1098–1115.
18. Chen, H-M.; Hwang, W-S. *Effect of EMBR on the Fluid Flow Inside a Thin Slab Continuous Casting Mold through Numerical Simulations*; Brno: Czeeh Repulic, EU, 2014.
19. Sarkar, S.; Jatav, S. K.; Singh, V.; Ajmani, S. K. Characterization of Single and Multiphase Hydrodynamics in Continuous Slab Casting Mold in the Presence of Applied Magnetic Field. *Jom* **2018,** *70* (12), 2980–2992.

CHAPTER 6

EXPERIMENTAL EVALUATION OF MECHANICAL AND METALLURGICAL PROPERTIES OF GRAY CAST IRON

M. VIJAY KUMAR[1*], B. S. AJAY KUMAR[2], and PRANITA MADALLI[1]

[1]*Department of Industrial Engineering and Management, JSS Academy of Technical Education, Bengaluru, India*

[2]*Department of Mechanical Engineering, K S Institute of Technology, Bengaluru, India*

[*]*Corresponding author. E-mail: mvkjss@gmail.com*

ABSTRACT

This paper is an experimental work carried out to evaluate the mechanical and metallurgical properties of gray cast iron that is used as a material for lathe bed. The gray cast iron is studied and analyzed for two quenching mediums such as polymer quenchant metal working oil (servo quench 11) and soluble refined mineral oil (MAK Sherol B) before and after the flame hardening process. The mechanical and microstructural behavior of the material are determined and analyzed for raw and treated specimens using metallurgical microscope.

6.1 INTRODUCTION

A lathe bed is positioned under the work space that covers the area from headstock to tailstock and it gives a stable and firm support to the tooling and

Modern Manufacturing Systems: Trends and Developments. Rajiv Kumar Garg, Ravi Pratap Singh, Rajeev Trehan, & Ramesh Singh (Eds.)

other components place on it. The top surface of the lathe bed is the ground for providing slide or guide ways (in the form of "V"). Head stock, tailstock, and carriage assembly are the three important components that are mounted on the lathe bed. The bed is subjected to heavy loads and vibrations; hence, the material used must possess very high vibration damping properties.

Generally, the bed is cast by using high grade cast iron because it exhibits good range of mechanical properties. To produce castings of high quality, the meehanite process is usually used. The bed must possess a hardness value of 210 BHN and tensile strength of 27 kg-f/mm^2 approximately. Hardness is an important factor in accepting or rejecting the bed. The bed is casted by following the regular process of pattern making and molding. It is followed by sand blasting and fettling. Finally, the finishing operations—machining and painting—are carried out. Then, the bed is heat treated by employing induction or flame-hardening process to impart the required hardness and other mechanical properties into it. After the heat treatment process, the bed attains a hardness of around 45 HRC (hardness measured on Rockwell CScale).

6.2 PROPERTIES OF GRAY CAST IRON

ASTM A48 class 35 (GD 250) gray cast iron is the most commonly used grade of cast iron. It is an alloy of iron-carbon-silicon generally called as gray cast iron. It is majorly composed of 2.5–4% carbon, 1–3% silicon, and, the rest, iron. Traces of manganese sulfur and phosphor are also present. They are hard, brittle, and nonmalleable in nature. The manufacturing method and composition determine their characteristics. When fractured, the surface is gray in color because of the presence of graphite; hence, the name gray cast iron. The microstructure of cast iron exhibits a ferritic or pearlitic matrix. It is in the form of flake graphites. The chemical composition of gray cast iron in %weight is tabulated in Table 6.1.

TABLE 6.1 Chemical Composition of Gray Cast Iron in %weight.

Metal	C	Mn	Si	Cr	Ni	Mo
Gray cast iron	3.2	0.7	2	1	1	0.4

Heat treatment can be defined as the process of obtaining the desired mechanical properties and microstructure by carrying out the combination of heating and cooling operations on the material or alloy in solid state.

Flame hardening is carried out using oxygen–acetylene gas for surface hardening by using high-intensity heating flame for a short period followed by quenching process. Polymer quenchant metal working oil (servo quench 11) and soluble cutting oil blended from refined mineral oil, emulsifier, and other additives (MAK Sherol B) are used as the quenching media. This localized hardening and heating improve the mechanical properties of the material.

The Figures 6.1 and 6.2 show a lathe bed and guide way subjected to flame-hardening process. Here, the guide ways of the stationary lathe bed is treated. It is a progressive-type flame-hardening process. The burner head or flame torch traverses over the length of the guide ways slowly. The quenching unit is attached along with the head of the torch; hence, a jet of water is sprayed simultaneously with the flame. Both the flame and water are impinged on the surface of the guide ways. The surface gets highly heated to a temperature around 1000°C and immediately cooled by quenching with water, resulting in change of state from austenite to martensite on the surface.

FIGURE 6.1 Flame hardening of lathe bed.

This increases the hardness on the surface and the core is soft. Induction and flame hardening are generally used to treat parts such as shafts, lathe beds, and other components of lathe, gears, crankshafts, cams, and rollers.

FIGURE 6.2 Flame hardening of guide ways.

The high temperature material is quenched resulting in phase transformation of the material from austenite to martensite state. This change of crystalline structure or phase transformation of the material affects its hardness and other properties.

6.3 EXPERIMENTATION AND METHODOLOGY

i. Preparation of Samples

A gray cast iron bar, which was subjected to turning operation in a lathe, is shown in the Figure 6.3. The obtained sample after turning and cutting is shown in Figure 6.4.

ii. Flame Hardening

The flame-hardening temperature was around 1000° and the standoff distance was about 8–10 mm.

iii. Quenching

The quenching process immediately follows flame hardening. The treated samples are quenched at room temperature using polymer and Sherol.

FIGURE 6.3 Turning operation.

FIGURE 6.4 Sample specimen.

iv. Tempering

The tempering was done in an induction furnace maintained at a temperature of 200°C for 2 h and allowed to cool at room temperature gradually.

v. Testing Mechanical Properties (IS: 1586–2012)

The hardness value serves as a guide for setting up tempering temperature and the time in order to achieve the hardness value of a work piece as mentioned in production blue print.

A primary load of 10 kgf was applied at the beginning followed by a major load of 140 kgf; the depth of penetration of the diamond was indicated and measured by a dial indicator provided on the machine, which gives direct value of hardness of specimen in Rockwell Hardness Number; the higher the hardness, the higher the reading and vice–versa. Three hardness tests were done and the average value of these three readings was taken as a final hardness value.

vi. Pin on Disc Wear Testing (ASTM G99-05(2010))

This method described the procedure to be followed to determine the material wear during sliding utilizing a pin-on-disk apparatus. During experimentation, the important concepts to be known while using the apparatus to measure wear were mentioned. Through this procedure, even the friction co-efficient could be obtained. For this wear test, two specimens were necessary, a pin with a flat circular disc. The pin was held perpendicular to the circular disk rigidly. The pin served as the specimen to be tested for wear. The disc specimen revolved around the center continuously.

6.3.1 CORROSION TEST (ASTM B117-2007)

The neutral salt spray (NSS) test is the standard test to determine the resistance offered by the samples against corrosion. Generally, coatings are applied to the components to enhance corrosion resistance and to increase the intended life of the part in use. The test is evaluated over a period of time till the oxidation takes place. The duration of the test depends upon the corrosion resistance offered by the coatings. Corrosion-resistant coating is directly proportional to the period in testing without showing any signs of corrosion.

6.3.2 MICROANALYSIS USING METALLURGICAL MICROSCOPES (IS 7739(P-5)1976)

Metallography is the study of components and their physical structure through a microscope. The metallurgical microscopes usually are optical or electron microscopes. It is used for detecting grain size, surface defects, and any inclusions.

The microanalysis is carried out on samples either mounted in a resin mold or cut to size. The magnification usually varies from 50× to 500×. The commonly available microscopes are 80× and 100×. Depending on the application, reflected or transmitted type of illumination is used.

If the obtained crystal structures are non-cubic (hexagonal-closed packed crystal structure), then microstructure can be seen without any etching agent using cross polarized light, else the microstructure of the specimen can be revealed by using suitable chemical etchant.

The surface of the specimen to be tested is prepared by different methods of grinding, polishing, and etching. The surface of the sample to be analyzed should be free from any dust and completely degreased. A flat mirror finish surface is required for testing the specimen. The sample is roughened first with belt polish using bench grinder and then fine polished manually with various silicon carbide emery papers from 80, 120, 400, and 600 grit sizes, respectively, and finally fine polished using a fine polishing machine. The specimen is polished with slurry of alumina or silica on the rotating disc of the machine and finally polished on a Naples cloth to produce a scratch free mirror finish and with minimal deformation remaining from the preparation process.

After preparation of the sample, a mould is prepared to mount the specimen to be tested. The mould is prepared by using thermosetting resins or epoxy. Epoxy is popularly used because of less shrinkage during curing, which results in good edge retention and mount. The samples are generally etched with appropriate chemical solution before it is analyzed using optical or electron microscope.

The type of etchant used for gray cast iron sample is nital (2–4% nitric acid + 96–98% alcohol). The solution is swabbed lightly over the required surface to be analyzed using cotton.

The microstructure of polymer and Sherol quenched sample are represented in Figures 6.5 and 6.6.

6.4 RESULTS

i. Hardness test: The Rockwell hardness results are tabulated in Table 6.2.

TABLE 6.2 Rockwell Hardness.

Quenching medium	Hardness (HRC)
Raw material	26
Polymer	48
Sherol B	41

FIGURE 6.5 Microstructure of polymer.

FIGURE 6.6 Microstructure of Sherol.

ii. Corrosion test: A red corrosion is found for the treated specimens.
iii. Pin-on-disc test: For a load of 20 N is tabulated in Tables 6.3 and 6.4.

TABLE 6.3 Weight of the Sample for 20 N Load.

Particulars	Initial weight (g)	Final weight (g)
Raw material	6.721	6.664
Polymer quenched	12.740	12.713
Sherol quenched	10.584	10.556

TABLE 6.4 Pin-on-Disc for 20 N Load.

Type	Speed (rpm)	Time (min)	Sliding velocity (m/s)	Weight loss (g)	Wear rate (mm^3/Nm)
Raw material	300	10	0.786	0.057	8.39×10^{-4}
Polymer quenched	300	10	0.786	0.027	3.98×10^{-4}
Sherol quenched	300	10	0.786	0.0298	4.39×10^{-4}

iv. For a load of 30 N: is tabulated in Tables 6.5 and 6.6.

TABLE 6.5 Weight of the Sample for 30 N.

Particulars	Initial weight (g)	Final weight (g)
Raw material	6.664	6.598
Water quenched	10.791	10.764
Polymer quenched	12.713	12.674
Sherol quenched	10.556	10.514

TABLE 6.6 Pin-on-Disc for 30 N Load.

Type	Speed (rpm)	Time (min)	Sliding velocity (m/s)	Weight loss (g)	Wear rate (mm^3/Nm)
Raw material	300	10	0.786	0.086	8.45×10^{-4}
Polymer quenched	300	10	0.786	0.041	4.03×10^{-4}
Sherol quenched	300	10	0.786	0.045	4.42×10^{-4}

v. Microstructure

The sample is roughened with belt polish and fine polished with various emery papers from 80, 120, 400, 600, and fine polished using Lavigated Alumina with 100× magnification and 5% Etchant. The microstructure structure appears in the form of a needle-shaped bainite (light gray) with pearlite (dark gray) and graphite (black). The hardness of this sample is 26 HRC. Microstructure structure for polymer quenched consists of flake graphite in a matrix of fine pearlite with 10% ferrite and 5% cementite. The hardness of this sample is 48 HRC. Microstructure structure for Sherol quenched consists of graphite in matrix of ferrite and pearlite and dark bands of pearlite at boundaries of the cell. The hardness value of this sample is 41 HRC.

6.5 CONCLUSION

The mechanical properties and their microstructure are studied and analyzed for two different quenching media. It is observed from the results that after the flame-hardening process, there is increase in hardness and wear resistance. Maximum wear resistance was obtained when polymer quenchant metal working oil (servo quench 11) was used as quenching medium compared to soluble cutting oil (Sherol). Thus, servo quench 11 can be used as quenching media during the flame-hardening process in order to increase the mechanical properties of gray cast iron.

KEYWORDS

- **flame hardening**
- **quenching medium**
- **metallurgical properties**
- **microstructure analysis**

REFERENCES

1. Silva, A. J. S. T.; Goldenstein, H.; Guesser, W. L.; de Campos, M. Quenching and Partitioning Heat Treatment in Ductile Cast Irons. *Mater. Res.* 2014, *17* (5), 1115–1123.
2. Edmonds, D. V.; Speer, J. G. Martensitic Steels with Carbide Free Microstructures Containing Retained Austenite. *Mater. Sci. Technol.* **2010,** *26* (4), 386–391.
3. Matlock, D. K.; Speer, J. G. Processing Opportunities for New Advanced High-Strength Sheet Steels. *Mater. Manuf. Proces.* **2010,** *25* (1–3), 7–13.
4. Dyson, D. J.; Holmes, J. B. Effect of Alloying Additions on Lattice Parameter of Austenite. *J. Iron Steel Inst.* **1970,** *208*, 469–470.

5. Liu, C.; Zhao, Z.; Northwood, D. O.; Liu, Y. A New Empirical Formula for the Calculation of MS Temperatures in Pure Iron and Super-Low Carbon Alloy Steels. *J. Mater. Proces. Technol.* **2001,** *113* (1–3), 556–562.
6. Edmonds, D. V.; Speer, J. G. Martensitic Steels with Carbide Free Microstructures Containing Retained Austenite. *Mater. Sci. Technol.* **2010,** *26* (4), 386–391.
7. Zhong, N.; Wang, X. D.; Wang, L.; Rong, Y. H. Enhancement of the Mechanical Properties of a Nb-microalloyed Advanced High-Strength Steel Treated by Quenching–Partitioning–Tempering Process. *Mater. Sci. Eng.* **2009,** *506* (1–2), 111–116.
8. De Moor, E.; Lacroix, S.; Clarke, A. J.; Penning, J.; Speer, J. G. Effect of Retained Austenite Stabilized via Quench and Partitioning on the Strain Hardening of Martensitic Steels. *Metal. Mater. Trans.: Phys. Metal. Mater. Sci.* **2008,** *39* (11), 2586–2595.

CHAPTER 7

A CRITICAL REVIEW ON BAINITIC STRUCTURE AFTER SHOT PEENING IN HIGH SILICON SPRING STEEL

KULDEEP DWIVEDI*, RAJAN KUMAR, and RAVI K. DWIVEDI

Department of Mechanical Engineering, Maulana Azad National Institute of Technology, Bhopal, India

Corresponding author. E-mail: kshudwivedi@gmail.com

ABSTRACT

The review work aims to examine the fatigue strength, wear-resistance, and surface roughness of the materials after shot peening and three-dimensional numerical modelling are used to predict the behavior of the material in severe conditions. Basic parameters used for fatigue life estimation are shot size, peening intensity, the number of shots, shot velocity, and coverage. Shot peening is mainly a cold working process, but water cavitation peening and laser shock peening are also popular for different applications. Each shot acts like a peen hammer and produces a residual compressive stress zone on the surface. The effect of temperature has been studied on residual compressive stress distribution on the surface as plastic deformation increases with temperature. The process of shot peening is performed in the camshaft, crankshaft, clutch springs, gear wheels, and connecting rods to increase their fatigue life. Higher strain rates resulting from the shot peening move the dislocations toward the grain boundaries to strengthen the material.

Modern Manufacturing Systems: Trends and Developments. Rajiv Kumar Garg, Ravi Pratap Singh, Rajeev Trehan, & Ramesh Singh (Eds.)

7.1 INTRODUCTION

Shot peening is known to be a cold working process used to produce the residual compressive stresses on the specimen surfaces and it also brings out the changes in the mechanical properties of metals and composites. Small spherical media called shot to behave like a small peen hammer strikes the material to impart a small indentation called as a dimple. The yielding of the surface fibers of the material leads to the creation of a dimple. Overlapping of the dimples develops a highly compressively stressed zone in which residual stress minimizes the chances of crack propagation. Almen numbers is a standard method to specify the shot peening intensity. These numbers indicate the arc height (curvature) of a steel strip exposed to the shot peening treatment. Almen strip is made up of spring steel with close tolerance in flatness and hardness peened on single side only. The arc height of almen strip varies with the mass and velocity of the shot. Surface treatment processes like milling, grinding, bending, and some heat treatments set up tensile residual stresses in the material. The life cycle of these parts is enormously reduced due to stresses. Shot peening is employed to convert these tensile stresses into residual compressive stress that results in an increase in the life cycle and load-carrying capacity of these parts. The fatigue life of a component is of great importance when the component is subject to cyclic loading running at high speeds such as shafts, bearing, and gears.[1] The hardness of the component can also be increased by shot peening as in the case of nitrogen austenitic stainless steel. Control over the hardness of the component depends on the intensity of the shot peening process. The shot-peening process is used in clutch springs, camshafts, crankshaft, gearwheels, and connecting rods to increase their fatigue life. Surface finishing of the cylinder heads and engine blocks in the casting process is also done by sandblasting or shot peening process.

7.2 PROCESS VARIABLES

1. Shot size
2. Peening intensity
3. Homogeneity of the material
4. Peening coverage percentage
5. Time of coverage
6. Shot angle

7.3 RELATIONSHIP BETWEEN THE DIFFERENT VARIABLES

In all sorts of industrial applications, the shot peening process is mainly controlled by the shot size, intensity, peening coverage, and velocity. The area that has been hit by the indent with a specified almen intensity is termed as coverage. It can either be visually estimated with the help of fluorescent paint or by using a loupe. Peening intensity can also be calculated by the exposure of the standard almen strip to shot stream for different periods. Peening intensity mainly depends on media velocity; however, it is also dependent on media hardness, impact angle, and material properties. However, the size of the almen strips is standardized. The experimental results show the correlation between velocity and almen intensity in the following figure[2] (Figure 7.1).

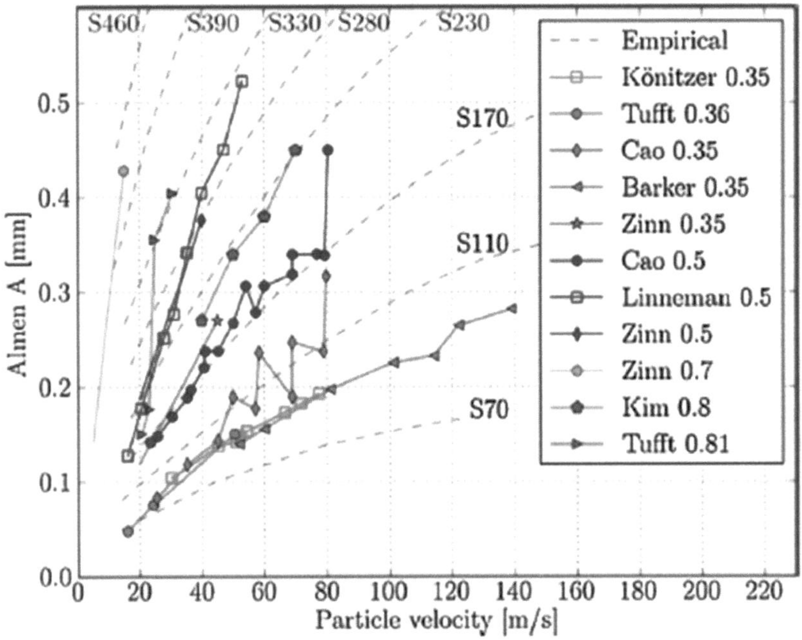

FIGURE 7.1 Relationship between particle velocity and almen intensity.

Source: Reprinted with permission from Ref. [2]. © 2016 Elsevier.

The curves have been generated by using different media sizes ranging from 0.1 to 0.5 mm, as we can see in the Figure 7.1. Some of the researchers use the nominal diameter of the shot despite the average diameter. Nominal diameter represents the amount of sieve opening in which at least 85% mass of the shot will remain. Konitzer and Polanetzki used a compressed air machine and also an optical-based velocity measurement system. S110 peening media was used with an impact angle of 70°. Tufft (1999) used electromagnetic velocity sensor to calculate media velocity. Three results have been measured with a mean diameter of 0.81 mm at an impact angle of 85° and the fourth result was obtained at the diameter of 0.36 mm.

Nanostructured surfaces from shot peening: It is not necessary to produce the nanofibers throughout the whole material; even the generation of the nanocrystalline layer on the upper surface will be extremely helpful to modify the properties. NC surfaces are obtained by deforming the material non-homogenously with large strain rates. The formation of nanosurface depends on the arrangement of dislocations during the shot peening process.

Surface nanocrystallization is obtained only through the high values of almen intensity and coverage percentage. The accumulated plastic strain on the surface during the process is the main factor for the generation of NC surfaces.

7.4 EFFECT OF TEMPERATURE ON SHOT PEENING PROCESS

A large amount of energy is required to carry out the shot peening process in high strength steels as in spring steels, which can fracture the shot or deteriorates the surface texture. Warm shot peening process is carried in the springs to decrease the flow stress. Keeping the working temperature constant during shot peening is quite a difficult task to attain. However, plastic deformation and the residual compressive stress increase with the increase in working temperature. Warm shot peening is to be carried out at an optimum temperature in the spring industry.[4]

Distribution of residual stress in ultra-shot peening plays a significant role to increase stiffness strength and fatigue life of the material. Moiré interferometry is a high-resolution optical method that is used to measure the residual stress. Specimen grating frequency is generally either 1200 or 2400 lines/mm. The maximum frequency, 4000 lines/mm, used in this method corresponds to the displacement resolution of 0.2 μm. The moiré interferometry set up (Figure 7.2) produces the fringe patterns by using the coherent beams interference. The specimen is bonded to the diffraction grating and moiré pattern is produced on it representing the displacement in the *x*-direction. Moiré pattern in *y*-direction can also be obtained by rotating the specimen through 90° (Figure 7.2).[5]

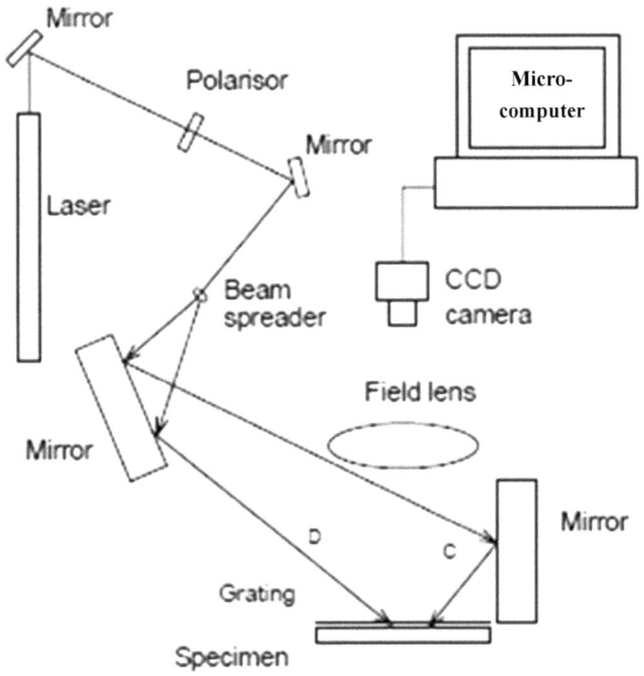

FIGURE 7.2 Moiré interferometry set up.

7.5 INFLUENCE OF NITRIDING ON SHOT PEENING TREATMENT

The thickness of the nitrided layer is generally kept between 0.3 and 0.4 mm; it is used to enhance the contact fatigue resistance and wear properties in aerodynamic industries. Depth of the nitrided surface layer followed by shot peening for 32Cr3MoVA steel is about 0.75 mm. It has been found that the fatigue life of nitrided and shot-peened specimen is greater in comparison to nitrided specimens. Scanning electron microscopy (SEM) test suggests that the area of failure and depth of the pit are small in case of nitrided shot-peened steel in comparison to nitrided steel.[6]

7.6 WATER CAVITATION PEENING

It is an advanced method of surface properties enhancement, which employs the use of ventilation nozzle to increase the impact pressure by inducing bubbles collapsing on the surface, and thus improves process capability.

The plunger pump is used to pressurized water jet. A test is conducted on the notched specimen of SAE 1045 steel and distribution of compressive residual stress in the superficial layer is measured by the XRD method. Maximum depth of compressive stress that can be attained is around 140 μm. A cavitating jet is developed by the researcher Soyama by mixing the high-speed and low-speed water jets by using concentric nozzle.[7,8] There are a lot of advantages of water cavitation peening over conventional shot peening:

1. Treatment of narrow and complicated surface is easier in water cavitation peening.
2. Material surface has greater thermal stability.
3. A smooth surface with less surface non-uniformities is obtained.

However, high-impact pressure induced in the process may also lead to stress corrosion and damage.[9] Aluminium structures of aircraft have direct exposure to adverse conditions. Samples of AA7050 high strength Al-alloy is subjected to the aqueous environment for a different period of 6, 12, 24, 48, and 72 h; they are shot-peened and compared with unpeeled samples. Laser displacement sensor measures the depth of corrosion. The result shows the increase in corrosion resistance of structure in aqueous conditions.[10]

Investigation of the steel corrosion behavior is done by using the polarization curve and pit corrosion test. Surface characterization of corroded samples is done by using SEM. Surface nanocrystallization increases the corrosion resistance in 1Cr18Ni9Ti stainless steel in chlorine ion solution.[11]

Fatigue experimentation and sensitivity analysis of shot-peened gears are done to focus on isotropic superfinishing and suitable choice of parameters is important to optimize the fatigue strength. Relative stress gradient theory and critical distance theory are used for fatigue life prediction of peened components. Design of experiment is a widely used method for shot peening optimization.[12]

Laser shock peening is a more advanced technique to improve fatigue life, wear resistance, and corrosive properties of materials. Laser shock peening induces compressive residual stress, which extends deeper in the surface than shot peening. A lot of models have been proposed for laser shock peening; a one-dimensional model is proposed, which assumed that shock wave propagation in confined medium and target is of a single dimension. This model is modified lately by using more conservation equations.[13]

7.7 CONCLUSION

The work investigates the effect of different process parameters that have been discussed on the fatigue strength of materials. Numerical models based on stress approach and other simulation models predict the fatigue strength of steel. However, a lot of work can still be done in the field of shot-peened spring steel specimens subjected to heat treatment. Heat treatment produces surface irregularities, which decreases the strength of the material; therefore, optimized conditions for heat treatment are required. Later on, S-N curves for different stress cycles are plotted after shot peening.

KEYWORDS

- **bainite**
- **steel**
- **shot peening**
- **microstructure**

REFERENCES

1. Takahashi, K.; Okada, H.; Ando, K. Effects of Shot Peening on the Torsional Fatigue Limit of High Strength Steel Containing an Artificial Surface Defect. *Int. J. Struct. Integ.* **2012,** *3*, 274–284.
2. Nordin, E.; Alfredsson, B. Measuring Shot Peening Media Velocity by Indent Size Comparison. J. Mater. Proces. Technol. **2016,** *235*, 143–148.
3. Bagherifard, S.; Fernandez-pariente, I.; Ghelichi, R.; Guagliano, M. *Handbook of Mechanical Nanostructuring*, 2015; p 14.
4. Harada, Y.; Mori, K. Effect of Processing Temperature on Warm Shot Peening of Spring Steel. *J. Mater. Proces. Technol.* **2005,** *162–163*, 498–503.
5. Liu, G.; Lu, J.; Lu, K. surface Nanocrystallization of 316 L Stainless Steel Induced by Ultrasonic Shot Peening. *Mater. Sci. Eng. A* **2000,** *286*, 91–95.
6. Gao, Y. K. Influence of Deep Nitriding and Shot Peening on Rolling Contact Fatigue Performance of 32Cr3MoVA Steel. *J. Mater. Eng. Perform.* **2007,** *17*, 455–459.
7. Soyama, H. High-Speed Observation of a Cavitating Jet in Air. *ASME J. Fluids Eng.* **2005,** *127*, 1095.
8. Soyama, H. Introduction of Compressive Residual Stress Using a Cavitating Jet in Air. *ASME J. Eng. Mater Technol.* **2004,** *126*, 123.

9. Han, B.; Ju, D. Y.; Jia, W. P. Influence of Water Cavitation Peening with Aeration on Fatigue Behaviour of SAE 1045 Steel. *Appl. Surf. Sci.* **2007,** *253*, 9342–9346.
10. Lv, S.; Cu, Y.; Zhang, W.; Tong, X.; Srivatsan, T. S.; Gao, X. Influence of Shot Peening on Failure of an Aluminium Alloy Exposed to Aggressive Aqueous Environments. *J. Mater. Eng. Perform.* **2012,** *22*, 1735–1743.
11. Wang, T.; Yu, J.; Dong, B. surface Nanocrystallization Induced by Shot Peening and Its Effect on Corrosion Resistance of 1Cr18Ni9Ti Stainless Steel. *Surf. Coatings Technol.* **2005,** *200*, 4777–4781.
12. Olmi, G.; Comandini, M.; Freddi, A. Fatigue on Shot Peened Gears: Experimentation, Simulation and Sensitivity Analysis. *Int. J. Exp. Mech.* **2010,** *46*, 382–395.
13. Hu, Y.; Yao, Z.; Hu, J. an Analytical Model to Predict Residual Stress Field Induced by Laser Shock Peening. *J. Manuf. Sci. Eng.* **2009,** *131*, 031017(1–6).

CHAPTER 8

VIRTUAL REALITY APPLICATIONS IN THE MANUFACTURING FIELD

MOHD JAVAID[1], ABID HALEEM[1], RAVI PRATAP SINGH[2*], and RAJIV SUMAN[3]

[1]*Department of Mechanical Engineering, Jamia Millia Islamia, New Delhi, India*

[2]*Department of Industrial and Production Engineering, Dr B R Ambedkar National Institute of Technology, Jalandhar, Punjab, India*

[3]*Department of Industrial & Production Engineering, G.B. Pant University of Agriculture & Technology, Pantnagar, Uttarakhand, India*

[*]*Corresponding author. E-mail: singhrp@nitj.ac.in*

ABSTRACT

Virtual reality (VR) is an innovative digital technology that uses glasses to substitute a virtual three-dimensional environment for the physical world and make users more interactive. As developed by researchers and designers, VR can accurately replicate the physical environments into a virtual world. In manufacturing, this technology is used for research and predictive analysis to modify and improve existing parts and effective learning processes. This immersive technology also changes the way production lines work for goods, reduces risk, and improves safety. It is used to assess a manufacturer's capability in making this process quicker and cost-effective, thereby enabling them to compete efficiently. The key benefits of immersive VR in manufacturing are

Modern Manufacturing Systems: Trends and Developments. Rajiv Kumar Garg, Ravi Pratap Singh, Rajeev Trehan, & Ramesh Singh (Eds.)

product development and refinement and designing production processes. Designers can test and refine a wide variety of configurations using VR at a much earlier stage in the process. As a result, there is less need for subsequent iterations when perfecting a concept in preparation for a product launch. This chapter briefs the significant applications of VR in the field of manufacturing. It elaborates capabilities and reduces time-to-market for testing, designing, and reviewing. It also offers a highly collaborative environment, allowing researchers, designers, and developers to interact effectively.

8.1 INTRODUCTION

Virtual reality (VR) opens a new way of developing innovative solutions because production units are equipped with intelligent machines, high connectivity, data intelligence platforms, and simulation tools. It helps to increase the manufacturing capability and satisfy the requirements of customers. It allows the corporation to give practically the actual world to its workers. The training sessions provided by using this technology will help tackle challenges in real-time.[1–3] During the development or upgrading of a new plant, massive efforts are required to design, test, and test the existing plant. In these situations, VR can help to upgrade the plant and have the capability to detect and repair faults in the whole device. It is possible to build and alter the whole plant from the beginning. Educated specialists will carry out protection and regular checks via virtual production processes. In VR, the experts can quickly take into account the minute data. VR can help operators to achieve advanced proficiency levels quicker, develop skills, and optimize the team training.[4,5]

VR Technology is highly used for handling machinery in the automotive industries. The high-tech industry is now recognizing the use of this technology. It helps processes from the space transportation to rail to medical infrastructure to make them more competitive, sustainable, cost-effective, and easy to build. VR has a crucial influence to encourage safety and health and environmental sustainability in the workplace. In manufacturing and processing, the most tactical use for VR can be seen as it satisfies the several criteria in industrial installations. It contributes significantly to the workforce's achievement by productive VR preparation, which is much needed today. It introduces the best field instruction options, where trainees can quickly adopt immersive learning with limited physical interaction in the real world.[6–8]

Companies worldwide are scaling up their efforts to increase the value of their data using new technologies. Even small companies are launching digital ventures to set the foundations, leverage energy, and drive expansion. VR is raising the standard of how production lines work and how products are produced, manufactured, and shipped to consumers in the manufacturing industry. A networked factory will use the correct technology as its base to make more automated and intelligent operations feasible through the use of sensor and Internet of Things (IoT) data from manufacturing machines and quality control inputs.[9–11]

VR offers a new platform for plant engineers to complete projects in a more synchronized and knowledgeable manner, whether on-site or at remote sites, leveraging real-time data and information from analytics. Industries owners, vendors, and consumers can monitor remote activities from anywhere globally and build trust or make decisions based on real-time data. Plant operators may be running complex machines, troubleshooting a fix, installing a system, or conducting safety checks, and VR can enhance their overall results and experience. Together, these systems improve workflows, offer extra knowledge and verification tools, minimize downtime for facilities, and maximize development schedules. They also improve the welfare of employees and factories.[12,13]

8.2 VIRTUAL REALITY: AN OVERVIEW

A VR experience occurs in a fully simulated, digital environment, which makes an ideal platform for gaming and 3D films. Nowadays, it is also used in reshaping industries all over the world, like manufacturing. Although the technology, like additive manufacturing and cobots before it, is currently used mostly by large manufacturers, increasing acceptance of the technology is expected to lower prices. Here, a consumer is made fully absorbed in the world or the situation created by a computer, and he is deprived of the actual perception of what is around us. Similarly, with VR controls, any type of object can be picked up and relocated. VR is a platform that streamlines the design process for car companies in the automotive industry. Engineers can imagine multiple models of vehicles, finishes, fabrics, colors, and a simulated prototype's function in real-time.[14–18] VR permits vehicles' safety checks by the simulations of individual conditions such as collisions during the vehicle's assembly process. Thus, design defects in cars can be identified with this technology before the prototype is made. In short, virtual truth is an excellent platform for the design team, manufacturers, and end-users to make agile improvements to the concept, achieve the expected product, evaluate those characteristics and see their effect on the car. VR technologies make it possible for the health industry to optimize the current systems and processes by countless means, thereby bringing value to the consumer. It empowers pharmaceutical practitioners to read, learn, and practice efficiently compared with the conventional learning methods and to achieve efficiency, effectiveness, and productivity. Visualizing a different reality offers useful perspectives and concepts and promotes human society.[19–21]

8.3 SIGNIFICANT ADVANCEMENTS THROUGH VIRTUAL REALITY

VR creates immersive experiences that facilitate the experience-based learning, which contributes to increased information sharing and retention. This technology is used in several sectors to construct digital twins of goods, machines, and assembly lines. This helps R&D teams imagine assembly line motions, automation procedures, and specialized equipment. The simulation of these in a simulated environment will help companies to prepare the deployment, testing, and preventive management of successful process modification vital to keep development systems online, streamlined, and highly efficient. Working in a collaborative virtual world with remote teams, partners, and consumers will strengthen collaboration and customer support. VR apps available on smartphones and tablets are more and more available; the obstacles to their implementation have decreased.[22–24]

Manufacturing companies understand the need for smarter digital tools through increased market pressures, increased profitability, and the need to minimize the risk. VR can increase coordination, teamwork, and production. Producers may use this method to maximize productivity, improve supply chains, and production output. VR provides many different ways of optimizing product and service presentation. It helps to train workers in real-life situations without the working world's risks, and employees can improve their skills. Virtual reality can help construct a popular virtual workspace that links many individuals in the same project. Users from multiple locations can display, view, and run on the same virtual model. This will facilitate contact between peers for validation processes without a physical meeting.[25–27]

Users can reach the virtual world through virtual technologies and communicate with virtual objects. It enables doctors to handle operations remotely and to contribute to industrial processes and quality management, etc. VR allows simulation of interactive and real-world knowledge. VR also makes virtual signals and directions that other employees can follow. This provides integration of digital information systems with modern production and process analysis. Smart links and data are the most distinctive characteristics of this new age. However, these massive data are not essential, but its analysis enhances decision-making, makes this data valuable. When intelligent ties become significant, the knowledge must be essential and shown in the correct location at the right time. Machines and people are still related, and manufacturing processes enhance this modern knowledge age.[28–31]

8.4 VIRTUAL REALITY TECHNOLOGY FOR ENGINEERING AND DESIGNERS

The automotive industry uses VR mainly in the maintenance, service, engineering, and production fields. VR utilization allows engineers and designers to develop and push the visual display of colors, shapes, and designs before the first samples are even built. During the production period, VR aids workers in maximizing assembly processes and manual manufacturing. Furthermore, AR will refine and replicate entire plants for processing and production by the proper overlay of devices.[32–35] VR plays a vital role in the typical earlier stages, where optimization and enhanced productivity are more critical than later stages of innovation and genuine business transformation. Manual inspections are one of the most time-consuming methods of industrial logistics. This method can be completely integrated into a decrease in errors and resource and monetary savings. It facilitates the repair process by presenting crews with up-to-date details, possible problems, and the equipment's comprehensive operating history.[36,37]

8.5 SAFER MANUFACTURING BY VIRTUAL REALITY

VR eliminates uncertainties by allowing engineers to find challenges, which can speed deployment rapidly. Protection and ergonomics at work are especially advantageous as both suppliers and service producers follow VR. Some manufacturers of complex industrial equipment are deploying VR to detect possible issues before design. It reduces the risk of shortages and expensive substitution. Complete immersion in the digital twin field is an ideal means of teaching and expertise for workers.[38–41] This includes system preparation, line transitions, and emergency reprocessing. VR avoids cost-intensive downtimes and reduces the chance of failures and accidents in making choices simultaneously. During vehicle development, many automobile firms still use VR. In that case, VR systems allow engineers and designers without incurring the expense of physical prototypes to display and debate key features. Faster time-to-market and better product design are among the rewards.[42–44] Today, we are witnessing the digitization of production and its complete transformation, which forces the industry to find new business models and creative solutions to improve their employees' productivity, profitability, and protection. VR is one of the most powerful ways to view and manipulate this content. It gives the automotive industry, in particular new opportunities.[45–47]

8.6 IMPROVED MONITORING OF PRODUCTION PROCESSES THROUGH VIRTUAL REALITY

VR provides the means of practically production and monitoring, using overlays on existing systems and finding failures, and fixing defects without major investment in equipment and workplace. These innovations make development more intelligent and save large and small enterprises massive time, effort, and expense. VR plays a vital role in proper processing during the product life cycles, from creation to development. However, at present, VR technology is affecting the CAD and CAM markets significantly. It helps producers test scenarios and prototypes before the goods are manufactured by improving efficiency and reducing development costs.[48–50] Today, VR technology will digitally recreate the shop floor's situation from traditional to digital manufacturing within immersive interactive decision platforms. This digital evolution discusses industrial engineers' future uses through workers' ergonomic tests, cinematic assessments, instrument validations, etc.[51,52]

In terms of product design and preparation, this technology also provides major developments. Welding techniques are learned in a calm and healthy atmosphere without fire, exposition of the eyes, or gases' inhalation. In that sense, the trainee is inspired to learn and perform harder. The time to learn is reduced, and the cost savings are tremendous. It is easier to incorporate VR technology in training since this is the reason for which it was planned. The teaching is quicker, better, more stable, and faster than to do live with a simulated welder. Increased output reality will take the operator a little longer to understand, but he will not succeed without it until he has mastered it. A substantial degree of preparation is required in architecture.[53,54]

The critical prerequisite for the use of VR technologies is the availability of 3D models. 3D modelling was embraced in the Metal Engineering industry, but 3D design and tool fixture design were only slowly implemented. When the industry understood the 3D need and began to embrace it for development, VR technology was inferior in affordability. The industry understands the importance of 3D modeling. It is possible to use VR technologies. VR developments in the gaming industry now experience a paradigm shift. In the industrial industry, too, this technology is being studied. The possibilities and implementations are enormous, but they are still developing.[55–59] This simulated complex manufacturing system helps users to analyze disparate project workflows and test new possibilities to create the most streamlined and better working production plan for what-if scenarios. The digital experience in building and adjusting the plant is improved, while the overall performance and output protection are enhanced.[60–64]

8.7 VIRTUAL REALITY APPLICATIONS IN CONTEMPORARY MANUFACTURING

There are various ways by which virtual reality can be used in manufacturing. In this scenario, virtual reality shows its extraordinary ability to immerse trainees in a realistic world while lowering training costs. It is the best system that can create a virtual atmosphere to either replicate the physical world or imagine phenomena. These innovations boost the efficiency of the business primarily or assist in information gathering and delivery. These modern business processes are focused on the highest data protection tools to record, analyze, and maximize their goods in any step of development in real-time. Today's VR software may be a backup device for repair, operation, and quality assurance for manufacturing firms. Besides, given that organizations can respond fast to business shifts to retain their competitive advantage, VR preparation warns the teams about the changes in operating procedures anticipated.[65–71] Significant applications of VR in manufacturing are as under:

8.7.1 MANUFACTURING PLANNING

Although manufacturing planning is performed in several industries, which is more complex and challenging, it is useful for proper planning. Multiple considerations can be addressed when designing space for assembly lines or manufacturing workshops, including equipment dimensions, connections between various systems, safety distances, power outlets, supply line inputs, and many others.

8.7.2 DESIGNING OF MANUFACTURING PLANT

VR makes it much easier to design factory floors. The manufacturers can check that everything is placed and linked correctly by recreating the actual areas in VR and putting the equipment models there. They can check in VR that all equipment is available, that all safety distances are preserved, and that the production sequence is designed in the proper order. Companies can maximize their floor area before physically moving equipment by using VR in manufacturing. This technology can assist in lowering construction costs and reducing the risk of equipment damage or worker injury.

8.7.3 PROCESS OF PRODUCTION

It is used to prepare for review, testing, and trial, after which it is determined whether it is ready for production or requires revision. VR can reduce the total of the product with the help of the proper process of production.

8.7.4 TRAINING

Digital reality applications for training are an excellent use of VR in manufacturing. VR training is used in a number of sectors, but manufacturing profits the most from it. The criteria for trainee protection in the preparation for a challenging job are very stringent. Such training is often difficult to coordinate because, on the one hand, the trainees lack practical skills, making them more vulnerable to risk. On the other hand, simulators have minimal capacities and, in the end, can only approximate the feel of actual work.

8.7.5 INSPECTION

VR has the potential to carry out a proper inspection and testing processes by minimizing the time. Companies can use this technology for the plant that manufactured parts can be a regular inspection. VR removes the need for the factory to assign an inspector to the unit.

8.7.6 TOWARD SMART MANUFACTURING

3D virtual models can track and analyze data in real-time in plant operations and construct a virtual representation of a mirror image of the physical environment. Machines, products, sensors, and people are part of this. It leads to downtime and quality enhancement for the machine.

8.7.7 WORKFORCE HEALTH

Health is an important application in which VR may have a big influence. Health practitioners are also using simulated models to train for human bodies, and VR was also used for pain relief of burn wounds. VR may also be used for mental health disorders, where immersion therapy from VR is considered to be particularly beneficial for the treatment.

8.7.8 DATA COLLECTION

VR drives manufacturing efficiently by collecting data and using that data to produce insights that will positively influence operational decisions. Process simulators will test and ensure the quality of these decisions. Manual decision processes are being re-imagined thanks to these new friendly techniques. With remote collaboration assistance, users can monitor all elements of their 3D plants. The interactive visualization of manufacturing installations enables knowledge to be gathered and solutions to issues without interrupting output.

8.7.9 COST REDUCTION

Often manufacture makes a prototype that can be tested and retested in order to manufacture products that meet requirements. This can be very costly and increase the cost of the research and development process. Manufacturers used this technology to eliminate the need for prototypes.

8.7.10 IMPROVING WORKERS SAFETY

Industrial protection has improved significantly over the years. Any accident or death is one too many. VR is also used to improve the safety of the workers.

This technology can simulate the manufacturing process and assembly line configurations and detect possible issues.

8.8 MAJOR CONTRIBUTIONS

VR technologies give users the ability to explore an expanded environment by overlaying real-world visual knowledge. In that way, both the physical and the virtual universe can be linked, and data or figures are collected in real-time. This will have a range of benefits for business 4.0. It may be the best way for technicians and staff to reflect important information and see the information they are doing in real-time. It also helps technicians to gain knowledge about the challenge a system has, access a user guide, or even call a professional to get real-time help. It is used for the production stage of devices or processes and concept evaluation so engineers can monitor the improvement accomplished by virtual simulation in a more visual and immersive manner. This will reduce errors and improve productivity in this process.

VR teaching helps to learn the details about the precise position where the tasks are to be done, with screen guidance. The consumer can use a flexible support system to execute the instructions before they are unable to do so. For the commissioning and maintenance of systems and facilities, the dynamics that can be applied in the VR training application can be successful. When travel time and costs are factored out of the equation, inspections, and remote maintenance can be performed much more efficiently in VR, resulting in fewer interruptions in the manufacturing cycle. VR blends digitally generated data with data obtained from the real world. It can be used on smartphones, and it's already finding applications in daily life. This 3D virtual technology allows us to recreate or replicate structures and existing computers digitally. It helps one experience high-risk situations, explosions, and injuries and learns the necessary protocols for emergencies and evacuation preparations.

8.9 FUTURE SCOPE OF THE TECHNOLOGY AND ITS APPLICATIONS

VR technology is expected to double in pixels over the next 5 years, allowing for better image processing, a broader field of view, and enhanced virtual sound amplification. These enhancements can help manufacturers already using VR drive efficiencies, improve efficiency, and save lives and resources.

VR allows users to explore various aspects of a product before committing to some sort of physical production in designing and prototyping. It provides staff with useful advanced knowledge about ergonomic specifications for particular parts and goods. It will become a real-time production line training tool that simulates a wide variety of environments and processes. This technology can speed up the whole production process, allowing for even quicker product releases. Manufacturers will put themselves at the forefront of industries experiencing profound transformations by embracing and exploring interactive VR possibilities. There will be some improvement in productivity, profitability, and better protection system in the industry.

8.10 CONCLUSION

VR can be a big aid in training and preparing workers for new product lines and processes. Employees can train in real-time while minimizing risk and optimizing the advantages of routine and self-directed learning. Since VR is adaptable and can be used in a wide variety of industries. This technology has great potential to increase flexibility in manufacturing. VR will help producers align themselves with these developments as supply chains become quicker and more effective, placing them in the best possible position to take advantage of emerging technology. Digital reality is a medium for industry transformation, enabling manufacturers to cross over into the world of modern industrial revolution. Manufacturing, like other sectors, needs to keep searching for new business models to help improve efficiency and profitability. Thereby, by using this technology, there are also innovative ways of equipping workers with the skills they need to meet evolving production demands.

KEYWORDS

- **applications**
- **digital technology**
- **manufacturing**
- **production planning**
- **virtual reality**

REFERENCES

1. Mujber, T. S.; Szecsi, T.; Hashmi, M. S. Virtual Reality Applications in Manufacturing Process Simulation. *J. Mater. Proces. Technol.* **2004,** *155*, 1834–1838.
2. Berg, L. P.; Vance, J. M. Industry Use of Virtual Reality in Product Design and Manufacturing: A Survey. *Virt. Reality* **2017,** *21* (1), 1–17.
3. Choi, S.; Jung, K.; Noh, S. D. Virtual Reality Applications in Manufacturing Industries: Past Research, Present Findings, and Future Directions. *Concurrent Eng.* **2015,** *23* (1), 40–63.
4. Abidi, M. H.; Al-Ahmari, A.; Ahmad, A.; Ameen, W.; Alkhalefah, H. Assessment of Virtual Reality-Based Manufacturing Assembly Training System. *Int. J. Adv. Manuf. Technol.* **2019,** *105* (9), 3743–3759.
5. Malik, A. A.; Masood, T.; Bilberg, A. Virtual Reality in Manufacturing: Immersive and Collaborative Artificial-Reality in Design of Human-Robot Workspace. *Int. J. Comput. Integ. Manuf.* **2020,** *33* (1), 22–37.
6. Aziz, F. A.; Mousavi, M. A Review of Haptic Feedback in Virtual Reality for Manufacturing Industry. *J. Mech. Eng.* **2009,** *40* (1), 68–71.
7. Jimeno, A.; Puerta, A. State of the Art of the Virtual Reality Applied to Design and Manufacturing Processes. *Int. J. Adv. Manuf. Technol.* **2007,** *33* (9–10), 866–874.
8. Sung, W. T.; Ou, S. C. Using Virtual Reality Technologies for Manufacturing Applications. *Int. J. Comput. Appl. Technol.* **2003,** *17* (4), 213–219.
9. Singh, R. P.; Javaid, M.; Kataria, R.; Tyagi, M.; Haleem, A.; Suman, R. Significant Applications of Virtual Reality for COVID-19 Pandemic. *Diabetes Metab. Syndrome: Clin. Res. Rev.* **2020,** *14* (4), 661–664.
10. Gonzalez-Franco, M.; Pizarro, R.; Cermeron, J.; Li, K.; Thorn, J.; Hutabarat, W.; Bermell-Garcia, P. Immersive Mixed Reality for Manufacturing Training. *Front. Robot. AI* **2017,** *4*, 3.
11. Bougaa, M.; Bornhofen, S.; Kadima, H.; Rivière, A. Virtual Reality for Manufacturing Engineering in the Factories of the Future. *Appl. Mech. Mater.* **2015,** *789*, 1275–1282.
12. Grajewski, D.; Górski, F.; Zawadzki, P.; Hamrol, A. Application of Virtual Reality Techniques in Design of Ergonomic Manufacturing Workplaces. *Procedia Comp. Sci.* **2013,** *25*, 289–301.
13. Salah, B.; Abidi, M. H.; Mian, S. H.; Krid, M.; Alkhalefah, H.; Abdo, A. Virtual Reality-Based Engineering Education to Enhance Manufacturing Sustainability in Industry 4.0. *Sustainability* **2019,** *11* (5), 1477.
14. Ong, S. K.; Mannan, M. A. Virtual Reality Simulations and Animations in a Web-Based Interactive Manufacturing Engineering Module. *Comput. Educ.* **2004,** *43* (4), 361–382.
15. Darmoul, S.; Abidi, M. H.; Ahmad, A.; Al-Ahmari, A. M.; Darwish, S. M.; Hussein, H. M. Virtual Reality for Manufacturing: A Robotic Cell Case Study. In *2015 International Conference on Industrial Engineering and Operations Management (IEOM)*; IEEE, March 2015; pp 1–7.
16. Hamid, N. S. S.; Aziz, F. A.; Azizi, A. Virtual Reality Applications in Manufacturing System. In *2014 Science and Information Conference*; IEEE, August 2014; pp 1034–1037.
17. Lipton, J. I.; Fay, A. J.; Rus, D. Baxter's Homunculus: Virtual Reality Spaces for Teleoperation in Manufacturing. *IEEE Robot. Autom. Lett.* **2017,** *3* (1), 179–186.

18. Aurich, J. C.; Ostermayer, D.; Wagenknecht, C. H. Improvement of Manufacturing Processes with Virtual Reality-Based CIP Workshops. *Int. J. Product. Res.* **2009,** *47* (19), 5297–5309.
19. Javaid, M.; Haleem, A. Virtual Reality Applications toward Medical Field. *Clin. Epidemiol. Global Health* **2020,** *8* (2), 600–605.
20. Novak-Marcincin, J. Selected Applications of Virtual Reality in Manufacturing. *J. Technol. Plasticity* **2011,** *36* (1), 25–34.
21. Dangelmaier, W.; Fischer, M.; Gausemeier, J.; Grafe, M.; Matysczok, C.; Mueck, B. Virtual and Augmented Reality Support for Discrete Manufacturing System Simulation. *Comput. Industry* **2005,** *56* (4), 371–383.
22. Novak-Marcincin, J.; Barna, J.; Janak, M.; Novakova-Marcincinova, L. Augmented Reality Aided Manufacturing. *Procedia Comput. Sci.* **2013,** *25*, 23–31.
23. Matsas, E.; Vosniakos, G. C. Design of a Virtual Reality Training System for Human-Robot Collaboration in Manufacturing Tasks. *Int. J. Interact. Design Manuf.* **2017,** *11* (2), 139–153.
24. Jezernik, A.; Hren, G. A Solution to Integrate Computer-Aided Design (CAD) and Virtual Reality (VR) Databases in Design and Manufacturing Processes. *Int. J. Adv. Manuf. Technol.* **2003,** *22* (11–12), 768–774.
25. Okulicz, K. Virtual Reality-Based Approach to Manufacturing Process Planning. *Int. J. Prod. Res.* **2004,** *42* (17), 3493–3504.
26. Javaid, M.; Haleem, A.; Vaishya, R.; Bahl, S.; Suman, R.; Vaish, A. Industry 4.0 Technologies and Their Applications in Fighting COVID-19 Pandemic. *Diabetes Metab. Synd.: Clin. Res. Rev.* **2020,** *14* (4), 419–422.
27. Büttner, S.; Mucha, H.; Funk, M.; Kosch, T.; Aehnelt, M.; Robert, S.; Röcker, C. The Design Space of Augmented and Virtual Reality Applications for Assistive Environments in Manufacturing: A Visual Approach. In *Proceedings of the 10th International Conference on PErvasive Technologies Related to Assistive Environments*, June 2017; pp 433–440.
28. Yang, X.; Malak, R. C.; Lauer, C.; Weidig, C.; Hagen, H.; Hamann, B.; Aurich, J. C. Virtual Reality Enhanced Manufacturing System Design. In *Proceedings of the 7th CIRP International Conference on Digital Enterprise Technology*, September 2011; pp 125–133.
29. Pappas, M.; Karabatsou, V.; Mavrikios, D.; Chryssolouris, G. Development of a Web-Based Collaboration Platform for Manufacturing Product and Process Design Evaluation Using Virtual Reality Techniques. *Int. J. Comput. Integ. Manuf.* **2006,** *19* (8), 805–814.
30. Dorozhkin, D. V.; Vance, J. M.; Rehn, G. D.; Lemessi, M. Coupling of Interactive Manufacturing Operations Simulation and Immersive Virtual Reality. *Virtual Reality* **2012,** *16* (1), 15–23.
31. Abidi, M. A.; Lyonnet, B.; Chevaillier, P.; Toscano, R. Contribution of Virtual Reality for Lines Production's Simulation in a Lean Manufacturing Environment. *Int. J. Comput. Theory Eng.* **2016,** *8* (3), 182.
32. Matsas, E.; Vosniakos, G. C.; Batras, D. Prototyping Proactive and Adaptive Techniques for Human-Robot Collaboration in Manufacturing Using Virtual Reality. *Robot. Computer-Integ. Manuf.* **2018,** *50*, 168–180.
33. Agethen, P.; Sekar, V. S.; Gaisbauer, F.; Pfeiffer, T.; Otto, M.; Rukzio, E. Behavior Analysis of Human Locomotion in the Real World and Virtual Reality for the Manufacturing Industry. *ACM Trans. Appl. Percep.* **2018,** *15* (3), 1–19.

34. Penttilä, S.; Lund, H.; Ratava, J.; Lohtander, M.; Kah, P.; Varis, J. Virtual Reality Enabled Manufacturing of Challenging Workpieces. *Procedia Manuf.* **2019,** *39*, 22–31.
35. Rubio, E. M.; Sanz, A.; Sebastián, M. A. Virtual Reality Applications for the Next-Generation Manufacturing. *Int. J. Computer Integ. Manuf.* **2005,** *18* (7), 601–609.
36. Haleem, A.; Javaid, M. Additive Manufacturing Applications in Industry 4.0: A Review. *J. Indust. Integ. Manage.* **2019,** *4* (04), 1930001.
37. Stratos, A.; Loukas, R.; Dimitris, M.; Konstantinos, G.; Dimitris, M.; George, C. A Virtual Reality Application to Attract Young Talents to Manufacturing. *Procedia CIRP* **2016,** *57*, 134–139.
38. Bellalouna, F. Virtual-reality-based Approach for Cognitive Design-review and FMEA in the Industrial and Manufacturing Engineering. In *2019 10th IEEE International Conference on Cognitive Infocommunications* (*CogInfoCom*); IEEE, October 2019; pp 41–46.
39. Calvo, I.; López, F.; Zulueta, E.; González-Nalda, P. Towards a Methodology to Build Virtual Reality Manufacturing Systems Based on Free Open Software Technologies. *Int. J. Interact. Des. Manuf.* **2017,** *11* (3), 569–580.
40. Liagkou, V.; Salmas, D.; Stylios, C. Realising Virtual Reality Learning Environment for Industry 4.0. *Procedia CIRP* **2019,** *79*, 712–717.
41. Gogouvitis, X. V.; Vosniakos, G. C. Construction of a Virtual Reality Environment for Robotic Manufacturing Cells. *Int. J. Computer Appl. Technol.* **2015,** *51* (3), 173–184.
42. Moreau, G.; Fuchs, P.; Stergiopoulos, P. Applications of Virtual Reality in the Manufacturing Industry: From Design Review to Ergonomic Studies. *Mécanique Industries* **2004,** *5* (2), 171–179.
43. Peniche, A.; Diaz, C.; Trefftz, H.; Paramo, G. Combining Virtual and Augmented Reality to Improve the Mechanical Assembly Training Process in Manufacturing. In *Proceedings of the 2012 American Conference on Applied Mathematics*, January 2012; pp 292–297.
44. Al-Ahmari, A. M.; Abidi, M. H.; Ahmad, A.; Darmoul, S. Development of a Virtual Manufacturing Assembly Simulation System. *Adv. Mech. Eng.* **2016,** *8* (3), 1687814016639824.
45. Haleem, A.; Javaid, M.; Khan, I. H. Virtual Reality (VR) Applications in Dentistry: An Innovative Technology to Embrace. *Indian J. Dental Res.* **2020,** *31* (4), 666.
46. Karvouniari, A.; Michalos, G.; Dimitropoulos, N.; Makris, S. An Approach for Exoskeleton Integration in Manufacturing Lines Using Virtual Reality Techniques. *Procedia CIRP* **2018,** *78*, 103–108.
47. Azizi, A.; Yazdi, P. G.; Hashemipour, M. Interactive Design of Storage Unit Utilising Virtual Reality and Ergonomic Framework for Production Optimisation in Manufacturing Industry. *Int. J. Interact. Des. Manufact.* **2019,** *13* (1), 373–381.
48. Barkokebas, R.; Ritter, C.; Sirbu, V.; Li, X.; Al-Hussein, M. Application of Virtual Reality in Task Training in the Construction Manufacturing Industry. In *ISARC. Proceedings of the International Symposium on Automation and Robotics in Construction*, Vol. 36; IAARC Publications, 2019; pp 796–803.
49. Jou, M.; Wang, J. Investigation of Effects of Virtual Reality Environments on Learning Performance of Technical Skills. *Comput. Hum. Behav.* **2013,** *29* (2), 433–438.
50. Nee, A. Y.; Ong, S. K. Virtual and Augmented Reality Applications in Manufacturing. *IFAC Proc.* **2013,** *46* (9), 15–26.
51. Damiani, L.; Demartini, M.; Guizzi, G.; Revetria, R.; Tonelli, F. Augmented and Virtual Reality Applications in Industrial Systems: A Qualitative Review towards the Industry 4.0 Era. *IFAC-PapersOnLine* **2018,** *51* (11), 624–630.

52. Robert, O.; Iztok, P.; Borut, B. Real-Time Manufacturing Optimisation with a Simulation Model and Virtual Reality. *Procedia Manuf.* **2019,** *38*, 1103–1110.
53. Gamlin, A.; Breedon, P.; Medjdoub, B. Immersive Virtual Reality Deployment in a Lean Manufacturing Environment. In *2014 International Conference on Interactive Technologies and Games*; IEEE, October 2014; pp 51–58.
54. Haleem, A.; Javaid, M.; Vaishya, R.; Khan, I. H. Virtual Reality Applications in Orthopaedics. *J. Orthopaed. Spine* **2019,** *7* (2), 83.
55. Damgrave, R. G. J.; Lutters, E.; Drukker, J. W. Rationalising Virtual Reality Based on Manufacturing Paradigms. *Procedia CIRP* **2014,** *21*, 264–269.
56. Marzano, A.; Friel, I.; Erkoyuncu, J. A.; Court, S. Design of a Virtual Reality Framework for Maintainability and Assemblability Test of Complex Systems. *Procedia CIRP* **2015,** *37*, 242–247.
57. Golda, G.; Kampa, A.; Paprocka, I. The Application of Virtual Reality Systems as a Support of Digital Manufacturing and Logistics. *IOP Conf. Ser. Mater. Sci. Eng.* Aug 2016, *145* (4), 042017.
58. Ong, S. K.; Yuan, M. L.; Nee, A. Y. C. Augmented Reality Applications in Manufacturing: A Survey. *Int. J. Product. Res.* **2008,** *46* (10), 2707–2742.
59. Whitman, L. A. R. R. Y.; Malzahn, D.; Madhavan, V.; Weheba, G. A. M. A. L.; Krishnan, K. R. I. S. H. N. A. Virtual Reality Case Study throughout the Curriculum to Address Competency Gaps. *Int. J. Eng. Educ.* **2004,** *20* (5), 690–702.
60. Pérez, L.; Rodríguez-Jiménez, S.; Rodríguez, N.; Usamentiaga, R.; Garcia, D. F. Digital Twin and Virtual Reality Based Methodology for Multi-Robot Manufacturing Cell Commissioning. *Appl. Sci.* **2020,** *10* (10), 3633.
61. Mousavi, M.; Aziz, F. A. State of the Art of Haptic Feedback in Virtual Reality in Manufacturing. In *2008 International Symposium on Information Technology*, Vol. 4; IEEE, Aug 2008; pp 1–7.
62. Novak-Marcincin, J.; Janak, M.; Barna, J.; Novakova-Marcincinova, L. Application of Virtual and Augmented Reality Technology in Education of Manufacturing Engineers. In *New Perspectives in Information Systems and Technologies, Volume 2*; Springer: Cham, 2014; pp 439–446.
63. Frontoni, E.; Loncarski, J.; Pierdicca, R.; Bernardini, M.; Sasso, M. Cyber-physical Systems for Industry 4.0: Towards Real Time Virtual Reality in Smart Manufacturing. In *International Conference on Augmented Reality, Virtual Reality and Computer Graphics*; Springer: Cham, June 2018; pp 422–434.
64. Hutabarat, W.; Oyekan, J.; Turner, C.; Tiwari, A.; Prajapat, N.; Gan, X. P.; Waller, A. Combining Virtual Reality Enabled Simulation with 3D Scanning Technologies towards Smart Manufacturing. In *2016 Winter Simulation Conference (WSC)*; IEEE, Dec 2016; pp 2774–2785.
65. Madathil, K. C.; Frady, K.; Hartley, R.; Bertrand, J.; Alfred, M.; Gramopadhye, A. An Empirical Study Investigating the Effectiveness of Integrating Virtual Reality-Based Case Studies Into an Online Asynchronous Learning Environment. *Computers Educ. J.* **2017,** *8* (3), 1–10.
66. Tahriri, F.; Mousavi, M.; Yap, H. J.; Siti Zawiah, M. D.; Taha, Z. Optimising the Robot Arm Movement Time Using Virtual Reality Robotic Teaching System. *Int. J. Simul. Model.* **2015,** *14* (1), 28–38.
67. Aurich, J. C.; Yang, X.; Schröder, S.; Hering-Bertram, M.; Biedert, T.; Hagen, H.; Hamann, B. Noise Investigation in Manufacturing Systems: An Acoustic Simulation and Virtual Reality Enhanced Method. *CIRP J. Manuf. Sci. Technol.* **2012,** *5* (4), 337–347.

68. Brunner, K. E.; Perey, C.; Rauschnabel, P. A.; Sage, M. Implementation of Augmented Reality in Manufacturing: A Case Study Exercise. In *Augmented Reality and Virtual Reality*; Springer: Cham, 2020; pp 85–97.
69. Dahl, M.; Albo, A.; Eriksson, J.; Pettersson, J.; Falkman, P. Virtual Reality Commissioning in Production Systems Preparation. In *2017 22nd IEEE International Conference on Emerging Technologies and Factory Automation (ETFA)*; IEEE, Sept 2017; pp 1–7.
70. Lawson, G.; Salanitri, D.; Waterfield, B. Future Directions for the Development of Virtual Reality within an Automotive Manufacturer. *Appl. Ergonom.* **2016,** *53*, 323–330.
71. Novak-Marcincin, J.; Fecova, V.; Barna, J.; Janak, M.; Novakova-Marcincinova, L. Using of the Virtual Reality Application with the Scanning Device Kinect for Manufacturing Processes Planning. *Manufact. Technol.* **2013,** *13* (2), 215–219.

CHAPTER 9

A SURVEY ON RESPONSES AND OPTIMIZATION OF PROCESS PARAMETERS IN MILLING: A REVIEW

ASHUTOSH KUMAR GUPTA and RAVI PRATAP SINGH*

Department of Industrial & Production Engineering, Dr. B R Ambedkar National Institute of Technology Jalandhar, Punjab, India

Corresponding author. E-mail: singhrp@nitj.ac.in

ABSTRACT

Computer numerical control (CNC) machining has growing in the modern manufacturing for accuracy and machining time. CNC milling is one of the most commonly used machining processes in manufacturing of different parts. The present paper reviews the responses and optimization technique used in different milling processes. This paper also highlights the progress of optimization in the area of milling machining and discussed the issue that need to be solved.

9.1 INTRODUCTION

Machining is one of the most popular manufacturing processes along with forming casting and welding process. Designing and optimization of process variables have attracted the attention of the research scholars.

Machining is simply defined as a materials removal process and materials are subtracted in the form of chips by using different shape cutting tool (single or multiple edge cutting tool). Metal cutting is oriented toward

Modern Manufacturing Systems: Trends and Developments. Rajiv Kumar Garg, Ravi Pratap Singh, Rajeev Trehan, & Ramesh Singh (Eds.)

minimizing machining cost and machining time and maximizing tool life and production rate. The responses and the optimization of machining can be measured for conventional machining process and newer machining process like EDM, ECM, WJM, and ECDM. But in this literature, review paper different responses and optimizations techniques which are commonly used are summarized. Among the different machining process, milling draws attention over the other three most widely used process turning, drilling, and grinding.

Milling: Milling is a manufacturing process in which a rotary multitooth cutter with or without inserts is used. Milling cutters are designed and used to cut gears, drill, make flat, shaped surfaces, bore, or do slotting work. Based on the axis of rotation, milling machines can be classified as horizontal and vertical milling. Based on the chips and cutting, milling can divided into up and down milling; in these processes down milling is preferable.

Turning: Turning is a manufacturing process in which axisymmetric bars of material are held in a chuck and rotated, while a is fed to the workpiece to remove material in form of chips to create the desired shape and size.

9.1.1 FACE MILLING

In a face milling cutting operation, the cutter is mounted on a spindle. The axis of feed of work-piece surface is perpendicular to the axis of rotation of the spindle. The cutting teeth, such as tungsten carbide or other inserts, are screwed on the cutter. The SR of the workpiece depends on the geometry of the insert mounted on cutter teeth and the feed per tooth. A schematic diagram of face milling is shown in Figure 9.1.

9.2 LITERATURE REVIEW

Muaz and Choudhury[2] used fuzzy logic along with the Taguchi orthogonal array L27 to maximize the material removal rate (MRR) and minimized the surface roughness. The experimentation was performed on the CNC end milling of Al 7075 with input parameter speed, depth of cut, and nose radius. The dimension of work-piece was 590 mm × 55 mm × 20 mm. Two orthogonal array L9 was prepared for experiment first with coolant ON condition and second with coolant OFF condition which was used to calculate the tool-wear. The confirmatory experiment showed that surface roughness decreased to 0.14 μm and MRR increased to 2304 mm^3/min. The most influencing parameters affecting the MRR and SR were nose radius and depth of cut.

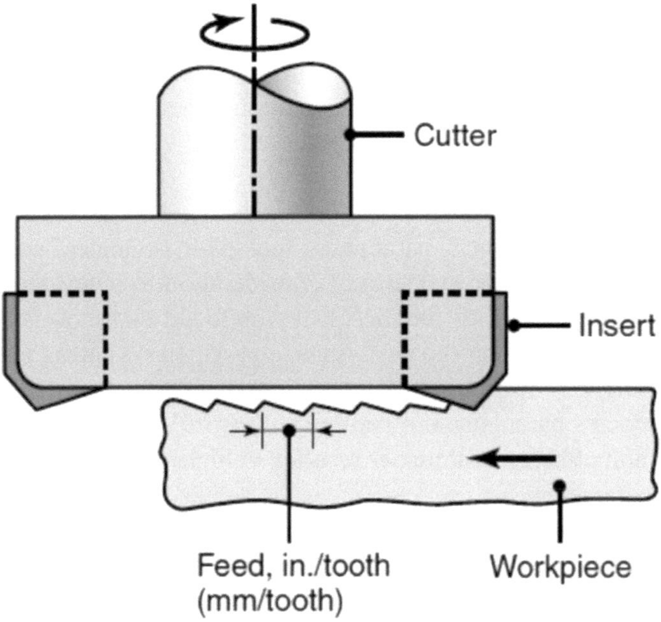

FIGURE 9.1 Face milling operation.

Perez et al.[10] investigated the cutting forces, surface integrities, residual stresses, and SR at different cutting speeds by keeping another parameter (like feed rate and depth of cut) constant. Face milled 7070-T7451 Al. alloy work-piece was used. The dimension of the specimen was taken as 100mm × 112mm × 40 mm and a cutter diameter of 50 mm with five uncoated inserts (APKTL93 1604 PDFR) was used. For measuring forces in all three-dimension KISTER dynamometer was used. The observation of the experiment was as follows: increase in cutting speed from 200 m/min to 800 mm/min. all forces decreased and at 1400 mm/min. cutting speed more reduction in F_x only but increase in F_y. Value of surface roughness was obtained at lowest cutting speed but effect of feed rate and tool geometry were more on SR than cutting speed. Residual stress was different at different cutting speed. The best optimum responses were found at 800 mm/min. cutting speed.

Khorasani et al.[11] concluded that tool vibration was a source of tool wear, chatter, tool breakage, and poor surface finish which was generated under the unsuitable cutting conditions during face milling. The experiments were formulated by using full factorial design and a modern technique ANN was used to predict tool vibration stability after the experimentation of Aluminum 7075 alloy and low carbon steel St 52. A modern optimization technique genetic algorithm (GA) was used to optimize the vibration that appeared during experiments. It was concluded that feed rate had the most influenced parameter on vibration. The error of 0.12365% was found between experimental result and predicted value by the optimization technique GA.

Pillai et al.[6] investigated the process variables for the end milling of Al 6005A alloy. The process parameters tool path strategies, stainless steel (SS) and feed rates were taken into consideration to study their effect on machine time and surface roughness. It was found that tool path strategies had maximum aspect on the performance parameter whereas feed rate had minimum aspect. Taguchi-Grey relation analysis was used to estimate the optimal process parameter and verified with ANOVA. It was also found that irregular and offset tool strategies resulted in higher SR in compare of raster tool path strategies.

Muaz and Choudhary[2] performed nano-finishing of AISI 4340 steel using nanopowder of TiCN/Al_2O_3/TiN that had been used during experimental investigation in the flat end milling machine. A set of experiments were undertaken using different parameters each time optimized through single performance Taguchi analysis as well as by a hybrid analysis (Gray relational analysis [GRA]). Then GRA results validated by GA confirmed that the method's produced values were within appropriate tolerance levels. Low viscosity fluids were found to provide better results during minimal quantity lubrication (MQL) application. This work was done with the aim of reducing the effects at reducing waste and making the process environment-friendly.

Pena et al. (2019) optimized the process parameters and coolant concentration for the milling process. In this study, nanoparticle (NP) was selected as cutting fluids. Montmorillonite (MMT) clay, a type of NP was used due to eco-friendly behavior, low cost, and natural existence. For the optimization of the input variables, cutting speed, feed rate, depth of cut, and NP concentration in the cutting fluid for milling of the AISI 1018 steel. A total of 27 experiments of different combinations of these process variables were performed by Box Behnken design of experiment method. The following responses were investigated: power consumption, wear, and surface roughness.

Ozsoy[9] optimized the condition of machining parameter feed by examining the SR of the aluminum 7075-T6 alloy in milling. The machining parameters spindle speed, feed per tooth, and cooling type were used for the experimentation. To validate the test result, Taguchi L_{16} orthogonal array DOE and signal to noise were used together. The percentage error between the calculated optimum value of the output parameter by using ANOVA and by experiment was around 0.59% which is very low it signifies that optimization was satisfactory. The minimum SR value was obtained at 3000 rev/min, 0.4 mm, and liquid for input parameter SS, feed, and cooling type, respectively.

Fabre et al. (2016) performed a series of experiments to study the influence of broaching parameter tool design the cutting conditions on the response of surface roughness along with forces and chip formation. The wide ranges of parameter taken into consideration were cutting speed tool design (rise per tooth angle) and workpiece material. $X_{12}Cr_{13}$ stainless steel, high-speed steel, and oil lubrication were used. It was concluded that cutting speed and rise per tooth are the most influencing parameter.

Carvalho et al. (2015) proposed that the planning process plays an important role in the productivity. An incomplete planning process leads to the loss of productivity that can be avoided by vibration analysis and energy efficiency in the process of face milling. A new mathematical model was built for the inclusion of vibration analysis and energy efficiency. If work was performed on the natural frequency of the tool, the energy consumption increased. The objective of this study was to achieve minimum power consumption and greater productivity by using vibration analysis and energy efficiency.

Zhang et al.[8] used Taguchi design of an experiment to optimize the surface roughness in face milling operation. The main parameter feed rate, spindle speed, and depth of cut were considered as control factors. An orthogonal array L_9 3 factor 4 variables (3^4) was selected to obtain the best surface roughness (response) and (S/N) signal to noise ratio. This paper also verified that Taguchi DOE successfully optimized the milling parameters for response, that is, surface roughness.

Achuthamenon et al. (2018) investigated the multi-response parameter surface roughness tool wear and depth of cut during the end milling process of C4. The optimization method used was response surface methodology (RSM). To fabricate the C4 AL 7068 (ductile) and SiC foam (brittle) of size 10 ppi and gravity infiltration were used. The RSM was used to optimize the SR and tool wear by varying the parameter speed depth of cut and feed rate. A 5 mm diameter uncoated carbide tool was taken. The result of the experiment was at 3000 rpm speed, 450 mm/rev feed, and 0.6 mm depth of cut removal rate and SR was 1752.89 mm^3/min and 2.6828 μm within 95% confidence level.

Masmiati and Sarhan (2014) investigated the surface integrity during end milling of S50C medium carbon steel using titanium-coated carbide ball end milling tool on 5 axis CNC milling machine. To find the greater significant factor, Taguchi optimization was carried out along with S/N ratio and ANOVA analysis. Surface inclination angle, axial depth of cut, spindle speed, and feed rate were taken as the input parameter while surface roughness was taken as response.

Karabulut (2015) first produced the aluminium 7039 by powder metallurgy process and then examined the effect of cutting speed, feed rate, and depth of cut in milling using uncoated carbide tool on the surface roughness and force. The experiments were designed using the Taguchi Orthogonal array and result was validated by S/N ratio analysis. Additionally, analysis of variance was performed to find the significance parameters. The result was found that material composition had a greater influence on the surface roughness but for the force, the feed rate was a dominating factor.

Lu et al. (2007) investigated the optimization techniques in the high-speed end milling of SDK tool steel. The responses selected for the evaluation were surface roughness, tool life, and MRR corresponding to the type of milling, axial, and radial depth of cut, feed per tooth, and SS. PCA was applied to find the weightage of responses for relative improvement of responses. The result was found that when GRA is added with PCA gives better results nearest to the experimental value.

Parihar et al. (2017) performed a finite element analysis to predict the various responses in turning of AISI H13 and then compared it with the experimental results. The result obtained by comparing data from simulation and experiments were found in good and valid accord. An approach was established to predict responses on the basis of cutting speed, feed, and depth of cut.

9.3 CUTTING PARAMETERS

The main cutting parameters on any milling machining are cutting speed, feed per tooth, depth of cut, cutting fluid (coolant type), and tool path strategy. Some other factors like workpiece material and tool material have also influenced on milling. But these five parameters are the only parameter that can be controlled right at the machine.

9.3.1 CUTTING SPEED

Cutting parameter has major influence on the output parameter. Cutting speed depends on spindle speed and diameter of milling cut may be calculated by using eq 9.1. Greater the cutting speed greater will be chatter suspension or it may cause greater vibration. The cutting speed can be selected by either workpiece material or by tool material. The cutting speed for Aluminium is

$$v_{\text{cutting}}(\text{mm/min.}) = \pi \text{Dn} \tag{9.1}$$

Where D= diameter of cutter, N= spindle speed in rpm.

9.3.2 FEED RATE

Feed rate is the distance of workpiece travels, horizontally (in X–Y plane) per unit revolution of the tool inserts. The feed measures in mm/rev. The rotation of tool along Z-axis and the movement of workpiece produce a chip by means of cutting the face of workpiece. The roughing of the surface in milling usually involves a high feed rate with the main purpose of getting high MRR with little regard for surface quality. But the finishing milling is executed at lower feed rate and the surface finish formed is a priority in spite of MRR.

9.3.3 DEPTH OF CUT

The depth of cut is the distance the tool travels, vertically (in Z-axis). The Depth of Cut (DOC) measures in mm or in. It is generally constant during machining. The roughing of surface in milling cuts usually involves large depths of cut with the main motive of eliminating large amounts of material with little regard for surface finish. But in the process of improving the quality of surface, milling is performed at a lower depth of cut and the surface quality formed is a priority in spite of MRR.

9.4 CUTTING RESPONSES

The list of process responses which affect the production costs and quality is discussed below. These variables are MRR, Surface roughness, tool wear, cutting forces, etc.

9.4.1 MATERIAL REMOVAL RATE

MRR is one of the essential responses which can be calculated by total material removed volume (mm^3) per unit cutting time (minutes). MRR is directly related to the production. For getting high production need for removal of extra material is more so, high MRR is desirable. MRR is usually measures in mm^3/min or mm^3/sec.

9.4.2 SURFACE ROUGHNESS

Irregularities and flaw in surface may primary site (nucleation) to start cracks or corrosion. The key factor of metal companies is to increase productivity and quality. The quality of a machined part is directly proportional to surface finish. The trial and error are not a cost-effective way to find out optimal combination of cutting parameter.

9.5 CONCLUSION

The main influence can be drawn from the past conducted study on milling of aluminium alloys, which are as follows:

- Most of the experiments are conducted on Al 2024, Al 6061, Al 7075, AISI metals, etc. But a lot of other material still untouched and need to study.
- Input parameters like spindle speed, feed Rate, and depth of cut are playing major role in the output responses. Increment or decrements in these input parameters can affect a lot of responses like MRR and Surface Roughness.
- Among all the investigations of Aluminium Alloys one response also affect the other response. Improvement in the surface roughness decreases material removal rate which is not desirable sometimes. So, optimization of Input parameter according to desired responses is needed.
- A lot of optimization techniques proposed by researches, that is, Taguchi-Grey, genetic Algorithm, neural network particle Swarm optimization teaching-learning based optimization, etc. Some optimization techniques are very popular and some need attention. Some are easy to use and fast but still not touched by researchers, that is, TLBO, PSO, ABC, etc.

TABLE 9.1

S. No.	Investigators	Work material(s) and tool(s)	Milling process type	Process parameters	Responses studied	Optimization technique	Research findings
1.	Chenxi wang (2019)	Al alloys 6061	Milling chatter suppression	Teeth No., helix angle, diameter and overhang	Tool life, MRR efficiency, workpiece surface quality	Comparative study of responses	Adaptive vibration reshaping which can precisely suppress chatter frequency effectively and do not change rotation frequency and its frequency multiplications. Adaptive vibration reshaping which can precisely suppress chatter frequency effectively and do not change rotation frequency and its frequency multiplications
2.	Muaz and Choudhury[2]	AISI 4340	End milling	MQL-type, SS, feed rate	Nano finishing	TGRA (Taguchi-Grey Relational Analysis) and GA (Genetic Algorithm)	Low viscosity fluids were found to provide better results during MQL application. This was done with the aim of reducing the effects at reducing/omitting waste and making the process environment-friendly
3.	Ozsoy[9]	Aluminum 7075-T6 alloy	End milling	Spindle speed, feed per tooth and cooling type	Surface roughness	Taguchi and ANOVA	The percentage error between calculated optimum value of output parameter by using ANOVA and by experiment was around 0.59% which is very low it signifies that optimization was satisfactory
4.	Pillai et al.[6]	Al 6005A	End milling	Tool path strategy, spindle speed and feed	Machining time and surface roughness	Taguchi-Grey And ANOVA	It was found that irregular and offset tool strategy resulted higher surface roughness in compare of raster tool path strategies

TABLE 9.1 *(Continued)*

S. No.	Investigators	Work material(s) and tool(s)	Milling process type	Process parameters	Responses studied	Optimization technique	Research findings
5.	Perez et al.[10]	7070-T7451 Aluminium	Face milling	Five different inserts and cutting speed	Force measurement	Comparative study	Value of surface roughness was obtained at lowest cutting speed but effect of feed rate and tool geometry were more on SR than cutting speed. Residual stress was different at different cutting speed. The best optimum responses were found at 800 mm/min. cutting speed
6.	Zhai et al.[13]	10L501	Milling	Operation type, axial DOC, cutting velocity and feed per tooth	Force and cost	Different TLBO	The result obtained by study is that Dynamic assignment teaching-learning based optimization algorithm gives better result than other proposed TLBO algorithms
7.	Mahesh and Rajesh[5]	Al 7075	End milling	Coolant condition speed, feed, depth of cut and nose radius	MRR and surface roughness	Taguchi and Fuzzy	Nose radius and Depth of cut had major contribution in MRR and SR Multi response performance index of fuzzy optimal variables was 36% better than initial process
8.	Wan et al. (2012)	Al 7050-T7451	End milling	Milling type, SS, *f*, radial DoC, axial DoC	Prediction of cutting forces	Comparative modelling	The suitability of the planned model is confirmed by relating the predicted cutting forces using mathematical modeling with the experimental ones under the specified cutting conditions. The study shows that parameters in experimental model and mathematical model are different but cutting force obtained using both models is approximately same

TABLE 9.1 *(Continued)*

S. No.	Investigators	Work material(s) and tool(s)	Milling process type	Process parameters	Responses studied	Optimization technique	Research findings
9.	Khorasani et al. (2011)	Al. 7075 and St.42	Face milling	Spindle speed deed rate, depth of CU and cooling condition	Tool vibration stability	ANN and GA	Feed rate had the most influenced parameter on vibration with 94% correlation. ANN was used to predict tool vibration and GA was used to optimize tool vibration stability. The genetic algorithm had calculated error of 0.124% in the measurement of vibration during machining
10.	Benardos and Vosniakos[7]	Aluminium alloy	Face milling	Cutting force, depth of cut, cutting fluid.	Surface roughness	Artificial neural network	ANN used for modeling and prediction of surface roughness was reliable and accurate.

KEYWORDS

- **responses**
- **optimization machining**
- **milling**
- **process models**

REFERENCES

1. Rao, R. V.; Savsani, V. J.; Vakharia, D. P. Teaching–Learning-Based Optimization: An Optimization Method for Continuous Non-Linear Large-Scale Problems. *Info. Sci.* **2012,** *183* (1), 1–15.
2. Muaz, M.; Choudhury, S. K. Experimental Investigations and Multi-Objective Optimization of MQL-Assisted Milling Process for Finishing of AISI 4340 Steel. *Measurement* **2019,** *138*, 557–569.
3. Shandilya, P.; Rouniyar, A. K.; Saikiran, D. Multi-Objective Parametric Optimization on Machining of Inconel-825 Using Wire Electrical Discharge Machining. *Proc. Inst. Mech. Eng. C J. Mech. Eng. Sci.* **2020**.
4. Siddiqui, F.; Thakur, P. Optimal Selection of Parameters in CNC End Milling of Al 7075 T6 Alloy by TBLO. *Int. J. Eng. Sci.* **2019,** *11* (4).
5. Mahesh, T.P.; Rajesh, R. Optimal Selection of Process Parameters in CNC End Milling of Al 7075-T6 Aluminium Alloy Using a Taguchi-fuzzy Approach. *Procedia Materials Science* **2014,** *5*, 2493–2502.
6. Pillai, U. J.; et al. Optimization of Multiple Response Characteristics on End Milling of Aluminium Alloy Using Taguchi-Grey Relational Approach. *Measurement* **2018,** *124*, 291–298.
7. Benardos, P. G.; Vosniakos, G. C. Prediction of Surface Roughness in CNC Face Milling Using Neural Networks and Taguchi's Design of Experiments. *Robot. Comput.-Integr. Manuf.* **2002,** *18* (5–6), 343–354.
8. Zhang, X.; Wang, C.; Liu, J.; Yan, R.; Cao, H.; Chen, X. Robust Active Control Based Milling Chatter Suppression with Perturbation Model via Piezoelectric Stack Actuators. *Mech. Syst. Signal Process.* **2019,** *120*, 808–835.
9. Özsoy, N. Experimental Investigation of Surface Roughness of Cutting Parameters in T6 Aluminum Alloy Milling Process. *Int. J. Comput. Exp. Sci. Eng.* **2019,** *5* (3), 105–111.
10. Perez, I.; Madariaga, A.; Cuesta, M.; Garay, A.; Arrazola, P. J.; Ruiz, J. J.; Sanchez, R. Effect of Cutting Speed on the Surface Integrity of Face Milled 7050-T7451 Aluminium Workpieces. *Procedia Cirp* **2018,** *71*, 460–465.
11. Khorasani, A. M.; Saadatkia, P.; Kootsookos, A. Tool Vibration Prediction and Optimisation in Face Milling of Al 7075 and St 52 by Using Neural Networks and Genetic Algorithm. *Int. J. Mach. Machinab. Mater.* **2012,** *12* (1–2), 142–153.

12. Wang, B.; Liu, Z.; Song, Q.; Wan, Y.; Shi, Z. Proper Selection of Cutting Parameters and Cutting Tool Angle to Lower the Specific Cutting Energy during High-Speed Machining of 7050-T7451 Aluminum Alloy. *J. Clean. Prod.* **2016,** *129*, 292–304.
13. Zhai, Z.; Li, S.; Liu, Y. Parameter Determination of Milling Process Using a Novel Teaching-Learning-Based Optimization Algorithm. *Math. Prob. Eng.* **2015,** *2015*.
14. Singh, R. P.; Singhal, S. Rotary Ultrasonic Machining: A Review. *Mater. Manuf. Process.* **2016,** *31*, 1795–1824.
15. Singh, R.P.; Singhal, S. Investigation of Machining Characteristics in Rotary Ultrasonic Machining of Alumina Ceramic. *Mater. Manuf. Process.* **2017,** *32*, 309–326.
16. Singh, R.P.; Tyagi, M.; Kataria, R. Selection of the Optimum Hole Quality Conditions in Manufacturing Environment Using MCDM Approach: A Case Study. In *Operations Management and Systems Engineering*; Springer: Singapore, 2019; pp 133–152.
17. Singh, R.P.; Singhal, S. Rotary Ultrasonic Machining of Macor Ceramic: An Experimental Investigation and Microstructure Analysis. *Mater. Manuf. Process.* **2017,** *32*, 927–939.
18. Singh, R.P.; Singhal, S. Experimental Investigation of Machining Characteristics in Rotary Ultrasonic Machining of Quartz Ceramic. *J. Mater. Des. Appl.* **2018,** *232,* 870–889.
19. Singh, R.P.; Kataria, R.; Kumar, J.; Verma, J. Multi-Response Optimization of Machining Characteristics in Ultrasonic Machining of WC-Co Composite through Taguchi Method and Grey-Fuzzy Logic. *AIMS Mater. Sci.* **2018,** *5*, 75–92.
20. Singh, R.P.; Kumar, J.; Kataria, R.; Singhal, S.: Investigation of the Machinability of Commercially Pure Titanium in Ultrasonic Machining Using Graph Theory and Matrix Method. *J. Eng. Res.* **2015,** *3*, 75–94.
21. Singh, R.P.; Singhal, S. An Experimental Study on Rotary Ultrasonic Machining of Macor Ceramic. *J. Eng. Manuf.* **2018,** *232,* 1221–1234.
22. Singh, R.P.; Singhal, S. Rotary Ultrasonic Machining of Alumina Ceramic: Experimental Study and Optimization of Machining Responses. *J. Eng. Res.* **2018,** *6*, 01–24.
23. Tyagi, M.; Panchal, D.; Singh, R. P.; Sachdeva, A. Modeling and Analysis of Critical Success Factors for Implementing the IT-Based Supply-Chain Performance System. In *Operations Management and Systems Engineering*; Springer: Singapore, 2019; pp 51–67.
24. Singh, R. P.; Kataria, R.; Singhal, S. Decision-Making in Real-Life Industrial Environment through Graph Theory Approach. In *Computer Architecture in Industrial, Biomechanical and Biomedical Engineering*. IntechOpen, 2019. DOI: 10.5772/intechopen.82011
25. Singh, R.; Singh, R. P.; Trehan, R. State of the Art in Processing of Shape Memory Alloys with Electrical Discharge Machining: A Review. *Proc. Inst. Mech. Eng. B J. Eng. Manuf.* **2020,** 0954405420958771.
26. Singh, R.; Singh, R. P.; Tyagi, M.; Kataria, R. Investigation of Dimensional Deviation in Wire EDM of M42 HSS Using Cryogenically Treated Brass Wire. *Mater. Today Proc.* **2019.**

PART II

Studies on Advanced/Matrix Composite Materials

CHAPTER 10

MICROSTRUCTURAL EVALUATION OF AL MATRIX REINFORCED Al_2O_3–C/SiC/TiO_2 HYBRID METAL MATRIX COMPOSITE

NASEEM AHAMAD[1*], AAS MOHAMMAD[1], and PALLAV GUPTA[2]

[1]*Department of Mechanical Engineering, Jamia Millia Islamia, New Delhi, India*

[2]*Department of Mechanical Engineering, A.S.E.T., Amity University Uttar Pradesh, Noida, India*

[*]*Corresponding author. E-mail: naseem.ahamad9412@gmail.com*

ABSTRACT

The aim of the present work is to study the phase and microstructure of prepared samples of Al–Al_2O_3–C/SiC/TiO_2 hybrid metal matrix composite. Three compositions of hybrid composite, that is, (Al–Al_2O_3–C), (Al–Al_2O_3–SiC), and (Al–Al_2O_3–TiO_2) were prepared by mechanical stir casting in equal proportion of reinforcement (2.5, 5, 7.5, and 10 wt.%). X-ray diffraction studies reported the phase identification and respective miller indices of Al matrix as well as of the reinforcement. Scanning electron microscopy (SEM) micrographs of prepared samples of different compositions have been investigated. SEM shows that ceramic Al_2O_3–C/SiC/TiO_2 was distributed uniformly in Al matrix at lowest percentage, after that varying the percentage of reinforcement in Al matrix, the reinforcement was distributed nonuniformly in Al matrix.

Modern Manufacturing Systems: Trends and Developments. Rajiv Kumar Garg, Ravi Pratap Singh, Rajeev Trehan, & Ramesh Singh (Eds.)

10.1 INTRODUCTION

Aluminum-based composite materials have lightweight, low wear rate, high strength, and maximum corrosion resistance in contrast with ordinary material. Aluminum-based composite material used in the fabrication of lower weight and superior strength structure of the vehicle, industries, and aerospace. Garg et al. investigated that aluminum-based composite used widely in engineering work and strength to weight ratio is high in these types of composite material. The various properties of aluminum matrix depend on reinforcement and matrix.[1] Jamwal et al. observed the mechanical behavior of Al–Al_2O_3–TiC hybrid composites. The hybrid composite Al–Al_2O_3–TiC has greater tensile strength, maximum–minimum hardness, and wear rate, respectively.[2] Jamwal et al. worked on various properties of hybrid composite (Cu–SiC–Gr). It is studied that SiC and graphite were spread evenly in Cu matrix and Cu-reinforcement has no intermediate reaction between each other.[3] Bandil et al. worked on various characteristic of hybrid composite (Al–Si–SiC). It is investigated that the SiC mixed evenly in Al–Si matrix, and Al–Si–SiC has higher wear and corrosion resistance as compared Al matrix.[4] Towhidul et al. found the various characteristics of Al–CNT–TiC hybrid composite. The Al–CNT–TiC composite material has lower weight, maximum hardness, and lower wear rate at minimum load.[5] Kumar et al. observed the different properties of hybrid composite Al–SiC–TiC and it is investigated that ceramic mixed uniformly in aluminum matrix. The density and wear rate of the prepared hybrid material was reduced with addition of reinforcements.[6] Zahid et al. found the various properties, such as wear, density, and hardness, of SiC–TiC reinforced with copper–aluminum alloy. It is observed that hardness & wear resistance increased as well as no intermediate reaction developed between ceramic and copper–aluminum alloy.[7] Towhidul et al. studied the optimization of electric discharge machining. It is observed that a quality product, quality of machining process, and prepared complicated shapes in minimum time by EDM.[8] Hossain et al. studied the epoxy fiber composite material, which is a natural material and it is biodegradable and ecofriendly composite material. It is found that epoxy composite has improved tensile strength, impact strength, and flexible strength.[9] Hossain et al. worked on mechanical and structural properties of Al_2O_3–SiC reinforced with aluminum alloy. Al–Al_2O_3–SiC hybrid composite has lightweight, higher hardness, and minimum wear rate.[10] Jamwal et al. found the different characteristics, that is, wear, density, and hardness of copper-based hybrid composite. It is observed that copper-based composite material has higher wear resistance and temperature resistance, respectively.[11]

Ahamad et al. found the various mechanical characteristics, that is, hardness, density, and tensile strength of Al–Al_2O_3–TiO_2 hybrid composite. It is found that composite material has high tensile strength, high impact strength, and uniform mixing of ceramic in aluminum matrix.[12] Ahamad et al. found the various characteristics, that is, density, hardness, toughness, and wear analysis of Al–Al_2O_3–C hybrid composite. It is observed that the Al–Al_2O_3–C composite has uniform mixing of reinforcement, lightweight, and no intermediate reaction developed between ceramic & matrix. Wear rate of the Al–Al_2O_3–C composite was reduced with addition of reinforcement.[13, 14] Ahamad et al. investigated the wear rate of Al–Al_2O_3–TiO_2 hybrid composite and optimizes the wear rate with help of ANOVA and TOPSIS tool. The wear rate of Al–Al_2O_3–TiO_2 composite decreased with the addition of Al_2O_3–TiO_2 reinforcement.[15]

In this study, three different types of hybrid composites, Al–Al_2O_3–C, Al–Al_2O_3–SiC, and Al–Al_2O_3–TiO_2, were fabricated by mechanical stir casting. The purpose of this experimental work is to find the phase identification and microstructural analysis of these hybrids composite with a variation of Al_2O_3–C/SiC/TiO_2 reinforcement in aluminum matrix. The prepared composite materials have lightweight and maximum strength with a broad application in the automotive and aerospace sector.

10.2 EXPERIMENTAL WORK

10.2.1 SAMPLE PREPARATION

In this study, aluminum (Al-6351) was used as a base Al alloy material having 0.15% Si, 0.39% Mg, 0.49% Mn and 1.89% O, and 97.08% Al. The reinforcement used for the different hybrid composite material is Al_2O_3, C, SiC, and TiO_2. The mesh size of reinforcements Al_2O_3, C, SiC, and TiO_2 are 70–230, 60, 350, and 120, respectively. The sample code of different hybrids composite samples is shown in Tables 10.1–10.3. The hybrid composite of three different compositions is fabricated with a mechanical stir casting process.

The melting temperature of aluminum (Al-6351) is 910°C. The reinforcement used in fabrication was preheated. First of all, the Al matrix melted in crucible at 910°C after that reinforcement Al_2O_3–C/SiC/TiO_2 in equal proportion put into the crucible and prepared the composite material. There is prepared three hybrid composite materials with four different internal compositions of every hybrid composite. So, total 12 samples were prepared by a machining process after stir casting process.

TABLE 10.1 Grading of AlAlC (Al–Al_2O_3-C) Composite Sample.

Sample number	Al-6351 (% by wt.)	Al_2O_3 (wt.%)	Carbon (wt.%)	Name of sample
1	95	2.5	2.5	AlAlC01
2	90	5	5	AlAlC02
3	85	7.5	7.5	AlAlC03
4	80	10	10	AlAlC04

TABLE 10.2 Grading of AlS (Al–Al_2O_3–SiC) Composite Sample.

Sample number	Al-6351 (% by wt.)	Al_2O_3 (wt.%)	Silicon carbide (wt.%)	Name of sample
1	95	2.5	2.5	AlS01
2	90	5	5	AlS02
3	85	7.5	7.5	AlS03
4	80	10	10	AlS04

TABLE 10.3 Grading of AlT (Al–Al_2O_3–TiO_2) Composite Sample.

Sample number	Al-6351 (% by wt.)	Al_2O_3 (wt.%)	Titanium dioxide (wt.%)	Name of sample
1	95	2.5	2.5	AlT01
2	90	5	5	AlT02
3	85	7.5	7.5	AlT03
4	80	10	10	AlT04

10.2.2 CHARACTERIZATION

X-ray diffraction (XRD) was the analysis of Bruker X-ray diffractometer for finding out phase identification. The reinforcements are in powder form for XRD analysis. The angle range of XRD is 20°–90° and operating voltage is 40 kV with Ni filter and Cu-Kα radiation. Samples were prepared for scanning electron microscopy (SEM) with the help of polishing machine. The sample has 14 mm diameter and 2 mm thickness in a round disc shape. For polishing, the different grit of emery paper was required and after polishing sample surface look like mirror. The morphological investigation of the hybrid composite was carried out by using SEM (SEM-Bruker and ZEISS).

10.3 RESULT AND DISCUSSION

10.3.1 X-RAY DIFFRACTION

Figure 10.1 shows the phase identification of Al, Al_2O_3, carbon, SiC, and TiO_2. XRD pattern plots between 2θ angles (*X*-axis) and intensity (*Y*-axis). In XRD, it is observed that maximum percentage of composition shows with high peaks, and minimum percentage of composition shows with low peak. The miller indices of different composition are shown in Figure 10.1 with respective peak. It is found that Al-6351 has Al, Si, Mg, and Mn composition as represented in Figure 10.1. The XRD pattern of ceramic reinforcement Al_2O_3, carbon, SiC, and TiO_2 with miller indices is shown in Figure 10.1.

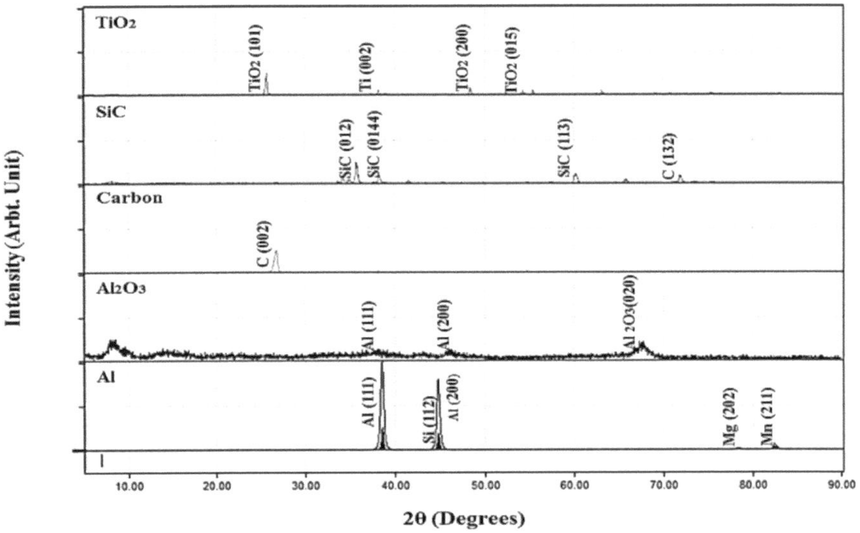

FIGURE 10.1 X-ray diffraction pattern of Al, Al_2O_3, carbon, SiC & TiO_2.

10.3.2 SCANNING ELECTRON MICROSCOPY

The morphological approach of hybrid composite was studied by SEM. Figures 10.2–10.4 show the microstructure of hybrid composites material, such as (Al–Al_2O_3–C), (Al–Al_2O_3–SiC), and (Al–Al_2O_3–TiO_2), respectively. It is observed that specimen AlAlC01 has uniform distributed of ceramics in Al matrix, as represented in Figure 10.4a.

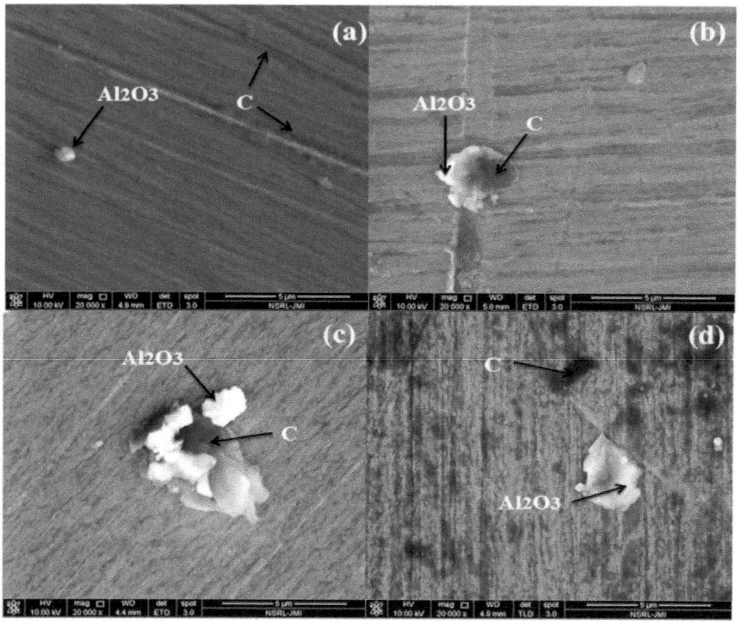

FIGURE 10.2 SEM of sample (a) AlAlC01, (b) AlAlC02, (c) AlAlC03, (d) AlAlC04 at 20,000× magnification.

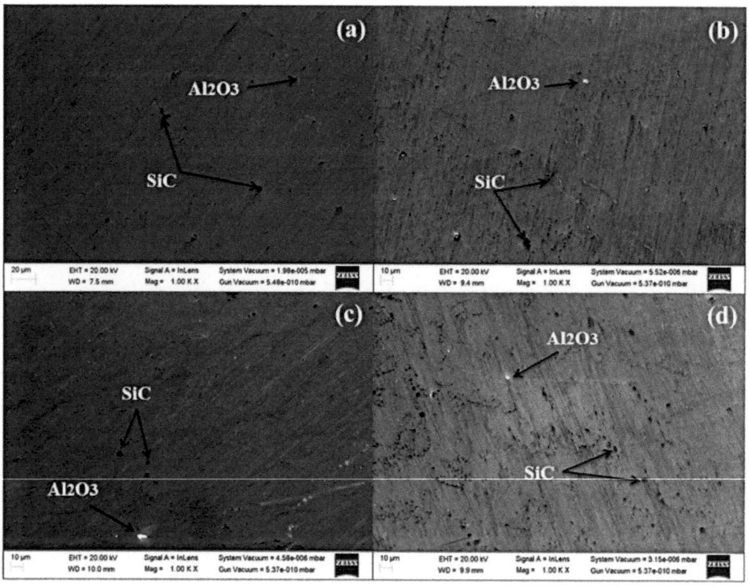

FIGURE 10.3 SEM of sample (a) AlS01, (b) AlS02, (c) AlS03, (d) AlS04 at 1000× magnification.

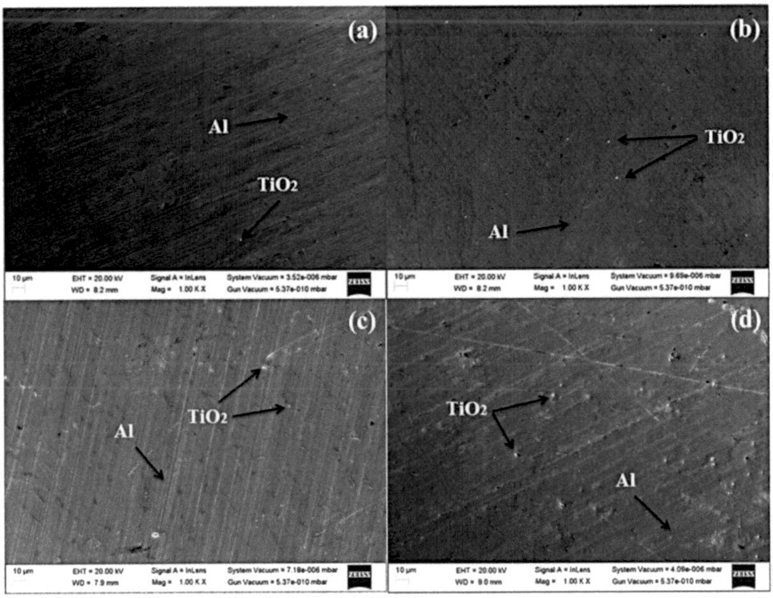

FIGURE 10.4 SEM of sample (a) AlT01, (b) AlT02, (c) AlT03, (d) AlT04 at 1000× magnification.

Specimen AlAlC02 has a higher concentration of ceramic with respect to AlAlC01. It is found that specimen AlAlC02 has more particles of reinforcement carbon and Al_2O_3 in Al matrix, as shown in Figure 10.2b. Specimen AlAlC03 in Figure 10.2c and specimen AlAlC04 in Figure 10.2d have a higher concentration of ceramic. It is observed that in specimen AlAlC04, the reinforcement particles were distributed uniformly. The agglomeration is happening in specimen AlAlC04 because this specimen has a higher concentration value of ceramic.

Similarly, Figure 10.3 represents the morphology of composite specimen Al–SiC–Al_2O_3. It is observed that specimens AlS01 and AlS02 have a minimum concentration value of ceramic with respect to specimen AlS03 and AlS04. Specimens AlS01 and AlS02 have a uniform spread of ceramic SiC-Al_2O_3 in Al matrix, as shown in Figure 10.3a and b. Due to the increase concentration of ceramic SiC-Al_2O_3 in sample AlS03 and AlS04, the agglomeration is found in these two samples that are shown in Figure 10.3c and d, respectively. The ceramic SiC-Al_2O_3 is spread evenly in the base material at a lower concentration after that the ceramic SiC-Al_2O_3 spread unevenly in the base Al matrix.

In Figure 10.4, the morphological approach of Al reinforced of Al_2O_3 and TiO_2 reinforcement is studied by SEM. Specimens AlT01 and AlT02 have a minimum percentage of reinforcement and respective microstructure, as shown in Figure 10.4a and b. The ceramics Al_2O_3 and TiO_2 are spread evenly in Al matrix at low concentration of ceramic. The concentration of reinforcement in specimen AlT03 and specimen AlT04 is maximum, due to higher concentration of Al_2O_3—TiO_2 reinforcement, the reinforcement is spread unevenly in the base material as represented in Figure 10.4c and d. The agglomeration is found as indicated in Figure 10.4c and d.

It is observed that these three hybrid composites, such as (Al–Al_2O_3–C), (Al–Al_2O_3–SiC), and (Al–Al_2O_3–TiO_2), have uniform distribution of reinforcement at lower percentage and nonuniform distribution of reinforcement of the higher percentage of reinforcement. The reinforcements Al_2O_3–C/SiC/TiO_2 particles are accumulated during stir casting process because reinforcements have insoluble nature in Al matrix. This is the region for unevenly spread of ceramic reinforcement in Al matrix at higher percentages.

10.4 CONCLUSION

Three compositions of hybrid composite, that is, (Al–Al_2O_3–C), (Al–Al_2O_3–SiC), and (Al–Al_2O_3–TiO_2) were prepared by mechanical stir casting in equal proportion of reinforcement (2.5, 5, 7.5, and 10 wt.%). XRD and SEM micrographs of the prepared sample were examined. The finalized comments of the above given research work are as follows:

1. XRD result of Al6351 matrix & reinforcements Al_2O_3–C/SiC/TiO_2 shows the different composition of material and respective miller indices. It is found that a higher percentage of composition shows with highest peak and dense peak.
2. Scanning electron microscopic is used for the investigation of microstructure of three compositions of hybrid composite, such as (Al–Al_2O_3–C), (Al–Al_2O_3–SiC), and (Al–Al_2O_3–TiO_2), at different concentration of ceramic reinforcements. It is found that at lower concentration of reinforcement approximate 15%, the ceramic reinforcement was distributed evenly in Al matrix. After that, the ceramic reinforcement was distributed unevenly in Al matrix due to nonsoluble nature in Al matrix. In the present study, it is observed that interfacial bonding between Al6351 & reinforcements Al_2O_3–C/SiC/TiO_2 is very good.

KEYWORDS

- **hybrid metal matrix composite**
- **mechanical stir casting**
- **XRD**
- **SEM**

REFERENCES

1. Garg, P.; Jamwal, A.; Kumar, D.; Sadasivuni, K. K.; Hussain, C. M.; Gupta, P. Advance Research Progresses in Aluminium Matrix Composites: Manufacturing & Applications. *J. Mater. Res. Technol.* **2019,** *8* (5), 4924–4939.
2. Jamwal, A.; Vates, U. K.; Gupta, P.; Aggarwal, A.; Sharma, B. P. Fabrication and Characterization of Al_2O_3–TiC-Reinforced Aluminum Matrix Composites. In *Advances in Industrial and Production Engineering*; Springer: Singapore, 2019; pp 349–356.
3. Jamwal, A.; Prakash, P.; Kumar, D.; Singh, N.; Sadasivuni, K. K.; Harshit, K.; Gupta, P. Microstructure, Wear and Corrosion Characteristics of Cu Matrix Reinforced SiC–Graphite Hybrid Composites. *J. Compos. Mater.* **2019,** *53* (18), 2545–2553.
4. Bandil, K.; Vashisth, H.; Kumar, S.; Verma, L.; Jamwal, A.; Kumar, D.; Gupta, P. Microstructural, Mechanical and Corrosion Behaviour of Al–Si Alloy Reinforced with SiC Metal Matrix Composite. *J. Compos. Mater.* **2019,** *53* (28–30), 4215–4223.
5. Nayim, S. T. I.; Hasan, M. Z.; Seth, P. P.; Gupta, P.; Thakur, S.; Kumar, D.; Jamwal, A. Effect of CNT and TiC Hybrid Reinforcement on the Micro-mechano-tribo Behaviour of Aluminium Matrix Composites. *Mater. Today: Proc.* **2020,** *21*, 1421–1424.
6. Kumar, A.; Arafath, M. Y.; Gupta, P.; Kumar, D.; Hussain, C. M.; Jamwal, A. Microstructural and Mechano-tribological Behavior of Al Reinforced SiC–TiC Hybrid Metal Matrix Composite. *Mater. Today: Proc.* **2019,** *21* (3), 1417–1420.
7. Sohag, M. A. Z.; Gupta, P.; Kondal, N.; Kumar, D.; Singh, N.; Jamwal, A. Effect of Ceramic Reinforcement on the Microstructural, Mechanical and Tribological Behavior of Al-Cu Alloy Metal Matrix Composite. *Mater. Today: Proc.* **2019,** *21* (3), 1407–1411.
8. Nayim, S. T.; Hasan, M. Z.; Jamwal, A.; Thakur, S.; Gupta, S. Recent Trends & Developments in Optimization and Modelling of Electro-discharge Machining Using Modern Techniques: A Review. In *AIP Conference Proceedings* (Vol. 2148, No. 1, p. 030051); AIP Publishing LLC, September 2019.
9. Hossain, S.; Rahman, M. M.; Jamwal, A.; Gupta, P.; Thakur, S.; Gupta, S. Processing and Characterization of Pine Epoxy Based Composites. In *AIP Conference Proceedings* (Vol. 2148, No. 1, p. 030017); AIP Publishing LLC, September 2019.
10. Hossain, S.; Rahman, M. M.; Chawla, D.; Kumar, A.; Seth, P. P.; Gupta, P.; Jamwal, A. Fabrication, Microstructural and Mechanical Behavior of Al–Al_2O_3-SiC Hybrid Metal Matrix Composites. *Mater. Today: Proc.* **2019**. DOI: 10.1016/j.matpr.2019.10.089

11. Jamwal, A.; Mittal, P.; Agrawal, R.; Gupta, S.; Kumar, D.; Sadasivuni, K. K.; Gupta, P. Towards Sustainable Copper Matrix Composites: Manufacturing Routes with Structural, Mechanical, Electrical and Corrosion Behaviour. *J. Compos. Mater.* **2020**. https://doi.org/10.1177/0021998319900655
12. Ahamad, N.; Mohammad, A.; Sadasivuni, K. K.; Gupta, P. Structural and Mechanical Characterization of Stir Cast Al–Al_2O_3–TiO_2 Hybrid Metal Matrix Composites. *J. Compos. Mater.* **2020,** *54* (21), 2985–2997.
13. Ahamad, N.; Mohammad, A.; Sadasivuni, K. K.; Gupta, P. Phase, Microstructure and Tensile Strength of Al–Al_2O_3–C Hybrid Metal Matrix Composites. *Part C: J. Mech. Eng. Sci.* **2020,** *234* (13), 2681–2693.
14. Ahamad, N.; Mohammad, A.; Gupta, P. Wear Characteristics of Al Matrix Reinforced with Al_2O_3-carbon Hybrid Metal Matrix Composites. *Mater. Today: Proc.* **2020**. https://doi.org/10.1016/j.matpr.2020.05.739.
15. Ahamad, N.; Mohammad, A.; Sadasivuni, K. K.; Gupta, P. Wear, Optimization and Surface Analysis of Al–Al_2O_3–TiO_2 Hybrid Metal Matrix Composites. *IMechE Part J: J. Eng. Tribol.* **2020,** *235* (1), 93–102.

CHAPTER 11

AN INSIGHT INTO CYCLIC OXIDATION BEHAVIOR OF ZrB_2-20 VOL.% $MoSi_2$ BASED ULTRAHIGH TEMPERATURE CERAMIC MATRIX COMPOSITE

MAINAK SAHA[1*], SOURAJIT PRAMANIK[1], RAJDEEP BANIK[1], POOJA MAURYA[1], SIMRAT SINGH[1], S. VANDAN RAJ[1], D. PREM MADHUR[1], TANAY RUDRA PAUL[1], RAHUL MITRA[2], MANAS KUMAR MONDAL[1], and MANAB MALLIK[1]

[1]*Department of Metallurgical and Materials Engineering, National Institute of Technology, Durgapur, India*

[2]*Department of Metallurgical and Materials Engineering, Indian Institute of Technology, Kharagpur, India*

[*]*Corresponding author. E-mail: mainaksaha1995@gmail.com*

ABSTRACT

In the context of hypersonic re-entry nosecones, subjected to repeated thermal shocks while descending through different layers of the atmosphere, with tremendously high velocities, it becomes extremely important to investigate the cyclic oxidation behavior of ultra-high temperature ceramics. The present study focuses on the cyclic oxidation law obeyed by ZrB_2-20 vol.% $MoSi_2$ composite during oxidation at temperatures ranging from 1100°C to 1300°C by carrying out six cycles for 6 h, at every individual temperature. The oxidation kinetic law has been estimated by measuring weight change at a

Modern Manufacturing Systems: Trends and Developments. Rajiv Kumar Garg, Ravi Pratap Singh, Rajeev Trehan, & Ramesh Singh (Eds.)

different time for each temperature. The curves plotted between weight gain per unit area and exposure time (in second) having the R^2 (goodness of fit) values more than 0.9 are considered to be the accurate fit curves for defining correct oxidation law. Cyclic oxidation of ZrB_2-$MoSi_2$ composite follows linear rate law up to 1250°C, and for 1300°C, it follows parabolic rate law, which is diffusion-controlled process. Furthermore, XRD of the oxide layers has also been performed to analyze the scale formed, and scanning electron microscopy-energy dispersive spectroscopy of the oxide layer has also been performed to determine its thickness and its precise chemical composition in order to characterize the oxide layer.

11.1 INTRODUCTION

Among UHTCs, ZrB_2-based ceramics have been reported to be potential candidates for the manufacture of reusable thermal protection systems in hypersonic re-entry nosecones, due to very high thermal conductivity and low density.[1–5] However, the low fracture toughness and poor thermal shock resistance of these ceramics,[6–8] coupled with the poor oxidation resistance of ZrB_2 at temperatures above 1200°C, due to formation of B_2O_3, and a porous scale of ZrO2Click or tap here to enter text.,[9,10] poses severe restrictions to its use at elevated temperatures, especially above 1200°C. This necessitates one to find materials, which may highly enhance the oxidation resistance of ZrB_2. A significant amount of work has already been done on reinforcing diborides like ZrB_2, and HfB_2 with SiC, $MoSi_2$, or Si_3N_4 for enhanced oxidation resistance beyond 800°C (Opeka et al., 1999),[11,12] since a few decades now. However, a limited amount of investigation has been made on correlating the oxidized microstructure with oxidation kinetics of ZM20 at temperatures exceeding 1100°C. Thus, the scope of the present work is to investigate the cyclic oxidation behavior of ZM20 between 1100°C and 1300°C and correlating the phase evolution in the oxidized microstructure with different laws of oxidation kinetics.

11.2 EXPERIMENTAL PROCEDURE

11.2.1 PREPARATION OF COMPOSITE

The composite (ZrB_2-20 vol.% $MoSi_2$-ZM20) used in the present investigation was obtained as a courtesy of the IIT Kharagpur, India. The powders of ZrB_2 and $MoSi_2$ (with average sizes of nearly 5.4 ± 2.2 μm, and 5.5 ± 1.0 μm, respectively, and purity: almost 99.5%) were procured from H.C. Starck, Germany. The raw powders were mixed in a ball Mill (Pulverisette 6, Fritsch). The milling speed and time were 250 rpm and 2 h, respectively. The powders were subsequently hot-pressed at 1800°C in a vacuum with a unidirectional pressure of 30 MPa for 30 min.

11.2.2 CHARACTERIZATION OF SINTERED PELLET

An isoMet diamond saw cutter did sectioning of hot-pressed pellets. These samples were subsequently polished using standard metallographic practices. Pellets were ground on SiC papers and finally polished on clothes spread with diamond paste. This was followed using the density measurement using the Archimedes principle. The XRD technique (Cu Kα radiation, λ= 0.154 nm) determined phases present in the composite. Microstructural imaging was performed using secondary electron and backscattered electron (BSE) imaging modes on a scanning electron microscopy (SEM) to investigate the dissemination of the critical phases. Besides, the chemical composition was identified using an energy dispersive X-ray analyzer (EDS).

11.2.3 CYCLIC OXIDATION STUDIES

The cuboidal samples were cut from the sintered pellet and subsequently polished using standard metallographic practices. Oxidation studies were then carried out for six cycles at various temperatures: 1100°C/1200°C/1250°C /1300°C in a furnace. The oxide layer formed on the surface was subjected to XRD analysis at a standard rate of 10/min between 20° and 90°. Oxide scale morphology and cross-section were examined under SEM coupled with EDS.

11.2.4 DAMAGE EVOLUTION STUDIES

Vickers-microhardness testing in Matsuzawa–Hibiki (Japan) was performed to determine the amount of degradation underneath the oxide scale. Measurements were carried out with a load of 50 gf and constructed a contour plot of the microhardness variation across the cross-section.

11.3 RESULTS AND DISCUSSIONS

11.3.1 MATERIAL CHARACTERIZATION

The Archimedes principle measures the ZM20 composite density. It is found to be 5.94 g/cm^3, which turns out to be 97% of the relative density using the rule of mixtures to calculate theoretical density. XRD analyses have characterized the constituent phases present in the composite, and the XRD pattern is shown in Figure 11.1. The SEM image (in BSE mode) and corresponding EDS spectrum of different phases are shown in Figure 11.2a and b. Based on Figures 11.1 and 11.2, it is observed that ZrB_2 looks bright; $MoSi_2$ appears gray, while the SiO_2 looks dark. This is due to the difference in atomic number contrast between two phases in SEM-BSE mode.[13]

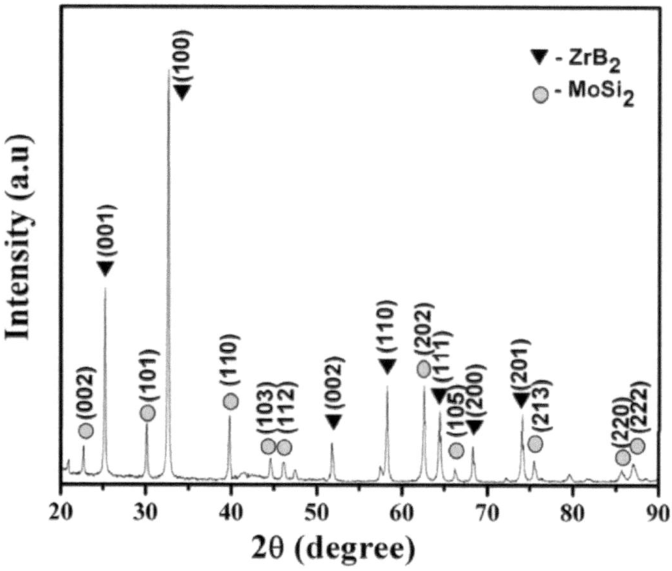

FIGURE 11.1 XRD pattern of ZM20 between 2θ = 20°–90° with step size: 10°/min.

11.3.2 OXIDATION KINETICS

The determination of the law of oxidation between 1100°C and 1300°C has been based on the Arrhenius kinetics[11] and the fact that oxide layer formation in ZM20 at these elevated temperatures obeys diffusion kinetics, in which only that law of oxidation is followed for which the goodness of fit (R^2) value of the weight change per unit area versus time (in second) plot is more than 0.9. Also, based on previous reports,[11,14] oxidation of ZM20 at elevated temperatures is expected to obey either linear or parabolic or cubic rate law of oxidation. However, the weight gain calculations (as a function of oxidation time) aid in determining whether the law of oxidation followed by ZM20 between 1100°C and 1300°C is linear, parabolic, inverse, logarithmic, or cubic. The various oxidation rate laws are presented in Table 11.1.

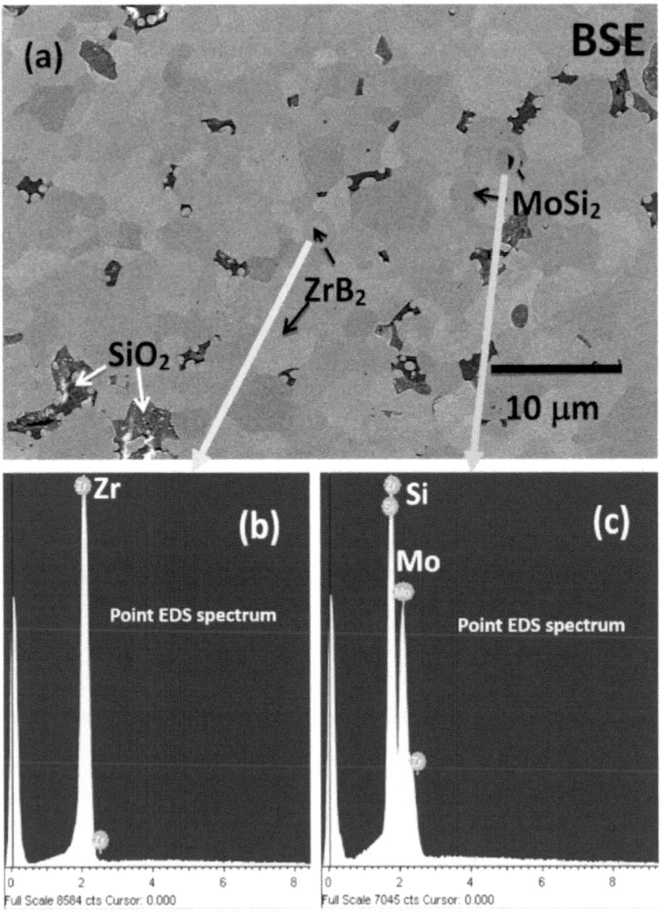

FIGURE 11.2 (a) SEM image of ZM20 composite shows ZrB_2 of bright phase, $MoSi_2$ of gray-colored phase, and SiO_2 phase with the darkest contrast. EDS spectrum from (b) ZrB_2 and (c) $MoSi_2$ grain shows Zr, Mo, and Si's peaks. The red-outlined circles in parts (a) and (b) are the regions where EDS has been performed.

TABLE 11.1 Various Rate Laws of Oxidation and Their Representative Equations.

Name of the rate law of oxidation	Equation for the rate law
Linear rate law	$W = K \times t + c$
Parabolic rate law	$W^2 = K \times t + c$
Cubic rate law	$W^3 = K \times t + c$
Direct logarithmic rate law	$W = K \times \ln t + c$
Inverse logarithmic rate law	$1/W = c - K \times \ln t$

W, weight change per unit surface area; K, rate constant; t, time of oxidation.

Figure 11.3a–e illustrates the variation of change in weight during oxidation as a function of oxidation (exposure) time for linear, parabolic, cubic, logarithmic, and inverse logarithmic rate laws, respectively. Moreover, from the X-ray diffractogram (Figure 11.4), it is observed that in addition to ZrO_2 peaks at all temperatures from 1100°C to 1300°C, peaks belonging to Mo_5Si_3 are observed only at 1200°C and 1300°C, whereas peaks belonging to MoO_3 are observed at 1100°C, 1200°C, and 1300°C.

From the SEM images (Figures 11.5–11.8a) and corresponding EDS spectra (Figures 11.5–11.8b), it is observed that oxygen (O), silicon (Si), zirconium (Zr), and molybdenum (Mo) are the principal elements present. Due to the presence of these elements, some compounds like MoO_3, ZrO_2, MoB, Mo_5Si_3, and SiO_2 have also been observed. From these images, it can be observed that the bright field is ZrO_2 while the dark field is SiO_2. When SiO_2 is less, the oxide layer developed on the specimen surface is coarse. On the contrary, for higher SiO_2 content, the oxide layer formed is smooth. The reason is that with a lesser concentration of SiO_2, from 1100°C to 1250°C, the specimen follows linear oxidation kinetics having a higher rate of oxidation while with a higher concentration of SiO_2 above 1250°C, it follows parabolic oxidation kinetics having a lower rate of oxidation as reported by Silvestroni et al.[11] From the EDS spectrum, Si has been observed to possess the highest intensity. But Si is present as SiO_2, which is amorphous; hence SiO_2 peaks cannot be detected by XRD analysis. The peaks of MoO_3, ZrO_2, MoB, and Mo_5Si_3 are present. Also, during the conversion of ZrB_2 to ZrO_2, B_2O_3 is formed, which is a volatile material. Hence, no distinct EDS spectrum for boron (B) is observed. In this context, it is important to mention that EDS is not a suitable spectroscopic technique to detect the exact concentration of light elements, such as B, which produces characteristic X-rays of very low energy, as reported by Newbury et al.[15] From Figures 11.9a and 11.10a, the average oxide layer thickness is found to be ~3.58 and ~5.2 μm for ZM20 specimens subjected to cyclic oxidation at 1200°C and 1300°C, respectively. The layer thickness for the latter is higher because of oxidation at a higher temperature. From Figures 11.9 and 11.10, elemental mapping of the oxide layer reveals the presence of B, O, Si, Zr, and Mo. The results from this mapping match with the results of XRD analysis and also with the EDS spectra.

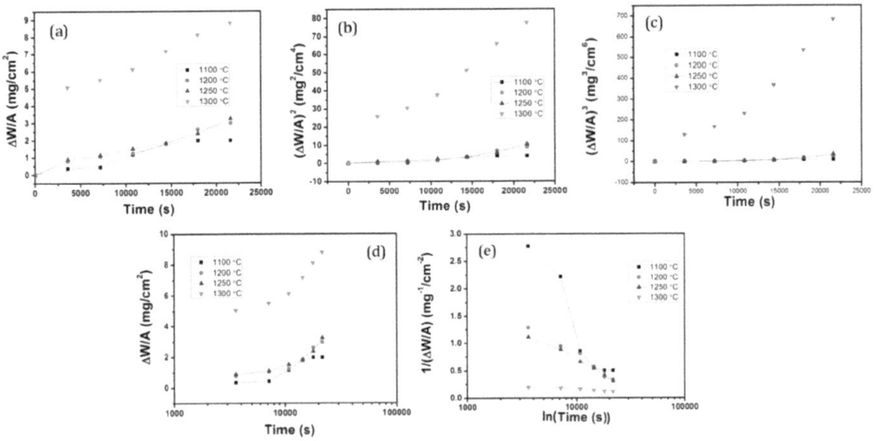

FIGURE 11.3 This figure shows the dependence of (a) weight change, (b) square of weight change, and (c) cube of weight change per unit surface area as a function of oxidation time. Further (d) weight change and (e) inverse weight change per unit surface area as a function of the logarithmic value of exposure time of ZrB_2-20 vol.% $MoSi_2$ at 1100°C–1300°C in air.

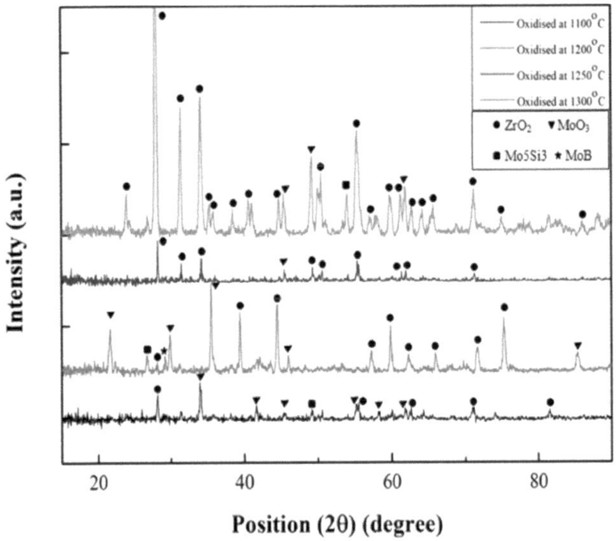

FIGURE 11.4 XRD patterns of oxide scale form on ZM-20 during oxidation at 1100°C, 1200°C, 1250°C, and 1300°C for 6 h in air.

It can also be observed from Table 11.2 that the values of goodness of fit (R^2) are more than 0.9 for linear rate law for temperature range: 1100°C–1250°C. But for 1300°C the value of R^2 is more than 0.9 for parabolic rate law. This conveys a quantitative information that oxidation in ZM20 composite follows a linear rate law ranging from 1100°C to 1250°C and a parabolic rate law at 1300°C, owing to the protective action of SiO_2 phase above 1250°C, as discussed previously.

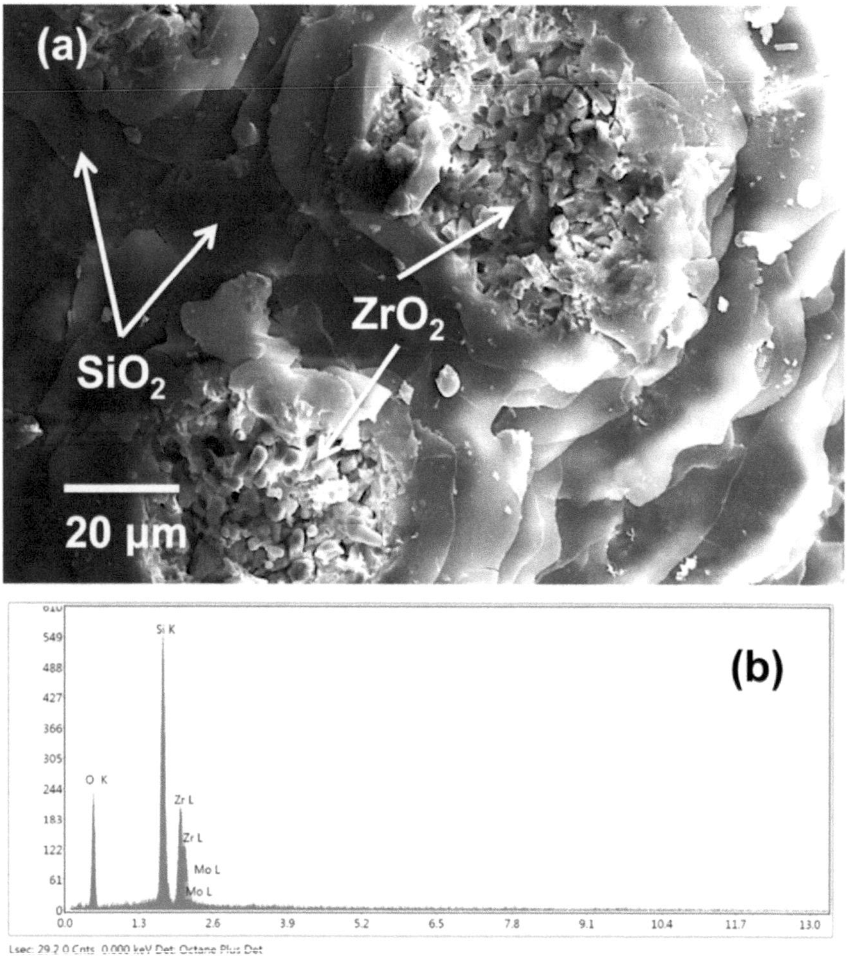

FIGURE 11.5 (a) SEM micrograph shows oxide surface formed on ZM20 composite after oxidation at 1100°C, and (b) typical EDS spectrum of oxide scale.

TABLE 11.2 Slopes and Goodness of Fit (R^2) for Weight Change Per Unit Area Versus Time Graphs.

Parameters	1100°C	1200°C	1250°C	1300°C	Kinetics
Kl (mg/cm^2 s)	1.046×10^{-4}	1.32×10^{-4}	1.33×10^{-4}	3.39×10^{-4}	Linear
R^2	0.92512	0.9632	0.95872	0.78871	
Kp (mg^2/cm^4 s)	2.22×10^{-4}	4.06×10^{-4}	4.32×10^{-4}	0.00331	Parabolic
R^2	0.88256	0.84221	0.80907	0.96534	
Kc (mg^3/cm^6 s)	4.42×10^{-4}	0.00117	0.00132	0.03035	Cubic
R^2	0.8472	0.73127	0.64173	0.95231	
Klg (mg/cm^2 lns)	0.38187	0.49824	0.52022	1.77591	Logarithmic
R^2	0.39344	0.45746	0.49971	0.87294	
Ki(cm^2/mglns)	−0.32077	−1.2533	−1.03527	−0.11188	Inverse logarithmic
R^2	0.88782	0.97656	0.98522	0.92557	

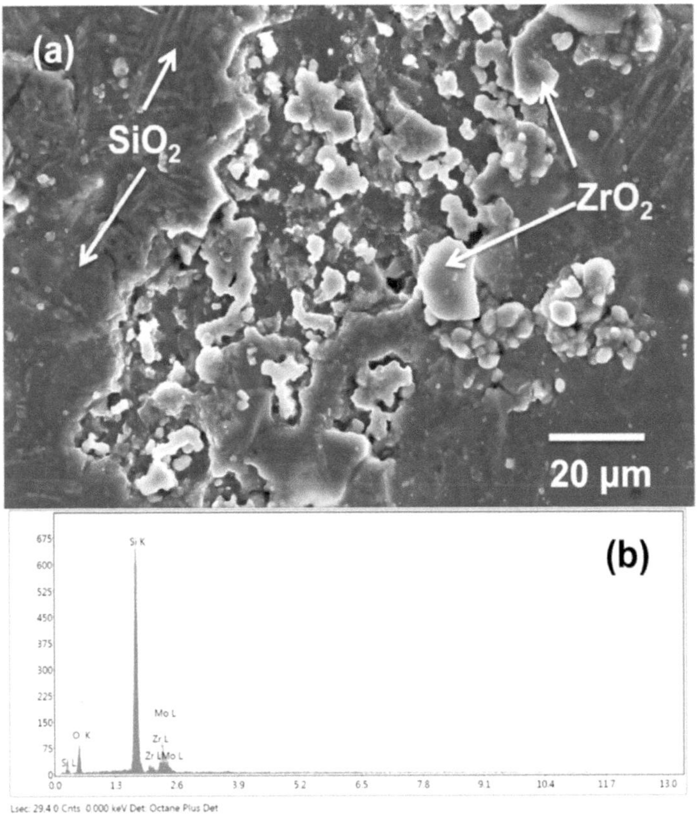

FIGURE 11.6 (a) SEM micrograph shows oxide surface formed on ZM20 composite after oxidation at 1200°C, and (b) typical EDS spectrum of oxide scale.

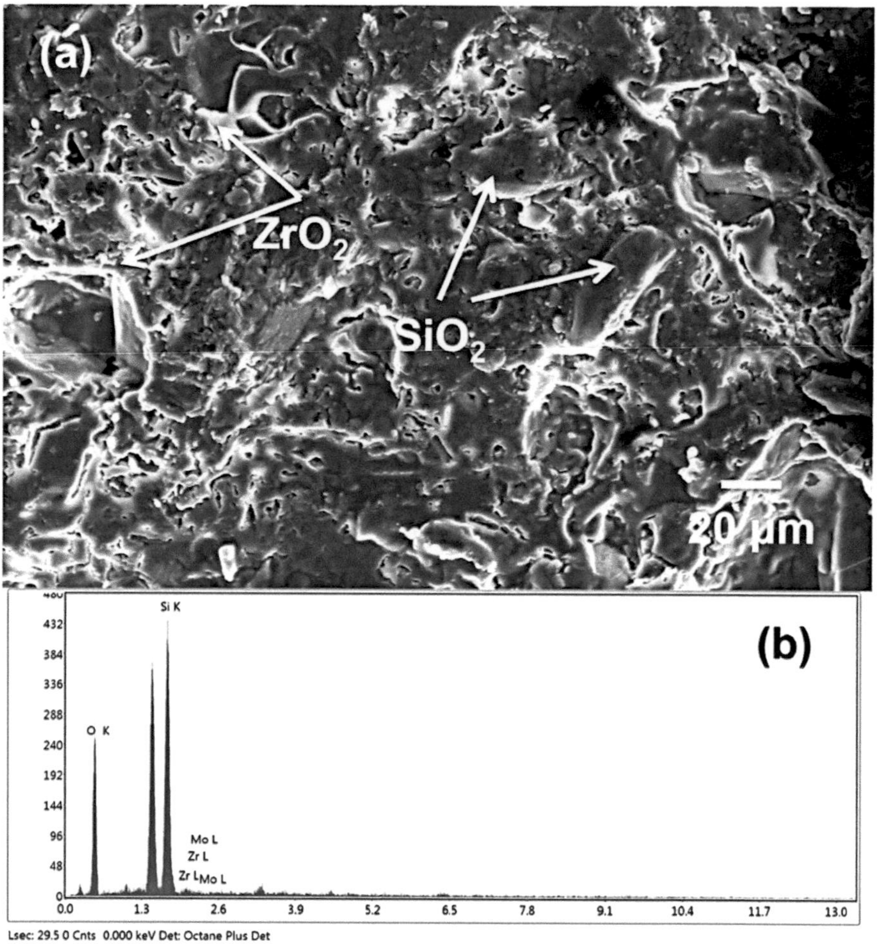

FIGURE 11.7 (a) SEM micrograph shows oxide layer formed on ZM20 composite after oxidation at 1250°C, and (b) typical EDS spectrum of oxide scale.

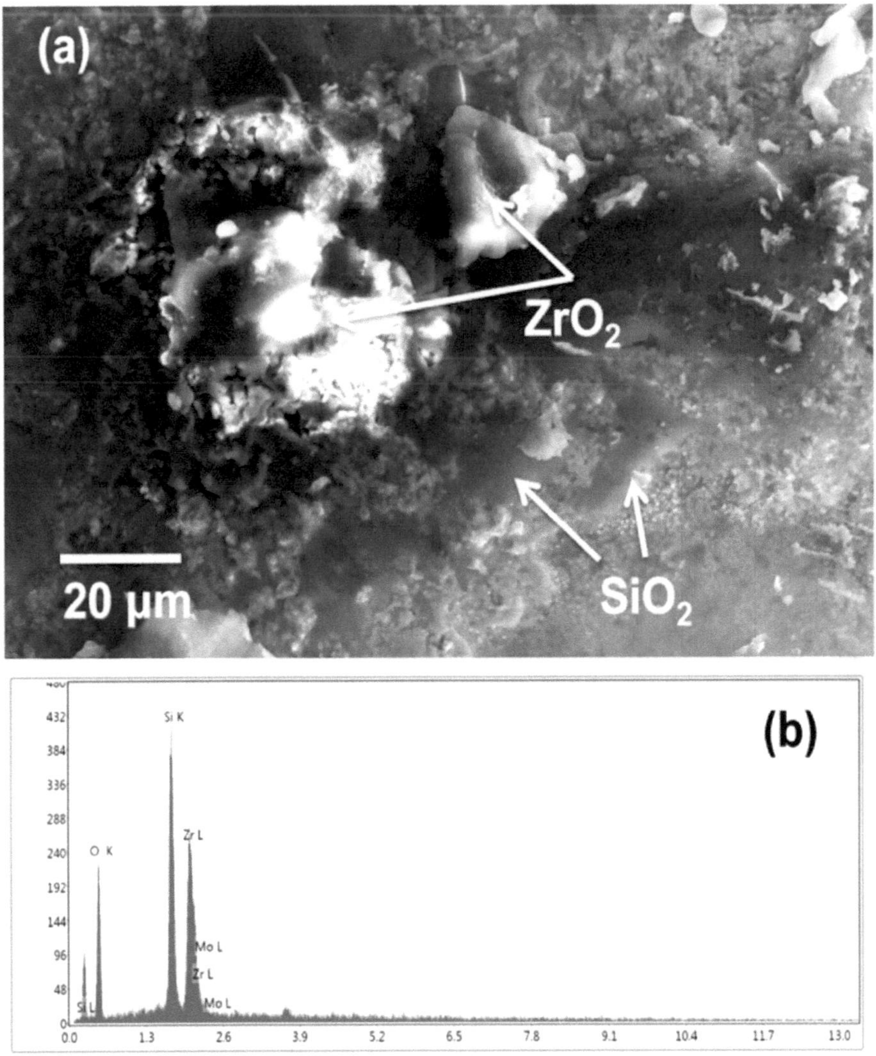

FIGURE 11.8 (a) SEM micrograph shows oxide layer formed on ZM20 composite after oxidation at 1250°C, and (b) typical EDS spectrum of oxide scale.

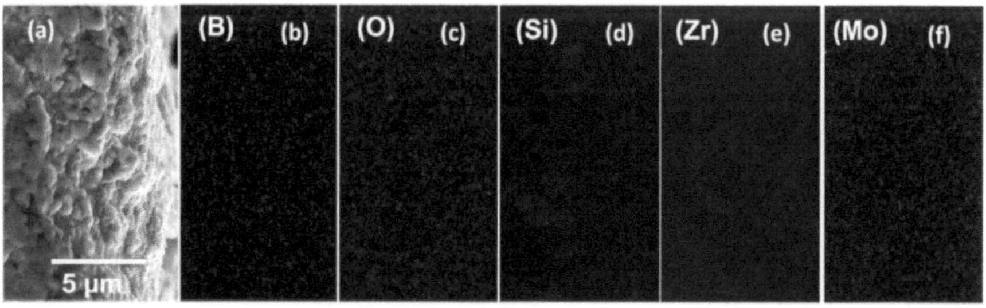

FIGURE 11.9 (a) SEM cross-sectional image of oxide layer formed after cyclic oxidation at 1200°C, 6 h (1 cycle/h); EDS maps of (b) B, (c) O, (c) Si, (d) Zr, and (e) Mo.

Possible reactions during cyclic oxidation at different temperatures are shown below:

General reactions:

$$ZrB_2(s) + 5/2O_2(g) = ZrO_2(s) + B_2O_3(l) \tag{11.1}$$

$$B_2O_3(l) = B_2O_3(s) \tag{11.2}$$

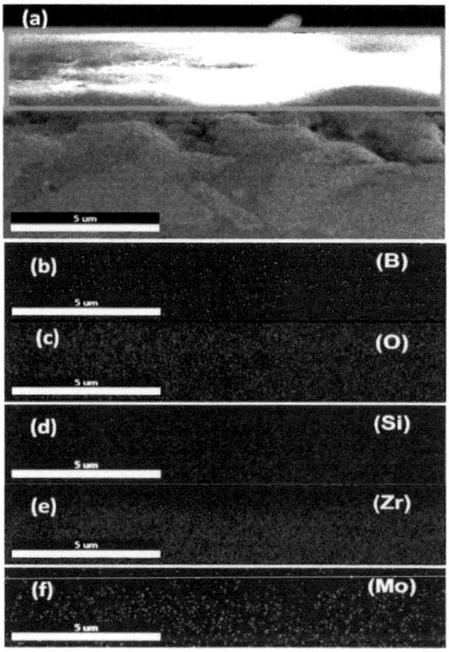

FIGURE 11.10 (a) SEM cross- sectional image of oxide layer formed after cyclic oxidation at 1200°C, 6 h (1 cycle/h); EDS maps of (b) B, (c) O, (d) Si, (e) Zr, and (f) Mo.

These reactions occur in the temperature range of (800°C–1200°C):

$$MoSi_2(s) + 7/2O_2(g) = SiO_2(s) + MoO_3(s) \quad (11.3)$$

$$ZrB_2(s) + 2MoSi_2(s) + 5O_2(g) = ZrO_2(s) + 2MoB(s) + 4SiO_2(s) \quad (11.4)$$

$$SiC + 3/2O_2(g) \rightarrow SiO_2(l) + CO(g) \quad (11.5)$$

These reactions occur in the temperature range of (1200°C–1500°C):

$$SiC + O_2(g) \rightarrow SiO(g) + CO(g) \quad (11.6)$$

$$SiO(g) + 12O_2(g) \rightarrow SiO_2(l) \quad (11.7)$$

So, what can be seen from these reactions is that after oxidation the constituents of the oxide layer should be ZrO_2, MoO_3, SiO_2, and MoB. From different analyses like XRD characterization and phase determination, SEM images and EDS spectrums can be observed that all the compounds formed according to these reactions are present in the oxide layer of the samples used for this present work.

The contour plots of Vickers microhardness variation across the cross-section of the sample exposed at 1200°C and 1300°C are shown in Figure 11.11a and b, respectively.

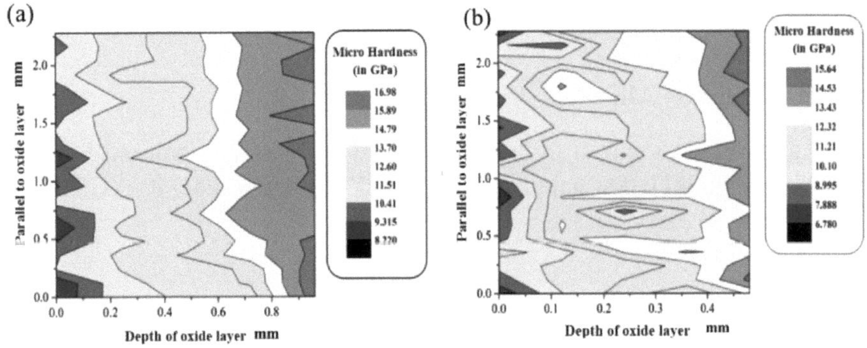

FIGURE 11.11 Hardness contour plots from oxide (left) to base material (right) for ZM20 composite oxidized at (a) 1200°C and (b) 1300°C.

In both cases, the hardness is decreasing from the center to the surface of the specimen, which has undergone oxidation. This is due to the formation of oxide layers on the surface of the specimen, which degraded its mechanical properties.

For the sample oxidized at 1200°C, it can be observed that the oxide layer is not too thick, and so the microhardness for that sample varies in a range of 9–17 GPa, and also the transition layer is not so prominent. For the second sample, which is oxidized at 1300°C, the oxide film is thicker than the former one. As a result, the microhardness data vary within a range of 7–16 GPa. The transition zone between the oxide layer and base metal is prominent; it possesses a hardness value of moderate level.

11.4 SUMMARY AND CONCLUSIONS

Cyclic oxidation behavior of ZM20 composite has been investigated at 1100°C, 1200°C, 1250°C, and 1300°C for 6 h. Monitoring weight change and examining oxide scales, the following conclusions may be drawn:

i) Weight gain for both the composites increases with increasing temperature and time.
ii) The main oxidation products formed between 1100°C and 1300°C are ZrO_2, MoO_3, and SiO_2.
iii) The presence of $MoSi_2$ particles leads to a significant improvement in the oxidation resistance of the composite due to the formation of borosilicate glass (especially at or above 1200°C).
iv) The cyclic oxidation of the samples follows linear oxidation kinetics from 1100°C to 1250°C while at 1300°C, it follows parabolic oxidation kinetics due to the protective action of SiO_2 above 1250°C.

ACKNOWLEDGMENT

The authors are grateful to the Department of Metallurgical and Materials Engineering, NIT Durgapur, and Central Research Facility (CRF), IIT Kharagpur, to carry out the work.

KEYWORDS

- **borides**
- **ceramic composites**
- **cyclic oxidation**
- **kinetics**
- **oxide layer**

REFERENCES

1. Golla, B. R. et al.;. Review on Ultra-High Temperature Boride Ceramics. In *Progress in Materials Science*; Elsevier Ltd, 2020.
2. Gasch, M. J.; Ellerby, D. T.; Johnson, S. M. Ultra High Temperature Ceramic Composites. In *Handbook of Ceramic Composites*; Springer US, 2006; pp 197–224.
3. Mitra, R. et al. Mechanical, Thermal, and Oxidation Behaviour of Zirconium Diboride Based Ultra-High Temperature Ceramic Composites. *Key Eng. Mater.* **2009,** *395*, 55–68.
4. Mallik, M. et al. Electrical and Thermophysical Properties of ZrB 2 and HfB 2 Based Composites. *J. Eur. Ceram. Soc.* **2012,** *32* (10), 2545–2555.
5. Mallik, M. et al. Effect of SiC Content on Electrical, Thermal and Ablative Properties of Pressureless Sintered ZrB2-Based Ultrahigh Temperature Ceramic Composites. *J. Eur. Ceram. Soc.* **2017,** *37* (2), 559–572.
6. Mallik, M.; Roy, H. *Fracture Toughness Measurement of Hot Pressed ZrB 2-MoSi 2 Composite*. researchgate.net, 2013.
7. Mallik, M. et al. Effect of SiC Content, Additives and Process Parameters on Densification and Structure-Property Relations of Pressureless Sintered ZrB_2-SiC Composites. *Ceram. Int.* **2013,** *39* (3), 2915–2932.
8. Paul, T. R.; Mondal, M. K.; Mallik, M. Microstructure Dependent Physical and Mechanical Properties of Spark Plasma Sintered ZrB2-MoSi2–SiCw Composites. *Int. J. Refract. Metals Hard Mater.* **2019,** *79*, 131–137.
9. Mitra, R.; Mallik, M.; Kashyap, S. High-Temperature Environmental Degradation Behavior of Ultrahigh-Temperature Ceramic Composites. In *Handbook of Advanced Ceramics and Composites*; Springer International Publishing, 2020; pp 1221–1256.
10. Mallik, M.; Ray, K. K.; Mitra, R. Oxidation Behavior of Hot Pressed ZrB2-SiC and HfB2-SiC Composites. *J. Eur. Ceram. Soc.* **2011,** *31* (1–2), 199–215.
11. Silvestroni, L. et al. Understanding the Oxidation Behavior of a ZrB_2–$MoSi_2$ Composite at Ultra-High Temperatures. *Acta Mate.* **2018,** *151*, 216–228.
12. Mallik, M.; Ray, K.; Mitra, R. Effect of Si3N4 Addition on Oxidation Resistance of ZrB2-SiC Composites. *Coatings* **2017,** *7* (7), 92.
13. Dingley, D. Progressive Steps in the Development of Electron Backscatter Diffraction and Orientation Imaging Microscopy. *J. Microsc.* **2004,** *213* (3), 214–224.
14. Sciti, D.; Brach, M.; Bellosi, A. Long-Term Oxidation Behavior and Mechanical Strength Degradation of a Pressurelessly Sintered ZrB2-MoSi2 Ceramic. *Scripta Mater.* **2005,** *53* (11), 1297–1302.
15. Newbury, D. E.; Ritchie, N. W. M. Is Scanning Electron Microscopy/Energy Dispersive X-ray Spectrometry (SEM/EDS) Quantitative? *Scanning* **2013,** *35* (3), 141–168.

CHAPTER 12

EXPERIMENTAL STUDY OF ALUMINUM METAL MATRIX COMPOSITE USING ELECTRODISCHARGE MACHINE: A REVIEW

SOURAV MAJUMDAR[1], NEERAJ KUMAR BHOI[2*], and HARPREET SINGH[3]

Department of Mechanical Engineering, PDPM Indian Institute of Information Technology Design and Manufacturing, Jabalpur, 482005, Madhya Pradesh, India

[*]Corresponding author. E-mail: neerajbitd@gmail.com

ABSTRACT

Aluminium metal matrix composites called as Al-MMCs are one of the foremost diverse set of materials possessing greater strength-to-weight ratio, excellent wear resistance, and mechanical and thermal properties. For catering the specified shape and advent of new class of Al-MMCs various nonconventional machining methods are utilized. Electrodischarge machining (EDM) is a widely utilized technique for developing fine end products through the means of electric spark. This work highlights the concept and focuses on properties of Al-MMCs machined using EDM. The various advancements made within the EDM and machining characteristics of Al-MMCs are discussed. This chapter aims at highlighting the viability and importance EDM possesses for machining Al-MMCs, associated difficulties and outcomes of the process.

Modern Manufacturing Systems: Trends and Developments. Rajiv Kumar Garg, Ravi Pratap Singh, Rajeev Trehan, & Ramesh Singh (Eds.)

12.1 INTRODUCTION

Generally traditional electrodischarge machining (EDM) process is restricted to handful industries because of low material removal rate (MMR), surface finish, and its poor productivity. The development in the field of EDM and electrodischarge coating is summarized in Figure 12.1.[1] With time, researchers have tried new variants and techniques to widen the application domain of EDM process by using methods such as insulating electrodes with coatings of high resistance, changing the orientation of electrodes, providing ultrasonic vibrations to the electrodes, and using powder mixed dielectrics.[2,3] Powder mixed EDM (PMEDM) came into light in late 1970s. Instead of using pure dielectric, particles of certain materials generally in the form of powder were suspended in dielectric and the powder and dielectric mix was used for the process. This resulted in a much better machining rate and surface finish. It was observed that different powder characteristics influenced the outcomes of the EDM process. Experimental works suggested that characteristics such as conductivities of both thermal and electrical, size, and density of powder material affected the process of PMEDM. It was found the spark gap for PMEDM was higher than the conventional EDM. Process variables including the likes of peak current, pulse on time, duty cycle, concentration of powder along with its material have considerable effect in PMEDM.[4–6] In PMEDM, an increase in both the spark gap and the electric conductivity of powder particle resulted in a better spark frequency and increased the material carrying capacity of the dielectric away from the machined zone.[4–6]

For improving MRR along with surface finish for Al-6061, Rouniyar and Shandilya used PMEDM along with magnetic field and noticed with an increment in magnitude of applied magnetic field, MRR highly improved.[6] When powder particles with high thermal conductivity were used larger amount of heat from the electrode gap was absorbed compared to traditional process resulting in a lower discharge density and hence shallower craters were formed on work surface. Moreover when smaller sized powder particles were used more particles could house themselves within inter electrode gap which led to an increase in overall discharge energy but reduced energy density. Hence in PMEDM, the intensity of discharge energy is less than the traditional EDM. This resulted in the formation of multiple craters of smaller size which resulted in higher MRR. Number of cracks along with their size produced on the machined surface was also found to be less compared to the traditional method which resulted in a much cleaner surface with lesser defects. When less dense powder material was used it was found that they were able to distribute themselves much better in the dielectric medium and reduced the settling phenomenon at the bottom of the tank resulting in minimization of the powder quantity initially required for the process. Characteristics such as low material density, size along with high thermal and electric conductivity provided better results. Most commonly used powder additives for this process include SiC, Si, and Al because of their low cost and easy availability. Kerosene, deionized water, and a few commercial oils are generally used as dielectrics.

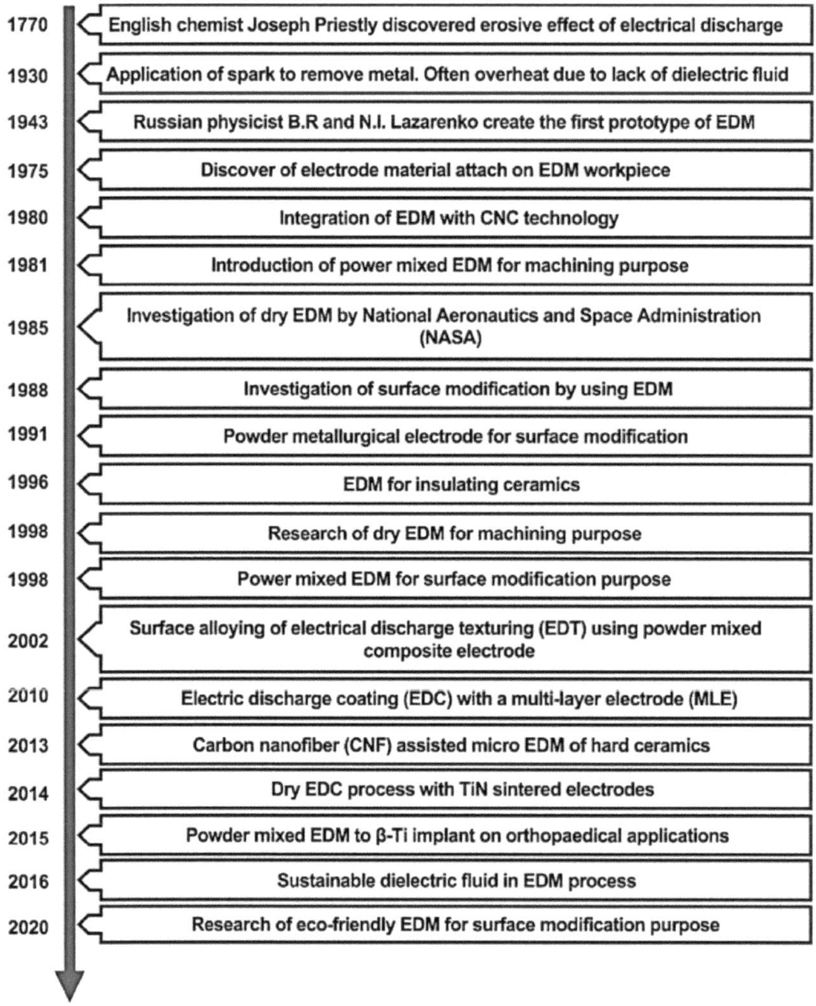

FIGURE 12.1 Chronology of EDM/EDC development.

Source: Reprinted from Ref. [1]. © 2020 The Author(s). Published by IOP Publishing Ltd on behalf of the IMMT https://creativecommons.org/licenses/by/3.0/

This chapter aims to review the process fundamentals related to the machining of Al-MMCs with different variants of EDM. The progressive development made in the timeline of the EDM is described in brief to illustrate the process variables affecting the machining conditions of Al-MMCs. The chapter will illustrate the recent developments for precision machining and sustainable variants of EDM process for Al-MMCs machining.

12.2 HISTORICAL BACKGROUND OF THE EDM PROCESS AND WORKING MECHANISM

EDM allows great precision and is generally employed for machining tough materials. EDM has played a major role, developing, and facilitating modern day advanced technology in industries like aerospace and automobile. The technological advancements in the 20th century led to rapid development in EDM. The working principle of EDM, first observed by the English scientist and philosopher Joseph Priestly in 1770, observed electrical discharges were capable of eroding materials and the process on which electric discharge machine works was discovered. EDM was independently developed first in the USSR and in the USA. B.R. and N.I. Lazarenko, observed in 1943 that controlled erosion was possible while the electrodes were still kept immersed within the dielectric while they were trying to reduce the erosion of tungsten electrodes. This resulted in the development of the earliest EDM machines. Jack Beaver, Harold Stark, and Victor Harding in the USA came up with an EDM machine which was capable of removing broken tools from cast aluminium. Subsequently, die-sink EDM method was used since the first machines were made, it was in the 1960s wire EDM technique was developed. Combination of new techniques with computer numeric controlled operation has opened new doors for EDM. Researchers are now focused on refining the existing techniques to further expand the application of EDM. Today EDM finds its use in industries such as aerospace, automobiles, mold making, die making, and tool-making industries all of which employ high degree of precision and use of super strength alloy materials to withstand elevated temperatures and high stress conditions, which are difficult to machine traditionally like drilling of holes in turbine blades or rocket nozzles, profiling and dressing of tools and dies.

EDM being one unconventional machining technique where energy as electric spark is provided which removes the material by melting and vaporization. A traditional EDM setup consists of a tank, workpiece, tool, dielectric like kerosene and commercial oils, and suitable potential difference. The workpiece is placed in the tank. Sparks are produced in the gap when the potential difference is applied and these results in erosion of the material both from tool and work piece. Due to high temperature produced in the process the work melts and vaporises and the mirror impression of the tool is obtained on the work piece. Similarly, in case of PMEDM conducting powders are added to the working tank for higher removal of the material along with better surface finish. At a certain applied voltage, an electric field of sufficient strength is created across the spark gap or inter electrode gap (IEG). This results in emission of electrons from cathode from the point of minimum IEG. These electrons accelerate towards the anode and strike the dielectric molecules and dissociate them producing electrons along with positive ions. These electrons further dissociate other dielectric molecules producing electrons as well as positive ions. The powder particle surface gets accumulated with these ions and a chain of powder particles is set up in this narrow column of ionized dielectric.[7,8]

When the discharge breakdown occurs, it is seen as a spark due to high conductivity of the ionized column. Discharge breakdown between the particles takes place where the field density is highest resulting in sparking between the two particles. The charges redistribute themselves and again the same phenomenon occurs, this time between different particles. In this manner, a series of discharges takes place in the ionized column and finally the discharge occurs between the powder particles and the electrodes.[9] The nature of these powder particles used in PMEDM also play a key role in determining the discharge gap. When metallic powder particles are used in the process their electrical as well as physical properties affect the discharge gap. Due to the series of discharges in the ionized column high temperatures are obtained in the IEG and free electrons are ejected from these metallic particles which decrease the resistance and increase the conductivity of the dielectric thereby allowing sparks to be generated and be effective at the same time even from a larger discharge gap as compared to the traditional method.[3,7,10] A combination of permanent magnets, use of steam in machining zone, different additives are highly used nowadays to improve the capabilities of the conventional EDM machining process. Figure 12.2 represents a hybrid variant of the EDM machining process for Al-MMCs materials. Along with the traditional EDM permanent magnets are used, abrasive powders are mixed with the dielectric medium to enhance the machining capability.[11]

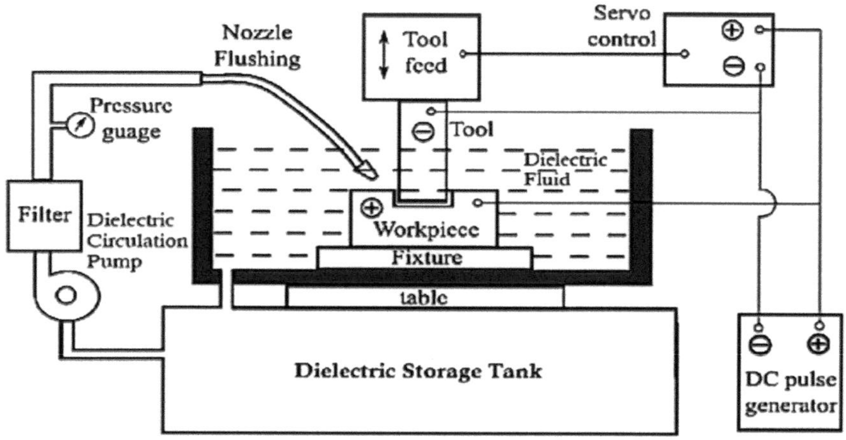

FIGURE 12.2 EDM setup.

Source: Reprinted from Ref. [29]. © 2019 Elsevier. https://creativecommons.org/licenses/by-nc-nd/4.0/

12.3 ALUMINUM METAL MATRIX COMPOSITE AND EDM

With the constantly changing scenarios and needs of the modern industries, new materials are being looked upon as an alternative or as an upgrade to the existing materials. Just like the needs and the demand of time forced the industries to move to alloys from pure metals, today's time calls for materials such as superalloys, nanomaterials, and composites. Composites have become a need of the industry and newer composites are being developed and researched for; they are far more superior and can be employed to a variety of applications as compared to the traditional materials. A composite is a combination and arrangement of more than two materials having indifferent properties compared to the parent material. These MMCs have high mechanical strength and toughness, better impact resistance, low sensitivity to the conditions of temperature and moisture and are more durable.[12–15] Due to its light weight, the components developed are lighter yet stronger, thereby making the product more effective as well as helping in cutting the cost of the material.[13–15] Al-MMCs are one class of MMCs which because of its numerous benefits like light weight, better strength and toughness, greater strength-to-weight ratio, superior fatigue resistance along with low sensitivity of hostile environment conditions make it an apt material and as a result is employed in manufacturing, automobile and electronics industries. Varying the percentage of reinforcement and matrix in Al-MMC, its properties can be altered as per the requirement making it both versatile and reliable. This affects the tensile and compressive strength along with composite hardness. Different reinforced particles like carbides, oxides, and nitrides also impart different properties to the composite such as for thermal applications Al–SiC, Al–TiC, and Al–Al_2O_3 are widely used.[16,17] Apart from improving mechanical strength and toughness, they impart heat and moisture resistance by lowering the coefficient of thermal and moisture expansion.

Machining of Al-MMCs is a daunting task because of harder reinforcing particles in the matrix. Conventional machining operations such as drilling, turning, milling are exposed to heavier tool wear due to higher hardness value. The nonhomogeneous and anisotropy nature of the composite material limits the conventional tooling material for machining applications. EDM is a widely used manufacturing operation for difficult to machine materials with zero cutting forces apart from minimal tool wear. Recent exploration made by Paswan et al.[18] a novel approach by combining steam-assisted machining process with EDM for Al-10SiC-3Cr MMCs (base material) they found that with the incorporation of the steam during EDM machining improved surface quality to a greater extent. On a similar note, the steam-assisted process in EDM results in thinner recast layer with comparatively lower defect compared to kerosene as a dielectric medium. In Figure 12.3, surface morphology of tungsten carbide machined zone shows different kinds

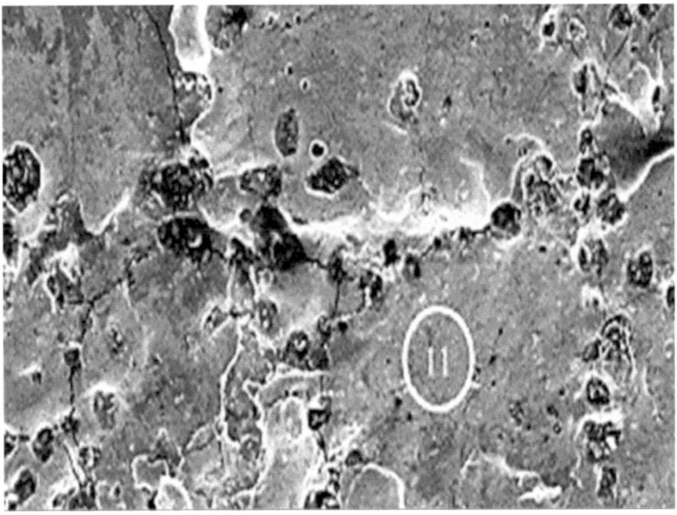

FIGURE 12.3 SEM micrograph showing the locations where the EDX analyses were taken. The places marked by the numbers 11 and 12 are the locations taken by the EDX analyses as shown in Figure 12.4a and Figure 12.4b, respectively.

Source: Reprinted with permission from Ref. [30]. © 2003 Elsevier.

of cracks and deposited machined debris in the surface. Higher current and larger pulse duration establish conditions of high thermal stresses resulting in the formation of cracks. Examination of machined surface depicts higher carbon composition at the craters formed as depicted in Figure 12.4a whereas in Figure 12.4b the EDX analysis shows larger traces of copper, main source being the copper electrode.

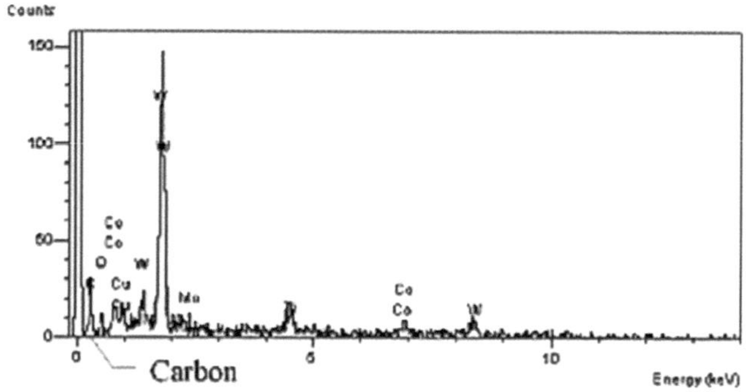

FIGURE 12.4a EDX analysis result for the EDMed surface, taken from the location marked by 11 as shown in Figure 12.3.

Source: Reprinted with permission from Ref. [30]. © 2003 Elsevier.

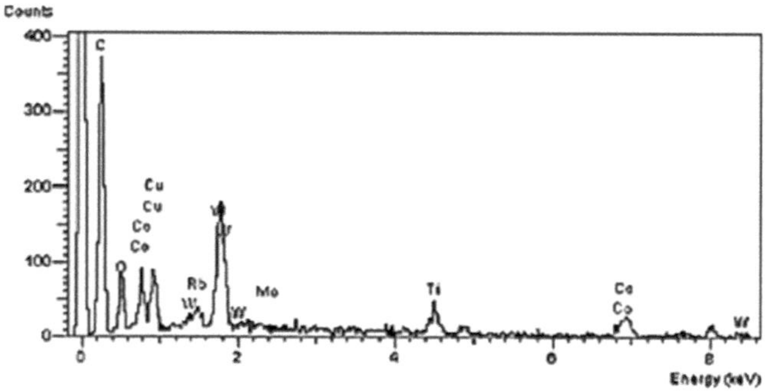

FIGURE 12.4b EDX analysis result for the EDMed surface, taken from the location marked by 12 as shown in Figure 12.3.

Source: Reprinted with permission from Ref. [30]. © 2003 Elsevier.

12.4 MATERIAL RESPONSES AND OUTCOME

The Al-MMCs have been researched with a variety of reinforced elements and were machined using various EDM processes. Input variables namely current *I*, pulse on time T_{on}, pulse off time T_{off}, gap voltage *V*, duty cycle DC, powder concentration was varied, their different combinations were used for

TABLE 12.1 Various Outcomes Reported by the Different Researcher in the Timeline for EDM Machining and Al-MMCs.

Sr. no	Workpiece material	Input parameter	Output parameter	Remark/summary	Reported by
1.	Al–SiC (10%)	Current I = 5, 10 and 15 A Voltage = 1, 1.5 and 2 V = 100,200,300 μs T_{on} = 10,20,30 μs Flushing pressure = 0.2.0.4,0.6 kg/cm^2	For copper tool MRR = 0.115 g/s EWR = 0.36% For brass tool MRR = 0.065 g/s EWR = 64.67%	• MRR is high at larger values of current and pulse on time along with medium voltage. • EWR is least for smaller values of current, high voltage with medium pulse off time.	Dhupal et al.[19]
2.	Al–SiC and Al–SiC–B_4C	Current I = 5, 10 and 15 A T_{on} = 50, 75 and 100 μs Voltage = 40, 45 and 50 V Flushing pressure = 1.5 kg/cm^2		• As MRR increases, crater diameter increases. • B_4C produces greater crater diameter of machined surface than Al–SiC. • As pulse current increases so is the surface roughness. • B_4C leads to greater surface roughness than Al-SiC.	Suresh et al.[20]
3.	AA6061/10% Al_2O_3	Pulse current I = 6, 10 and 14 A T_{on} = 75, 100 and 200 μs Duty cycle DC = 50, 60 and 70%	For Max MRR Current I = 14 A T_{on} = 200 μs Duty cycle = 50%	• MRR increases as current and T_{on} sincreases. • Primary influencing factor is Pulse current	Kandpal et al.[21]
4.	Al-SiC$_p$ 12%	Peak current I = 6, 12 A T_{on} =50, 150 μs T_{off} = 4, 8 μs	Max MRR= 0.0566 g/min at Peak current I = 12 A T_{on} = 150 μs T_{off} = 8 μs	• As current increases so does the MRR. • As peak current increases so is the tool wear rate.	Mohanty et al.[22]

TABLE 12.1 *(Continued)*

Sr. no	Workpiece material	Input parameter	Output parameter	Remark/summary	Reported by
5.	Al7075-10% SiC–Mg	Current = 3, 5, 7, 9 and 11 A Pulse on time = 50, 100, 150, 200 and 250 μs Duty cycle DC = 0.5, 0.55, 0.6, 0.65 and 0.70	MRR decreased as pulse duration increased. Increment in pulse current along with duty cycle MRR increased.	• MRR of rotary EDM is greater than MRR of conventional EDM. • EWR of rotary EDM is less than EWR of conventional EDM.	Malhotra et al.[23]
6.	Al-10% SiC-2.5% TiB_2	Current $I = 2$, 5 and 8 A $T_{on} = 30$, 50 and 70 μs Powder concentration C = 0, 2 and 4 g/L	MRR max = 171.41 mg/min at I =8 A $T_{on} = 70$ μs $C = 0$ g/L R_a min= 3.79 μm at $I = 2$ A $T_{on} = 30$ μs $C = 4$ g/L	• High MRR was observed for larger current and T_{on}. • Low TWR and surface roughness was observed for smaller current and T_{on} with higher Al powder concentration.	Roy et al.[24]
7.	Al–SiC 12% Al_2O_3	Current $I = 6$, 12 A $T_{on} = 50$, 150 μs $T_{off} = 4$, 8 μs	MRR max in die sink EDM= 0.0396 g/min MRR max in PMEDM= 0.0566 g/min	• Nanopowder EDM provides better surface finish and MRR • As peak current increases so does the TWR and decreases for increase in pulse duration.	Mohanty et al.[23]

TABLE 12.1 *(Continued)*

Sr. no	Workpiece material	Input parameter	Output parameter	Remark/summary	Reported by
8.	AA7075-B_4C	Current I = 3, 6, 9, 12 and 15 A T_{on} = 30, 60, 90, 120 and 150 µs T_{off} = 8, 56, 104, 152 and 200 µs Voltage V= 30, 40, 50, 60 and 70 V	MRR max = 0.2134 g/min at I =12 A T_{on} =120 µs T_{off} = 56 µs V = 40 V	• PMEDM gives better MRR and surface finish. • Current along with T_{on} is most influential input parameter for MRR.	Dubey and Singh[26]
9.	AA8011/15 wt% ZrO_2	Current I = 5, 10 and 15 A T_{on} = 300, 600 and 900 µs T_{off} = 30, 60 and 90 µs	R_a min = 4.38 µm at I = 5 A T_{on} = 300 µs T_{off} = 60 µs	• Pulse current was primary process input parameter affecting the SR followed by T_{on}. • As T_{on} and pulse current increased so did the SR.	Alagarsamy et al.[27]
10.	Al-SiC 20% SiC	Input Current I = 0–50 A Pulse on time = 0.5–4000 µs Gap voltage V= 0–250 V Flushing pressure = 0–2 bar	Optimum value was found at I = 3 A T_{on} = 75 µs DC= 80% V= 40 V	• Maximum MRR observed at combination of larger input current along with duty cycle, smaller T_{on} and medium voltage. • For least TWR condition, MRR was almost near to its least value.	Satpathy et al.[28]

DoE and the output parameters in terms surface roughness, MRR, tool wear rate (TWR) were recorded and the following conclusions were drawn.

12.5 CONCLUSION

This chapter reviews the range of advancement made in the direction of machining of Al-MMCs with EDM and its variants. It was observed that PMEDM produces better surface finish along with higher MRR as compared to traditional methods. Hard ceramic content present within the material greatly affects the product quality. Use of hybrid variant in the EDM such as permanent magnets, abrasive powders, steams, and any additives improve the overall productivity and service life of the components. Selection of proper dielectric medium, input process parameters, effect of tooling sizes and shapes must be explored further. The machining of nonconducting materials and the presence of soft phase in the composite requires intensive research with EDM. The use of different liquid medium and gases during the machining process can be explored for the feasibility and improving the product quality.

KEYWORDS

- **Al-MMCs**
- **mechanical behavior**
- **nonconventional machining**
- **EDM**
- **surface morphology**

REFERENCES

1. Liew, P. J.; Yap, C. Y.; Wang, J.; Zhou, T. Surface Modification and Functionalization by Electrical Discharge Coating : A Comprehensive Review. *Int. J. Extrem. Manuf.* **2020**. https://doi.org/https://doi.org/10.1088/2631-7990/ab7332.
2. Talla, G.; Gangopadhayay, S. State of the Art in Powder-Mixed Electric Discharge Machining: A Review. *J. Eng. Manuf.* **2017,** *231*, 2511–2526. https://doi.org/10.1177/0954405416634265.
3. Sahu, A. K.; Mahapatra, S. S. Performance Analysis of Tool Electrode Prepared through Laser Sintering Process during Electrical Discharge Machining of Titanium. *Int. J. Adv. Manuf. Technol.* **2020,** 1017–1041.

4. Bains, P. S.; Sidhu, S. S.; Payal, H. S. Fabrication and Machining of Metal Matrix Composites : A Review Fabrication and Machining of Metal Matrix Composites: A Review. *Mater. Manuf. Process.* **2016,** *31*, 553–573. https://doi.org/10.1080/10426914.2015.1025976.
5. Taylor, P.; Kumar, S. S.; Uthayakumar, M.; Kumaran, S. T.; Parameswaran, P. Electrical Discharge Machining of Al (6351)– SiC – B 4 C Hybrid Composite Electrical Discharge Machining of Al (6351) – SiC – B 4 C Hybrid Composite. *Mater. Manuf. Process.* **2014,** 37–41. https://doi.org/10.1080/10426914.2014.952024.
6. Rouniyar, A. K.; Shandilya, P. Fabrication and Experimental Investigation of Magnetic Field Assisted Powder Mixed Electrical Discharge Machining on Machining of Aluminum 6061 Alloy. *J. Eng. Manuf.* **2019,** *233*, 2283–2291. https://doi.org/10.1177/0954405419838954.
7. Paswan, K.; Pramanik, A.; Chattopadhyaya, S.; Basak, A. K. A Novel Approach towards Sustainable Electrical Discharge Machining of Metal Matrix Composites (MMCs). *Int. J. Adv. Manuf. Technol.* **2019.**
8. Uthayakumar, M.; Babu, V. K.; Kumaran, S. T.; Kumar, S. S.; Jappes, J. T. W., D. R. T. P. A Study on the Machining of Al – SiC Functionally Graded Metal Matrix Composite Using Die – Sinking EDM. *Part Sci. Technol.* **2017,** 6351. https://doi.org/10.1080/02726351.2017.1346020.
9. Singh, T.; Dvivedi, A. Developments in Electrochemical Discharge Machining: A Review on Electrochemical Discharge Machining, Process Variants and Their Hybrid Methods. *Int. J. Mach. Tools Manuf.* **2016,** *105*, 1–13. https://doi.org/10.1016/j.ijmachtools.2016.03.004.
10. Taylor, P.; Sidhu, S. S.; Batish, A.; Kumar, S. Materials and Manufacturing Processes Study of Surface Properties in Particulate-Reinforced Metal Matrix Composites (MMCs) Using Powder-Mixed Electrical Discharge Machining (EDM) Study of Surface Properties in Particulate-Reinforced Metal Matrix Compo. *Mater. Manuf. Process.*, **2014,** 37–41. https://doi.org/10.1080/10426914.2013.852211.
11. Singh, S.; Yeh, M. F. Optimization of Abrasive Powder Mixed EDM of Aluminum Matrix Composites with Multiple Responses Using Gray Relational Analysis. *J. Mater. Eng. Perform.* **2012,** *21*, 481–91. https://doi.org/10.1007/s11665-011-9861-z.
12. Singh, H.; Jain, P. K.; Bhoi, N.; Pratap, S. Experimental Study Pertaining to Microwave Sintering (MWS) of Al-Metal Matrix Composite—A Review. *Mater. Sci. Forum* **2018,** *928*, 150–155. https://doi.org/10.4028/www.scientific.net/MSF.928.150.
13. Pattnaik, S. K.; Bhoi, N. K.; Padhi, S.; Sarangi, S. K. Dry Machining of Aluminum for Proper Selection of Cutting Tool: Tool Performance and Tool Wear. *Int. J. Adv. Manuf. Technol.* **2018,** *98*, 55–65. https://doi.org/10.1007/s00170-017-0307-0.
14. Bhoi, N. K.; Singh, H.; Pratap, S. Developments in the Aluminum Metal Matrix Composites Reinforced by Micro/Nano Particles—A Review. *J. Compos. Mater.* **2020,** *54*, 813–833. https://doi.org/10.1177/0021998319865307.
15. Bhoi, N. K.; Singh, H.; Pratap, S. Synthesis and Characterization of Zinc Oxide Reinforced Aluminum Metal Matrix Composite Produced by Microwave Sintering. *J Compos Mater*. **2020**. https://doi.org/10.1177/0021998320918646.
16. Bhoi, N. K.; Singh, H.; Pratap, S. Synthesis and Characterization of Alumina Nanoparticles: A Case Study. *J. Inst. Eng. India Ser. C* **2019,** *101*, 407–413. https://doi.org/10.1007/s40032-019-00542-w.
17. Dasari, B. L.; Morshed, M.; Nouri, J. M.; Brabazon, D.; Naher, S. Mechanical Properties of Graphene Oxide Reinforced Aluminium Matrix Composites. *Compos Part B Eng* **2018,** *145*, 136–144. https://doi.org/10.1016/j.compositesb.2018.03.022.

18. Paswan, K.; Pramanik, A.; Chattopadhyaya, S. Machining Performance of Inconel 718 Using Graphene Nanofluid in EDM. *Mater. Manuf. Process.* **2020,** *00*, 1–10. https://doi.org/10.1080/10426914.2020.1711924.
19. Dhupal, D.; Naik, S.; Das, S. R. Modelling and Optimization of Al–SiC MMC through EDM Process Using Copper and Brass Electrodes. *J. Mater. Today Proc.* **2017**. https://doi.org/10.1016/j.matpr.2018.02.096.
20. Suresh, S. K.; Uthayakumar, M.; Kumaran, S. T.; Varol, T.; Canakci, A. Investigating the Surface Integrity of Aluminium Based Composites Machined by EDM. *J Defence Tech.* **2019**. https://doi.org/10.1016/j.dt.2018.08.011.
21. Kandpal, B. C.; Kumar, J.; Singh, H. Optimisation of Process Parameters of Electrical Discharge Machining of Fabricated AA 6061/10% Al_2O_3 Aluminium Based Metal Matrix Composite. *J. Mater. Today: Proc.* **2018**. https://doi.org/10.1016/j.matpr.2017.12.009.
22. Mohanty, S.; Routara, B. C.; Nanda, B. K.; Das, D. K.; Sahoo, A. K. Study of Machining Characteristics of Al-SiCp12% Composite in Nano Powder Mixed Dielectric Electrical Discharge Machining Using RSM. *J. Mater. Today: Proc.* **2018**. https://doi.org/10.1016/j.matpr.2018.10.365.
23. Malhotra, P.; Tyagi, R. K.; Singh, N. K.; Sikarwar, B. S. Experimental Investigation and Effects of Process Parameters on EDM of Al7075/SiC Composite Reinforced with Magnesium Particles. *J. Mater. Today: Proc.* **2019**. https://doi.org/10.1016/j.matpr.2019.11.069.
24. Roy, C.; Syed, K. H.; Kuppan, P. Machinability of Al/10%SiC/2.5%TiB2 Metal Matrix Composite with Powder-Mixed Electrical Discharge Machining. *J. Procedia Tech.* **2016**. https://doi.org/10.1016/j.protcy.2016.08.206.
25. Mohanty, S.; Singh, S. S.; Routara, B. C.; Nanda, B. K.; Nayak, R. K. A Comparative Study on Machining of AlSiCp Metal Matrix Composite Using Electrical Discharge Machine with and without Nano Powder Suspension in Dielectric. *J. Mater. Today: Proc.* **2019**. https://doi.org/10.1016/j.matpr.2019.07.386.
26. Dubey, V.; Singh, B. Study of Material Removal Rate in Powder Mixed EDM of AA7075/B4C Composite. *J. Mater. Today: Proc.* **2018**. https://doi.org/10.1016/j.matpr.2017.11.418.
27. Alagarsamy, S. V.; Ravichandran, M.; Sakthivelu, S.; Kumar, S. D.; Chanakyan, C.; Meignanamoorthy, M. Optimization of Electric Discharge Machining Parameters on Surface Roughness for Al/ZrO2 Composite through Response Surface Methodology. *J. Mater. Today: Proc.* **2020**. https://doi.org/10.1016/j.matpr.2020.01.344.
28. Satpathy, A.; Tripathy, S.; Senapati, N. P.; Brahma, M. K. Optimization of EDM Process Parameters for AlSiC- 20% SiC Reinforced Metal Matrix Composite with Multi Response Using TOPSIS. *J. Mater. Today: Proc.* 2017. https://doi.org/10.1016/j.matpr.2017.02.187.
29. Joshi, A. Y.; Joshi, A. Y. A Systematic Review on Powder Mixed Electrical Discharge Machining. *Heliyon* **2019,** 5 (12). https://doi.org/10.1016/j.heliyon.2019.e02963
30. Lee, S. H.; Li, X. Study of the Surface Integrity of the Machined Workpiece in the EDM of Tungsten Carbide. *J. Mater. Process. Tech. 139* (1–3). https://doi.org/10.1016/S09240136(03)00547-8

CHAPTER 13

ROBUST COMPARISON OF Al/Mg ALLOYS BASED HYBRID METAL MATRIX COMPOSITES: A REVIEW

NIKHILESH SINGH*, R. M. BELOKAR, and R. S. WALIA

Production and Industrial Engineering Department, Punjab Engineering College (Deemed To Be University), Chandigarh 160012, India

**Corresponding author. E-mail: nikhileshsingh.phdprod@pec.edu.in*

ABSTRACT

In the present scenario, most of the natural resources are going to be vanished due to extensive overuse by mankind and their needs. To overcome the man-made challenges regarding resources or materials, most researchers all over the world are working on the materials, which are formed by mixing two or more materials that give superior properties to naturally obtained monolithic materials, called composite materials. Composite materials have their properties such as robustness, light in weight, economical, etc., which enables large-scale production in many esteemed manufacturing industries like automobiles and avionics sector, biomedical field, etc. The mixture of two or more reinforcements in the desired metal matrix enhances the mechanical as well as tribological properties of the desired material termed as hybrid metal matrix composites. This review article enlightens the current trends for the fabrication of aluminum/magnesium-based hybrid metal matrix composites by stir casting, that is, vortex technique incorporated with process parameters, material properties, associated technological problems, and some recent applications.

Modern Manufacturing Systems: Trends and Developments. Rajiv Kumar Garg, Ravi Pratap Singh, Rajeev Trehan, & Ramesh Singh (Eds.)

13.1 INTRODUCTION

In a modern era, there is a huge demand for advanced engineering materials that act as a multifunctional catalyst like a light in weight, economical as well as eco-friendly for the various application sectors such as automotive, avionics, biomedical, civil-construction, electronics, sports industries, etc., shown in Figure 13.1. To achieve this, various conventional metallic alloys have been replaced by metal matrix composites (MMCs) acts as the superior class for various manufacturing industries because of their excellent material properties.[1] Aluminum/magnesium-based metallic alloys are being widely used as the base matrix for MMCs because of low density and high strength-to-weight ratio make them highly potential alternatives for lightweight applications.[2] The various elements present in the composition of base alloys provide tremendous properties for the fabrication of HMMCs shown in Table 13.1. As per the typical classification of the matrix composites, such as metal matrix composites (MMCs), ceramics matrix composites (CMCs), and polymer matrix composites (PMCs), MMCs have capabilities to reinforced with other material, that is, ceramics, which have higher temperature capacity, strength, and thermal conductivity.[3,4] MMCs typically used Al/Mg base matrix along with various ceramics reinforcement to improve the mechanical well as tribological properties because the soft nature of Al/Mg alloys gets easily mixed with some carbides and oxides.[5] As we discussed, the various properties of Al-based MMCs showed higher stiffness and higher mechanical damping as compared to pure alloys.[6] On the other hand of Mg-based MMCs, widely used because of their attractive mechanical properties than their monolithic alloy, but Mg is highly reactive at high temperature, which is controlled by either using surface coatings or maintained an inert atmosphere to avoid oxidation phenomenon.[7] Although Al/Mg-based alloys have a low melting point (about 650°C–660°C), usually not recommended for long-time service above 180°C.

The processing route for Al/Mg MMCs can be classified as solid-state or powder metallurgy (spark plasma sintering, vacuum/gas sintering, microwave sintering) liquid state (infiltration method, stir casting, vacuum die casting), and tertiary state (compocasting, rheocasting, in-situ casting). Among all, the stir casting or vortex technique is considered as the simplest way, economical, and capable of mass production.[8] Some researchers have been also reported that the enhanced stir casting method has the capability for mass production in lightweight applications.[5,9] Solid-state processing route, that is, powder metallurgy needs a sufficient diffusion to ensure uniformity throughout the microstructure of HMMCs and their properties solely depend on the characteristics of the matrix and reinforcements used.[10,11] Most of the researchers under review report for the production of HMMCs, bottom pouring stir casting with electromagnetic and ultrasonic stirrer along with squeeze attachment is being recommended because of the significant reduction of porosity and uniform distribution of particles.[8,12] The schematic setup of a vacuum-assisted bottom pouring stir casting is depicted in Figure 13.2.

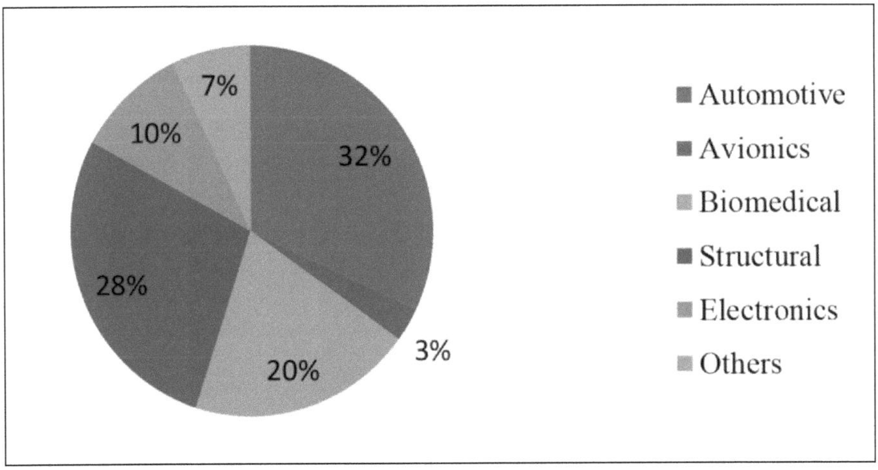

FIGURE 13.1 Applications of Al/Mg-based MMC in various sectors reviewed.

TABLE 13.1 Composition of Widely Used Al/Mg-Based Alloys for MMC's (in wt.%).

Alloy series		Si	Mn	Mg	Zn	Al
Al-based	6061	0.40–0.8	0.15	0.8–1.2	0.25	Balance
	7075	0.40	0.30	2.1–2.9	5.1–6.1	Balance
Mg-based	AZ91D	0.034	0.03	Balance	0.775	8.40

Properties of HMMCs fabricated by various processing routes can be altered by varying different process parameters—pouring temperature, preheat temperature of reinforcement, stirring speed, pressure holding time, reinforcement weight percentage, reinforcement size particle, size and position of the stirrer, etc. along with the addition of additives such as degasser, wetting agents that are also responsible for enhancing HMMCs on the qualitative basis.[13] According to core researchers suggested that types of reinforcements played a vital role in the mechanical as well as tribological properties of hybrid composites. Further classification of reinforcements can be categorized into two broad types, that is, organic (such as red mud, fly-ash, bagasse-ash, rice-husk, coir, eggshell, groundnut shell, etc.) and inorganic (such as SiC, B_4C, Al_2O_3, Gr, CNT, etc.).[14–17]

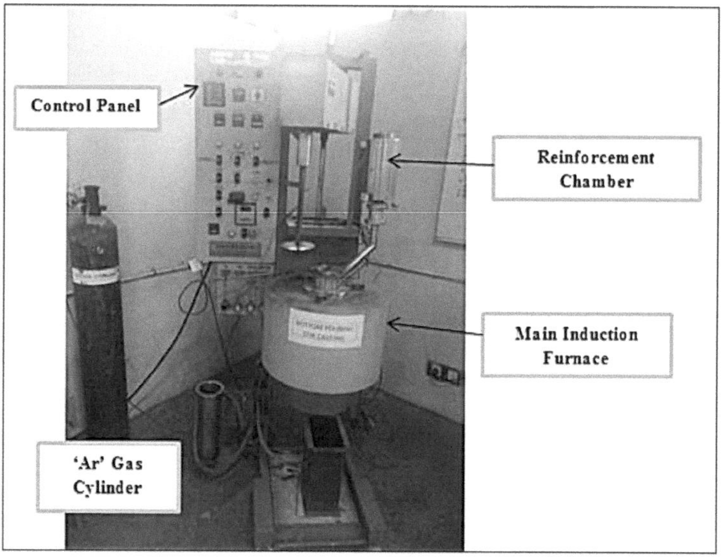

FIGURE 13.2 Schematic setup of vacuum-assisted bottom pouring stir casting.

Technical specifications and applications of some commonly used reinforcement are shown in Table 13.2. Salient features of used reinforcements with Al/Mg-based HMMCs discussed by some researchers, that is, silicon carbide (SiC) having high specific strength, stiffness, hardness, resistance to acids,[18,19] Boron Carbide (B_4C) showed better chemical stability, neutron absorption capability, high strength, and hardness,[17] Alumina or aluminum oxide (Al_2O_3) has a high strength to weight ratio, high hardness,[20] and Graphite (Gr) showed a high modulus of elasticity, high specific stiffness, and thermal conductivity properties being used by various automotive applications.[21] By using the appropriate proportion of particular reinforcement in Al/Mg-based hybrid composites can enhance their required mechanical and tribological properties accordingly.[22]

Martin et al. have reported for the fabrication of Al-Mg-Si/Al_2O_3/Rice husk ash (as organic reinforcement with 0–10%) via. the stir casting route resulted in decreased density, hardness, and tensile strength, on the other end, specific strength, % elongation, and fracture toughness increased.[24] Devaraju et al. studied Al6061-T6/SiC/Gr/Al_2O_3 MMC fabricated by friction

TABLE 13.2 Technical Specifications and Applications of Commonly Used Reinforcement.[23]

Reinforcement	Melting point, °C	Density, g/cm3	Tensile strength, MPa	Thermal conductivity, W/m K	Applications
Silicon Carbide (SiC)	2730	3.21	240–1625	3.8–20.7	Pistons, calipers, propeller shaft, connecting rod, brake rotors, driveshaft, engine
Boron Carbide (B_4C)	2763	2.52	261–569	17–42	Automobile Sectors- intake manifolds, chassis, suspension components
Alumina (Al_2O_3)	2072	3.95	69–665	12–38.5	Brake discs, pistons, cylinder heads, connecting rods of Engine
Graphite (Gr)	3600	2.26	4.8–76	8.7–114	Truss element, bus panel, waveguides

stir processing that decreased microhardness, increased wear resistance, provides excellent dimensional stability, reduced their properties.[25] Zhou et al. studied the different size and wt.% of SiC on Mg-based alloy, that is, AZ91, and results show that submicron is more favorable to grain refinement as compared to the micron of SiC due to this average grain size reduces and boundaries of submicron particles get dense, which affects interface bonding and dimples may occur in materials.[26] Sujan et al. investigated SiC and Al_2O_3 reinforcement on composite materials to provide a low coefficient of thermal expansion, high tensile strength, high impact stress, hardness, and lower wear rate.[27] Park et al. fabricated 6061 Al–Mg–Si alloy reinforced with Al_2O_3 (5–30 wt.%), found that composites prepared via the powder metallurgy route having more fatigue strength than prepared by the liquid metallurgy process.[28] Yilmaz et al. studied the abrasive wear resistance properties on Al alloys with varying particle size and volume fraction of Al_2O_3, shows the larger size of Al_2O_3 particles is more effective for wear resistance.[29] Thirugnanasambandham et al. reported that there is a 28% increase in wear resistance of MMCs as compared to Mg base alloys during the investigation of wear characteristics of Mg-based MMCs reinforced with Al_2O_3 by applying loads (10–20 N) with varying sliding speed (200–800 rpm).[30] Hashim et al. studied A359 by using SiC and Mg as a wetting agent to improve the wettability phenomenon.[31]

Veeresh et al. reported Al6061/SiC and Al7075/Al_2O_3 (2–6%) via. liquid metallurgy enhanced tensile, wear properties, increased hardness along with decreasing density. In Al6061/SiC composites—thermal conductivity, thermal capacity, and thermal expansivity decreased.[32] Aattisugan et al. experimentally investigated the microstructural and mechanical properties of base alloy AZ91D, AZ91D/B_4C, and AZ91D/B_4C/Gr hybrid composites fabricated via. stir casting showed density increases with the addition of B_4C in AZ91D and further slightly increases with the embedment of graphite as AZ91D/B_4C/Gr hybrid composite. The other outcomes of AZ91D/B_4C composite and AZ91D/B_4C/Gr hybrid composite showed superior hardness and ultimate tensile strength (UTS) than base alloy AZ91D magnesium, while hardness and UTS decrease for hybrid composite for AZ91D/B_4C composites. Uniformity distribution of B_4C/graphite particles throughout the AZ91D matrix phase with a lack of cracks has been found.[33] Reddy et al. suggested a stir casting route on Al7075/fly-ash/E-glass short fibers, increased tensile strength, hardness, and wear resistance.[34] Zhou et al. proposed a novel two-step mixing method with gravity casting applied on Al536 and Al6061 reinforced with 10–20% wt.% or volume fraction of SiC particles to improve the wettability and uniformity of the particle throughout the matrix, the author found that SiC particles are in the interdendrite regions and also proposed a thermal lag model to explain the concentration of particles.[35] Lloyd et al. investigated Al7075/B_4C particles through stir casting processing route showed increased hardness and increased wear resistance, on another composition of Al7075/TiB_2 showed increased microhardness, tensile strength as well as yield strength via the same processing route.[36] Keshavamurthy et al. fabricated Al7075/ TiB_2 in-situ composite via. stir casting revealed that microhardness, yield strength, and ultimate tensile strength of Al7075/ TiB_2 were higher than an unreinforced alloy of Al7075.[37] Habibi et al. reported the enhancement of strength and ductility of magnesium by incorporating Al/CNT hybrid reinforcement through the powder metallurgy route (with microwave sintering followed by hot extrusion). In hybrid nano, Mg/Al-CNT composites content of CNT was kept constant at about 0.18 wt.% and Al content was varied at the range of 0.5, 1, and 1.5 wt.%. Results revealed that mechanical properties such as elastic modulus, yield strength, ultimate tensile strength, and failure strain were improved by 3.6%, 38%, 36%, and 42%, respectively, than unreinforced magnesium alloy.[38]

Wei et al. investigated the effect of sliding velocity on the wear behavior of bare Mg, Mg/SiC, and Mg/SiC/MWCNTs HMMCs, via. Powder metallurgy observed that hybridizing thermal stable and high modulus SiC and MWCNTs in Mg revealed improved hardness and wear resistance under high load conditions. Also, Hybrid composites disclosed a different dominating mechanism at varying sliding velocity (i.e., abrasion and delamination at low sliding velocity and adhesion under higher velocity).[39] Li et al. examined the microstructural, mechanical, and fractural behavior of AZ91D, AZ91D/$Mg_2B_2O_5$W, and AZ91D/$Mg_2B_2O_5$W/B_4C HMMCs fabricated by the squeeze casting technique, results revealed that reinforcements Magnesium borate whisker ($Mg_2B_2O_5$W) and Boron carbide (B_4C) improved the flexural properties of hybrid composite by 29% as compared to base alloy composite and B_4C was the major contributor in the improvement phenomenon. This improvement may contribute to higher dislocation density in the matrix caused by the coefficient of linear thermal expansion (CTE) and elastic modulus difference between reinforcement and matrix. The microstructure and fracture study disclosed a random and isotropic distribution of reinforcement signifies the strong interfacial bonding between matrix and reinforcement.[40] Reinforcements such as silicon carbide, alumina, boron carbide, carbon nanotubes, graphite, etc. were used in Al/Mg-based HMMCs under this review report as shown in Figure 13.3, whereas the rigorous comparative study of Al/Mg alloys based metal matrix composites as elaborated in Table 13.3.

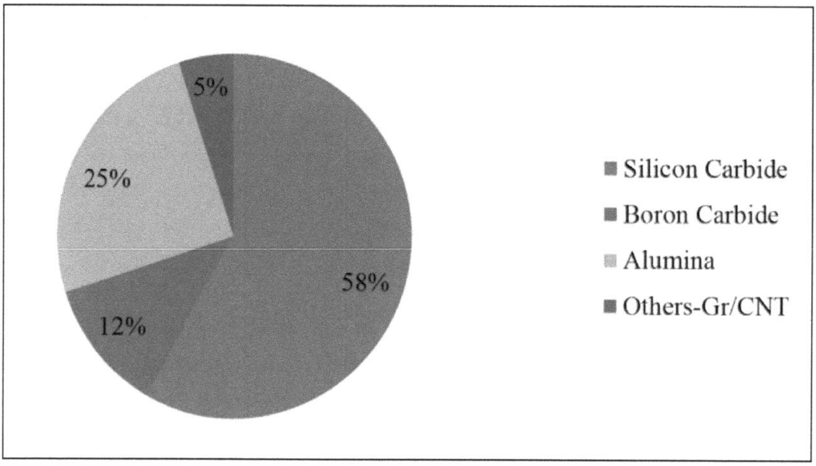

FIGURE 13.3 Reinforcements used in Al/Mg-based HMMCs reviewed.

13.2 CONCLUSIONS

From the above study, discussions about technological problems and conclusions can be obtained as follows:

1. Uniformity in reinforcement distribution with different metal matrix played a significant issue and severely affects the mechanical as well as tribological properties of the fabricated hybrid MMCs.
2. Wettability influences the interfacial bonding between the matrix and reinforcement of hybrid MMCs. 'Mg' is considered the most widely used wetting agent in the range of 1.5–2% to improve wettability.
3. Porosity (caused due to either gas formation or solidification shrinkage) is found to be a huge problem that directly weakens the casting. This problem can control by either making a protected mold design by introducing dry N_2 gas.
4. To avoid agglomeration during the stirring process, the coating of reinforcement and inert gas environment with argon gas may provide better results.
5. The optimum range of reinforcement of SiC/Al_2O_3/B_4C/Gr in Al/Mg-based HMMCs can enhance the composite material properties as per their required applications.

TABLE 13.3 Comparative Analysis of Al/Mg Alloys Based Metal Matrix Composites.

Matrix with reinforcement	Processing route	wt.%	Hardness	Strength, MPa	Wear loss	References
A356/Al_2O_3	SC	2.5	96 HR	182 (U)	–	[41]
AA2024/Al_2O_3	SC	1	–	215 (T)	–	[42]
AA6061/Al_2O_3	SC	20	38 BHN	–	–	[43]
AA7075/SiC	SC	20	50 HB	–	–	
AA6061/Al_2O_3	SC	12	–	193.47 (T)	–	[44]
AA2024/Al_2O_3	SC	10, 20 and 30	135 BHN	112 (U)	–	[45]
AA7075/Al_2O_3	SC	1.2	160 HV	400 (U)	–	[46]
AA6061/Al_2O_3	SC (with Ultrasonic stirring)	3.5	74.25 HRB	254.72 (T)	–	[47]
AA6063/SiC	SC	12	–	158.5 (T)	–	[48]
A356/SiC	SC (with electromagnetic stir)	15	104.66 BHN	309.83 (T)	–	[49]
AA1100/SiC	SC	12	–	181.2 (T)	–	[50]
AA6061/SiC	SC (with squeeze attachment)	30	84 HB	200 (U)	–	[51]
AA6061/SiC	SC	6	98 HV	270 (U)	–	[52]
Al base alloy/SiC	SC	10	67 HB	205 (U)	–	[53]
AA7075/SiC	SC	6	118 HB	269 (U)	–	[54]
AA6061/B_4C	SC	12	80.8 HV	215 (T)	–	[55]
Al base alloy/B_4C	SC	2	–	197 (T)	–	[56]
A356/B_4C	SC	10	74 BHN	265 (U)	–	[57]
AA6061/B_4C	SC	15	80 VHN	260 (U)	–	[58]
AA6061/B_4C	SC	15	97 VHN	270 (U)	–	[59]
AA7075/ B_4C	SC	20	210 BHN	305 (U)	–	[60]
Mg alloy-AZ91D	–	–	20.5 BHN	62.1 (T)	0.017 g	[61]
AZ91D/B_4C	SC	1.5	27.1 BHN	99.2 (T)	0.014 g	

TABLE 13.3 *(Continued)*

Matrix with reinforcement	Processing route	wt.%	Hardness	Strength, MPa	Wear loss	References
AZ91D/B_4C/Gr	SC	1.5, 1.5	22.5 BHN	86.3 (T)	0.013 g	
Mg alloy- AZ91	–	–	–	200 (T)	–	[62]
AZ91/Gr	SC	5	–	250 (T)	–	
AZ91/Gr	SC	10	–	240 (T)	–	
AZ91/Gr	SC	15	–	200 (T)	–	
AZ91/Gr	SC	20	–	175 (T)	–	
Mg/GNF/Al_2O_3	IM	10 GNF	142 HV	240 (T)	–	[63]
Mg/GNF/Al_2O_3	IM	15 GNF	136 HV	238 (T)	–	
Mg/GNF/Al_2O_3	IM	20 GNF	120 HV	233 (T)	–	
Mg/GNF/Al_2O_3	IM	10 Al_2O_3	130 HV	236 (T)	–	
Mg/GNF/Al_2O_3	IM	15 Al_2O_3	135 HV	237 (T)	–	
Mg/GNF/Al_2O_3	IM	20 Al_2O_3	140 HV	239 (T)	–	
AZ91/SiC/Gr	SC	1,1	62.8 HV	–	0.0045 mm^3/km	[64]
AZ91/SiC/Gr	SC	2,2	64.3 HV	–	0.0045 mm^3/km	
AZ91/SiC/Gr	SC	3,3	66.2 HV	–	0.0045 mm^3/km	

Abbreviations: AA, aluminum alloy; Mg, magnesium; AZ91, magnesium-based alloy; HR, Rockwell hardness; HB, Brinell hardness; HRB, Rockwell hardness B scale; HV, Vickers hardness; BHN, Brinell hardness number; VHN, Vickers hardness number; U, ultimate strength; T, tensile strength; SC, stir casting; GNF, graphite nanofibers; IM, infiltration method.

6. Vacuum-Assisted Bottom Pouring Stir Casting with electromagnetic or ultrasonic stirring with the proper squeeze attachment played a tremendous role in producing good quality of Al/Mg-based HMMCs as compared to other processing routes.

KEYWORDS

- **Al/Mg alloys HMMC's**
- **reinforcements**
- **stir casting**

REFERENCES

1. Rohatgi, P. K. In *Tribology for Scientists and Engineers: From Basics to Advanced Concepts*; Khorshid, M. T., Omrani, E., Lovell, M. R., Menezes, P. L., Eds.; Springer Science: London, 2013; p 233.
2. Campbell, F. C. *Lightweight Materials—Understanding the Basics*; ASM International: Ohio, 2012.
3. Sheshan, S.; Guruprasad, A.; Parbha, M.; Sudhakar, A. Fibre-reinforced Metal Matrix Composites—A Review. *J. Indian Inst. Sci.* **1996,** *76*, 1–14.
4. Rosso, M. *12th International Scientific Conference on Achievements in Mechanical & Materials Engineering*; Gliwice, Poland, 2003.
5. Jeevan, V.; Rao, C. S. P.; Selvaraj, N. Development of a New Aluminum Matrix Composite Reinforced with Silicon Carbide (SiC). *Int. J. Mech. Eng. Robot. Res.* **2013,** *1* (2), 34–38.
6. Rojas, J. I.; Siva, B. V.; Sahoo, K. L.; Crespo, D. Viscoelastic Behavior of a Novel Aluminum Metal Matrix Composite and Comparison with Pure Aluminum, Aluminum Alloys, and a Composite Made of Al–mg–Si Alloy Reinforced with SiC Particles. *J. Alloys Compounds* **2018,** *744*, 445–452.
7. Jayakumar, J. Recent Development and Challenges in Synthesis of Magnesium Matrix Nano Composites—A Review. *Int. J. Latest Res. Sci. Technol.* **2012,** *1*, 164–171.
8. Arunachalam, R.; Krishnan, P. K.; Muraliraja, R. A Review on the Production of Metal Matrix Composites through Stir Casting—Furnace Design, Properties, Challenges, and Research Opportunities. *J. Manuf. Proces.* **2019,** *42*, 213–245.
9. Gopalakrishnan, S.; Murugan, N. Production and Wear Characterisation of AA 6061 Matrix Titanium Carbide Particulate Reinforced Composite by Enhanced Stir Casting Method. *Compos. Part B: Eng.* **2012,** *43* (2), 302–308.
10. Moosa, I. S. *Int. J. Res. Develop.* **2014,** *2*, 18.
11. Tsutsui, T. Recent Technology of Powder Metallurgy and Applications. Hitachi Chemical Technical Report, 12, 2012.

12. Singh, N.; Belokar, R. M. Tribological Behavior of Aluminum and Magnesium-Based Hybrid Metal Matrix Composites: A State-of-art review. *Mater. Today: Proceedings*, **2020**.
13. Moona, G.; Walia, R. S.; Rastogi, V.; Sharma, R. Aluminium Metal Matrix Composites: A Retrospective Investigation. *Indian J. Pure Appl. Phys. (IJPAP)* **2018,** *56* (2), 164–175.
14. Prabu, S. B.; Karunamoorthy, L.; Kathiresan, S.; Mohan, B. Influence of Stirring Speed and Stirring Time on Distribution of Particles in Cast Metal Matrix Composite. *J. Mater. Process. Technol.* **2006,** *171* (2), 268–273.
15. Prasad, D. S.; Krishna, R. A. Production and Mechanical Properties of A356. 2/RHA Composites. *Int. J. Adv. Sci. Technol.* **2011,** *33*, 51–58.
16. Alaneme, K. K.; Bodunrin, M. O.; Awe, A. A. Microstructure, Mechanical and Fracture Properties of Groundnut Shell Ash and Silicon Carbide Dispersion Strengthened Aluminium Matrix Composites. *J. King Saud Univ. Eng. Sci.* **2018,** *30*, 96–103.
17. Singh, N. Influence of Stir Casting Technique on HMMCs- A Literature Survey. *J. Emerg. Technol. Innov. Res.* **2020,** *7* (6), 1636–1642.
18. Reddy, P. S.; Kesavan, R.; Ramnath, B. V. Investigation of Mechanical Properties of Aluminium 6061-silicon Carbide, Boron Carbide Metal Matrix Composite. *Silicon* **2018,** *10* (2), 495–502.
19. Prasad, S. V.; Asthana, R. Aluminium Metal-Matrix Composites for Automotive Applications: Tribological Considerations. *Tribol. Lett.* **2004,** *17*, 445–453.
20. Sajjadi, S. A.; Ezatpour, H. R.; Parizi, M. T. Comparison of Microstructure and Mechanical Properties of A356 Aluminum Alloy/Al_2O_3 Composites Fabricated by Stir and Compo-Casting Processes. *Mater. Des.* **2012,** *34*, 106–111.
21. Nturanabo, F.; Masu, L.; Kirabira, J. B. Novel Applications of Aluminium Metal Matrix Composites, Open Access Peer- Reviewed Chapter- Aluminium Alloys and Composites, 2019.
22. Kumar, N.; Soren, S. Selection of Reinforcement for Al/Mg Alloy Metal Matrix Composites. *Mater. Today: Proceedings* **2019,** 2214–7853.
23. Singh, N.; Belokar, R. M.; Walia, R. S. A Critical Review on Advanced Reinforcements and Base Materials on Hybrid Metal Matrix Composites. *Silicon* **2020,** 1–24.
24. Martin, B.; Cecilia, P.; Frantisek, S. *J. Alloys Comp.* **2011,** *509S*, 235.
25. Devaraju, A.; Kumar, A.; Kotiveerachari, B. Influence of Addition of Gr_p/ Al_2O_{3p} with SiC_p on Wear Properties of Aluminium Alloy 6061-T6 Hybrid Composites via Friction Stir Processing. *Trans. Nonferrous Metals Soc. China* **2013,** *23*, 1275.
26. Shan, S. Z.; Deng, K. K.; Li, J. C.; Shang, S. J.; Liang, W.; Fan, J. F. Effects of Volume Ratio on the Microstructure and Mechanical Properties of Particle Reinforced Magnesium Matrix Composite. *Mater. Des.* **2014,** *63*, 672–677.
27. Sujan, D.; Oo, Z.; Rahman, M. E.; Maleque, M. A.; Tan, C. K. Physio-mechanical Properties of Aluminium Metal Matrix Composites Reinforced with Al_2O_3 and SiC. *Int. J. Mater. Metal. Eng.* **2012,** *6* (8), 678–681.
28. Park, B. G.; Crosky, A. G.; Hellier, A. K. Fracture Toughness of Microsphere Al_2O_3– Al Particulate Metal Matrix Composites. *Compos. Part B: Eng.* **2008,** *39*, 1270–1279.
29. Yilmaz, O.; Buytoz, S. Abrasive Wear of Al_2O_3 Reinforced Aluminium Based MMCs. *Compos. Sci. Technol.* **2001,** *61*, 2381–2392.
30. Thirugnanasambandham, T.; Chandradass, J.; Sethupathi, P. B.; Martin, M. L. J. Experimental Study of Wear Characteristics of Al_2O_3 Reinforced Magnesium Based Metal Matrix Composites. *Mater. Today: Proc.* **2019,** *14*, 211–218.

31. Hashim, J.; Looney, L.; Hashmi, M. S. J. The Enhancement of Wettability of SiC Particles in Cast Aluminium Matrix Composites. *J. Mater. Process. Technol.* **2001,** *119* (1–3), 329–335.
32. Veeresh, K. G. B.; Rao, C. S. P.; Selvaraj, N.; Bhagyashekar, M. S. Studies on Al6061-SiC and Al7075-Al_2O_3 Metal Matrix Composites. *J. Minerals Mater. Characteriz. Eng.* **2010,** *9*, 43.
33. Aatthisugan, I.; Rose, A. R.; Jebadurai, D. S. Mechanical and Wear Behaviour of AZ91D Magnesium Matrix Hybrid Composite Reinforced with Boron Carbide and Graphite. *J. Magnes. Alloys* **2017,** *5* (1), 20–25.
34. Reddy, M. S.; Chetty, S. V.; Premkumar, S.; Reddappa, H. N. Influence of Reinforcements and Heat Treatment on Mechanical and Wear Properties of Al 7075 Based Hybrid Composites. *Procedia Mater. Sci.* **2014,** *5*, 508–516.
35. Liu, G. W.; Muolo, M. L.; Valenza, F.; Passeronea, A. Survey on Wetting of SiC by Molten Metals. *Ceram. Int.* **2010,** *36* (4), 1177–1188.
36. Bhushan, R. K.; Kumar, S.; Das, S. Fabrication and Characterization of 7075 Al Alloy Reinforced with SiC Particulates. *Int. J. Adv. Manuf. Technol.* **2013,** *65*, 611–624.
37. Keshavamurthy, R.; Mageri, S.; Raj, G.; Naveenkumar, B.; Kadakol, P. M.; Vasu, K. Microstructure and Mechanical Properties of Al7075-TiB2 in-situ Composite. *Res. J. Mater. Sci.* **2013,** *2320*, 6055.
38. Habibi, M. K.; Paramsothy, M.; Hamouda, A. M. S.; Gupta, M. Using Integrated Hybrid (Al+CNT) Reinforcement to Simultaneously Enhance Strength and Ductility of Magnesium. *Compos. Sci. Technol.* **2011,** *71*, 734–741.
39. Wei, T. Z.; Shamsuri, S. R. B.; Yee, C. S.; Rashid, M. W.; Ahsan, Q. Effect of Sliding Velocity on Wear Behavior of Magnesium Composite Reinforced with SiC and MWCNT. *Procedia Eng.* **2013,** *68*, 703–709.
40. Li, J.; Wang, F.; Weng, W.; Zhang, Y.; Wang, M.; Wang, H. Characteristic and Mechanical Properties of Magnesium Matrix Composites Reinforced with $Mg_2B_2O_5$w and B4Cp. *Mater. Des.* **2012,** *37*, 533–536.
41. Kongshaug, D. R.; Ferguson, J. B.; Schultz, B. F.; Rohatgi, P. K. Reactive Stir Mixing of Al- Mg/Al_2O_3np Metal Matrix Nanocomposites: Effects of Mg and Reinforcement Concentration and Method of Reinforcement Incorporation. *J. Mater. Sci.* **2014,** *49*, 2106–2116.
42. Su, H.; Gao, W.; Feng, Z.; Lu, Z. Processing, Microstructure and Tensile Properties of Nano-Sized Al_2O_3 Particle Reinforced Aluminum Matrix Composites. *Mater. Des.* **2012,** *36*, 590–596.
43. Lakshmipathy, J.; Kulendran, B. Reciprocating Wear Behavior of 7075Al/SiC in Comparison with 6061Al/Al_2O_3 Composites. *Int. J. Refractory Metals Hard Mater.* **2014,** *46*, 137–144.
44. Bharath, V.; Nagaral, M.; Auradi, V.; Kori, S. A. Preparation of 6061Al-Al_2O_3 MMC's by Stir Casting and Evaluation of Mechanical and Wear Properties. *Procedia Mater. Sci.* **2014,** *6*, 1658–1667.
45. Kok, M. Production and Mechanical Properties of Al_2O_3 Particle-Reinforced 2024 Aluminium Alloy Composites. *J. Mater. Process. Technol.* **2005,** *161*, 381–387.
46. Ezatpour, H. R.; Torabi Parizi, M.; Sajjadi, S. A.; Ebrahimi, G. R.; Chaichi, A. Microstructure, Mechanical Analysis and Optimal Selection of 7075 Aluminium Alloy Based Composite Reinforced with Alumina Nanoparticles. *Mater. Chem. Phys.* **2016,** *178*, 119–127.
47. Kumar Kohli, D. K.; Agnihotri, G.; Purohit, R. *Mater. Today: Proc.* **2015,** *2*, 3017–3026.

48. Alaneme, K. K.; Akulo, A. O. *Scientia Tractions A: Civil Eng.* **2012,** *19* (4), 992–996.
49. Dwivedi, S. P.; Sharma, S.; Mishra, R. K. Microstructure and Mechanical Properties of A356/SiC Composites Fabricated by Electromagnetic Stir Casting. *Procedia Mater. Sci.* **2014,** 1524–1532.
50. Mohanakumara, K. C.; Rajashekar, H.; Ghanaraja, S.; Ajitprasad, S. L. Development and Mechanical Properties of SiC Reinforced Cast and Extruded Al Based Metal Matrix Composite. *Procedia Mater. Sci.* **2014,** *5*, 934–943.
51. Reihani, S. S. Processing of Squeeze Cast Al6061–30vol% SiC Composites and Their Characterization. *Mater. Des.* **2006,** *27* (3), 216–222.
52. Kumar, G. V.; Rao, C. S. P.; Selvaraj, N.; Bhagyashekar, M. S. Studies on Al6061-SiC and Al7075-Al_2O_3 Metal Matrix Composites. *J. Minerals Mater. Characteriz. Eng.* **2010,** *9* (1), 43–55.
53. Sozhamannan, G. G.; Prabu, S. B. Effect of Processing Parameters on Metal Matrix Composites: Stir Casting Process. *J. Surface Eng. Mater. Adv. Technol.* **2012,** *2*, 11–15.
54. Kumar, G. B. V.; Rao, C. S. P.; Selvaraj, N. Mechanical and Dry Sliding Wear Behavior of Al7075 Alloy-Reinforced with SiC Particles. *J. Compos. Mater*. **2012,** *46*, 1201–1209.
55. Kalaiselvan, K.; Murugan, N.; Parameswaran, S. Production and Characterization of AA6061–B4C Stir Cast Composite. *Mater. Des.* **2011,** *32* (7), 4004–4009.
56. Alizadeh, A.; Taheri-Nassaj, E.; Hajizamani, M. Hot Extrusion Process Effect on Mechanical Behavior of Stir Cast Al Based Composites Reinforced with Mechanically Milled B4C Nanoparticles. *J. Mater. Sci. Technol.* **2011,** *27* (12), 1113–1119.
57. Mazahery, A.; Shabani, M. O.; Rahimipour, M. R.; Tofigh, A. A.; Razavi, M. Effect of Coated B4C Reinforcement on Mechanical Properties of Squeeze Cast A356 Composites. *Kovove Materialy* **2012,** *50* (2), 107–113.
58. Shabani, M. O.; Mazahery, A. Good Bonding between Coated B_4C Particles and Aluminum Matrix Fabricated by Semisolid Techniques. *Russian J. Non-Ferrous Metals* **2013,** *54* (2), 154–160.
59. Taylor, P.; Mazahery, A.; Shabani, M. O. Existence of Good Bonding between Coated B_4C Reinforcement and Al Matrix via Semisolid Techniques: Enhancement of Wear Resistance and Mechanical Properties. *Tribol. Trans.* **2013,** *3*, 37–41.
60. Mazahery, A.; Shabani, M. O. Sol–gel Coated B_4C Particles Reinforced 2024 Al Matrix Composites. *Proc. Inst. Mech. Eng. Part L: J. Mater.: Des. Appl.* **2012,** *226* (2), 159–169.
61. Aatthisugan, I.; Rose, A. R.; Jebadurai, D. S. Mechanical and Wear Behaviour of AZ91D Magnesium Matrix Hybrid Composite Reinforced with Boron Carbide and Graphite. *J. Magnes. Alloys* **2017,** *5* (1), 20–25.
62. Akhlaghi, F.; Mahdavi, S. Effect of the SiC Content on the Tribological Properties of Hybrid Al/Gr/SiC Composites Processed by in situ Powder Metallurgy (IPM) Method. *Adv. Mater. Res.* **2011,** *264*, 1878–1886.
63. Babu, J. S. S.; Nair, K. P.; Unnikrishnan, G.; Kang, C. G.; Kim, H. H. Fabrication and Properties of Magnesium (AM50)-based Hybrid Composites with Graphite Nanofiber and Alumina Short Fiber. J. Compos. Mater. **2010,** *44* (8), 971–987.
64. Girish, B. M.; Satish, B. M.; Sarapure, S. Optimization of Wear Behavior of Magnesium Alloy AZ91 Hybrid Composites Using Taguchi Experimental Design. *Metal. Mater. Trans. A* **2016,** *47* (6), 3193–3200.

CHAPTER 14

INVESTIGATION ON FIBER-REINFORCED CEMENTITIOUS COMPOSITE CONTAINING WASTE PRODUCTS

HARSH YADAV*, ANKESH KUMAR, PAWAN KUMAR, MANINDER SINGH, and BABITA SAINI

Department of Civil Engineering, National Institute of Technology Kurukshetra, India

Corresponding author. E-mail: harshuy062@gmail.com

ABSTRACT

The increase in the cost of cement-based construction material is basically due to use of cement and steel in construction work. Manufacturing process of cement and steel causes a lot of environmental degradation. Production of Portland cement and steel has a lot of negative impact on environment. Therefore, to reduce the degradation of environment and cost of work, waste products can be used in cementitious materials. During the past several decades, concrete technology has undergone rapid development, and fiber-reinforced cementitious composite (FRCC) is one of those newly developed cement-based material. In the present research, investigation thermal (fly ash) waste was used as a supplementary cementitious material. The coconut fibers have been used to reinforce the FRCC, which are generally thrown as waste. Combined use of both waste products (thermal and coconut) make the cement-based structural component economical and eco-friendly to a great extent. To evaluate the performance of FRCC the flexural and compressive

Modern Manufacturing Systems: Trends and Developments. Rajiv Kumar Garg, Ravi Pratap Singh, Rajeev Trehan, & Ramesh Singh (Eds.)

properties were analyzed and compared with normal concrete. Laboratory results showed that the behavior of FRCC was better as compared to normal concrete.

14.1 INTRODUCTION

Construction industry has a great contribution in the development of any nation. But, it consumes a huge part of natural resources to complete the demand of cementitious products. The processing of cementitious constituents and industrial products generated a big amount of waste products (i.e., solid and slurry waste)[1,2] Therefore, the surrounding environment of industries and construction sites are degraded through the direct disposal of these wastes and some other activities. Industries, cement plants, and construction sites create nuisance, emit large amount of harmful gases into the environment and generate various types of wastes like fly ash, kiln dust, stone dust, slag, hyposludge, jarosite, etc.[3–6] Thermal power plants, steel, and iron industries are the leading sources to generate solid wastes into the open land environment.[7] Therefore, to reduce the degradation of environment and cost of work, addition of these waste products into the cementitious products is the best alternative solution.

Nowadays to complete the demand of shelter, the cement-based products are the most used material due to its durable nature.[8] In the last few decades, the rapid advances have been made in the development of cement matrix with the introduction of various types of cementitious products like high-performance concrete (HPC), reactive powder concrete (RPC), fiber-reinforced concrete (FRC), engineered cementitious composite (ECC), fiber-reinforced cementitious composite (FRCC), etc.[9] FRCC is a new category of cementitious product in which short fibers are randomly distributed into the matrix.[10] Numerous researchers reported that the overall performance of the FRCC is better than conventional concrete.[11–14] Various types of short-length fibers such as steel, polypropylene, polyethylene, polyvinyl alcohol, carbon, etc. have been used to reinforce the FRCC.[15] Addition of these fibers suppress the formation of wider cracks and enhance the performance of the cementitious products.[16–18] The usage of these fibers increases the overall cost and limit the use of FRCC in place of normal concrete.[19,20] The cost of the FRCC needs to be reduced for its widespread use in construction work. Therefore, in the present study with the supplementary cementitious material (SCM) coconut fibers were used to prepare the FRCC. Coconut fibers are generally thrown as a waste, which are easily available in India as well as other countries. Therefore, the usage of coconut fibers and SCM helps in reducing the cost of matrix, save the natural resources, and contribute in making the environment ecofriendly.

14.2 EXPERIMENTAL PROGRAM

14.2.1 CONSTITUENTS AND MIX PROPORTIONS

To prepare the normal concrete and FRCC, the solid constituents selected during this study were Portland cement (PC), class F-type fly ash, river sand (passing from 1.18 mm IS Sieve), fine aggregates (FA), and coarse aggregates (CA). The physical properties of PC and aggregates have been provided in Tables 14.1 and 14.2. The coconut fibers were used to reinforce the FRCC, which are taken from coconut thrown waste. To achieve the stirred state, locally available freshwater was used.

TABLE 14.1 Physical Properties of Cement (PC).

Sr. no.	Physical properties	Experimental value	IS values
1.	Consistency	28	26–33%
2.	Initial setting time	30 min	Not less than 30 min
3.	Final setting time	480 min	Not more than 600 min
4.	Fineness of cement	2%	10% Max

TABLE 14.2 Physical Properties of Aggregates.

Sr. no.	Physical properties	Value
1.	Specific gravity	3.01
2.	Fineness modulus of fine aggregate	3.89
3.	Fineness modulus of coarse aggregate	7.08

14.2.2 MIX PROPORTIONS

After the selection and procurement of constituents, the compositions of materials for the design of normal concrete (M-25 grade) and FRCC have been decided. The details of the mix proportions for normal concrete (NC) and FRCC have been provided in Tables 14.3 and 144 respectively.

TABLE 14.3 Mix Proportions of Normal Concrete.

Mix Id	PC	FA	CA	w/c ratio
Normal concrete (NC)	1	1	2	0.35

TABLE 14.4 Mix Proportions of Different Mix of FRCC.

Mix Id	PC	Fly ash	River sand	w/c	Coconut fiber (%)
FRCC 2	1.0	1.84	1.2	0.35	2
FRCC 3	1.0	1.84	1.2	0.35	3
FRCC 4	1.0	1.84	1.2	0.35	4
FRCC 5	1.0	1.84	1.2	0.35	5

14.2.3 MIXING TECHNIQUE AND SPECIMEN CASTING

The mix of NC was prepared with the help of concrete mixer by a simple method. However, the mix of FRCC has been prepared in a power-driven mortar mixer. To prepare the FRCC firstly, the determined amount of river sand, PC, and fly ash were added into the drum of mixer and rotated for 2–3 min. After this, 2/3rd part of total water content was slowly poured into the mixer and rotated for another 2–3 min. After this stage, the remaining part of water content was poured into the mixed constituents and rotated for 3 min. At last, a determined amount of coconut fibers was added into the mortar mixer and uniformly mixed with the mortar matrix to form homogenous mix.

After the completion of mixing process, the prepared cementitious composite was poured into the molds of beam (500 × 100 × 100 mm) and cube (70.6 × 70.6 ×70.6 mm) and put on vibrator for proper compaction. After compaction, the cube and beam specimens as shown in Figures 14.1 and 14.2 were kept for 48 h at room temperature. After demolding from molds, the specimens were cured for 28 and 56 days. Testing was done after 28 and 56 days of water curing.

14.3 TEST METHODOLOGIES

To evaluate and compare the performance of NC and FRCC, the compression and flexural testing was done on prepared specimens. Fully automatic compression testing machine (Figure 14.3) was used to record the compressive strength and testing was done as per IS 516:1959 and BS-EN-12390 part 3 specifications. Two-point loading technique was used to find the failure load and corresponding flexural strength was recorded as shown in Figure 14.4 and tested as per IS 516:1959 and BS-EN-12390 part 5 specifications.

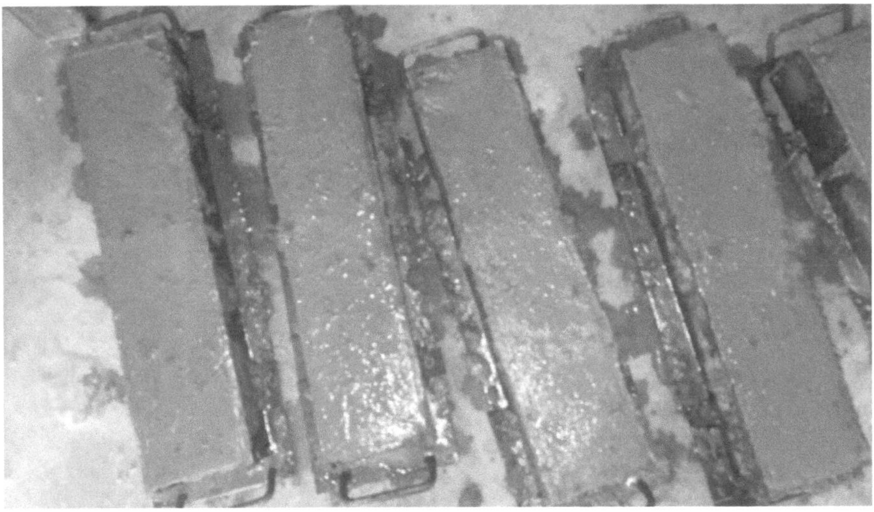

FIGURE 14.1 Beams after casting and compaction.

FIGURE 14.2 Cubes after casting and compaction.

14.4 RESULTS AND DISCUSSION

14.4.1 COMPRESSIVE STRENGTH

The compression behavior of various mixes has been recorded after 28 and 56 days of water curing to analyse the impact of coconut fiber percentage on

FIGURE 14.3 Fully automatic compression testing machine.

FIGURE 14.4 Flexural testing machine.

FRCC. It has been noticed from Figure 14.5 that the compressive strength of all FRCC mixes was less than NC mix at both curing ages (28 and 56 days) as compared to normal mix. The increment in the percentage of coconut fibers decreased the compressive strength values, maybe due to presence of large number of fiber filaments, which created the voids in the internal structure of the matrix. From all the FRCC mixes, the optimum value of the compressive strength was observed with the usage of 2% fibers. Compressive strength (CS) values of FRCC mixes increased with increase in the age of the specimens, maybe due to long term pozzolanic reaction of fly ash.[21–23] The 56 days CS of FRCC 2 mix increased by 81.46% as compared to 28 days of FRCC 2 mix. From the CS result, it has been concluded that FRCC 2 mix was the most efficient mix w.r.t other FRCC mixes.

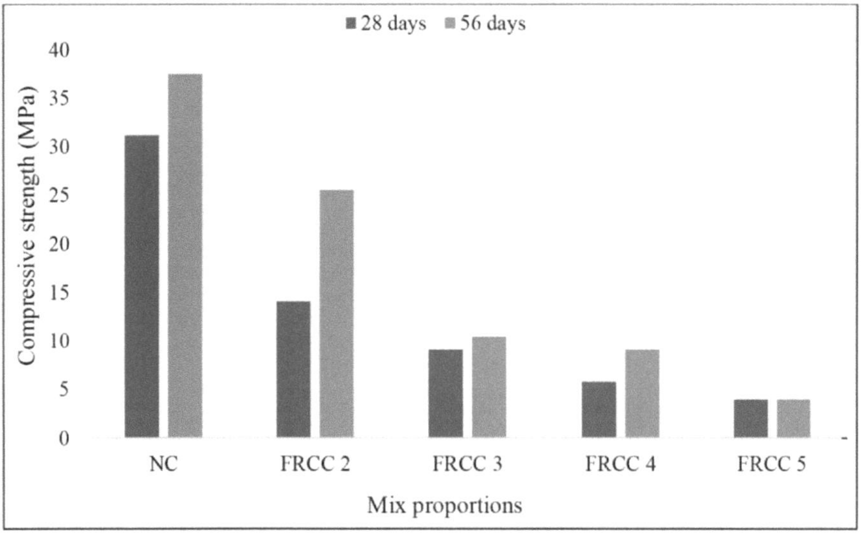

FIGURE 14.5 Compressive strength of various mixes.

14.4.2 FLEXURAL STRENGTH

Beams were casted in order to get the flexural strength of different mixes. The flexural performance of various mixes has been recorded after 28 and 56 days of water curing. It has been noticed from Figure 14.6 that the flexural strength of FRCC 2 mix was higher than other FRCC mixes and normal concrete mix. Flexural strength of all the mixes increased with increase in the curing period. Maximum increase in flexural strength at 56 days is for

FRCC 5 mix, that is, 131.89% as compared to 28 days flexural strength. The recorded values confirm that the mixes containing fly ash attained the strength in the long term, which may be due to long-term pozzolanic activities of fly ash.[24,25] The optimum flexural strength was observed in FRCC 2 mix after 56 days of water curing.

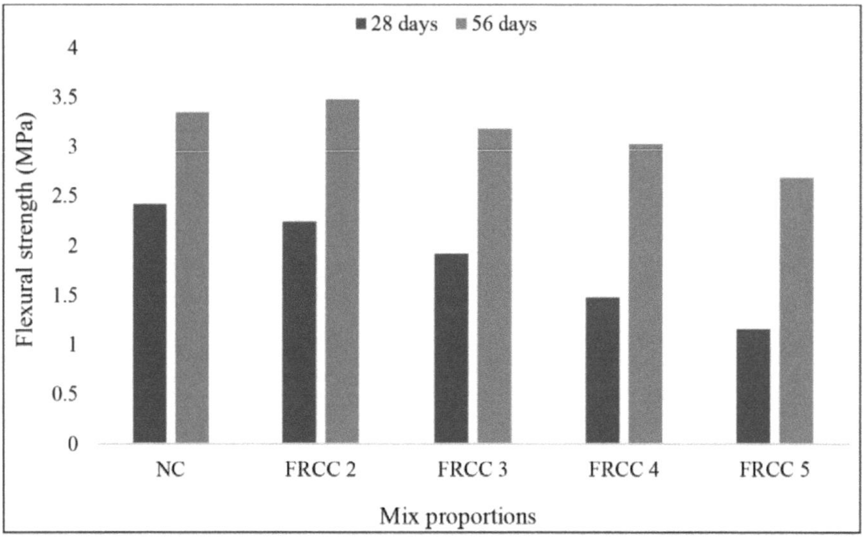

FIGURE 14.6 Flexural strength of various mixes.

14.5 CONCLUSIONS AND SUGGESTIONS

In the current investigation, the impact of coconut fiber percentages on fly ash containing FRCC was evaluated and compared with normal concrete mix. The major-specific conclusions from the current study have been mentioned below:

- Compression behavior of various mixes depicted that the CS of all the FRCC mixes was lesser as compared to normal concrete mix.
- Increment in the percentage of coconut fiber decreased the strength parameters owing to presence of large number of fiber filaments which create the voids at the internal structure of the cement matrix.
- Flexural behavior of various mixes depicted that the flexural strength of FRCC mix was slightly better than normal concrete mix. Coconut fiber in cementitious matrix acts as bridge and suppress the formation of wider cracks

- From the above results and evaluations, it can be concluded that FRCC 2 is most effective composition as compared to other mixes.
- With increase in the curing period, the maximum enhancement in strength parameters was in fly ash containing mixes which may be due to long-term pozzolanic activities of fly ash.
- To recommend the better guidelines on the usage of coconut fibers in cementitious products more research will be needed with some lesser percentages of coconut fibers in long-term curing period.

KEYWORDS

- **cementitious**
- **coconut fiber**
- **environmental degradation**
- **flexural**
- **waste products**

REFERENCES

1. Naik, T. R. Sustainability of Concrete Construction. Pract. Period. Struct. Des. Construct. **2008,** *13* (2), 98. https://doi.org/10.1061/(ASCE)1084-0680(2008)
2. Singh, M.; Saini, B.; Chalak, H. D. An Overview on Utilization of Stone Waste in Construction Industry. In *3rd International Conference on Innovative Technologies for Clean and Sustainable Development*, 2021; pp 569–583.
3. Gautam, N.; Krishna, V.; Srivastava, A. Sustainability in the Concrete Construction. *Int. J. Environ. Res. Dev.* **2014,** *4* (1), 81–90.
4. Gambhir, M. L. *Concrete Technology*; The McGraw Hill Companies, 2004.
5. Siddique, R. Utilization of Industrial By-Products in Concrete. *Proc. Eng.* **2014,** *95*, 335–347.
6. Woodson, R. D. Concrete Materials. In *Concrete Portable Handbook*; Elsevier Inc.: Oxford, 2012; pp 5–18.
7. Pappu, A.; Saxena, M.; Asolekar, S. R. Solid Wastes Generation in India and Their Recycling Potential in Building Materials. *Build. Environ.* **2007,** *2* (6), 2311–2320.
8. Singh, M.; Saini, B.; Chalak, H. D. Properties of Engineered Cementitious Composites: A Review. In *ICSWMD, LNCE 21*, 2018; pp 1–11, https://doi.org/10.1007/978-3-030-02707-0_54
9. Singh M.; Saini B.; Chalak H. D. An Overview on Waste Materials Used in Engineered Cementitious Composite. In *International Conference on Smart Materials and Techniques for Sustainable Development*, 2019; pp 213–222.

10. Ali, M. A. E. M.; Soliman, A. M.; Nehdi, M. L. Hybrid-fiber Reinforced Engineered Cementitious Composite under Tensile and Impact Loading. *Mater. Des.* **2017,** *117*, 139–149.
11. Ahmed, S. F. U.; Mihashi, H. A Review on Durability Properties of Strain Hardening Fibre Reinforced Cementitious Composites (SHFRCC). *Cement Concrete Comp.* 2007, *29* (5), 365–376.
12. Singh, M.; Saini, B.; Chalak, H. D. Performance and Composition Analysis of Engineered Cementitious Composite (ECC) – A Review. *J. Build. Eng.* **2019,** *26*, 100851. https://doi.org/10.1016/j.jobe.2019.100851
13. Fanghong, W.; Lihua, X.; Yin, C.; Yanqin, Z.; Fangqian, D.; Qian, C. Compressive and Flexural Properties of Ultra-High Performance Fiber-Reinforced Cementitious Composite: The Effect of Coarse Aggregate. *Compos. Struct.* 2020, *236*, 111810.
14. Linzhu, S.; Qiang, H.; Junliang, Z.; Dongyan, W.; Fang, Y. Stress Strain Behavior of Hybrid Steel-PVA Fiber Reinforced Cementitious Composites under Uniaxial Compression. *Construct. Build. Mater.* 2018, *188*, 349–360.
15. Singh, M.; Saini, B.; Chalak, H. D. Flexural Response and Durability Aspects of ECC Containing Stone Slurry Powder. *Asian J. Civil Eng.* **2020.**
16. Rathod, J. D.; Patodi, S. C. Interface Tailoring of Polyester-Type Fiber in Engineered Cementitious Composite Matrix against Pullout. *ACI Mater. J.* **2010,** *107*, 114–122.
17. Siddique, R.; Kapoor, K.; Kadri, E. H.; Bennacer, R. Effect of Polyester Fibres on the Compressive Strength and Abrasion Resistance of HVFA Concrete. *Construct. Build. Mater.* **2012,** *29*, 270–278.
18. Singh, M.; Saini, B.; Chalak, H. D. Performance Characteristics of ECC Using Micro Fibers in Hybridization. *Indian Concrete J.* **2020,** *94*, 66–73. https://www.icjonline.com/explore_journals/2020/06/
19. Zuanfeng, P.; Chang, W.; Jianzhong, L.; Wei, W.; Jiwei, L. Study on Mechanical Properties of Cost-Effective Polyvinyl Alcohol Engineered Cementitious Composites (PVA-ECC). *Construct. Build. Mater.* **2015,** *78*, 397–404.
20. Pakravan, H. R.; Jamshidi, M.; Latifi, M. The Effect of Hybridization and Geometry of Polypropylene Fibers on Engineered Cementitious Composites Reinforced by Polyvinyl Alcohol Fibers. *J. Compos. Mater.* **2015,** *50*, 1007–1020.
21. Al-Akhras Nabil, M.; Ayman, A.; Alaraji Wail A. Using Burnt Stone Slurry in Mortar Mixes. *Construct. Build. Mater.* **2010,** *24*, 2658–2663.
22. Almeida, N.; Branco, F.; De Brito, J.; Santos, J. R. High-performance Concrete with Recycled Stone Slurry. *Cement Concr. Res.* **2007,** *37*, 210–220.
23. Singh, M.; Saini, B.; Chalak, H. D. Long Term Evaluation of Engineered Cementitious Composite Containing Stone Slurry Powder. *Construct. Build. Mater.* **2020,** *264*, 120183.
24. Goel, S.; Singh, S. P.; Singh, P. Flexural Fatigue Strength and Failure Probability of Self-Compacting Fiber Reinforced Concrete Beams. *Eng. Struct.* **2012,** *40*, 131–140. https://doi.org/10.1016/j.engstruct.2012.02.035.
25. Siddique, R. Utilization of Industrial By-products in Concrete. *Procedia Eng.* **2014,** *95*, 335–347. https://doi.org/10.1016/j.proeng.2014.12.192.
26. IS 516-1959. *Methods of Tests for Strength of Concrete*, 18th ed.; Bureau of Indian Standards: New Delhi, India, 2006.
27. BS-EN-12390-3. Testing Hardened Concrete, Compressive Strength of Test Specimens, British Standard Institution, 2009.
28. BS-EN-12390-5. Testing Hardened Concrete, Flexural Strength of Test Specimens, British Standard Institution, 2009.

CHAPTER 15

ANALYSIS OF TRIBOLOGICAL PROPERTIES OF ALUMINUM METAL MATRIX COMPOSITE FABRICATED BY STIR CASTING METHOD

FAIM KHAN, N. YUVARAJ, LAKSHAY TYAGI, LUCKSHAYA KEM, and RAVI BUTOLA*

Department of Mechanical, Production & Industrial and Automobile Engineering, Delhi Technological University, Delhi 110042, India

Corresponding author. E-mail: ravibutola33855@gmail.com

ABSTRACT

Composites are being widely used in our manufacturing industry. Reinforcement is added to metal to enhance the properties like strength, toughness, ductility, hardness, tensile strength, etc. In the present work, Al6063 is reinforced with varying wt.% (0%, 2%, 4%) yttrium oxide. The composite is fabricated using Stir Casting. Tribological behavior of all the fabricated was analyzed and it was found that with increasing composition of Y_2O_3 its tribological properties were enhanced, that is, weight loss decreased, when Y_2O_3 was increased. But, for each sample weight loss and wear rate increases on increasing load.

15.1 INTRODUCTION

Metal matrix composites (MMCs) are occupying great interest in automobile and aerospace sector because of properties like improved ductility,

Modern Manufacturing Systems: Trends and Developments. Rajiv Kumar Garg, Ravi Pratap Singh, Rajeev Trehan, & Ramesh Singh (Eds.)

high strength, toughness, etc.[1–4] Addition of reinforcement to MMC increases mechanical and tribological properties like specific strength, stiffness, creep and fatigue, and wear. All these properties are increased on comparison with their parent engineering materials.[5] There are some ceramic reinforcements used in metal matrix to enhance the properties.[6] Some of the reinforcements are alumina (Al_2O_3), boron carbide (B_4C), silica (SiO_2), silicon carbide (SiC), graphite (Gr), tungsten carbide (WC),[7] and yttrium oxide.[8] Butola et al.[9] reinforced AA7075 with SiC particles via FSP. They reported that with reinforcement addition, materials' wear resistance, toughness, and tensile strength were enhanced. Worn surfaces were also examined using SEM and it was observed that adhesive wear was prominent wear mechanism among the wear-tested samples. Tyagi et al.[10] fabricated aluminum HMMCs using aloevera and SiC as reinforcement and they observed that addition of reinforcement enhances the material wear performance. Butola et al.[11] via FSP reinforced aluminum base material with nano B_4C and they found that microhardness and tensile strength of the material enhanced significantly.

There are many methods for mechanical alloying but casting method is mostly used due to its less cost. There are two types of fabrication methods, first one is liquid state processing and second one is solid-state processing.[12] The method that is commercially practiced is stir casting.[13] Stir casting method is very simple like it is very flexible, simple, and good for large production.[14] Aravindan et al.[15] reinforced Al6063 with SiC and studied its mechanical behavior and microstructure properties. Composites were fabricated by stir casting and as a result mechanical property of the material improved, also there is less particle distribution when SiC weight percentage is low. Dewangan et al.[16] studied the graphite reinforced Al6063, they observed that hardness and brittleness of the reinforced composite were enhanced. Rozhbiany et al.[17] observed the influence of various reinforcement like MA, SA, NFC, etc on aluminum matrix. They found that with the reinforcement addition, tensile strength and hardness improved. Lokesh et al.[18] used TiO_2–Cu as a reinforcement with Al6063 matrix and it is observed that there was minor improvement in hardness on increasing the weight percentage of Cu reinforcement and tensile strength also increased, but ductility and impact strength decreased. Hemalatha et al.[19] used Al_2O_3 as a reinforcement with Al6063 matrix and prepared composite via stir casting. It was found that hardness and tensile strength enhanced with the increasing the wt.% of reinforcement but the percentage elongation decreased on increasing the content of Al_2O_3. Moosa et al.[20] observed the effect of rice husk ash—yttrium oxide on mechanical and tribological properties of aluminum and the results found that on increasing the reinforcement (Y_2O_3) ratio hardness increased and wear rate decreased.

Butola et al.[21] analyzed the effect on mechanical properties of Al6061 composite. Al6061 matrix was reinforced with three different reinforcement, that is. Banana Ash, Coconut Ash, and Sugarcane Ash, composite was fabricated via stir casting method. UTS was observed and was found to be maximum in the case of aluminum composite having banana ash as reinforcement and it was minimum for the base. Coyal et al.[22] studied the tensile, hardness, and wear properties of AA6061 matrix reinforced with SiC and jute ash. Composite samples were fabricated via stir casting. Observations were made that SiC as reinforcement turned out to have the best UTS and hardness value of all tested samples. Bharat et al.[23] analyzed mechanical and wear properties on Al6061 reinforced with aluminum oxide prepared by using stir casting. It was found that tensile strength and hardness increased with the increase in concentration of reinforcement. Regarding wear loss, it was observed that as the wt.% of Al_2O_3 increased, wear loss decreased. Bhaumik et al.[24] observed the effect on mechanical properties of Al6063 reinforced with 5% ZrO_2/5% Al_2O_3. Fabrication was done by using stir casting. It was observed that fabricated samples showed more strength under compressive load.

Emmanual et al.[25] analyzed the wear behavior of Al6063 reinforced with SiC and granite. Composite was prepared by stir casting method with constant concentration of silicon carbide (3 wt%) and varying concentration of granite (2%, 4%, 6%) for a total of three samples. It was observed that the wear resistance was maximum in sample having 6 wt.% of granite. Joshi et al.[26] compared the mechanical performance of cast and heat-treated Al6063, reinforced with MWCNT. The fabrication was done using stir casting. It was observed that the tensile strength of heat-treated composite was more compared to un-heat-treated composite. It was also observed that with the increasing content of reinforcement in composite, impact strength increased. Pavani et al.[27] observed the mechanical behavior of Al6063-SiC_p MMCsfabricated via stir casting. It was found that on increasing the concentration of SiC in composite, mechanical properties like hardness, tensile strength enhanced. Hindi et al.[28] studied the mechanical behavior of AA6063 reinforced with SiC, fabricated via stir casting. It was found that as the wt.% of SiC increased, tensile strength improved. Maximum UTS of all tested samples was recorded for composite with 6 wt.% SiC.

In this study, an attempt is made to fabricate AMMCs having Yttrium oxide as reinforcement. The wear properties of the fabricated aluminum composites are studied and analyzed to govern the effect of reinforcement added.

15.2 METHODOLOGY

15.2.1 MATERIALS

Al6063 is the most common alloy used for the extrusion of alumina. It has good mechanical properties and has good heat treatable and welding properties. Table 15.1 shows the composition of Al 6063. Yttrium Oxide (Y_2O_3) is used as reinforcement for this study. It is an air-stable white solid substance. It is widely used as additives in steel, iron, and nonferrous alloys.

TABLE 15.1 Al6063 Composition.

Elements	Si	Fe	Cu	Mn	Mg	Zn	Al
Wt% composition	0.6	0.35	0.1	0.1	0.45	0.1	Balance

15.2.2 FABRICATION OF COMPOSITE

Stir casting process was used to fabricate the composite in a setup involving electric furnace and stirrer. Solid AA6063 was kept in the furnace and was heated at a melting temperature of 900°C for 90 min, Simultaneously, Yttrium oxide powder of size 25 μm was preheated at 300°C for 30 min. All the process parameters are mentioned in Table 15.2, the preheated reinforcement particles and melted Al 6063 were poured together and stirred at 200 rpm for 15 min in order to ensure proper dispersion of reinforcement particles in matrix. A die of diameter 80 mm and length 60 mm was casted out of the mixture. Similarly, all other Specimens mentioned in Table 15.3 were casted. Figure 15.1 shows the stir casting setup.

15.2.3 WEAR TESTING

Wear test was carried out on a DUCOM manufactured pin on disc rotatory tribometer. For the testing samples of diameter 10 mm and length 40 mm, shown in Figure 15.2, were cutted by wire EDM (electric discharge machining) from the casted cylinders. Wear was tested on the disc made of diameter 100 mm and thickness of 8 mm made of EN 24 steel and heat treatment was carried out on it in order to enhance its hardness. All the process parameters for the experiment are mentioned in Table 15.4, and model of tribometer is shown in Figure 15.3.

FIGURE 15.1 Stir casting setup.

TABLE 15.2 Stir Casting Process Parameters.

Process parameters	Values
Melting temperature (alloy)	900°C
Melting time (alloy)	90 min
Preheating temperature (reinforcement)	300°C
Preheating time (reinforcement)	30 min
Stir rpm	200
Stir time	15 min

TABLE 15.3 Composition of Specimens.

Casting samples	Al 6063 (%)	Yttrium oxide (%)
Sample 1	100	0
Sample 2	98	2
Sample 3	96	4

FIGURE 15.2 Specimen prepared for wear testing.

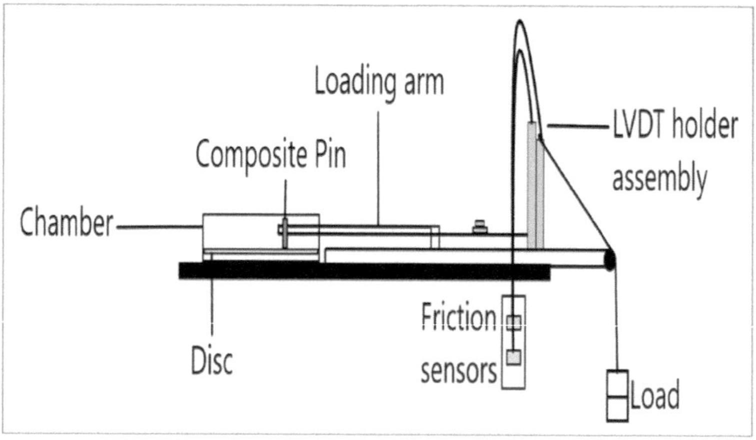

FIGURE 15.3 Rotatory tribometer model.

TABLE 15.4 Wear Test Process Parameters.

S. no.	Test conditions	Values
1.	Load	20 N, 40 N, 60 N
2.	Pin diameter	10 mm
3.	Pin length	40 mm
4.	Sliding distance	2000 m
5.	Sliding speed	2 m/s
6.	Track diameter	40 mm, 50 mm, 60 mm
7.	Rpm	478, 382, 318
8.	Disc roughness	58 HRC
9.	Disc hardness	0.2 μm

15.3 RESULTS AND DISCUSSION

15.3.1 WEIGHT LOSS AT DIFFERENT APPLIED LOAD

Figure 15.4 shows the weight loss of all tested samples at different loads. It can be concluded that weight loss increases with the increasing load. This is because at lower load the pressure acting is low and it increases gradually as the load is increased, that is, at larger load there is more pressure applied compared to the pressure applied at lower loads.[29–33] Similar type of scenario was seen when Alam et al. analyzed the wear properties of AMMC reinforced with SiC at different loads.[34] It can be observed that the reinforced sample showed less weight loss compared to nonreinforced sample at all applied load. This can be attributed to more load-bearing capacity of well-dispersed Y_2O_3 particles. As most of the load applied on the matrix is transferred on these well bonded well-dispersed reinforcement particles hence increasing the wear resistance of the reinforced samples. Similar type of results were obtained by Ahmed et al.[35] when they studied wear polymer matrix composite reinforced with Y_2O_3 particles.

15.3.2 EFFECT OF LOAD ON WEAR RATE

Wear rate for a composite sample is calculated using eq 15.1

$$W = \frac{W1 - W2}{T} \tag{15.1}$$

W_1 = Initial wear in μm
W_2 = Wear at 2000 m in μm
T = Time in min
W = Wear rate in μm/min

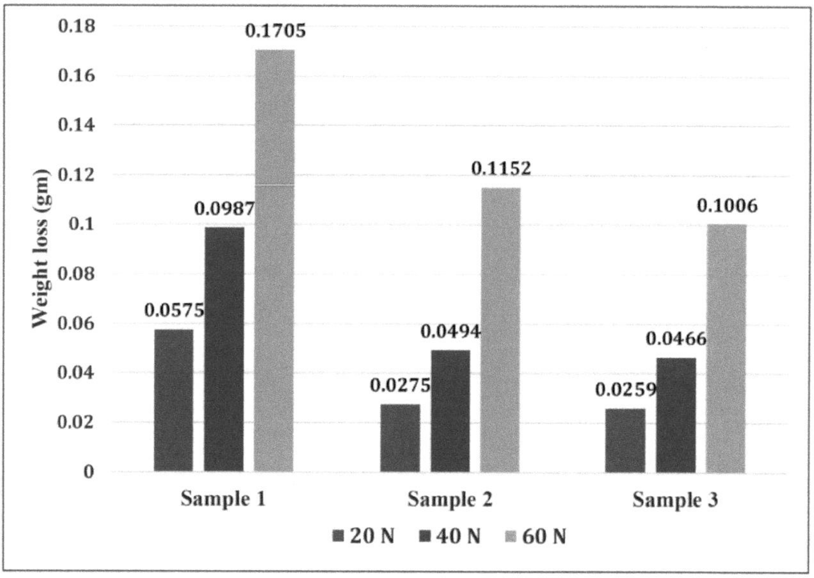

FIGURE 15.4 Weight loss at different loads.

The relationship between wear rate and load can be seen in Figure 15.5. It can be concluded that with increasing load, wear rate increases. Stating that wear rate is maximum at 60 N load and minimum at 20 N load for any composite. With escalating load effect of friction and contact between the metal disc and sample tested increases hence increasing the wear rate.[36] It is also observed that wear rate decreases with increasing the reinforcement at a specific load. This is due to the increase in the number of hard reinforcement particles present in the matrix. As these hard particles bear more fraction of load, slowing the deformation of the composite and reducing the wear rate.

15.4 CONCLUSION

AMMCs were successfully fabricated using the stir casting method. Stir casting is found to be quite efficient and flexible for producing composites. Uniform distribution of yttrium oxide particles is there in the aluminum matrix, which enhanced the wear performance of the material, all other observations made from the study are as follows

- Weight loss of composite decreases on increasing the content of Y_2O_3 and it increases with the increasing load.
- Wear rate was escalated with increasing load. With increasing reinforcement content in the matrix wear rate decreased.
- Sample 3 having the maximum reinforcement content of all fabricated composite showed the optimum result having minimum weight loss at all tested loads.

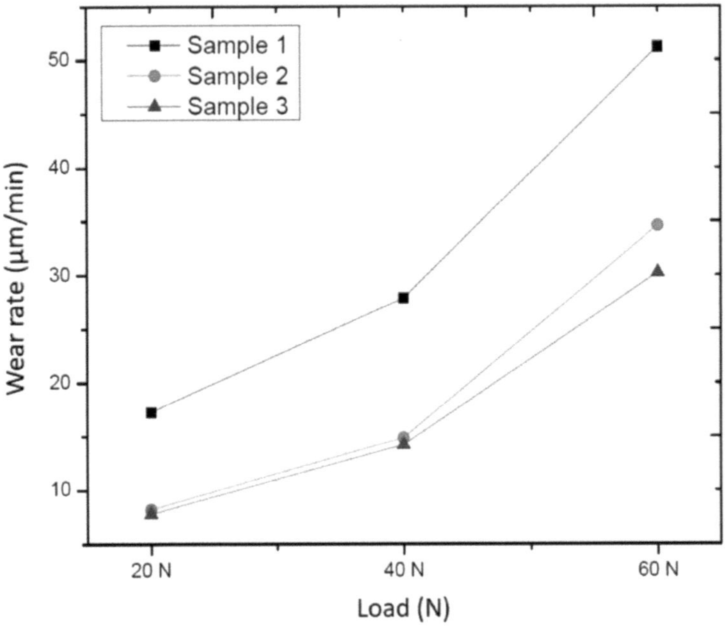

FIGURE 15.5 Wear rate at different load.

KEYWORDS

- **stir casting**
- **yttrium oxide**
- **wear**
- **composite**

REFERENCES

1. Reddy, M.; Shakoor, R.; Parande, G.; Manakari, V.; Ubaid, F.; Mohamed, A.; Gupta, M. Enhanced Performance of Nano-sized SiC Reinforced Al Metal Matrix Nanocomposites Synthesized through Microwave Sintering and Hot Extrusion Techniques. *Prog. Nat. Sci.: Mater. Int.* **2017,** *27* (5), 606–614.
2. Butola, R.; Kanwar, S.; Tyagi, L.; Singari, R.; Tyagi, M. Optimizing the Machining Variables in CNC Turning of Aluminum Based Hybrid Metal Matrix Composites. *SN Appl. Sci.* **2020,** *2* (8).
3. Butola, R.; Tyagi, L.; Kem, L.; Ranganath, M.; Murtaza, Q. *Mechanical and Wear Properties of Aluminium Alloy Composites: A Review*. Lecture Notes on Multidisciplinary Industrial Engineering, 2020; pp 369–391.
4. Butola, R.; Singari, R.; Murtaza, Q. Mechanical and Wear Behaviour of Friction Stir Processed Surface Composite through Self-Assembled Monolayer Technique. *Surf. Topogr.: Metrol. Prop.* **2020,** *8* (4), 045007.
5. Al-Salihi, H.; Mahmood, A.; Alalkawi, H. Mechanical and Wear Behavior of AA7075 Aluminum Matrix Composites Reinforced by Al_2O_3 Nanoparticles. *Nanocomposites* **2019,** *5* (3), 67–73.
6. Iqbal, A.; Nuruzzaman, D. Effect of the Reinforcement on the Mechanical Properties of Aluminium Matrix Composite: A Review, 2016.
7. Grover, M.; Sharma, S.; Kataria, T.; Samdani, S.; Agarwal, S.; Singh, S. Soft Tissue Reactions Following Cochlear Implantation. *Eur. Arch. Oto-Rhino-Laryngol.* **2018,** *276* (2), 343–347.
8. Kruk, A.; Mrózek, M.; Domagała, J.; Brylewski, T.; Gawlik, W. Synthesis and Physicochemical Properties of Yttrium Oxide Doped with Neodymium and Lanthanum. *J. Electr. Mater.* **2014,** *43* (9), 3611–3617.
9. Butola, R.; Tyagi, L.; Singari, R.; Murtaza, Q.; Kumar, H.; Nayak, D. Mechanical and Wear Performance of Al/SiC Surface Composite Prepared through Friction Stir Processing. *Mater. Res. Express* **2021,** *8* (1), 016520.
10. Tyagi, L.; Butola, R.; Jha, A. Mechanical and Tribological Properties of AA7075-T6 Metal Matrix Composite Reinforced with Ceramic Particles and Aloevera Ash via Friction Stir Processing. *Mater. Res. Express* **2020,** *7* (6), 066526.
11. Butola, R.; Murtaza, Q.; Singari, R. Formation of Self-Assembled Monolayer and Characterization of AA7075-T6/B4C Nano-ceramic Surface Composite Using Friction Stir Processing. *Surf. Topogr.: Metrol. Prop.* **2020,** *8* (2), 025030.
12. Dwivedi, A.; Joshi, U. Evaluation of Mechanical Properties of Aluminium Reinforced with Al_2O_3 Nano Particles: A Review. *J. Emerg. Technol. Innov. Res.* **2017,** *4* (6).
13. Rizam, S.; Harun, M.; Hafizal, Y.; Mazlee, M.; Fitri, M.; Shamsul, J. Fabrication of A356 Al Alloy - SiCp Composites by Stir Casting Method, 2006.
14. Hashim, J.; Looney, L.; Hashmi, M. Metal Matrix Composites: Production by the Stir Casting Method. *J. Mater. Proces. Technol.* **1999,** 92–93, 1–7.
15. Aravindan, M.; Balamurugan, K.; Murali, G. Effect of Reinforcement of AL-6063 with SiC on Mechanical Behavior and Microstructure of Metal Matrix Composites. *Carbon – Sci. Technol.* **2014,** *6* (2), 388–394.
16. Dewangann, S.; Arora, D. Mechanical Properties of Aluminum 6063 Alloy based Graphite Particles Reinforced Metal Matrix Composite. *Int. J. Sci. Res. Develop.* **2015,** *3* (5).

17. Rozhbiany, F.; Jalal, S. Influence of Reinforcement and Processing on Aluminum Matrix Composites Modified by Stir Casting Route. *Adv. Compos. Lett.* **2019,** *28*, 2633366 X1989658.
18. Lokesh, N.; Manoj, B.; Srikanth, K.; kumar, P.; Ramanayya, P.; Rao, M. Mechanical Characterization of Stir Cast Al 6063 TiO_2-Cu Reinforced Hybrid Metal Matrix Composites. *Mater. Today: Proc.* **2018,** *5* (9), 18383–18392.
19. Hemalatha, K.; Venkatachalapathy, V.; Alagumurthy, N. Processing and Synthesis of Metal Matrix Al 6063/Al_2O_3 Metal Matrix Composite by Stir Casting Process. *Int. J. Eng. Res. Appl.* **2013,** *3* (6), 1390–1394.
20. Moosa, A.; Yass Awad, A. Influence of Rice Husk Ash-Yttrium Oxide Addition on the Mechanical Properties Behavior of Aluminum Alloy Matrix Hybrid Composites. *Int. J. Curr. Eng. Technol.* **2016,** *6* (3).
21. Butola, R.; Pratap, C.; Shukla, A.; Walia, R. Effect on the Mechanical Properties of Aluminum-Based Hybrid Metal Matrix Composite Using Stir Casting Method. *Mater. Sci. Forum* **2019,** *969*, 253–259.
22. Coyal, A.; Yuvaraj, N.; Butola, R.; Tyagi, L. An Experimental Analysis of Tensile, Hardness and Wear Properties of Aluminium Metal Matrix Composite through Stir Casting Process. *SN Appl. Sci.* **2020,** *2* (5).
23. Bharath, V.; Nagaral, M.; Auradi, V.; Kori, S. Preparation of 6061Al-Al 2 O3 MMC's by Stir Casting and Evaluation of Mechanical and Wear Properties. *Procedia Mater. Sci.* **2014,** *6*, 1658–1667.
24. Bhaumik, M.; Maity, K. Fabrication and Characterization of the Al6063/5%ZrO2/5% Al_2O_3composite. *IOP Conf. Series: Mater. Sci. Eng.* **2017,** *178*, 012011.
25. Emmanual, L.; Ragunath, L.; Karthikeyan, T. Wear Behaviour on Composite of Aluminium 6063 Alloy with Silicon Carbide and Granite by Stir Casting Method. *Int. Conf. Mater. Manuf. Mach.* **2019**.
26. Joshi, A.; Kumaraswamy, H.; Anilkumar, Mithun Reddy, A.; Ashwin Raj, A. A Comparative Study of Mechanical Properties of As Cast and Heat Treated Al6063-MWCNT Composite Fabricated by Stir Casting Process. *IOP Conf. Series: Mater. Sci. Eng.* **2018,** *376*, 012075.
27. Pavani, M. Preparation and Mechanical Behaviour of Al6063-SICPR RMMC by Using Stir Casting Technique. *Int. J. Innov. Sci. Eng. Technol.* **2015,** *2* (8).
28. Hindi, J.; Kini, U.; Sharma, S.; Gurumurthy, B.; Shankar, M. Mechanical Characterization of Stir Cast Al 6063 Matrix SiC Reinforced Metal Matrix Composites. In *5th International Conference on Automotive, Mechanical and Materials Engineering*, 2015.
29. Radhika, N.; Raghu, R. Investigation on Mechanical Properties and Analysis of Dry Sliding Wear Behavior of Al LM13/AlN Metal Matrix Composite Based on Taguchi's Technique. *J. Tribol.* **2017,** *139* (4).
30. Haiter Lenin, A.; Vettivel, S.; Raja, T.; Belay, L.; Singh, S. A Statistical Prediction on Wear and Friction Behavior of ZrC Nano Particles Reinforced with Al Si Composites Using Full Factorial Design. *Surf. Interf.* **2018,** *10*, 149–161.
31. Dorri Moghadam, A.; Omrani, E.; Menezes, P.; Rohatgi, P. Mechanical and Tribological Properties of Self-Lubricating Metal Matrix Nanocomposites Reinforced by Carbon Nanotubes (CNTs) and Graphene – A Review. *Compos. Part B: Eng.* **2015,** *77*, 402–420.
32. Dama, K.; Prashanth, L.; Nagaral, M.; Mathapati, R.; Hanumantharayagouda, M. Microstructure and Mechanical Behavior of B4C Particulates Reinforced ZA27 Alloy Composites. *Mater. Today: Proc.* **2017,** *4* (8), 7546–7553.

33. Tyagi, L.; Butola, R.; Kem, L.; Singari, R. Comparative Analysis of Response Surface Methodology and Artificial Neural Network on the Wear Properties of Surface Composite Fabricated by Friction Stir Processing. *J. Bio- and Tribo-Corrosion* **2021,** *7* (2).
34. Alam, M.; Arif, S.; Ansari, A. Wear Behaviour and Morphology of Stir Cast Aluminium/SiC Nanocomposites. *Mater. Res. Express* **2018,** *5* (4), 045008.
35. Ahmed, A.; Irhayyim, S.; Hammood, H. Effect of Yttrium Oxide Particles on the Mechanical Properties of Polymer Matrix Composite. *IOP Conf. Series: Mater. Sci. Eng.* **2018,** *454*, 012036.
36. Mahmood, I. A.; Soud, W. A.; Abdullah, O. S. Effects of Different Types of Fillers on Dry Wear Characteristics of Carbon-Epoxy Composite. *Al-Khwarizmi Eng. J.* **2013,** *9* (2), 85–93.

CHAPTER 16

IMPACT BEHAVIOR OF GLASS AND CARBON-FIBER-BASED REINFORCED POLYMER COMPOSITES

MAHENDRA SINGH TOMAR* and SATNAM SINGH

Department of Mechanical Engineering, National Institute of Technology, Kurukshetra 136119, India

**Corresponding author. E-mail: mahendra_31806209@nitkkr.ac.in*

ABSTRACT

Fiber-based composites have extensive application in various fields. Fiber-based composite materials are used widely as a substitute for metallic material because of their large strength-to-weight ratio. Light-weight structure is the important factor in improving the fuel efficiency of transport vehicles. In this research work, Izod impact test and hardness of carbon and glass fiber composites are studied. Specimens were prepared as per ASTM D265 standard. Unidirectional glass and carbon-fiber-based sheets with 400 GSM, epoxy resin YD535 LV, and hardener TH7255 were used for the preparation of sheets by vacuum-assisted resin transfer molding process. Three combinations like glass–glass, carbon–carbon, and glass–carbon were tested under the impact tester and their properties are studied.

16.1 INTRODUCTION

Fiber-based composites are special type of materials that are made up of fibers of different materials like metals, glass, carbon embedded in a matrix

Modern Manufacturing Systems: Trends and Developments. Rajiv Kumar Garg, Ravi Pratap Singh, Rajeev Trehan, & Ramesh Singh (Eds.)

of different materials. Composites are made in order to take advantage of mechanical properties provided by different type of materials. Composites based on glass and carbon are widely used because of their large strength-to-weight ratio. Various research works have been done to test the mechanical properties of different types of composites in order to check the applications of different composites known. A study conducted by Sohi and Hahn[1] on composite made up of carbon fiber bundles embedded on epoxy resin casting. This study was done to calculate the compressive strength of the composite and to study the modes of failure of the mentioned composites. Another study was conducted by Chaplin[2] to investigate the effect of manufacturing defect in the mechanical properties of the glass epoxy based composite. In his experiment, he found that if a crack initiates due to compressive load failure, the crack expanded through the entire material. The failure initiates from a notch produced due to manufacturing defect. An experiment was conducted by Knauss et al.[3] to study the mechanical properties of composites in the presence of holes. The mechanical properties studied included harness, tensile strength, and compressive strength. Khan et al.[4] in their experiment compare the interfacial and mechanical properties of E-glass and jute fiber reinforced polypropylene composite. They found that the mechanical properties like tensile strength, tensile modulus, bending strength shown by the E-glass-based composite were double that as shown by the jute-based composite. Another experiment was conducted by Ramchandra Reddy et al.[5] to study the properties such as hardness, impact strength, chemical resistivity, and frictional coefficient of a hybrid composite developed with the amalgamation of glass and sisal fibers in epoxy matrix. The impact strength and hardness of the composite was found to be a function of length of the fiber and optimum length of the fiber was found as 2 cm so that it can give the maximum impact strength and hardness to the composite. The composite was found to be chemically resistive to all corrosive chemicals except toluene and sodium carbonate. Devendra[6] did an investigation to find the effect of mechanical properties due to different filler fluid used to fill glass fiber reinforced composites. The composite was filled with a combination of different fluids in varying concentration which included magnesium hydroxide, aluminum oxide, ash, and hematite powder solution. The experiment showed that the maximum tensile strength was shown when the composite was filled with 10% concentration by volume of magnesium hydroxide while maximum impact strength was shown when the composite was filled with 10% concentration by volume of ash solution and maximum hardness was found when the composite was filled with magnesium hydroxide alone.

This chapter studies the mechanical properties of hardness and impact strength of unidirectional glass and carbon-fiber-based sheets with 400 GSM, epoxy resin YD535 LV, and hardener TH7255. The test is done using the combination of composites, such as like glass–glass, carbon–carbon, and glass–carbon. The experiment performed for this study includes Izod impact test and hardness test.

16.2 COMPOSITE MATERIALS

Unidirectional carbon fiber sheets and glass fiber are fabricated with the help of epoxy resin YD535 LV and hardener TH7255 by vacuum-assisted resin transfer molding (VARTM) process. Dimensions of the specimen are taken according to ASTM codes.

There are three types of combination of carbon fibers and glass fibers used for mechanical testing.

The combination of fiber composites are:

i. Carbon–carbon–carbon fiber.
ii. Carbon–glass–carbon fiber.
iii. Glass–glass–glass fiber.

16.2.1 GLASS FIBER 400 GSM

Unidirectional E-glass fiber of 400 GSM having fabric design thickness of 0.172 mm and fiber density of 2.56 g/cm^3 manufactured by AB Chemicals Ltd. is being used to prepare specimens. The material, that is, glass fiber is supplied by the supplier named as CF composites (Figure 16.1).

Properties of Glass Fiber

- Density (g/cc): 2.56.
- Modulus (N/mm^2): 76,000.
- Failure strain (%): 2.8.
- Strength(N/mm^2): 3400.

Advantages of using glass fiber as a reinforcement material are:

➢ Strong in tensile loading
➢ Easily available
➢ Corrosion resistant

- Easy to form in different shapes
- Low density
- Nonflammability
- High thermal insulation
- High electrical insulation and sound insulation.

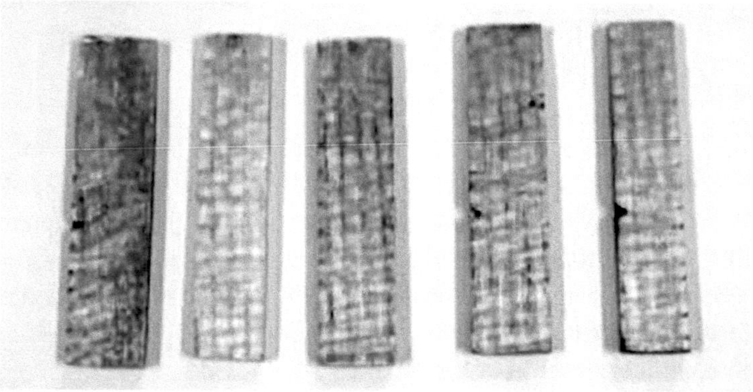

FIGURE 16.1 Glass fiber composite.

16.2.2 CARBON FIBER 400 GSM

Unidirectional carbon fiber of grade 400 GSM manufactured by AB Chemicals Ltd. was purchased by CF Composites, Delhi. The fabric thickness of the carbon fiber was 0.255 mm and fiber density was 1.82 g/cm^3 (Figure 16.2).

Properties of carbon fiber

- Density (g/cc): 1.82
- Modulus (N/mm^2): 230,000
- Failure strain (%): 1.7
- Strength (N/mm^2): 4000

The advantages of carbon fiber as reinforcement material are listed below:

- High stiffness
- High tensile strength
- High compressive strength
- Low weight
- High temperature tolerance
- Low thermal expansion

FIGURE 16.2 Carbon fiber composite.

16.2.3 CARBON AND GLASS MIX FIBER

Unidirectional carbon and glass mix fiber of grade 400 GSM manufactured by AB Chemicals Ltd. was purchased by CF Composites, Delhi. The fabric thickness of the carbon and glass mix fiber was 0.225 mm. Mixing different fibers in the same load path can lead to a component being weaker than it would be if only one type of fiber had been used (Figure 16.3).

FIGURE 16.3 Glass and carbon mixed fiber composite.

16.2.4 EPOTEC EPOXY

A laminated epoxy system is made up of one resin and six hardeners, of choice, to provide wider range of processing properties and to make it suitable for laminate fabrication techniques with various environmental conditions. The working time of EPOTEC TH 7258 hardener can provide less than 12 h at room temperature with low exothermic reactions. EPOTEC TH 7253 hardener is used in order to produce relatively smaller components that can be transported in a very short period of time, in as little as a few hours of time, at room temperature, and it has also related to rapid curing character.

16.3 SAMPLE GEOMETRY AND FABRICATION PROCESS

According to the standard ASTM D265, test samples for impact testing having dimension of 64 × 12.7 × 3.2 mm for each combination of fiber layer are being used.

The three possible stacking sequences are:

CCCC: eight layers
GGGG: eight layers
CGCG: eight layers

16.3.1 VACUUM-ASSISTED RESIN TRANSFER MOLDING PROCESS AND STEPS

For the fabrication of composites, the process used is VARTM, which is a variation of the fabrication method of resin transfer molding. The basic difference in this method from other methods is that the upper mold is replaced by a vacuum bag that helps in assisting the resin flow over the fiber layup. Then the part is left to cure at room temperature. The ratio of resin to fiber is an important factor in deciding the final strength and performance parameters. The mechanical properties are being influenced mostly by the type of fiber reinforcement used and the properties such as corrosion resistance and surface finish are determined by the type of resin used in the fabrication process. In order to produce the high-quality parts, it is very essential that air leakage is avoided otherwise there are the chances of improper deposition of resin on the fiber layup, which can lead to the formation of air bubbles in part. Various methods are available for the detection of air leakage in vacuum bag.

16.3.2 GET THE EQUIPMENT AND CONSUMABLES

The required equipment for the purpose of fabrication are a pump, valve, and some barbed tube fitting that can be joined by the screws onto the pump. Apart from that tacky tape, plastic t-connectors, green, plastic tubing, spiral tubing wrap, vacuum bagging film, and flow medium are also required.

16.3.3 SETUP

Once all the equipment are ready, the infusion is made from a distance of 12 inches after arranging all the necessary setups.

- Wax and tacky tape
- Mold is properly waxed in layers over the mold.

- Carbon and glass fiber
  Fiber is made lay down. And then by calculation of the weight of fiber, the amount of resin to be used can be estimated. Now the resin can be manually squeezed into the fiber.

- Peel ply
  A few layers of peel ply are used to absorb the unnecessary amount of the resin.

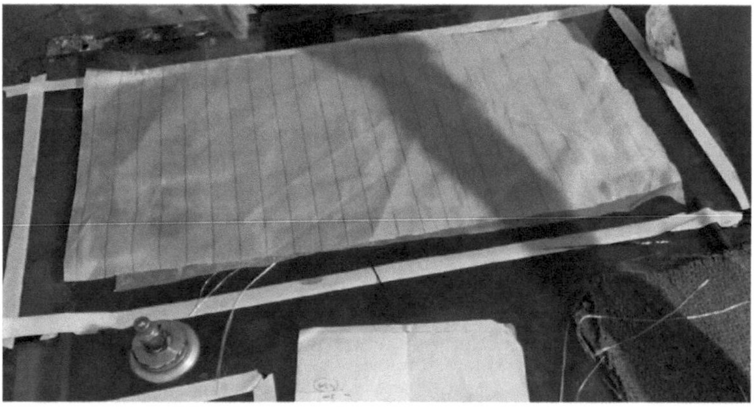

- Flow medium
  Over the peel ply green floe medium is used, which helps the resin to flow properly across the surface.
- Resin & vacuum ports
  For the purpose of infusion the ports and connections are also provided.

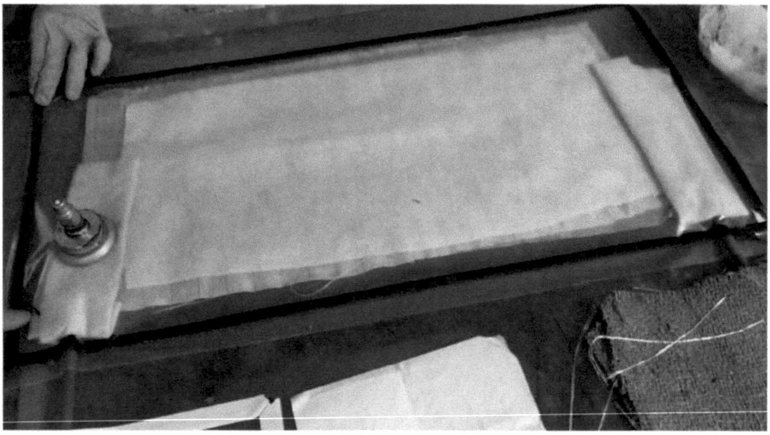

- Vacuum bagging: vacuum bagging is the last layer applied over the other layers.
- Vacuum, catch pot, and tubing.

16.3.4 PULL VACUUM

Before the resin is made flow, it is highly crucial that the vacuum can be pulled perfectly. Now slowly pull the vacuum with the help of vise grips by clamping down the input tubes of resin. Ball valve on the pump allows the desirable speed for pulling the vacuum. It is very important that the spiral-wrapped tubes are placed correctly so it preferable to start slowly in the beginning. Input is made to touch the green flow medium and release cloth and it is ensured by taking high precautions. It is, also, ensured that the output is off the carbon part and the release cloth or it can be simply removed from the flow medium. The resin flow will be slowed down if there is any space left.

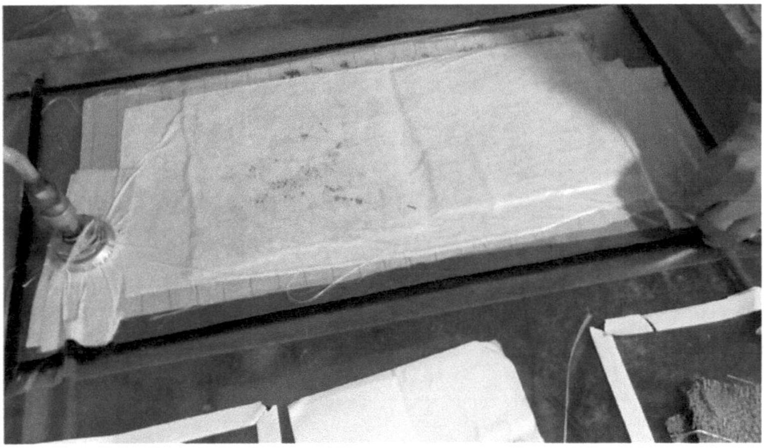

16.3.5 WEIGH OUT THE RESIN AND INFUSE THE RESIN

To start the infusion process, the weight of the resin is to be calculated on the basis of the weight of fiber used. The weight of fiber used initially is measured and combined with the some other factor the required amount of resin is calculated. The addition amount of resin is generally absorbed by the release cloth and flow medium.

16.3.6 REMOVE THE PART

By allowing some time to cure it in room temperature, the part is ready to be removed and it is done. All the connections and ports are removed before the removal of the part to ensure the safe and proper removal.

16.3.7 SAMPLE CUTTING

Sample cutting was done at central workshop of NIT, Kurukshetra.

16.4 RESULTS AND DISCUSSION

16.4.1 IZOD IMPACT TEST RESULT

Results obtained in Izod impact test are as follows:

TABLE 16.1 Izod Impact Strength of Composites.

Fiber	Sample	Rise angle	Absorbed energy	Potential energy	Compensated energy
	C1	128.20	38.67	216.83	37.81
Carbon fiber	C2	127.90	39.22	216.27	38.36
	C3	131.60	32.57	222.98	37.70
	M1	125.00	44.68	210.82	43.83
Carbon and glass fiber	M2	129.90	35.58	219.91	34.72
	M3	119.30	55.96	199.54	55.12
	G1	118.60	57.39	198.11	56.56
Glass fiber	G2	121.90	50.72	204.77	49.88
	G3	121.60	51.32	204.18	50.48

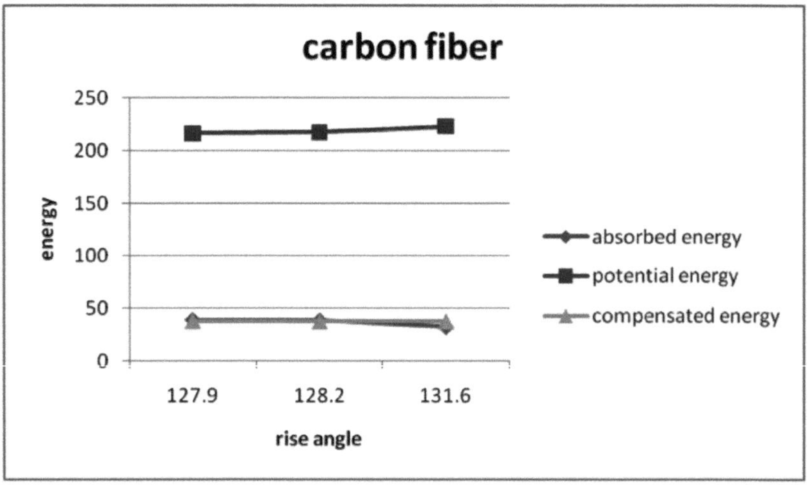

FIGURE 16.4 Rise angle versus energy (J) graph of carbon fiber.

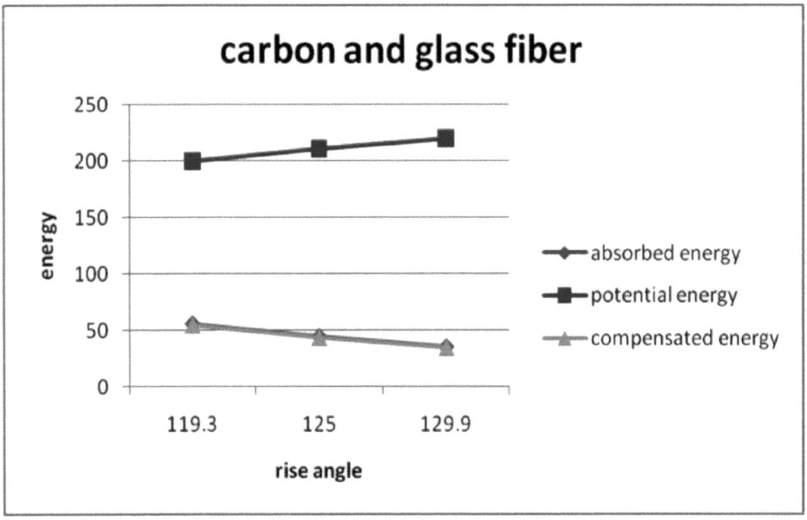

FIGURE 16.5 Rise angle versus energy (J) graph of carbon and glass fiber.

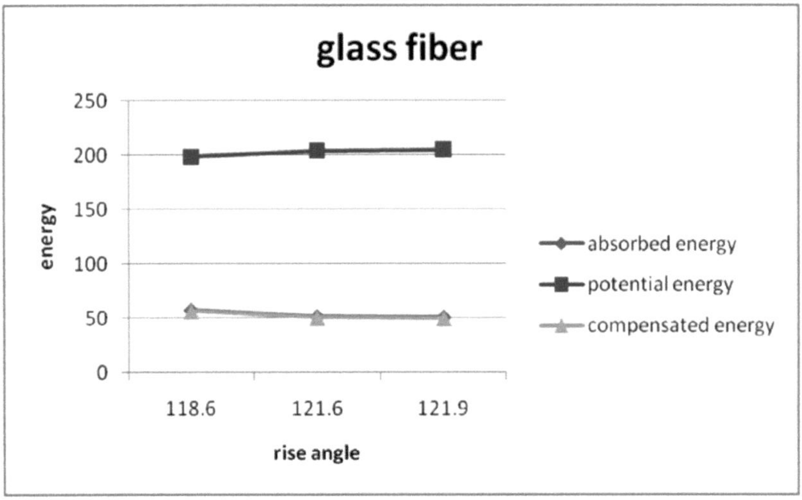

FIGURE 16.6 Rise angle versus energy (J) graph of glass fiber.

16.4.2 ROCKWELL HARDNESS TEST

The results found for Rockwell hardness test are given in Table 16.2.

TABLE 16.2 Hardness of Composite.

Type of composite	Hardness	Load
Carbon fiber	56.8	150 kgf
Glass fiber	56.9	150 kgf
Carbon and glass mixed fiber	50.8	150 kgf

16.5 CONCLUSION

Glass fiber composite materials showed an increase in impact strength and hardness while the weight is reduced by applying VARTM process for composite material fabrication.

Various plots of rise angle versus energy have been shown in Figures 16.4–16.6 for the carbon fiber, carbon and glass fiber, and glass fiber, respectively, after the performance of the Izod test. The potential energy in the case of carbon fiber, carbon and glass fiber, and glass fiber is found out to be 216.83, 210.82, and 199.54 J, respectively, at different rise angles. This can be noted from Table 16.1 that carbon fiber has 2.85% and 8.66% more potential energy than carbon and glass fiber, and glass fiber, respectively. The compensated energy in the case of carbon fiber, carbon and glass fiber, and glass fiber is found out to be 37.81, 43.83, and 56.56 J, respectively, at different rise angles. This can be noted from Table 16.1 that carbon fiber has 13.73% and 33.15% less compensated energy than carbon and glass fiber and glass fiber, respectively.

Above all the absorbed energy is the most important parameter. The absorbed energy in the case of carbon fiber, carbon and glass fiber, and glass fiber is found out to be 38.67, 44.68, and 57.39 J, respectively, at different rise angle. This can be noted from Table 16.1 that carbon fiber has 13.45% and 32.61% less absorbed energy than carbon and glass fiber and glass fiber, respectively. Carbon fiber has the lowest absorbed energy.

The Rockwell hardness values for carbon fiber, glass fiber, and carbon and glass fiber are 56.8, 56.9, and 50.8, respectively. So the carbon fiber has the highest value of Rockwell hardness.

KEYWORDS

- **carbon fiber**
- **epoxy resin**
- **glass fiber**
- **Izod impact test**

REFERENCES

1. Hahn, H. T.; Sohi, M. M. Buckling of a Fiber Bundle Embedded in Epoxy. *Compos. Sci. Technol.* **1986,** *27* (1), 25–41.
2. Chaplin, C. R. Compressive Fracture in Undirectional Glass-Reinforced Plastics. *J. Mater. Sci.* **1977,** *12* (2), 347–352.
3. Waas, A. M.; Babcock Jr, C. D.; Knauss, W. G. An Experimental Study of Compression Failure of Fibrous Laminated Composites in the Presence of Stress Gradients. *Int. J. Solids Struct.* **1990,** *26* (9–10), 1071–1098.
4. Khan, R. A.; Khan, M. A.; Zaman, H. U.; Pervin, S.; Khan, N.; Sultana, S.; Mustafa, A. I. Comparative Studies of Mechanical and Interfacial Properties between Jute and E-glass Fiber-Reinforced Polypropylene Composites. *J. Reinf. Plast. Compos.* **2010,** *29* (7), 1078–1088.
5. Ramachandra Reddy, G.; Ashok Kumar, M.; Karthikeyan, N.; Mahaboob Basha, S. Tamarind Fruit Fiber and Glass Fiber Reinforced Polyester Composites. *Mech. Adv. Mater. Struct.* **2015,** *22* (9), 770–775.
6. Devendra, K.; Rangaswamy, T. Strength Characterization of E-glass Fiber Reinforced Epoxy Composites with Filler Materials. *J. Minerals Mater. Characteriz. Eng.* **2013,** *1* (6), 353–353.

PART III

Case Studies on Welding/Joining Processes

CHAPTER 17

EXPERIMENTAL INVESTIGATIONS ON EFFECT OF TOOL WELDING SPEED ON MICROSTRUCTURE AND TENSILE PROPERTIES OF FSW DISSIMILAR JOINTS REINFORCED WITH SiC NANOPARTICLES

K. S. ANIL KUMAR[1], H. RAJNEESH[2], and H. C. MADHU[3]

[1]*Department of Mechanical Engineering, J.S.S. Academy of Technical Education, Bangalore 560060, India*

[2]*Department of Mechanical Engineering, Indian Institute of Science, Bangalore 560012, India*

[3]*Department of Mechanical Engineering, Siddaganga Institute of Technology, Tumakuru 572103, India*

[*]*Corresponding author. E-mail: anilkumar_aks@rediffmail.com*

ABSTRACT

Aluminum alloy plates of 2024-T351 and 7075-T651 reinforced with nano-sized particles of silicon carbide (SiC_{NP}) in the groove cut of the rectangular shape made on the adjoining surfaces of the plates and are welded in the butt–joint combination using friction stir welding (FSW) technique. FSW tool of shape taper threaded cylindrical is employed to produce composite at the weld nugget zone (WNZ) of the FSW dissimilar joint. During FSW, the dissimilar aluminum alloy plates were welded by varying the tool welding

Modern Manufacturing Systems: Trends and Developments. Rajiv Kumar Garg, Ravi Pratap Singh, Rajeev Trehan, & Ramesh Singh (Eds.)

speed (TWS) from 20 to 200 mm/min, while the tool rotation speed (TRS) and tool plunge depth (TPD) were kept at a constant of 1000 rpm and 6.20 mm, respectively. The effectiveness of varying the TWS on the SiC_{NP} distribution in the WNZ of the joint specimens is investigated by conducting microstructure analyses using optical microscopy and scanning electron microscopy. The tensile tests for the FSW joint specimens reinforced with SiC_{NP} at the WNZ are conducted, and the weld joint strength is investigated. The results indicated that FSW dissimilar joint produced for TWS range 40–50 mm/min with constant TRS of 1000 rpm and TPD of 6.20 mm results in the higher weld joint tensile properties, observes a defect-free and uniform distribution of SiC_{NP} at the WNZ. Increasing the TWS above 75 mm/min results in SiC_{NP} being stretched toward the retreating side of the weld region and improper distribution of SiC_{NP}.

17.1 INTRODUCTION

Friction stir welding (FSW) technique was developed in the year 1991 at the welding Institute (TWI) and initially applied to aluminum alloys.[1] It is a kind of solid-state joining technique. In the FSW process, a nonconsumable tool rotating in either clockwise (CW) or counter CW direction consists of pin and shoulder design that is plunged from the faying surface of the joints in the form of the sheets or plates. After developing a sufficient amount of heat, the tool is made to traverse in a particular direction.[2]

In the last few years, the FSW is adopted to develop a friction stir processing (FSP) technique used for the purpose of surface modification by refining the microstructure of the exposed surface area of the materials. Further, the FSP techniques are also used to produce the surface composites by reinforcing micro- or nano-particles into the grooves cut on the faying surface of the plates. The cross-sectional view of the weld region of the FSW/FSP consists of different distinct regions, namely, weld nugget zone (WNZ), thermo-mechanically affected zone (TMAZ), and heat-affected zone (HAZ). These distinct regions are present in advancing side (AS) and retreating side (RS). The WNZ consists of fully recrystallized, equiaxed, and fine grains of the materials, which are processed in a better way, compared to the other zone. FSW joints fabricated from similar or dissimilar alloys show better mechanical and metallurgical properties compared to the joints produced from conventional fusion welding techniques.[2]

17.2 RESEARCH BACKGROUND

In recent years, aluminum-based metal matrix composites (MMCs) are gaining importance for their applications as structural and functional materials. Several studies[3–7] have reported the fabrication of MMC and metal matrix nanocomposites using FSW and FSP techniques by reinforcing the micro- or nano-sized particles.

Byung-Wook et al.[3] in their study, fabricated AA5083 and silicon carbide nanoparticles (SiC_{NP}) composite using the FSW technique. SiC_{NP} are inserted into the rectangular grooves that are cut on the adjoining surfaces of the plates. They investigated the effect of FSW passes (one and two pass) on the tensile, hardness properties, and the microstructure characteristics of the fabricated AA5083/SiC_{NP} composite using constant tool rotation speed (TRS) of 1800 rpm and tool welding speed (TWS) of 22 mm/min. Further, they report that compared to FSW one pass the two-pass results in WNZ consisting of higher hardness of HV90 and uniform distribution of SiC_{NP}. Further, they report that the presence of SiC_{NP} in the WNZ facilitates the grain refinement by the pinning effect and results in the reduced grain size compared to the joints produced without SiC_{NP}. Similar studies are carried out by Abbasi et al.[4] in their experimental study on the effect of TRS (600, 800, and 1000 rpm) and TWS (25, 75, 125, and 175 mm/min) in fabricating AZ31 magnesium alloy FSW joints produced by inserting the SiC_{NP} in the rectangular grooves using FSW single pass. They report that TRS of 800 rpm and TWS of 75 mm/min results in proper distribution of SiC_{NP} and yields higher weld joint strength. Further, they report that higher strength in the weld joint of the FSW joints is due to the low grain size and interparticle spacing (λ) observed in the WNZ.

Bahrami et al.[5] fabricated an MMC of 7075/SiC_{NP} using FSW technique using a single pass. They report higher UTS and percentage elongation (%EL) for the FSW joint specimens fabricated for the combination of TRS of 1250 rpm with TWS of 40 mm/min compared to the specimen without SiC_{NP}. Similarly, Tabasi et al.[6] fabricated FSW dissimilar joint of AA7075-O to AZ31 magnesium alloy by reinforcing SiC_{NP} in the cut groove. TRS of 450, 560, 710, 900, 1100 rpm and TWS of 11.2, 22.4, 35.5, and 45 mm/min are used to weld the joints. They report that the FSW dissimilar joint specimen prepared with TRS of 560 rpm and TWS of 22.4 mm/min yields maximum UTS of 206 MPa and grain size of 4.36 μm. Further, they report that the pinning effect of SiC_{NP} reduces the grain size and has significant effect on weld joint properties. In addition, the combination of higher TRS and lower TWS results in a more uniform distribution of SiC_{NP}. Pantelis et al.[7] in their investigations to study the effect of TRS (600–1180 rpm), TWS (60–85 mm/min) fabricated the FSW dissimilar joints of AA5083-111 and AA6082-T6 by reinforcing SiC_{NP} in the rectangular groves. Further, they report the FSW second pass given in the opposite direction to the first pass results in uniform distribution of SiC_{NP}, higher joint strength, %EL and hardness of 81HV compared to the joint without SiC_{NP}. Kumar et al.[8] in their experimental investigation in finding the effect of TRS range 400–1800 rpm on tensile properties and microstructure of FSW joints of AA2024/AA7075 reinforced with SiC_{NP} reports that TRS of 1000 rpm, constant TWS of 40 mm/min, tool plunge depth (TPD) of 6.20 mm yields UTS, 380 MPa, yield strength (YS), 150 MPa.

The present study aims in fabricating composite at the WNZ of the FSW joints of AA2024-T351 to AA7075-T651 by reinforcing SiC_{NP} in the WNZ. The effect of TWS range of 20–200 mm/min on the tensile properties, microstructure is influenced by the distribution of SiC_{NP} inside the weld region of the weld joint specimens has been investigated. The FSW experiments and joints are produced by varying TWS (20–200 mm/min) along the weld joint line using FSW single pass. The optimum TWS is selected based on the higher UTS, YS, and %EL.

17.3 RESEARCH METHODS

17.3.1 MATERIALS AND MACHINE

The aluminum alloy individual plates of AA2024-T351 and AA7075-T651 each with thickness of 6.35 mm are cut to dimensions of each plate with a width of 75 mm and length of 220 mm using electrical discharge machine (EDM). The adjacent surfaces that are used to produce the butt–joint of each plate were machined to have a rectangular groove cut of size 0.2 mm width and depth of 5 mm. The total dimension of the groove width would be 0.4 mm when the two plates are fixed in butt–joint configuration. Figure 17.1a shows the two plates arrangement in butt–joint configuration. SiC_{NP} of size 50 nm (supplier Alfa Aesar, UK) having a purity ~98.5% were reinforced in the groove and pressed tightly to ensure that unfilled region in the weld joint line was occupied with SiC_{NP} before the FSW process. FSW tool used in fabricating the joints consists of a taper threaded cylinder with a pin probe height of 5.80 mm tapered from top-end diameter of pin, d_2 of 7.50 mm to bottom end diameter of pin, d_1 of 6 mm, and the remaining length is of cylinder with shoulder diameter of 20 mm. The tool is having a hardness of 55 HRC and machined out of hot die steel. A five-axis FSW machine (BISS Pvt. Ltd., Bengaluru) capable of varying the TRS, TWS, and TPD along the weld length is used for joining the plates reinforced with SiC_{NP}. After the fabrication of FSW dissimilar joints for the varying TWS experiments range 20–200 mm/min, the tensile specimens are cut and prepared following ASTM code of E8M-04 standards, as shown in Figure 17.1b. The specimens are evaluated for the UTS, YS, and % EL using 100 kN capacity universal testing machine (Make SHIMADZU and type AGS-X series). The stress–strain graphs are plot, and UTS, YS, %EL are evaluated. Further, the microstructure analysis was carried out using scanning electron microscopy (SEM) machine (JEOL type JSM-6380 LA model), and the grain size, microstructure modification, and distribution of SiC_{NP} are investigated. A low magnification (10X) stereo microscope is used to study the shape of the weld nugget, the joint quality of the fabricated, and defects in the weld region of dissimilar joints.

17.3.2 VARYING TOOL WELDING SPEED EXPERIMENT TO PRODUCE FSW DISSIMILAR JOINT REINFORCED WITH SIC_{NP}

Figure 17.1b presents the varying TWS experiment trials carried out to produce FSW joint reinforced with SiC_{NP}. The FSW tool is made to rotate in CW direction and traversed along the weld joint line for a 160 mm length. The FSW single pass is used for joining dissimilar joint by varying the TWS range 20–120 mm/min along the weld-joint line for the first set of experiment trials. The TPD of 6.20 mm, TRS of 1000 rpm, and tool tilt angle of 2° are considered for the varying TWS experiment. Similarly, the dissimilar joint is produced in the second set of FSW experiment by varying the TWS from 120 to 200 mm/min.

In Figure 17.1b, "*T*" and "*M*" indicate extraction locations of the tensile and microstructure specimens, respectively. In the first set of FSW experiment, total eight tensile specimens of T1–T8, and in the second set for varying TWS range 120–200 mm/min the tensile specimens of T9–T16 are extracted using wire EDM process. Similarly, 16 microstructure specimens are prepared according to the dimension of width 40 mm and thickness of 10 mm from both the first and second set of FSW experiments. The tensile test is conducted to find UTS, YS, and %EL for the specimens produced for different conditions of varying TWS. Similarly, for the extracted microstructure specimens, the metallographic analyses are carried out in order to investigate the microstructure and distribution of SiC_{NP} at the WNZ for different conditions of varying TWS used to obtain dissimilar joints produced with SiC_{NP}.

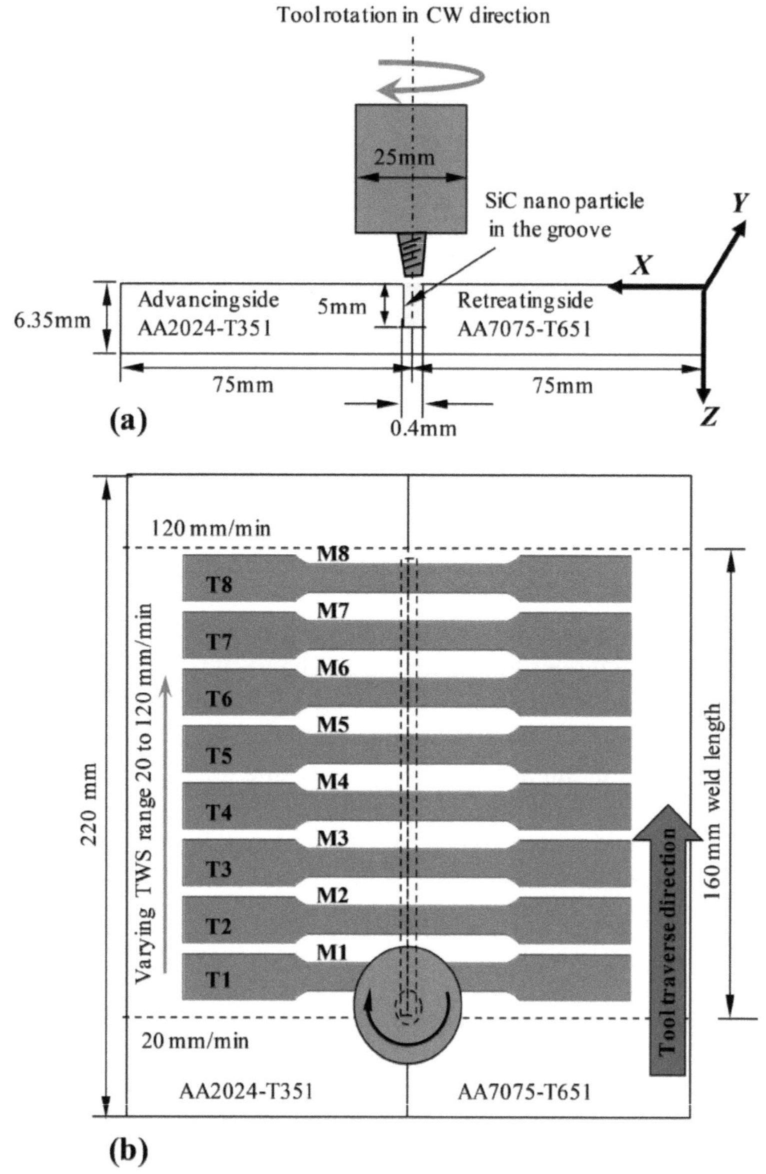

FIGURE 17.1 Schematic view of AA2024-T351 and AA7075-T651 showing (a) rectangular cut groove 0.4 mm width and 5 mm depth reinforced nanoparticles and (b) tensile and microstructure specimen extraction location in the FSW dissimilar joint.

17.4 RESULTS AND DISCUSSION

17.4.1 MACRO- AND MICROSTRUCTURE

Macro- and microstructures of the fabricated FSW joints reinforced with SiC_{NP} for varying TWS condition range 20,200 mm/min, constant TRS of 1000 rpm, and TPD of 6.20 mm by two sets of FSW experiments are discussed in the proceeding section.

Effect of varying TWS on microstructure

Figure 17.2 shows the FSW joints fabricated with SiC_{NP} reinforcement by varying TWS experiment for 20,200 mm/min, constant TRS of 1000 rpm, and TPD of 6.20 mm. One can observe from Figure 17.2a for TWS range 80–120 mm/min, the weld crown surface has a smooth surface finish. However, for a TWS from 20 to 70 mm/min, the weld crown surface has a moderate flash. This flash observed is due to lower TWS that results in the increased heat input to the weld region and tool shoulder contact with the material. Figure 17.2b shows excessive flash for the TWS of 120 mm/min. This is due to the initial tool dwell time of 20 s. Further increase in the TWS, above 120 mm/min observes smooth weld crown surface with absence of flash.

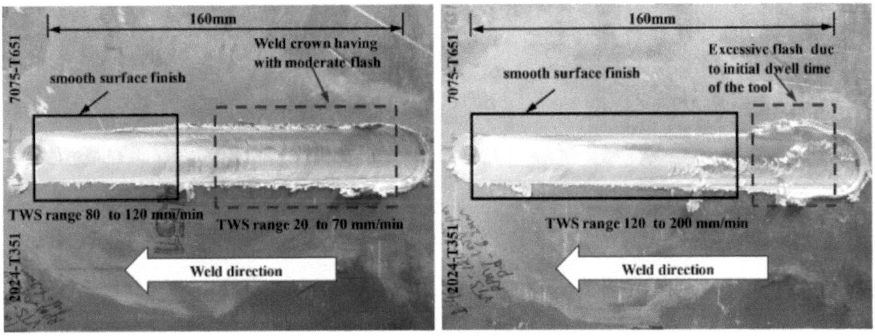

FIGURE 17.2 Photographs showing the FSW joints reinforced with SiC_{NP} for varying TWS experiment for (a) 20–120 mm/min (b) 120–200 mm/min.

Table 17.1 illustrates the macrostructure of the transverse sections of the weld joints specimens fabricated for the varying TWS range 20–200 mm/min. The macrostructural changes in the WNZ, TMAZ, and HAZ in terms of size and shape resulting from the varying TWS condition can be observed in Table 17.1. For the lower TWS of 30–75 mm/min, the WNZ exhibits

TABLE 17.1 Optical Macrostructure for Varying TWS Range 20–200 mm/min.

Tool welding speed	Macrostructure AS RS	Tool welding speed	Macrostructure AS RS
30	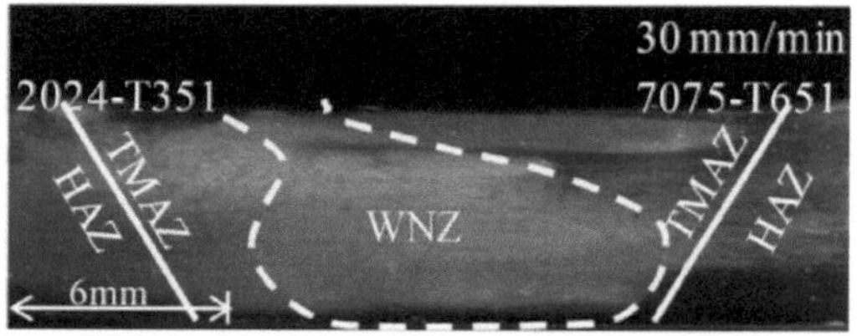	135	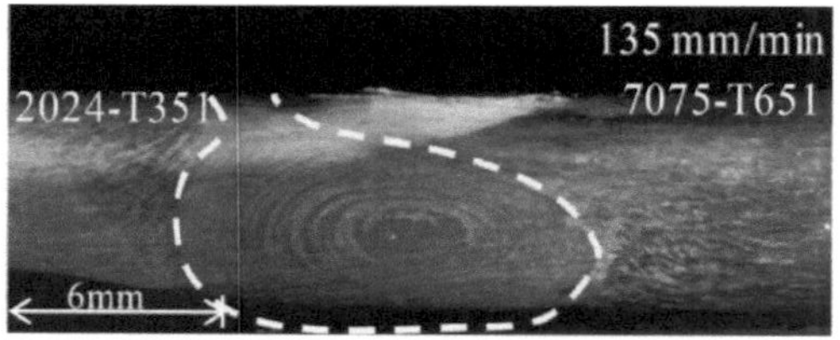
40	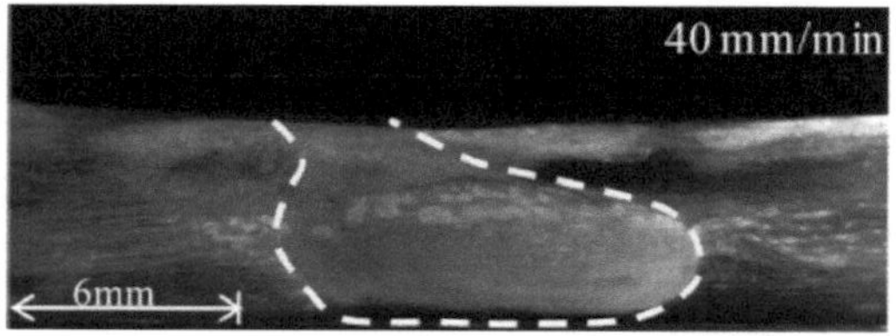	145	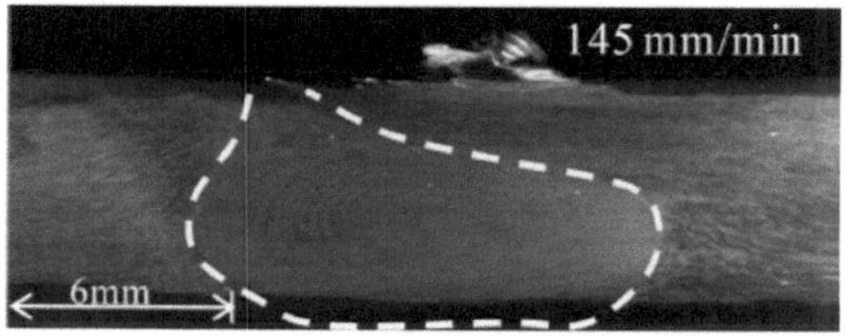
55	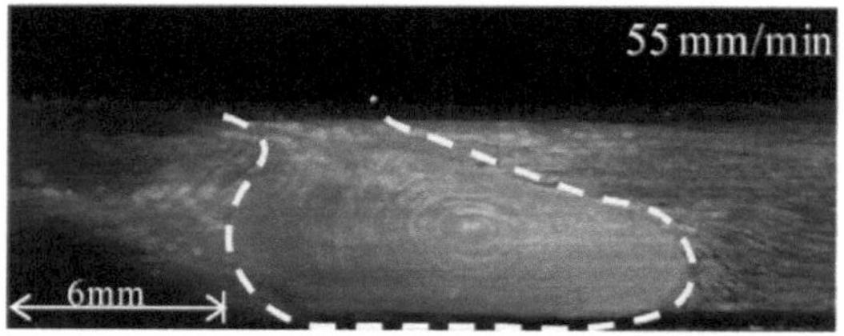	155	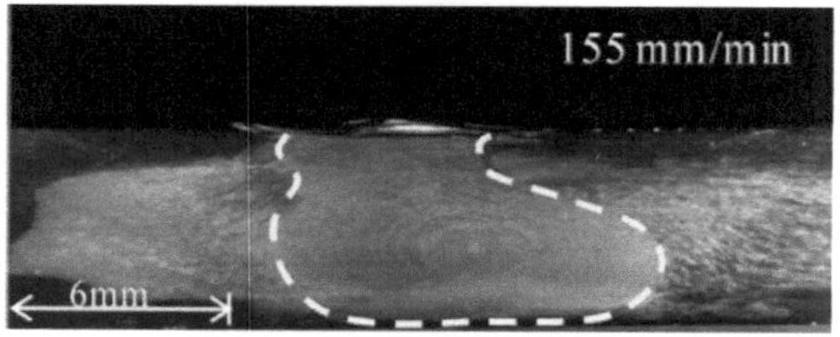

TABLE 17.1 *(Continued)*

Tool welding speed	Macrostructure (AS – RS)	Tool welding speed	Macrostructure (AS – RS)
75	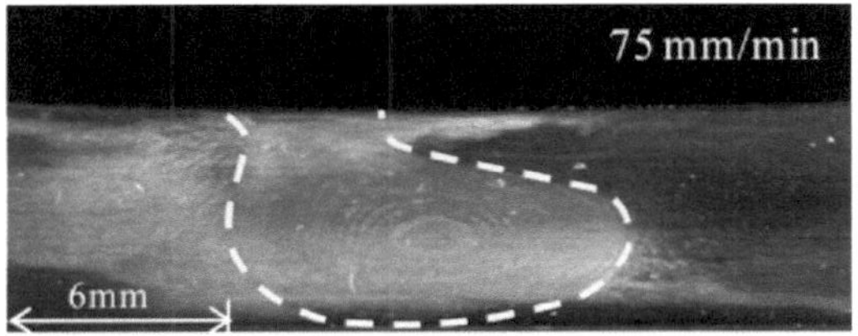	165	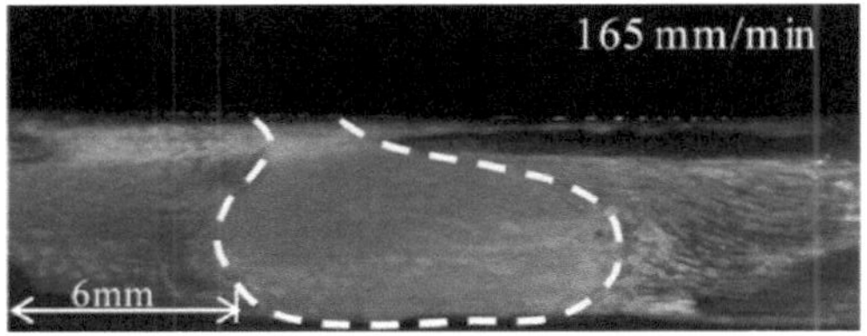
90	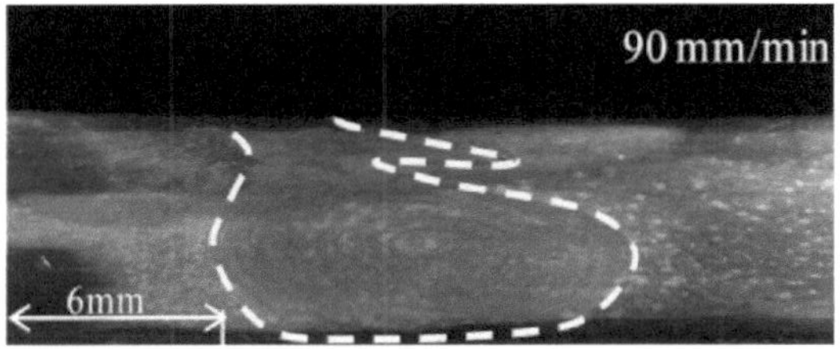	175	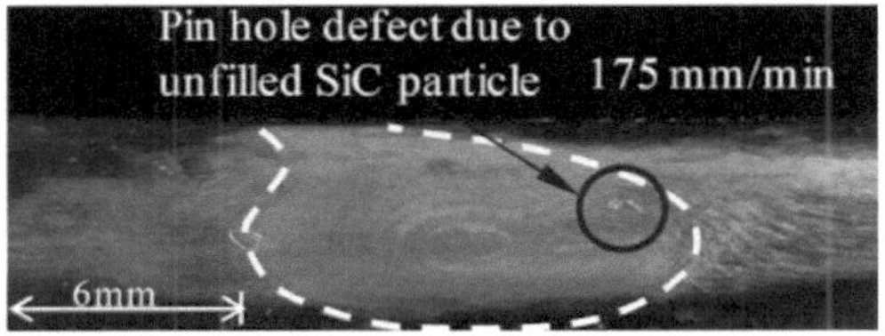
105	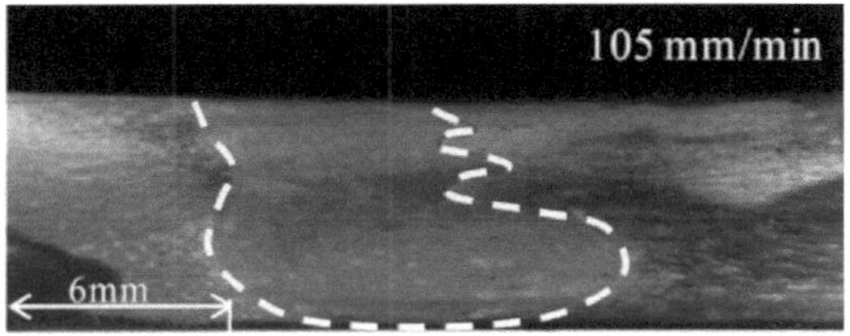	185	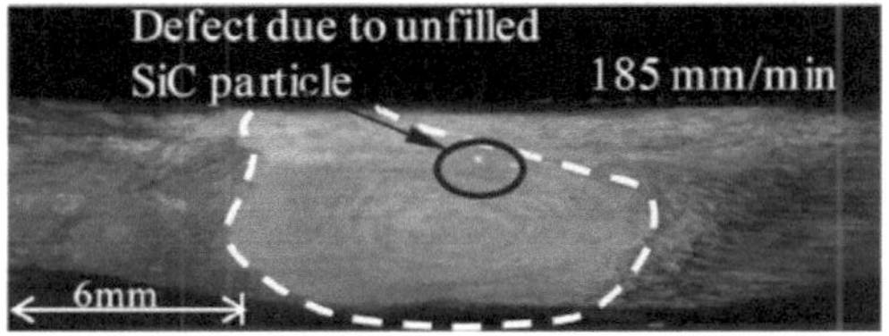

"onion ring" pattern, a typical characteristic of FSW. In addition, the width of the WNZ was much larger compared to the WNZ for higher TWS. This variation in the width of the WNZ is due to the change in the heat supply to the weld region for varying TWS. The heat supply to the weld region increases for lower TWS and decreases or higher TWS.[9] Further, increase in the TWS above 165 mm/min leads to the formation of pinhole defects in the WNZ. This is due to the low heat input to the weld region that results in cold processing condition.[9,10] From the macrostructure, it is observed that the changes in the heat input with varying TWS condition affect the WNZ size and shape, and presence of the defects in the weld region.

The SiC_{NP} distribution in the WNZ of the microstructure specimens for the varying TWS was carried out using SEM. Figure 17.3a–i presents the microstructural features of WNZ observed for TWS range 20–200 mm/min. For the TWS 30 mm/min, in Figure 17.3a, only the top region of the WNZ observes agglomeration of the SiC_{NP}. This is may be due to the increased tool residence time during the initial plunge as a result SiC_{NP} stretch was observed at the top of the WNZ. However, the middle portion of the WNZ observes uniform distribution of SiC_{NP}. Further, the TWS to 40 mm/min results in increased frictional force induced by the tool shoulder and observes uniform distribution of SiC_{NP} in the entire weld region.[11] The distribution of SiC_{NP} is influenced by the varying TWS. Further, increase in the TWS above 55 mm/min results in less heat generation and reduced tool residence time causing the SiC_{NP} agglomeration defects near to the top region of the WNZ. From Figure 17.3d–i, higher TWS speed range 75–185 mm/min results in SiC_{NP} stretching toward RS. Thus, from the above varying TWS experiment, it can be concluded that in the present study, the TWS of 40 mm/min results in defect-free WNZ and observes homogeneously distributed SiC_{NP}. For TWS of 40 mm/min, the SEM images of Figure 17.4a observe a grain size range 3.5–5.5 μm at the WNZ, and Figure 17.4b is the back scattered electron (BSE) images, which presents the homogeneous distribution of nanoparticles along the grain boundaries at the WNZ.

17.4.2 EFFECT OF VARYING TWS EXPERIMENT ON TENSILE STRENGTH

Figure 17.5 shows the graph plot for the tensile properties of the FSW joints fabricated with SiC_{NP} reinforcement for the varying TWS range 20,200 mm/

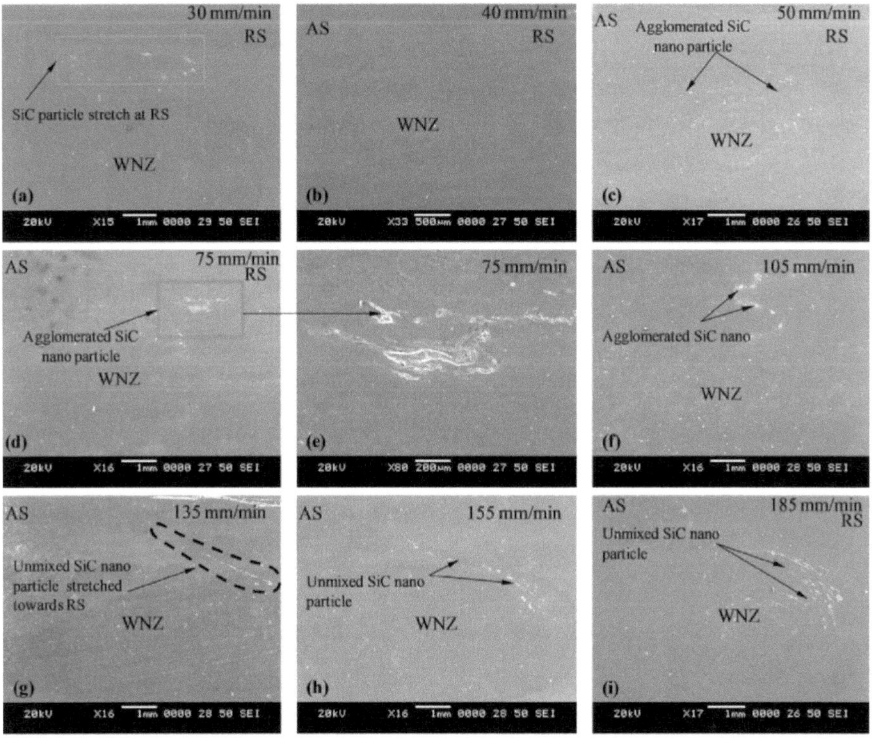

FIGURE 17.3 Low magnification SEM images of FSW joint reinforced with SiC_{NP} for varying TWS range 20–200 mm/min.

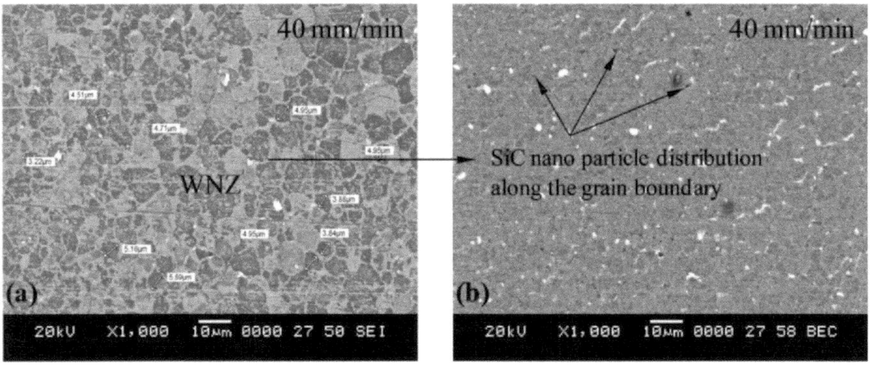

FIGURE 17.4 SEM images showing (a) grain size range 3.5–5.5 µm at the WNZ, (b) BSE images showing SiC_{NP} distribution at WNZ.

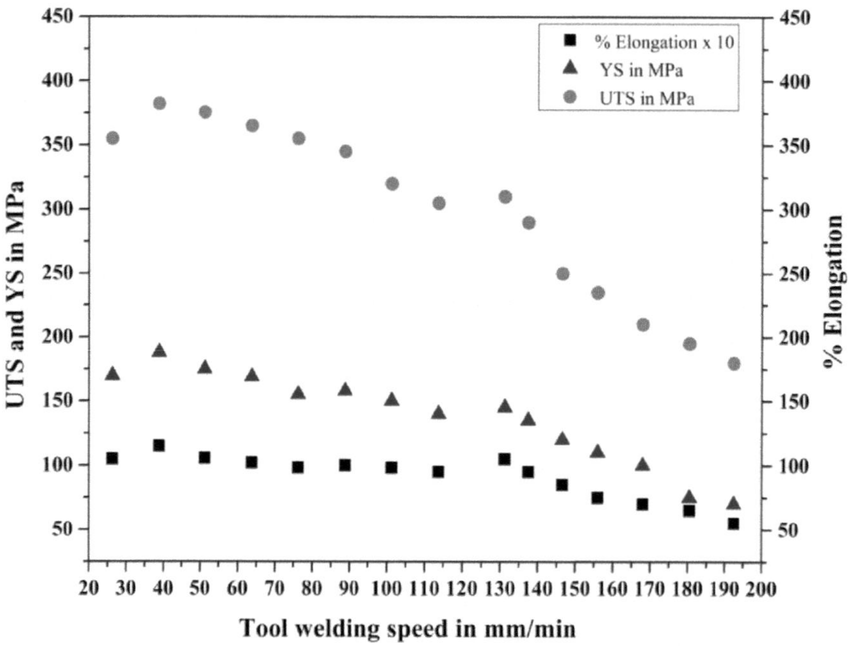

FIGURE 17.5 Tensile properties of the FSW joints for the varying TWS range 20–200 mm/min.

min. From Figure 17.5, a higher UTS, YS, and %EL is observed for lower TWS range 30–50 mm/min. UTS varies from 355 to 382 MPa, YS from 170 to 185 MPa, and %EL from 10.5 to 11.5 was observed for lower TWS. This is due to the sufficient heat input and stirring of the SiC_{NP} in the weld region that results in homogeneous distribution of SiC_{NP}. Further, the pinning pressure increases due to the uniform distribution of secondary phase particles SiC_{NP} in the weld region. Humphreys's[12] study reports about the Zener pinning pressure (Z) that arises due to the effect of pinning along the grain boundaries by the secondary-phase particles. The parameter, Z, is given by[12]

$$Z=\frac{3\dot{F}_v R}{d}. \tag{17.1}$$

where d, R, F_v represent the diameter of the second phase particles, average radius of the grain, and volume fractions, respectively. Thus, it can be stated that the lower TWS results in uniform distribution secondary phase particles SiC_{NP} in the weld region improve the strength of the joint due to load transfer and Orowan strengthening mechanisms. Orowan-Ashby's[13] equation is given by:

$$\sigma_{\text{orowan}} = \frac{0.13 G_m b}{\lambda} ln \frac{r}{b} \tag{17.2}$$

where b is the burgers vector, G_m is the shear modulus of the matrix, λ is the mean interparticle spacing, and r is the radius of reinforcement particles.

Further increasing the TWS above 75 mm/min observes decrease in UTS, YS, % EL, as a result of lack of heat input and inadequate stirring of the SiC_{NP} at the WNZ for higher TWS. This results in the agglomeration defects in the WNZ, as observed in Figure 17.3d–i, and leads to the lower tensile properties. Thus, the FSW joints fabricated with TWS of 40 mm/min yield higher UTS of 382 MPa, YS of 188 MPa, and %EL of 11.5. Hence, TWS of 40 mm/min is considered as optimum TWS for joining FSW joint reinforced with SiC_{NP}.

17.5 CONCLUSION

In this present study, the varying TWS experiment is carried out to fabricate the FSW joint by reinforcing SiC_{NP}, and the results are summarized as follows:

- The SiC_{NP} distribution in the weld region, tensile properties, and microstructure of the FSW joint specimens are influenced by the TWS.
- The homogeneously distributed SiC_{NP} in the WNZ, higher tensile strength, and better microstructure characteristics were observed for lower TWS. FSW joints produced with TWS of 40 mm/min yields higher UTS, 382 MPa, YS, 188 MPa, and %EL, 11.5.
- Increase in the TWS above 75 mm/min results in lack of heat generation and inadequate stirring of the SiC_{NP} in the WNZ results in agglomeration defects in the WNZ. Thus, higher TWS yields decrease in UTS, YS, and % EL.

KEYWORDS

- **friction stir welding**
- **aluminum alloys**
- **dissimilar joints**
- **nanoparticles**
- **microstructure**
- **composite**

REFERENCES

1. Thomas, W. M.; Nicholas, E. D.; Needam, J. C.; Murch, M. G.; Templesmith, P.; Dawes, C. J. GB Patent Application No. 9125978.8, Dec 1991 and US Patent No. 5460317, Oct 1995.
2. Mishra, R.S.; Ma, Z.Y.;. Friction Stir Welding and Processing. *Mater. Sci. Eng. R Rep.* **2005,** *50* (1), 1–78.
3. Byung-Wook, A. H. N.; Don-Hyun, C. H. O. I.; Yong-Hwan, K. I. M.; Seung-Boo, J.U.N.G.;. Fabrication of SiCp/AA5083 Composite via Friction Stir Welding. *Trans. Nonferrous Metals Soc. China* **2012,** *22*, 634–638.
4. Abbasi, M.; Abdollahzadeh, A.; Bagheri, B.; Omidvar, H. The Effect of SiC Particle Addition during FSW on Microstructure and Mechanical Properties of AZ31 Magnesium Alloy. *J. Mater. Eng. Perform.* **2015,** *24* (12), 5037–5045.
5. Bahrami, M.; Dehghani, K. Givi, M.K.B. A Novel Approach to Develop Aluminum Matrix Nano-Composite Employing Friction Stir Welding Technique. *Mater. Design* **2013,** *53*, 217–225.
6. Tabasi, M.; Farahani, M.; Givi, M.B.; Farzami, M.; Moharami, A. Dissimilar Friction Stir Welding of 7075 Aluminum Alloy to AZ31 Magnesium Alloy Using SiC Nanoparticles. *Int. J. Adv. Manuf. Technol.* **2016,** *86* (1), 705–715.
7. Pantelis, D.I.; Karakizis, P.N.; Daniolos, N.M.; Charitidis, C.A.; Koumoulos, E.P.; Dragatogiannis, D.A. Microstructural Study and Mechanical Properties of Dissimilar Friction Stir Welded AA5083-H111 and AA6082-T6 Reinforced with SiC Nanoparticles. *Mater. Manuf. Process.* **2016,** *31* (3), 264–274.
8. Kumar, K. A.; Murigendrappa, S. M.; Kumar, H.; Shekhar, H. Effect of Tool Rotation Speed on Microstructure and Tensile Properties of FSW Joints of 2024-T351 and 7075-T651 Reinforced with SiC Nano Particle: The Role of FSW Single Pass. *AIP Conf. Proc.* **2018,** *1943* (1), 020056.
9. Liu, H. J.; Hou, J.C.; Guo, H. Effect of Welding Speed on Microstructure and Mechanical Properties of Self-Reacting Friction Stir Welded 6061-T6 Aluminum Alloy. *Mater. Design* **2013,** *50*, 872–878.
10. Rajakumar, S.; Muralidharan, C.; Balasubramanian, V. Influence of Friction Stir Welding Process and Tool Parameters on Strength Properties of AA7075-T6 Aluminium Alloy Joints. *Mater. Design* **2011,** *32* (2), 535–549.
11. Tutunchilar, S.; Haghpanahi, M.; Givi, M.B.; Asadi, P.; Bahemmat, P. Simulation of Material Flow in Friction Stir Processing of a Cast Al–Si Alloy. *Mater. Design* **2012,** *40*, 415–426.
12. Humphreys, F. J. A Unified Theory of Recovery, Recrystallization and Grain Growth, Based on the Stability and Growth of Cellular Microstructures—I. The Basic Model. *Acta Mater.* **1997,** *45* (10), 4231–4240.
13. Dieter, G. E.; Bacon, D. *Mechanical Metallurgy*, Vol. 3; McGraw-Hill: New York, 1986.

CHAPTER 18

EFFECT OF TOOL PIN PROFILE ON MECHANICAL AND WEAR PROPERTIES OF FRICTION STIR WELDING OF DISSIMILAR AA6061 AND AA5052 ALUMINUM ALLOYS

PRADYUMN K. ARYA[1], BHAVESH CHAUDHARY[1], NEELESH KUMAR JAIN[1], and M JAYAPRAKASH[2]

[1]*Discipline of Mechanical Engineering, Indian Institute of Technology Indore, Madhya Pradesh, India*

[2]*Discipline of Metallurgy Engineering and Materials science, Indian Institute of Technology Indore, Madhya Pradesh, India*

**Corresponding author. E-mail: phd1801103006@iiti.ac.in*

ABSTRACT

The main objective for the existing research needs to identify the appropriate friction stir weld tool pin profile for welding distinct aluminum alloys AA6061 and AA5052. To investigate their effect on tensile strength, microhardness and wear characteristics of dissimilar joints, four separate tool pin forms were utilized: straight cylindrical (SC), straight threaded cylindrical (STC), tapered cylindrical (TC), and straight square (SS). Joints were produced with different tool pin profiles even at a constant rotational speed of 1000 rpm and two different table traverse speed of 60 and 80 mm/min. A SC tool pin design has been found to generate surface galling defects, while

Modern Manufacturing Systems: Trends and Developments. Rajiv Kumar Garg, Ravi Pratap Singh, Rajeev Trehan, & Ramesh Singh (Eds.)

a TC tool pin profile generates inadequate heat diffusion as well as defective morphology of the surface. The results showed that tool pin profiles had an important influence on the mechanical characteristics of dissimilar joints. Threaded cylindrical tool pin profile yielded a higher tensile strength of 231 MPa besides a joint efficiency around 74.5 %, while the square tool pin profile obtained a maximum microhardness of 94 HV in the stir region. Grain size in the stir zone was enhanced through adjustments in the design of the tool pin profile and the average grain size for the square and threaded cylindrical tool pin profile was 35 and 39 μm. The highest hardness value was observed in the weld NZ near to advancing side of AA6061, and the minimum hardness value was observed throughout the HAZ near retreating side of AA5052 Al Alloys. Among all pin profiles, STC and SS pin profiles had been identified to be appropriate tool pin profiles of friction stir welding of AA6061–AA5052 Al alloy.

18.1 INTRODUCTION

Aluminum and its alloys provide an extensive variety of application areas in the manufacturing, aerospace, automotive, defense, and medical industries because of that strong strength to weight proportion.[1] In industrial application fields, the requirement for merging distinct metals also emerges due to the necessities for a wide range of metals for weight reduction and enhanced engineering structure efficiency.[2] The current practices of joining aluminum alloys by fusion-based processes are not amenable due to their long solidification range that induces hot cracking, especially in AA6061 alloys.[3] The alternative approach is the use of FSW, which is an autogenously new solid-state welding process, particularly welding of aluminum alloys invented by TWI.[4] To present, several welding methods have been established to interact with various categories of alloys where welding of multiple metals remained a challenging issue.[5] For welding of numerous metals, friction stir welding (FSW) is already introduced and has already been implemented on a number of joint combinations namely butt joint, lap joint, and T joint.[6] In FSW, heat is generated by friction between the nonconsumable moving tool as well as the workpieces, causing the materials to plasticize and subsequently material bonding.[7] However, defects such as tunnel, flash, and kissing bonds are generally observed during FSW due to improper process parameters or tool geometry.[11,12] Through FSW, the tool pin profile performs an important function in joint success.[13] It is a major consideration in deciding the morphology of the weld, localized heat, and stirring activity.[9] The behavior of flow of metal furthermore, being mainly affected either by morphology of tool pin, the specifications of tool pin as well as the variables of the FSW operation.[10]

The effects of tool morphology on different features of FSW had currently been reviewed by several researchers. Palanivel et al.[14] observed that the influence of rotation speed of the tool as well as the pin profile greatly affected the weld quality and microstructure. They also stated that difference in tensile strength being attributed to metal transfer behavior, the removal of cold working influence on heat-affected zone (HAZ) of AA5083, as well as the dissolution of second phases precipitates in alloy AA6351. Ilangovan et al.[15] analyzed the influence of pin feature on a microstructural and weld characteristics of distinct AA6061–AA5086 aluminum alloys of friction stir welded. They stated that throughout the nugget region, the straight cylindrical (SC) pin feature produces cross-sectional macro-level faults. On the other side, threaded and tapered cylindrical (TC) pin profile tools yield defect-free joints. Similarly, owing to excellent joint achievement they preferred the threaded cylindrical pin features over the TC pin features. Rajakumar et al.[16] reported the suitable tool dimension to weld AA7075-T6 alloys. Tool specifications of 15 mm shoulder diameter, 5 mm pin diameter with 45 HRC hardness were recommended to achieve stronger strength characteristics according to optimal process specifications. Kalemba et al.[17] reported an improvement in table traverse speed, leading to an improvement in weld strength to a higher amount whereas a further raise in tool traverse speed contributes in a decline in the weld strength of FSWed joints. Optimum tensile and fatigue characteristics had been obtained for welding with AA6082 on advancing side for joining of dissimilar AA6082 and AA2024 al alloy. Venkateshkannan et al.[18] noted the maximum tensile strength and defects free welds on plates that were welded using the stepped pin tool for the welding of AA2024 and AA5052 al alloy. Sivachidambaram et al.[19] studied the FSW of dissimilar AA5383 and AA7075 aluminum alloy and found that the lower tool rotational speed with low welding speed yields maximum tensile strength. Owing to incorrect mixing, they reported reduced strength at higher rotational speed of tool and table speed. Khan et al.[20] owing to inadequate preference of instrument pin deviation and plunge depth, recorded tunneling and kissing bonding faults in FSWed unsimilar AA5083-H116 and AA6063-T6 alloys. Ahmed et al.[21] studied the welding of dissimilar AA7075–AA5083 alloys. They mentioned here that grain shape and form of precipitates get a significant effect on recrystallization operation, which is necessary for the finished grain configuration within the NZ, in addition to process specifications.

Though several studies reported the influence of various process parameters on microstructure and mechanical properties of FSWed dissimilar aluminum alloys, the influence of various tool pin profiles on FSWed of dissimilar AA6061–AA5052 Al alloy is still missing and needs to be addressed. In the context of this requirement, the goal of such a research is being to explore the impact on mechanical and wear properties of various tool pin features of dissimilar FSWed AA6061–AA5052 aluminum alloy joints.

18.2 EXPERIMENTAL PROCEDURE

Aluminum alloy sheets AA6061 and AA5052 having a size of 100 × 50 × 3 mm had been involved in the present work. The chemical composition and mechanical properties of the alloys AA6061 and AA5052 are presented in Tables 18.1 and 18.2. FSW tools had been manufactured from H13 tool steel; each of them has a 46 HRC hardness. The should of each tool was designed with a concavity of 10°. The shoulder diameter was 14 mm for each tool, described in Figure 18.1. Table 18.3 presents the variables of the welding procedure.

TABLE 18.1 Chemical Composition of AA6061 and AA5052.

Alloy	Mg	Si	Mn	Fe	Zn	Ti	Al
AA6061	0.92	0.61	0.06	0.33	0.03	0.02	Bal.
AA5052	2.30	0.15	0.02	0.20	0.01	0.01	Bal.

TABLE 8.2 Mechanical Properties of AA6061 and AA5052.

Base Material	UTS (MPa)	% Elongation	Microhardness (HV)
AA6061	308	12	106
AA5052	260	10	81

The aluminum alloys plates were cleaned by the steel brush and acetone before the weld was initiated. As stated by too many reports, the high-strength alloy must be positioned mostly on the advancing side, which exhibits intense mixing compared to the retreating side. Therefore, mostly on the advancing side, Al Alloy AA5052 is being placed, while along the retreating side, the AA6061 plate is being positioned.

TABLE 18.3 Welding Parameters.

Process parameters	Values
Tool rotational speed (rpm)	1000
Welding speed (mm/min)	40, 60
Tool tilt angle	0
Tool shoulder diameter	14
Tool pin profile	SC, TC, SS, STC
Shoulder surface	Concave
Tool material	H13 tool steel

SC, straight cylindrical; TC, tapered cylindrical; SS, straight square; STC, straight threaded cylindrical.

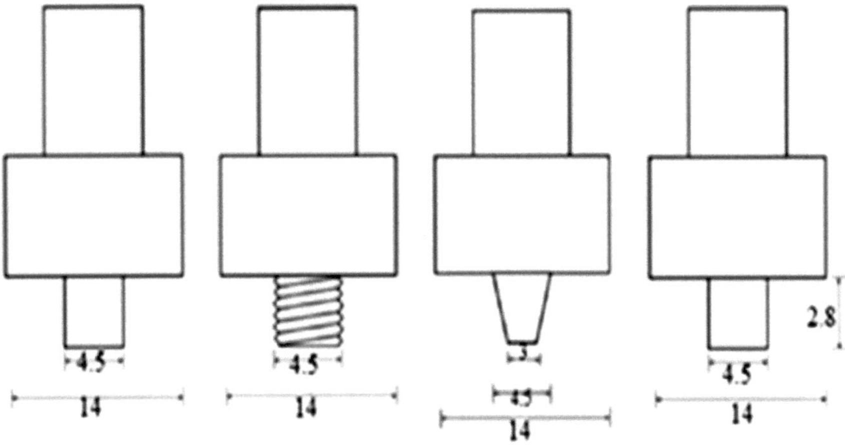

FIGURE 18.1 Different tool pin profile (a) straight cylindrical (SC), (b) straight threaded cylindrical (STC), (c) tapered cylindrical (TC), and (d) straight square (SS).

After welding, metallographic samples were cut from each welded cross-section using wire electric discharge machining. To reveal grain boundaries, samples were finely polished, and Keller's etchant was applied. Tensile specimens being extracted throughout the transverse direction from the cross-section of the weld according to the ASTM E-08 standard, as shown in Figure 18.2. In a Mitutoyo HM-112 Vickers microhardness tester, microhardness indentations had been carried out under a load of 200 g with a dwelling time of 15 s. Samples for wear were cut from the welded zone in rectangular form according to the standard of ASTM G99-04, as shown in

Figure 18.3. Wear tests were carried out on a sliding wear test rig ED-201 wherein a hardened wear disc was made of EN-31 steel. The data on the wear test are as follows:

$$\text{Sliding wear rate, } W_r = \Delta m/L$$

where, W_r = wear rate in gm/mm, Δm = loss in weight in gm, L = sliding distance in mm = $\pi DNT/60\times10^3$ = 2714.336 m (for present work), D = diameter of track in mm = 60 mm, N = Disc rotational speed = 480 rpm, T = Test duration in sec. = 30 × 60 s (for present work).

FIGURE 18.2 Tensile sample of FSWed dissimilar AA6061–AA5052 alloy.

18.3 RESULTS AND DISCUSSION

In this section, surface morphologies and mechanical properties in terms of tensile strength and microhardness, and wear properties of friction stir welded AA6061 and AA5052 concerning different tool pin profile has been discussed.

18.3.1 SURFACE MORPHOLOGIES OF WELD

The surface morphologies of the welded sample produced by different tool pin profiles are shown in Figure 18.4. The cylindrical tool pin profile showed surface galling defects as well as the generation of flash, as shown in Figure 18.4a. The straight threaded cylindrical (STC) pin profile showed the cleaned surface with flash, as shown in Figure 18.4b. TC tool pin profile showed insufficient heat diffusion and defective surface morphology, as shown in Figure 18.4c. By utilizing a square and threaded cylindrical tool pin design, defect-protected welding was developed.

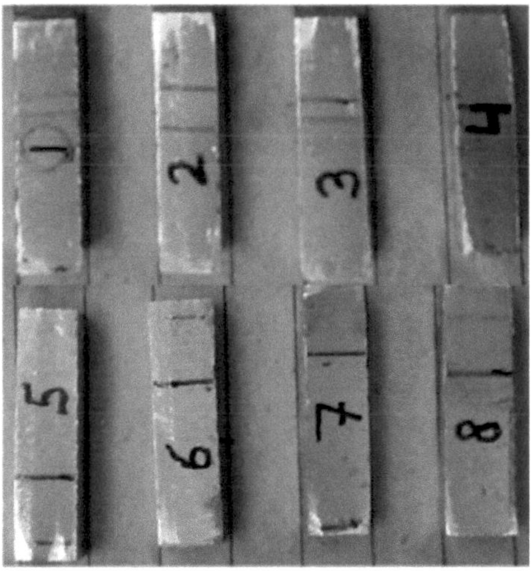

FIGURE 18.3 Wear sample of FSW.

18.3.2 TENSILE PROPERTIES OF THE JOINTS

The effect of the various tools profiles of pin on the tensile strength of dissimilar FSWed AA6061 and AA5052 Al alloy has been illustrated in Figure 18.5. It is indeed observable that, owing to the existence of surface faults throughout the weld region, the weld created by using TC tool pin profile showed a minimum tensile strength of 147 MPa. On the other side, a joint fabricated with threaded cylindrical tool pin profile showed the maximum tensile strength of 231 MPa due to proper mixing and better materials flow between two dissimilar alloys. Consequently, the maximum joint efficiency of 74.5% with the threaded cylindrical tool and a minimum efficiency of 47.4% with the TC tool was obtained. Fracture of the dissimilar Al Alloy joints was occurred in the NZ and near to TMAZ of AA6061 side, as depicted in Figure 18.6. Table 18.4 shows the ultimate tensile strength (UTS), joint efficiency, and position of fracture point for FSWed dissimilar AA6061–AA5052 Al alloys.

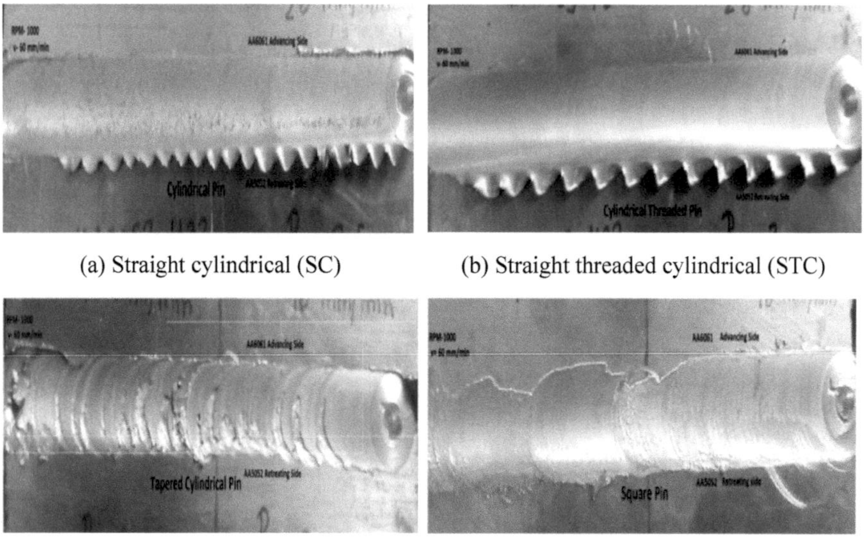

(a) Straight cylindrical (SC) (b) Straight threaded cylindrical (STC)

(c) Tapered cylindrical (TC) *(d) Straight square (SS)*

FIGURE 18.4 Surface morphologies of the welded samples carried out at 1000 rpm and 60 mm/min with different pin profile.

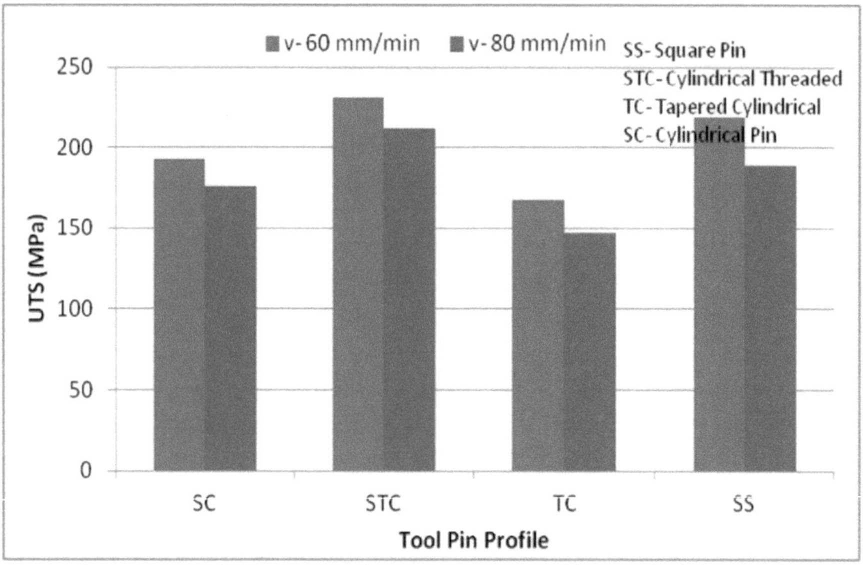

FIGURE 18.5 Shows the UTS at different pin profile for FSWed dissimilar AA6061–AA5052 alloy.

TABLE 18.4 Ultimate Tensile Strength (UTS), Joint Efficiency, and Position of Fracture Point for FSWed Dissimilar AA6061–AA5052 Al Alloys.

S. No.	Tool pin profile	Welding speed (mm/min)	UTS (MPa)	Joint efficiency	Fracture spot
1	Cylindrical	60	193	62.2	NZ
2	Cylindrical	80	176	56.7	NZ
3	Cylindrical threaded	60	231	74.5	TMAZ
4	Cylindrical threaded	80	212	68.4	TMAZ
5	Square	60	219	70.6	TMAZ
6	Square	80	189	60.9	TMAZ
7	Tapered cylindrical	60	168	54.2	NZ
8	Tapered cylindrical	80	147	47.4	NZ

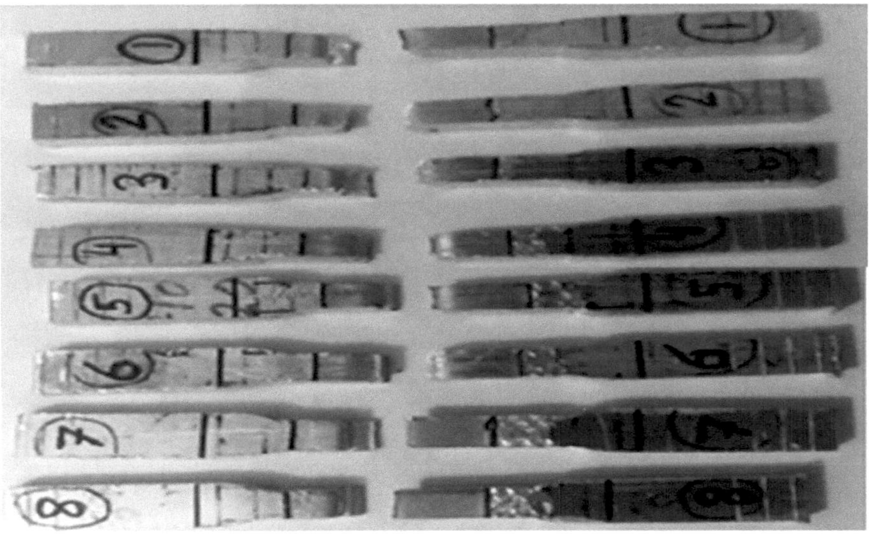

FIGURE 18.6 Tensile fracture sample and corresponding fracture spot position for FSW dissimilarAA6061-AA5052 alloy.

18.3.3 MICROHARDNESS DISTRIBUTION

Figure 18.7 presents the hardness variation for each tool geometry, whereas Figure 18.8 depicts the variation of microhardness for square and cylindrical threaded pin along the traverse cross-section of the weld. From the hardness distribution curve of the square and cylindrical threaded pin, the highest hardness value observed in the weld NZ near to advancing side of AA6061, and the lowest hardness value found in the HAZ near retreating side of AA5052 Al Alloys as shown in Figure 18.7. Weld joints implemented using a straight square (SS) pin profile with the tool rotation speed of 1000 rpm, table speed of 60 mm/min showed the high hardness value in the NZ due to the profound generation of the vorticity in nugget region, as shown in Figure 18.8. Weld joints implemented utilizing a TC profile with the tool rotation speed of 1000 rpm, welding speed of 60 mm/min showed the lowest hardness value in the NZ due to the presence of macroscopic defects. At high welding speed, dissolution and over-aging of precipitates occurs, reducing the hardness in the weld region.

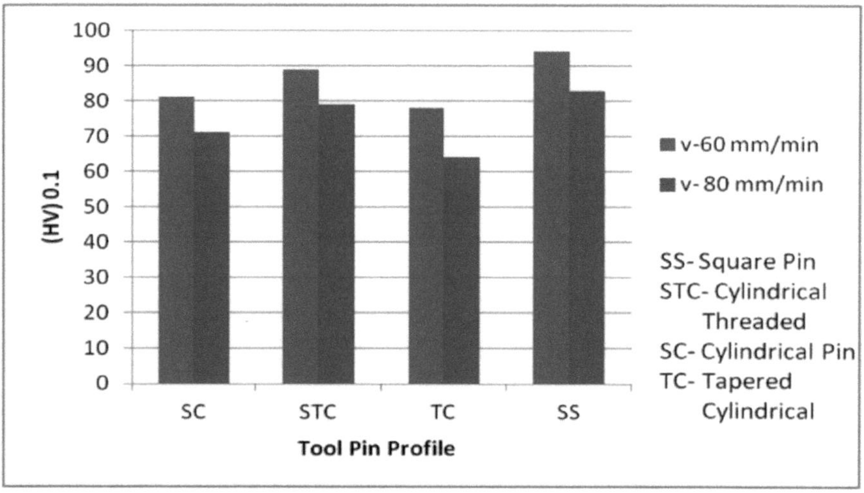

FIGURE 18.7 Graph showing the maximum microhardness at NZ for different pin profile in FSWed dissimilar AA6061–AA5052 alloy.

18.3.4 MICROSTRUCTURE ANALYSIS

Each weld obtained by different tool geometry was observed under DM 2500 M optical microscope. Here, Figure 18.9 shows the optical microstructural diagrams of FSW of distinct Al alloy AA6061 and AA5051 at various tool pin profiles, that is, square, threaded cylindrical, cylindrical, and TC, generated using tool rotation speed of 1000 rpm. Fine-grain size was improved by square and threaded cylindrical tool pin profile; however, under cylindrical and TC tool pin profile, the lesser fine grain structure is identified.

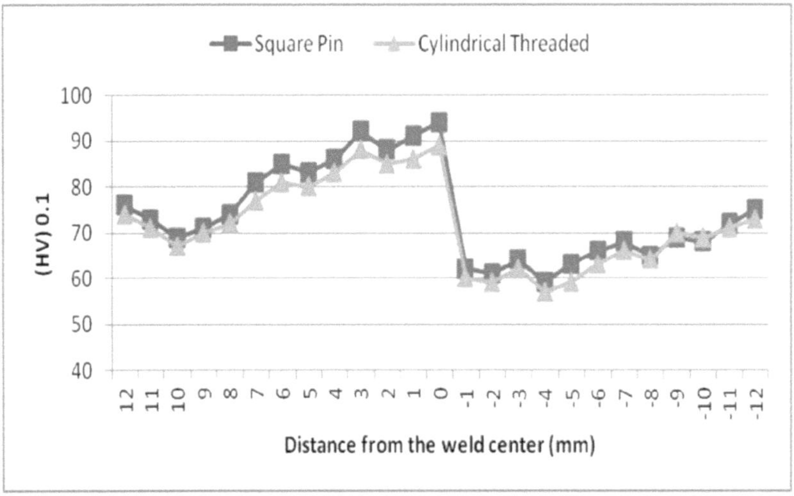

FIGURE 18.8 Graph showing the microhardness distribution across the cross-section of weld fabricated by square pin and a cylindrical threaded pin in FSWed of dissimilar AA6061–AA5052 alloy.

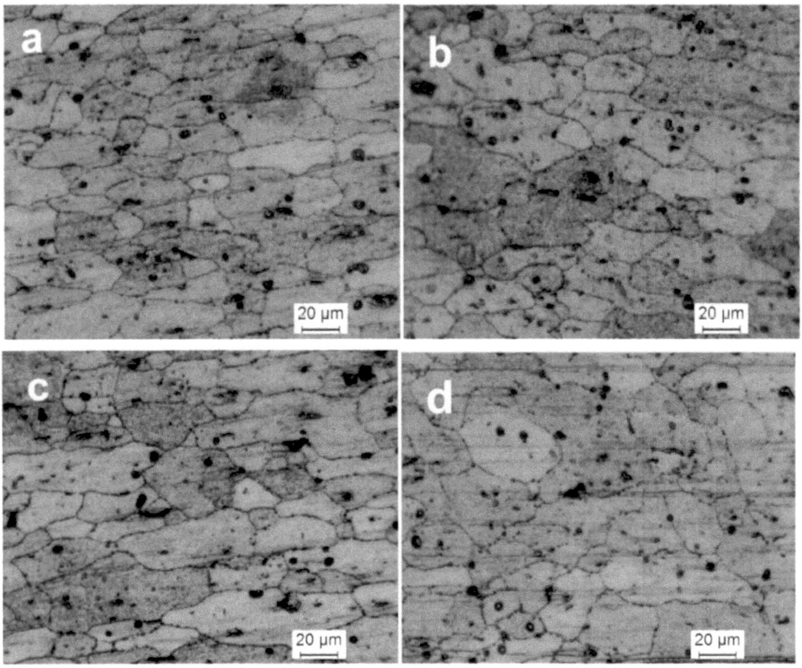

FIGURE 18.9 Microstructure in NZ at different tool pin profile (a) Square (b) Threaded cylindrical (c) Cylindrical (d) Tapered cylindrical.

Grain size in the stir zone enhanced through adjustments in the design of the tool pin profile and the average grain size for the square, and threaded cylindrical tool pin profile was 35 and 39 μm; however, the average grain size was 43 and 49 μm for the cylindrical and TC profiles, respectively. Finer grains arrangement suggests how NZ underwent dynamic recrystallization throughout welding. Two portions were being identified in most of the microstructure, that is, unmixed zone and mechanically mixed flowing zone; however, for cylindrical and TC tool pin profile, such zones were unavailable. A considerable portion of intermixing amongst two distinct alloys has not been observed for the cylindrical and TC tool pin profile, but appropriate mixing has been observed with the use of threaded cylindrical and square tool pin profile.

18.3.5 WEAR PROPERTIES OF JOINTS

Table 18.5 presents the amount of wear rate for friction stir welded samples welded using different tool pin profile, whereas Figure 18.10 demonstrates the influence of tool pin feature on wear rate of the distinct joints. It is evident from the figure that the joints fabricated using SS pin profile exhibit higher wear resistance compared to other tool pin profiles. The lower wear rate for the SS pin was due to proper mixing and extreme plastic flowing of distinct Al Alloy. The joints fabricated using a TC profile showed a higher wear rate in the weld zone due to macroscopic defects present in the weld zone.

TABLE 18.5 Wear Data at Different Tool Pin Profile of FSW Sample.

S. No	Tool pin profile	Welding speed (mm/min)	Initial mass m_1 (kg)	Final mass m_2 (kg)	Change in mass Δm (kg)	Wear Rate Wr (gm/mm) $\times 10^{-3}$
1	Square	60	2.7237	2.6634	0.06026	22.20
2	Square	80	2.3241	2.2559	0.06821	25.13
3	Cylindrical threaded	60	2.8276	2.7560	0.07160	26.38
4	Cylindrical threaded	80	2.5117	2.4330	0.07866	28.98
5	Cylindrical	60	2.5751	2.4979	0.07717	28.43
6	Cylindrical	80	2.9272	2.8415	0.08566	31.56
7	Tapered cylindrical	60	2.7853	2.6986	0.08669	31.94
8	Tapered cylindrical	80	2.4634	2.3715	0.09190	33.86

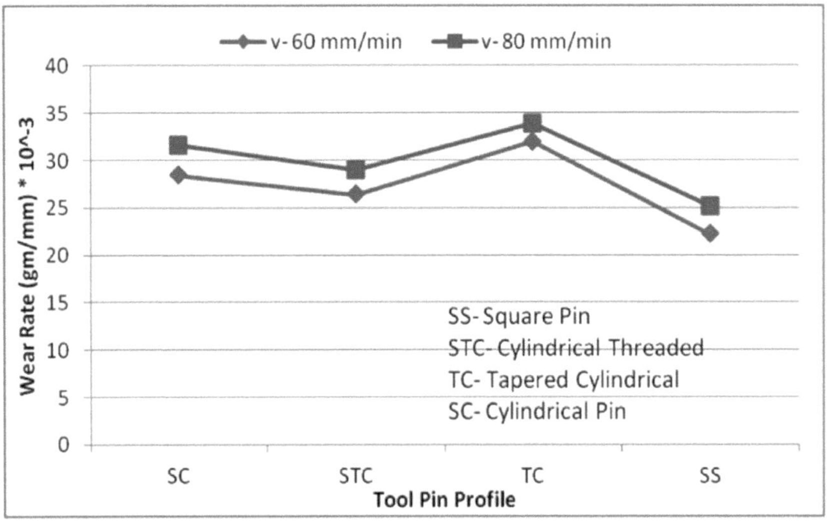

FIGURE 18.10 Graph showing the wear rate in gm/mm for different tool pin profile in welding of dissimilar AA6061–AA5052 Al alloy.

18.4 CONCLUSIONS

The effect of tool pin feature on mechanical and wear characteristics of the FSWed dissimilar aluminum alloys AA6061–AA5052 been analyzed in the current research. Following is the inference considered in the existing research:

- STC and SS pin profile was found to be suitable tool pin for FSWed joining of dissimilar AA6061–AA5052 Al alloy.
- Of the four tool pin profile, SC tool pin profile showed surface galling defects as well as the generation of flash and TC tool pin profile showed insufficient heat diffusion and defective surface morphology. Therefore, STC and SS pin profile should be preferred over SC and TC tool pin profiles.
- The results showed significant effects of tool pin profiles on the mechanical properties of dissimilar joints. Threaded cylindrical tool pin profile yielded the maximum tensile strength of 231 MPa with the joint efficiency of 74.5%, whereas maximum microhardness of 94 HV in the nugget zone was obtained with square tool pin profile. Fracture of the dissimilar Al Alloy joints was occurred in the weld nugget zone and near to TMAZ of AA6061 side.

- The highest hardness value observed in the weld NZ was near to advancing side of AA6061, and the lowest hardness value found in the HAZ was near retreating side of AA5052 Al alloys.
- The joints fabricated using SS pin profile exhibit best wear properties compared to other tool pin profiles owing to proper mixing and intense plastic flow of dissimilar Al Alloys, which might have reduced grain size and consequently higher hardness.
- Grain size in the stir zone enhanced through adjustments in the design of the tool pin profile and the average grain size for the square, and threaded cylindrical tool pin profile was 35 and 39 μm; however, the average grain size was 43 and 49 μm for the cylindrical and TC profiles, respectively.

KEYWORDS

- **friction stir welding**
- **distinct aluminum joints**
- **tool pin shape**
- **mechanical and wear characteristics**

REFERENCES

1. Thomas, W. M.; Nicholas, E. D.; Needham, J. C.; Murch, M. G.; Templesmith, P.; Dawes, C. J. Friction Stir Welding. International Patent Application no. PCT/GB92102203 and Great Britain Patent Application, Dec 6, 1991, 9125978.
2. Khan, N. Z.; Siddiquee, A. N.; Khan, Z. A. *Friction Stirs Welding: Dissimilar Aluminium Alloys*; CRC Press, Jul 28, 2017.
3. Kulkarni, N.; Mishra, R. S.; Yuan, W. *Friction Stir Welding of Dissimilar Alloys and Materials*; Butterworth-Heinemann, Mar 5, 2015.
4. Chaudhary, B.; Patel, V.; Ramkumar, P. L.; Vora, J. Temperature Distribution during Friction Stir Welding of AA2014 Aluminium Alloy: Experimental and Statistical Analysis. *Trans. Indian Inst. Metals* Apr 1, **2019,** *72* (4), 969–981.
5. Dawes, C.; Thomas, W. Friction Stir Joining of Aluminium Alloys. *TWI Bull.* 1995, *6*, 124.
6. Patel, V. K.; Kumar, P.; Bhattacharya, S. Mechanical, Microstructural and Sliding Wear Properties of Friction Stir Welded AA6063-T6 and AA5052-H32 Aluminium Alloys. *Mater. Focus* Feb 1, **2018,** *7* (1), 50–58.
7. Meilinger, Á.; Török, I. The Importance of Friction Stirs Welding Tool. *Prod. Process. Syst.* **2013,** *6* (1), 25–34.

8. Bahuguna, S.; Arya, P. K.; Patel, V. K. Mechanical and Abrasive Wear Properties of Friction Stir Welded Joints of Aluminium Alloy AA6061-T6 with/Without Nickel Coating. *Strojnícky časopis-J. Mech. Eng.* Nov 1, **2020,** *70* (2), 21–36.
9. Arya, P. K.; Gupta, G.; Rajput, A. K. A Review on Friction Stir Welding for Aluminium Alloy to Steel. *Int. J. Sci. Eng. Res.* **2016,** *7* (5), 119.
10. Dawood, H. I.; Mohammed, K. S.; Rahmat, A.; Uday, M. B. Effect of Small Tool Pin Profiles on Microstructures and Mechanical Properties of 6061 Aluminium Alloy by Friction Stir Welding. *Trans. Nonferrous Metals Soc. China* Sept 1, **2015,** *25* (9), 2856–2865.
11. Kim, Y. G.; Fujii, H.; Tsumura, T.; Komazaki, T.; Nakata, K. Three Defect Types in Friction Stir Welding of Aluminum Die Casting Alloy. *Mater. Sci. Eng. A* Jan 15, **2007,** *415* (1–2), 250–254.
12. Song, S. W.; Lee, S. H.; Kim, B. C.; Yoon, T. J.; Kim, N. K.; Kim, I. B.; Kang, C. Y. Liquation Cracking of Dissimilar Aluminum Alloys during Friction Stir Welding. *Mater. Trans* Feb 1, **2011,** *52* (2), 254–257.
13. Kumar, N.; Patel, V. K. Effect of SiC/Si 3 N 4 Micro-Reinforcement on Mechanical and Wear Properties of Friction Stir Welded AA6061-T6 Aluminum Alloy. *SN Appl. Sci.* Sept **2020,** *2* (9), 1–1.
14. Palanivel, R.; Mathews, P. K.; Murugan, N.; Dinaharan, I. Effect of Tool Rotational Speed and Pin Profile on Microstructure and Tensile Strength of Dissimilar Friction Stir Welded AA5083-H111 and AA6351-T6 Aluminum Alloys. *Mater. Design* Sept 1, **2012,** *40*, 7–16.
15. Ilangovan, M.; Boopathy, S. R.; Balasubramanian, V. Effect of Tool Pin Profile on Microstructure and Tensile Properties of Friction Stir Welded Dissimilar AA 6061–AA 5086 Aluminium Alloy Joints. *Defence Technol.* Jun 1, **2015,** *11* (2), 174–184.
16. Rajakumar, S.; Muralidharan, C.; Balasubramanian, V. Influence of Friction Stir Welding Process and Tool Parameters on Strength Properties of AA7075-T6 Aluminium Alloy Joints. *Mater. Design* Feb 1, **2011,** *32* (2), 535–549.
17. Kalemba-Rec, I.; Kopyściański, M.; Miara, D.; Krasnowski, K. Effect of Process Parameters on Mechanical Properties of Friction Stir Welded Dissimilar 7075-T651 and 5083-H111 Aluminum Alloys. *Int. J. Adv. Manuf. Technol.* Jul 1 **2018,** *97* (5–8), 2767–2779.
18. Venkateshkannan, M.; Rajkumar, V.; Sadeesh, P.; Arivazhagan, N.; Narayanan, S.; Ramkumar, K. D. Influences of Tool Geometry on Metallurgical and Mechanical Properties of Friction Stir Welded Dissimilar AA 2024 and AA 5052. *Procedia Eng.* Jan 1, **2014,** *75*, 154–158.
19. Sivachidambaram, S.; Rajamurugan, G.; Amirtharaj, D. Optimizing the Parameters for Friction Stir Welding of Dissimilar Aluminium Alloys AA5383/AA7075. *ARPN J. Eng. Appl. Sci.* Jul **2015,** *10*, 5434–5437.
20. Khan, N. Z.; Siddiquee, A. N.; Khan, Z. A.; Shihab, S. K. Investigations on Tunneling and Kissing Bond Defects in FSW Joints for Dissimilar Aluminum Alloys. *J. Alloys Compounds* Nov 5, **2015,** *648*, 360–367.
21. Ahmed, M. M.; Ataya, S.; Seleman, M. E.; Ammar, H. R.; Ahmed, E. Friction Stir Welding of Similar and Dissimilar AA7075 and AA5083. *J. Mater. Process. Technol.* Apr 1, **2017,** *242*, 77–91.

CHAPTER 19

TAGUCHI-GRA-BASED MULTIOBJECTIVE OPTIMIZATION OF UNDERWATER FRICTION STIR WELDING OF AA 6061-T6 ALLOY

KIRAN WAKCHAURE* and AJAYKUMAR THAKUR

Sanjivani College of Engineering, SPPU Pune University, Kopargaon, Maharashtra 423603, India

**Corresponding author. E-mail: kiran.wakchaure@yahoo.com*

ABSTRACT

In this work, aluminum alloy AA6061-T6 of 6 mm thickness fabricated with butt type weld configuration using underwater friction stir welding with water as cooling medium. Experimentations were performed by considering L9 Taguchi method followed by Grey Relational Analysis (GRA) multiobjective optimization. Welding parameters tool rotation speed, welding speed, and tilt angle were considered to study its influence on Tensile strength, yield strength impact strength, and % elongation of welded joint. Results showed that Tool rotation speed plays a significant role in mechanical properties of submerged friction stir welded as compared with welding speed and tilt angle. Optimum parameters obtained from Taguchi-GRA multiobjective optimization were considered for comparison of conventional and submerged friction stir welding. Due to refined grains in the nugget zone and strengthening of heat affected zone (HAZ), underwater friction stir welded sample shows highest strength with conventional in air friction stir welded joints.

Modern Manufacturing Systems: Trends and Developments. Rajiv Kumar Garg, Ravi Pratap Singh, Rajeev Trehan, & Ramesh Singh (Eds.)

19.1 INTRODUCTION

Friction stir welding (FSW) is a solid-state machine tool welding process. Earlier FSW was limited to laboratory experiments, but now it emerged as a cutting-edge technology that can weld almost all the types of similar and dissimilar engineering materials like steels, aluminum alloys, composites, plastics, etc.[1] In friction stir welding, a nonconsumable rotating tool consisting shoulder and pin is allowed to insert in abutting edges till shoulder touches the top of base material to be welded. Frictional heat between rotating tool shoulder and the top surface of base material allowed to plasticize the material.[1] Rotating pin transfers material from advancing side to retreating side and forged at the back of pin, below shoulder with the action of axial force. Continuous transverse motion of tool along weld line creates strong weld with good appearance having dimensional accuracy, fine grains in nugget zone, minimum heat-affected zone, and minimum residual stresses. Due to its enormous advantages over conventional welding processes. FSW is gaining popularity in defense technology, automobile industry, aircraft industry, railway industry, transportation, and many more.[2,3]

AA 6061-T6 is the moderate strength and heavily used heat treatable alloy in the 6XXX series consisting Al-Mg-Si are the main alloying elements. In 6XXX series, Mg_2Si precipitates are responsible for the strengthening of alloy. AA6061 possesses excellent welding characteristics but they suffer loss in strength due to dissolution of strengthening precipitates and softening of HAZ due to conventional fusion welding processes. Friction stir welding refines the grain in nugget zone and improves the mechanical properties as compared with other welding processes.

Underwater friction stir welding (UFSW) emerged as an innovative welding practice, which avoids welding defects occurred in conventional FSW and provides improved mechanical properties.[4] In underwater friction stir welding, the whole welding process is carried out in water. Hofmann et al.[5] have successfully generated ultra-fine grains in the nugget zone of friction stir processed AA6061-T6 using water as a cooling medium. Wnag et al.[6] revealed that underwater friction stir welding of spray-formed 7075 Al alloy improves tensile properties, corrosion resistance, and reduces corrosion rate due to fine grains and well-distributed precipitates in weld region because of water cooling compared with conventional FSW. Hosseini et al.[7] observed that underwater friction stir welding of ultrafine-grained AA1050 provides improved tensile, yield strength, and narrower HAZ as compared with conventional FSW. Also, authors proposed that underwater friction stir welding can be used as a method for surface hardening of materials. Liu et al.[8] investigated the effect of water as cooling medium on microstructural and mechanical properties of friction stir welded lap joint off 2 mm ME20M and reported that submerged friction stir welded joint shows refined grains, improved plasticity, and elongation compared with conventional friction stir welded joint. Hui-jie et al.[9] found that underwater friction stir welding of 2219 aluminum alloy shows improved tensile strength and low hardness value while deteriorated plasticity as compared with conventional friction stir welding. Also significant improvement was noticed in HAZ of underwater friction stir welded joint. Fathi et al.[10] observed that tensile strength of SFSW of AA6061 T6 increased with tool rotation speed and decreased with welding speed. Submerged friction stir welded joint shows improved tensile strength, hardness, and residual stress as compared with conventional friction stir welded joint.

Fratini et al.[11] investigated the effect of in-process cooling with in normal air, forced air, and in water on mechanical properties of friction stir welding for AA7075-T6. Results show that, water acts as a stronger cooling medium, improves tensile strength of joint due to narrows weakest TMAZ and HAZ. FSW samples with water as a cooling medium shows increasing microhardness values and low hardness value in the nugget zone of water FSW sample as compared with normal air and forced air FSW samples. Sharma et al.[12] studied the effect of liquid nitrogen, normal air, and compressed air on friction stir welding of AA7039 and reported that cooling during FSW absorb excess heat during welding, which is responsible to restrict grain growth in nugget zone and coarsening in HAZ. Author also concluded that water is the better cooling medium to give maximum tensile strength and percentage elongation as compared with compressed air and liquid nitrogen. Sinhmar et al.[13] observed that friction stir welded joint of AA2014 with water cooling shows improved yield strength, tensile strength, hardness in HAZ, and corrosion resistance due to more refined microstructure and the existence of small precipitates as compared with conventional FSW with natural air cooling. Cylindrical and conical tool pins are simple in shape and easy to manufacture tool pin profiles used in FSW. Conical pin tool gives improved mechanical properties as compared with cylindrical tool[14,15] and square tool.[16]

Taguchi method is a powerful statistical tool to produce high-quality products at minimum cost and time due to limited number of experiments. It gives a systematic approach to design and analysis of experiments to obtain optimum design parameters.[17] Vijayan et al.[18] used Taguchi-based grey

Relation analysis for multiobjective optimization of friction stir welding of AA 5083 with L9 orthogonal array using process parameters tool rotation, welding speed, and axial force to obtain maximum tensile strength weld joint and minimum power consumption of machine tool during process. Kasman[19] obtained optimum process parameters tool type, rotation speed, and welding speed for dissimilar friction stir welding of AA6082 and AA7075 to get maximum ultimate tensile strength and percentage elongation using Taguchi L19 orthogonal array and grey relation analysis. Sahu et al.[20] studied the feasibility of GRA with Taguchi technique (L18 array) for friction stir welded AM20 magnesium alloy by considering plunging depth, shoulder diameter, welding speed, and tool rotation speed as a process parameters to maximize tensile strength, yield strength, compressive strength, and bending angle and to minimize percentage elongation and hardness values in welded zone. Chien et al.[21] proposed grey-based Taguchi method to obtain optimum process parameters tool rotation speed, transverse speed (feed rate), tool tilt angle of friction stir welding of aluminum alloy AA5083 to get maximum tensile strength and elongation rate of welded joint.

Previous research has highlighted the importance of water as a cooling medium for the improvement of mechanical properties of friction stir welded joint. Very limited research is conducted on the optimization of underwater friction stir welded joint of AA 6061-T6. Hence underwater friction stir welding of AA 6061-T6 was performed using Taguchi L9 array and multiobjective optimization was performed using grey relation analysis by considering Tensile strength and % elongation as response parameters. After obtaining optimum input parameters, conventional in air, and water friction stir welded joints were compared in respect with microstructural and mechanical properties.

19.2 EXPERIMENTAL PROCEDURE

In this research, experiments were carried out on rolled plates of aluminum alloy AA 6061-T6 of 6 mm thickness. The chemical composition and mechanical properties of AA 6061-T6 are shown in Tables 19.1 and 19.2, respectively. Aluminum alloy sheets were cut into sample size of 200(L) × 75(W) × 6(T) mm with abrasive cutting machine. Edges are machined with milling machine and edge of 200 mm length cut to transverse direction rolling line was considered to form a weld. Plates were washed with acetone to avoid the contamination during welding procedure. Conical tool with 18 mm shoulder diameter and pin with 6 mm diameter, 14° semicone angle and 5.6 mm length was considered to manufacture the tool. Hot worked die steel (AISI- H13) was used as a tool material with hardening of 50–55 HRC. Main constituent of chemical composition of H13 are C-0.4, Mn-0.4, Si-1, Mo-1.35, Cr-0.25, V-1, and Fe-balance (all elements in wt.%). New fixture was developed to carry out welding trials, which can hold the workpiece properly and withstand load acting on it. Special vessel type arrangement was made with acrylic sheet to store water in the fixture.

TABLE 19.1 Chemical Composition of AA6061-T6 Alloy.

Elements	Mg	Si	Mn	Zn	Fe	Cu	Al
Alloying elements (in %)	1.1	0.6	0.12	0.25	0.35	0.3	Rest

TABLE 19.2 Mechanical Properties of AA6061-T6 Alloy.

Properties	Ultimate tensile strength	Tensile yield strength	Modulus of elasticity	Percentage elongation	Poisson's ratio	Melting point	Hardness, Vickers	Density
Values	303 MPa	240 MPa	68.9 GPa	20%	0.33	582°C–652°C	107Hv	2.7 g/cc

All the experiments were carried out on a universal milling machine (G.DUFOUR make) 624D model having vertical spindle attachment with DRO. Machine consist of main spindle motor of 15 HP with spindle speed ranges between 24 and 1180 rpm. Also two feed motors with 5 HP power with feed in the range of 22–1180 mm/min. Experimental setup of underwater friction stir welding is shown in Figure 19.1. Friction stir welding tool mounted with collet on spindle of universal milling machine. Fixture was rigidly mounted on bed of the milling machine. Specimens to be welded were fixed in the slot made in the fixture with clamps to avoid lateral and transverse movement of specimens during trials. Firstly, rotating tool shoulder is allowed to touch top surface of specimen and then water is allowed to pour in the fixture vessel to make 30 mm water head. Arrangement was made to pour the water and drain hot water from vessel. K-type thermocouples were employed at a distance of 15 mm from either sides of weld line and 100 mm from the start of weld line. Arduino-based data acquisition system was developed to measure the temperature and store data in excel format (Figure 19.2).

Taguchi method with L9 array was adopted to perform experimental trials. In friction stir welding tool geometry, tool dimensions, tool material, tool rotation speed, welding speed, tilt angle, plunge depth, axial load, and submerged conditions are the main parameters, which affects the mechanical and microstructural properties of welded joint.

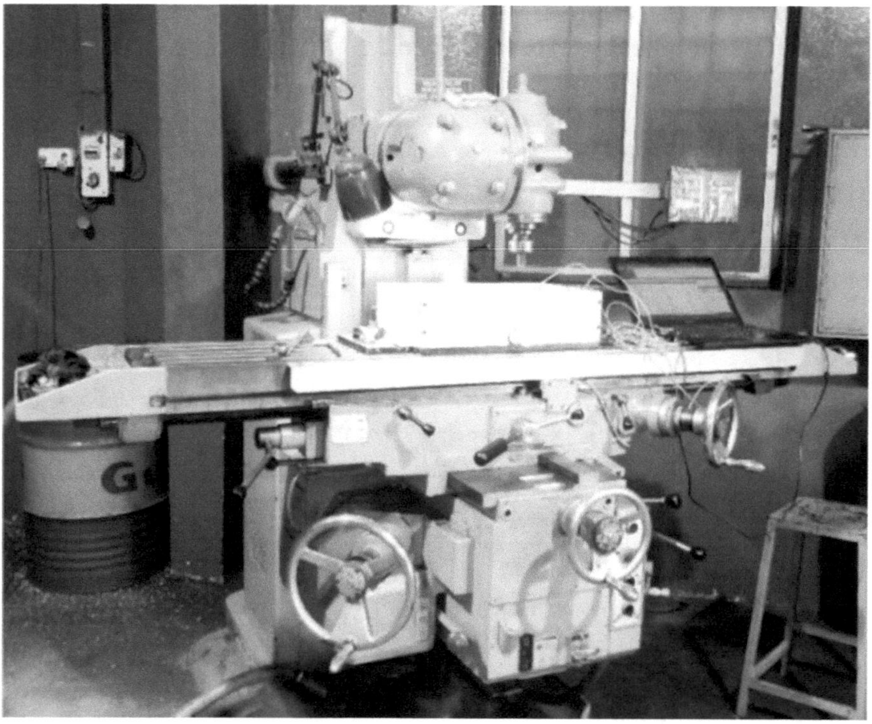

FIGURE 19.1 Experimental setup.

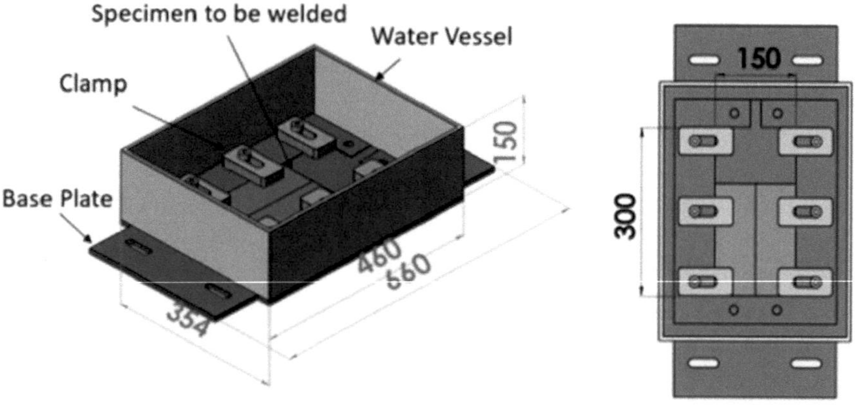

FIGURE 19.2 Underwater friction stir welding fixture (a) Isometric view (b) Top view.

In this research, tool rotation speed, welding speed, and tilt angle were chosen as process parameters by assuming all parameters constant. Initial trials were performed to derive levels of process parameters by considering defects in the weld zone. Levels chosen to perform Taguchi L9 are shown in Table 19.3.

TABLE 19.3 Process Parameters Levels.

Parameter (notation)	Unit	Level		
		Low (1)	Middle (2)	High (3)
Tool rotation speed (TRS)	RPM	710	900	1180
Welding speed (WS)	mm/min	29	64	108
Tilt Angle (TA)	degrees	1°	2°	3°

After finishing welding trials, all-welded samples were cut into the required size as per American Society for Testing of Materials (ASTM) E-8M-08 [ASTM-2008] guidelines using abrasive water jet machine. For the characterization three samples for tensile strength, one sample for impact strength and one common sample for microstructure and microhardness were considered. Tensile tests were performed on 2000 KN capacity universal testing machine (UTM) (Make: Universal Motion Inc, Model: MUTCS-20 model). Crosshead speed of 1 mm/min of UTM was selected during tensile testing. Selected samples were considered for macro and microstructure evaluation. Optical microscopy has been carried out using a light optical microscope (Make: Conation technologies) with image analyzing system. All the microstructure samples were polished to different grades of emery paper followed by diamond paste of 1 μm particle size to evolve different zone of welded specimen SZ, TMAZ, HAZ, and base material. Microhardness testing was carried on Vickers Microhardness taster of (Economet make) VH-1MD. Mid of weld zone (transverse to weld direction) was considered a testing area with 1mm interval using 100 g indentation load. Sub-size specimen was considered for Charpy impact test as per ASTM E23 standards.

19.3 TAGUCHI–GREY RELATION ANALYSIS

Taguchi is a part of design of the experiment, which gives optimum result with minimum number of experiment trials. In this research, three parameters with three levels were selected. Taguchi method, proposed L9 array with nine experiments trial with different combination of levels of factors with corresponding response variables shown in Table 19.4. Taguchi method can do analysis of system with single response variable. For multiple responses/ criteria, Taguchi method is always accompanied with grey relation analysis, TOPSIS, AHP, etc. multicriteria decision-making techniques.

TABLE 19.4 L9 Orthogonal Array with Experimental Results.

Sr. no.	TRS	WS	TA	UTS (MPa)	EL (%)	YS (MPa)	IS (J)
1.	710	29	1°	195.12	9.50	101.2	15.85
2.	710	64	2°	208.94	13.68	123.43	14.09
3.	710	108	3°	219.30	12.96	109.94	15.46
4.	900	29	2°	198.51	12.03	110.98	18.95
5.	900	64	3°	210.36	13.03	125.41	15.55
6.	900	108	1°	232.12	12.84	130.00	12.00
7.	1180	29	3°	198.54	14.50	117.69	15.46
8.	1180	64	1°	206.48	12.43	114.83	16.78
9.	1180	108	2°	222.52	12.09	138.00	11.30
10.	C1	C2	C3	C4	C5	C6	C7

Grey relation analysis is multicriteria decision-making technique founded by Chinese professor Julong Deng in 1980s. GRA is an effective tool for solving interrelationships among multiple responses/criteria. In GRA, multiple responses are converted into a single grey relation grade, which is convenient for decision making. GRA gives the appropriate good solution instead of the best solution for real-life complex problems. Three important steps in GRA are normalization, generation of grey relation coefficient (GRC), and grey relation grade (GRG). In normalization, responses are normalized between values 0 and 1 according to three different performance characteristics—larger the better, smaller the better, and nominal the better. In this research, all the response variables ultimate tensile strength, yield strength, impact strength, and % elongation should be maximum. Hence both the variable were normalized with larger the better performance characteristics (Ref. eq 19.1). Then normalized data converted in grey relation coefficient (Ref. eq 19.2) to represent relationship between experimental and desired data. In the last step (Ref. eq 19.3), the overall grey relation grade is generated by averaging weighed grey relation coefficients. Table 19.5 shows grey relation analysis.

Steps in grey relation analysis are as follows:

Step 1

Normalization [18–20]

$$x_i^*(k) = \frac{x_i^0(k) - \min x_i^0(k)}{\max x_i^0(k) - \min x_i^0(k)} \quad \textbf{(Table 5 C8 -C11)} \tag{19.1}$$

$x_i^*(k)$ is the sequence after the data preprocessing,

$x_i^0(k)$ is the reference sequence,

min $x_i^0(k)$ is the smallest value in the reference sequence **(Table 5 C4-C7)**

max $x_i^0(k)$ is is the largest value in the reference sequence, **(Table 5 C4-C7)**

$i = 1, 2, .., m; k = 1, 2,.., n;$

m is the number of experiment and n is the number of experimental data

Step 2

Grey Relation coefficient [18–20]

$$\xi\left(x_0^*(k), x_i^*(k)\right) = \frac{\Delta_{\min}(k) + \zeta\Delta_{\min}(k)}{\Delta_{0i}(k) + \zeta\Delta_{\max}(k)} \quad \textbf{(Table 5 C16-C19)} \tag{19.2}$$

$$\Delta_{0i}(k) = \left\| x_0^*(k) - x_i^*(k) \right\| \quad \textbf{(Table 5 C12-C15)} \quad (2.1)$$

$$\Delta_{\min}(k) = \min_{\forall j \in i} \min_{\forall k} \left\| x_0^*(k) - x_j^*(k) \right\| \quad \textbf{(Table 5 C12-C15)}$$

$$\Delta_{\max}(k) = \max_{\forall j \in i} \max_{\forall k} \left\| x_0^*(k) - x_j^*(k) \right\| \quad \textbf{(Table 5 C12-C15)}$$

where $\Delta_{0i}(k)$ is the deviation sequence of the comparability sequence $x_i^*(k)$ and reference sequence $x_0^*(k) = 1$

ζ is the distinguishing coefficient having value between 0-1, and the value of 0.5 is generally used

Step 3

Grey Relation Grade [18–20]

$$\gamma_i\left(x_0^*(k), x_i^*(k)\right) = \frac{1}{n}\sum_{i=1}^{n} w_i \xi\left(x_0^*(k), x_i^*(k)\right) \quad \textbf{(Table 5 C20)} \tag{19.3}$$

Where, W_i is the weighting value of the ith performance characteristic, here 0.25 is the weight consider for each response variable. $\gamma_i\left(x_0^*(k), x_i^*(k)\right)$ the GRG for an ith experiment and n is the number of performance characteristics,

19.4 RESULT AND DISCUSSION

Underwater friction stir welding experimentation was carried out by using grey assisted Taguchi method. Combine response of ultimate tensile strength, yield strength, impact strength, and % elongation are recorded into the single response form of GRG Table 19.5.

19.4.1 ANOVA ANALYSIS

In order to do analysis of significance of each input parameter and its percentage contribution on response the analysis of variance (ANOVA) was carried out, shown in Table 19.6. ANOVA analysis shows that tool rotation speed (37.17%) contributes significantly as compared with welding speed (28.73%) and tilt angle (10.69%). In friction stir welding, the tool rotation speed controls the heat generation and material flow during welding.

TABLE 19.5 Grey Relation Analysis.

Normalised				1- Normalized				GRC				GRG	Rank
UTS	EL	YS	IS	UTS	EL	YS	IS	UTS	EL	YS	IS		
0.00	0.00	0.00	0.65	1.00	1.00	1.00	0.35	0.33	0.33	0.33	0.59	0.40	**9**
0.40	0.86	0.64	0.43	0.60	0.14	0.36	0.57	0.45	0.78	0.58	0.47	0.57	**6**
0.67	0.73	0.27	0.61	0.33	0.27	0.73	0.39	0.61	0.65	0.41	0.56	0.56	**7**
0.10	0.56	0.30	1.00	0.90	0.44	0.70	0.00	0.36	0.53	0.42	1.00	0.58	**5**
0.44	0.75	0.69	0.62	0.56	0.25	0.31	0.38	0.47	0.66	0.62	0.57	0.58	**4**
1.00	0.71	0.81	0.12	0.00	0.29	0.19	0.88	1.00	0.63	0.72	0.36	0.68	**1**
0.10	1.00	0.49	0.61	0.90	0.00	0.51	0.39	0.36	1.00	0.50	0.56	0.60	**3**
0.33	0.64	0.41	0.76	0.67	0.36	0.59	0.24	0.43	0.58	0.46	0.68	0.54	**8**
0.76	0.57	1.00	0.00	0.24	0.43	0.00	1.00	0.67	0.54	1.00	0.33	0.64	**2**
C8	C9	C10	C11	C12	C13	C14	C15	C16	C17	C18	C19	C20	**C21**

TABLE 19.6 ANOVA Analysis.

Parameters	DOF	SS	MS	F Ratio	% Contributions
TR	2	0.0181	0.0090	1.5870	37.17%
WS	2	0.0140	0.0070	1.2267	28.73%
TA	2	0.0052	0.0026	0.4563	10.69%
ERROR	2	0.0114	0.0057		23.42%
TOTAL	8	0.0487	0.0243		100%

19.4.2 EFFECT OF PROCESS PARAMETERS

Process parameters control microstructural and mechanical properties of welded joint. Figure 19.3 shows mechanical properties of welded samples are increases from tool rotation speed 710 –900 rpm and then decreases at 1180 rpm. As rotation speed increases, temperature at processing zone increases, which eases material flow. But increment in temperature lead to reduce coefficient of friction, which limits material flow and deteriorates mechanical properties of welded joint. Also high temperature in weld zone is responsible for grain coarsening.

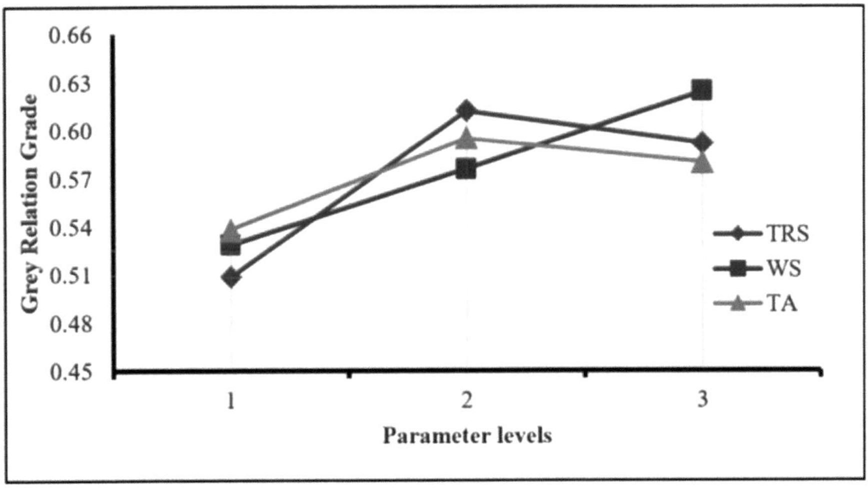

FIGURE 19.3 Influence of process parameters on grey relation grade.

As shown in Figure 19.3 for welding speed, as speed increases from 29 to 108 mm/min mechanical properties of welded joint increases. In FSW at low welding speed, temperature in weld nugget zone increases, which provides space for grain coarsening during cooling of weld. As the welding speed increases, temperature in the nugget zone decreases lead to decrease grain growth, which shows finer grains in the nugget zone. Finer grains are responsible for the improvement of mechanical properties in the nugget zone. Optimum temperature in nugget zone provides restrictions to dissolution of strengthening precipitates and coarsening of grains leads to increase mechanical properties of welded joint. Tilt angle is responsible to constraint material in weld zone. GRG increase as increase in tilt angle from 1° to 2° and decreases again by changing tilt angle to 3°.

The mean effect of parameter was calculated by averaging of gray relation grade for input parameters with their three levels (refer Table 19.7). The mean response gives information of influence of variation in factor from its low level to high level on performance characteristics, that is, grey relation grade. This gives an optimum combination of input parameters for multiobjective optimization. Table 19.7 shows TRS2 (910 rpm) WS3 (108 mm/min) TA2 (2°) will be the optimum process parameters.

TABLE 19.7 Influence of Process Parameters at Different Levels on GRG.

Parameters	Low (1)	Medium (2)	High (3)	Max–Min
TRS	0.5085	0.6121	0.5918	0.1036
WS	0.5284	0.5757	0.6241	0.0957
TA	0.5379	0.5947	0.5798	0.0568

19.4.3 SELECTION OF OPTIMUM PROCESS PARAMETERS

The optimum parameters combination observed from Table 19.7 is not the part of L9 array. Optimum performance characteristics, that is, GRG is calculated by considering optimum process parameters using eq 19.4,

$$\hat{\gamma} = \gamma_m + \sum_{i=1}^{l} \left(\bar{\gamma}_j - \gamma_m\right) \tag{19.4}$$

where γ_m is total mean grey relation grade, $\bar{\gamma}_i$ is the mean GRG at the optimal level, and l is the number of the main design parameters. Welding parameters tool rotational speed, tilt angle, and welding speed were set at optimum levels and grey relation grade calculated using obtained average tensile strength and % elongation Table 19.8.

TABLE 19.8 Confirmation Test.

Method	Optimum process conditions				UTS (MPa)	% EL	YS (MPa)	IS (J)	GRG	% Improvement in GRG
	Tool	TR	WS	TA						
TAGUCHI-GRA	2	900	108	1°	232.12	12.84	130.00	12.00	0.68	—
TAGUCHI-GRA (Pred)	2	900	108	2°					0.69	1.47%
TAGUCHI-GRA (Experimental)	2	900	108	2°	237.05	12.62	140.00	12.57	0.75	8.69%

After predicting grey relation grade for optimum parameters, confirmation test experiment has been carried out by keeping all conditions same. Experiment is carried out by taking process parameters tool rotation speed 900 rpm, welding speed 108 mm/min, and tilt angle 2 °. Results of confirmation test are shown in Table 19.8. Experimental result shows GRG improved by 8.69% in comparison with value predicted by Taguchi-GRA analysis.

19.4.4 MICROSCOPIC EVALUATION

Macroscopic images of all L9 samples are free from any volumetric defect except flash over top surface of the weld. Figure 19.4 Shows macroscopic image of underwater friction stir welding sample, which welded with optimum process parameters obtained with Taguchi- GRA analysis. For the micro- and macrostructural investigations, specimens processed with optimum process parameters were used. Macrographs shown in Figure 19.4 shows specimens are free from any welding effects like porosity, crack, kissing bond, etc.

Macrostructure of the weld sample shows well-defined onion rings consisting bright and dark field regions, which denotes uniform material flow in nugget zone shown in Figure 19.4. In UFSW, due to restriction on material flow width of weld is marginal small as compared with FSW. Exposure of water in UFSW makes weld smoother compared with FSW. Friction stir welding that redefines grains in nugget zone.

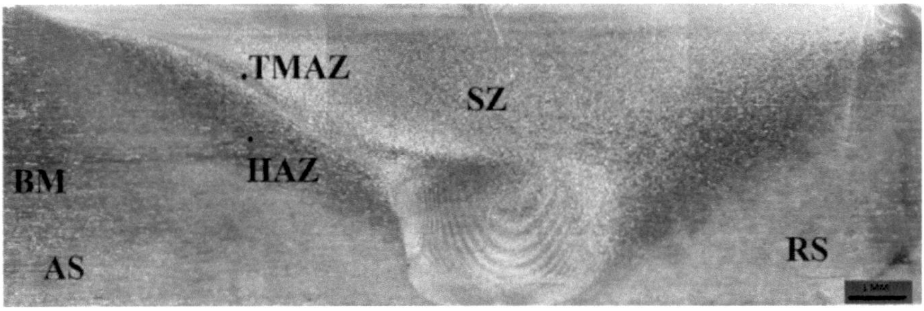

FIGURE 19.4 Macroscopic image of UFSW sample with optimum parameter.

Figure 19.5 shows EBSD base metal and compared with weld nugget zone of UFSW analyzed using ATEX software tool.[23] In FSW, base material of 84 μm grain size undergoes severe plastic deformation and grain refinements provide equiaxed grain size. Average grain size in UFSW is 2.56 μm. Water environment in UFSW restricted grain growth in nugget zone after processing which provides finer grain UFSW as compared with FSW.[5,7]

Vickers microhardness profiles are measured along mid thickness of Base metal and UFSW shown in Figure 19.6a. Base material shows average hardness value of 107 Hv. The microhardness of UFSW shows "W" shaped asymmetrical profile. In friction stir welding, advancing side undergoes severe plastic deformation compared with retreating side. Due to which advancing side undergoes higher temperature as compared with retreating side which reduces microhardness in the region. In UFSW, low-temperature exposure in welding region, value of minimum hardness increases shown in Figure 19.6a. It was found that in UFSW narrowing TMAZ and HAZ lead to reduce softening regions in nugget zone.[7]

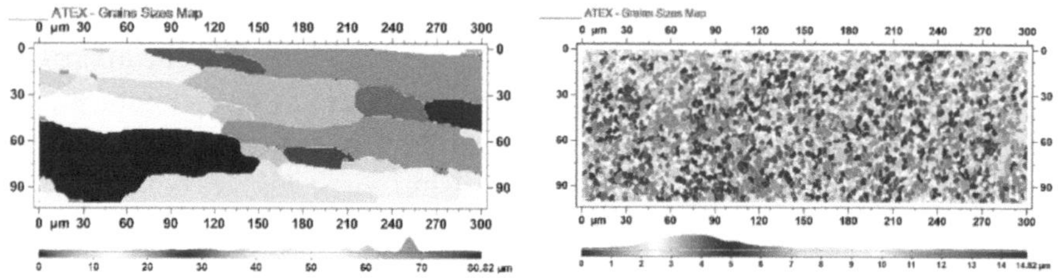

FIGURE 19.5 EBSD Maps (a) Base Metal (b) UFSW sample.

19.4.5 COMPARISON OF CONVENTIONAL AND UNDERWATER FRICTION STIR WELDING

Figure 19.6b shows tensile test result of Base metal, underwater friction stir welding, and conventional in air fraction stir welding (CFSW) sample. Both samples, CFSW and UFSW, were processed with optimum process parameters TRS 900 rpm, WS 108 mm/min, and TA 2°. Friction stir welding undergoes thermal heating that results in reduction in mechanical properties as compared with base metal consisting Tensile strength 303.5 MPa and % elongation of 20%. Results show that underwater samples show tensile strength of 237.25 MPa (78.17% of BM) with elongation 12.62% (63.2% of BM). FSW samples processed in air (CFSW) shows tensile strength of 216.88 MPa (71.45% of BM) with elongation 12.5% (63.2% of BM).

Tensile strength of underwater friction stir welding sample undergoes rapid cooling due to high heat absorption capacity of water compared with air in conventional friction stir welding. Due to cooling effect in UFSW, restrictions on grain growth are imposed which gives finer grains compared with CFSW. Also less exposure of heat in UFSW controls dissolution of strengthening precipitates lead to improve mechanical properties of welded joints compared with CFSW.[9]

19.5 CONCLUSION

The multiobjective optimization was carried out using Taguchi-based grey relation analysis for underwater friction stir welding of AA 60660-T6 aluminum alloy using conical tool. Response variables, ultimate tensile strength, yield strength, % elongation, and impact strength were considered to obtain optimum process parameters tool rotation speed, welding speed, and Tilt angle. The following conclusions were drawn from this study are:

- Taguchi-based grey relation analysis is very effective tool for multi-objective optimization of underwater friction stir welding.
- Optimum process parameter to obtain defect-free joint with highest GRG 0.75 are: tool rotation speed-910 rpm, welding speed 108 mm/min, and tilt angle 2° using conical tool.
- ANOVA result shows tool rotation speed (37.17%) and welding speed (28.18%) have major influence on mechanical and microstructural properties of UFSW of AA 6061-T6.
- Tensile test result shows ultimate tensile strength UFSW joint is about 7% higher than conventional CFSW joint.

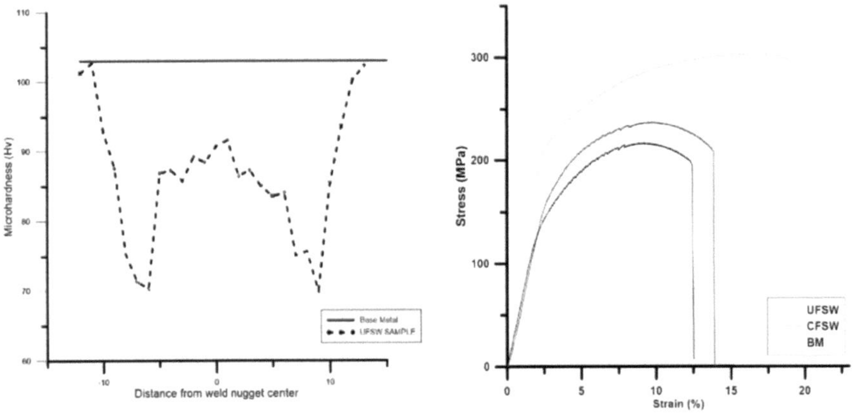

FIGURE 19.6 (a) Microhardness profile of UFSW sample. (b) Comparison of BM, UFSW and CFSW samples.

KEYWORDS

- **underwater friction stir welding**
- **AA6061-T6**
- **Taguchi-GRA multioptimization**
- **mechanical properties**

REFERENCES

1. Mishra, R. S.; Mahoney, M. W. *Friction Stir Welding and Processing*, 2007; p 368. DOI: 10.1361/fswp2007p001
2. Ma, Z. Y. Friction Stir Processing Technology: A Review. *Metal. Mater. Trans. A* **2008,** *39*, 642–658. DOI: 10.1007/s11661-007-9459-0
3. Mendes, N.; Neto, P.; Loureiro, A.; Moreira, A. P. Machines and Control Systems for Friction Stir Welding: A Review. *Mater. Des.* **2016**. DOI: 10.1016/j.matdes.2015.10.124
4. Paramaguru, D.; Pedapati, S. R.; Awang, M. *A Review on Underwater Friction Stir Welding (UFSW)*, 2016. DOI: 10.1007/978-981-10-9041-7
5. Hofmann, D. C.; Vecchio, K. S. Submerged Friction Stir Processing (SFSP): An Improved Method for Creating Ultra-Fine-Grained Bulk Materials. *Mater. Sci. Eng. A* **2005,** *402*, 234–241. DOI: 10.1016/j.msea.2005.04.032
6. Wang, Q.; Zhao, Y.; Yan, K.; Lu, S. Corrosion Behavior of Spray Formed 7055 Aluminum Alloy Joint Welded by Underwater Friction Stir Welding. *Mater. Des.* **2015,** *68*, 97–103. DOI: 10.1016/j.matdes.2014.12.019
7. Hosseini, M.; Danesh Manesh, H. Immersed Friction Stir Welding of Ultrafine Grained Accumulative Roll-Bonded Al Alloy. *Mater. Des.* **2010,** *31*, 4786–4791. DOI: 10.1016/j.matdes.2010.05.007
8. Liu, W.; Shen, Y.; Guo, C. Microstructures and Mechanical Properties of Submerged Friction Stir Welding of ME20M Magnesium Alloy. In *2019 16th International Bhurban Conference on Applied Sciences and Technology (IBCAST)*, 2019; pp 1–6.
9. Liu, H. J.; Zhang, H. J.; Huang, Y. X.; Yu, L. Mechanical Properties of Underwater Friction Stir Welded 2219 Aluminum Alloy. *Trans. Nonfer. Metals Soc. China (English Edition)* **2010,** *20*, 1387–1391. DOI: 10.1016/S1003-6326(09)60309-5
10. Fathi, J.; Ebrahimzadeh, P.; Farasati, R.; Teimouri, R. Friction Stir Welding of Aluminum 6061-T6 in Presence of Watercooling: Analyzing Mechanical Properties and Residual Stress Distribution. *Int. J. Lightweight Mater. Manuf.* **2019,** *2*, 107–115. DOI: 10.1016/j.ijlmm.2019.04.007
11. Fratini, L.; Buffa, G.; Shivpuri, R. In-process Heat Treatments to Improve FS-welded Butt Joints. *Int. J. Adv. Manuf. Technol.* **2009,** *43*, 664–670. DOI: 10.1007/s00170-008-1750-8
12. Sharma, C.; Dwivedi, D. K.; Kumar, P. Influence of In-process Cooling on Tensile Behaviour of Friction Stir Welded Joints of AA7039. *Mater. Sci. Eng. A* **2012,** *556*, 479–487. DOI: 10.1016/j.msea.2012.07.016
13. Sinhmar, S.; Dwivedi, D. K. Enhancement of Mechanical Properties and Corrosion Resistance of Friction Stir Welded Joint of AA2014 Using Water Cooling. *Mater. Sci. Eng. A* **2017,** *684*, 413–422. DOI: 10.1016/j.msea.2016.12.087
14. Ugender, S.; Kumar, A.; Reddy, A. S. Experimental Investigation of Tool Geometry on Mechanical Properties of Friction Stir Welding of AA 2014 Aluminium Alloy. *Proc. Mater. Sci.* **2014,** *5*, 824–831. DOI: 10.1016/j.mspro.2014.07.334
15. Elangovan, K.; Balasubramanian, V. Influences of Tool Pin Profile and Welding Speed on the Formation of Friction Stir Processing Zone in AA2219 Aluminium Alloy. *J. Mater. Proces. Technol.* **2008,** *200*, 163–175. DOI: 10.1016/j.jmatprotec.2007.09.019
16. Suresha, C. N.; Rajaprakash, B. M.; Upadhya, S. A Study of the Effect of Tool Pin Profiles on Tensile Strength of Welded Joints Produced Using Friction Stir Welding Process. *Mater. Manuf. Process.* **2011,** *26*, 1111–1116. DOI: 10.1080/10426914.2010.532527

17. Aydin, H.; Bayram, A.; Esme, U. et al. Application of Grey Relation Analysis (Gra) and Taguchi Method for the Parametric Optimization of Friction Stir Welding (Fsw) Process. *Materiali in Tehnologije* **2010,** *44*, 205–211.
18. Vijayan, S.; Raju, R.; Rao, S. R. K. Multiobjective Optimization of Friction Stir Welding Process Parameters on Aluminum Alloy AA 5083 Using Taguchi-Based Grey Relation Analysis. *Mater. Manuf. Process.* **2010,** *25*, 1206–1212. DOI: 10.1080/10426910903536782
19. Kasman, S. Multi-response Optimization Using the Taguchi-based Grey Relational Analysis: A Case Study for Dissimilar Friction Stir Butt Welding of AA6082-T6/AA5754-H111. *Int. J. Adv. Manuf. Technol.* **2013,** *68*, 795–804. DOI: 10.1007/s00170-012-4720-0
20. Sahu, P. K.; Pal, S. Multi-response Optimization of Process Parameters in Friction Stir Welded AM20 Magnesium Alloy by Taguchi Grey Relational Analysis. *J. Magnes. Alloys* **2015,** *3*, 36–46. DOI: 10.1016/j.jma.2014.12.002
21. Chien, C-H.; Lin, W-B.; Chen, T. Optimal FSW Process Parameters for Aluminum Alloys AA5083. *J. Chin. Inst. Eng.* **2011,** *34*, 99–105. DOI: 10.1080/02533839.2011.553024
22. Elangovan, K.; Balasubramanian, V.; Babu, S. Predicting Tensile Strength of Friction Stir Welded AA6061 Aluminium Alloy Joints by a Mathematical Model. *Mater. Des.* **2009,** *30*, 188–193. DOI: 10.1016/j.matdes.2008.04.037
23. Beausir, B.; Fundenberger, J.-J. Analysis Tools for Electron and X-ray diffraction, ATEX—software. www.atex-software.eu, Université de Lorraine—Metz, 2017.

CHAPTER 20

IMPROVED MECHANICAL AND ABRASIVE WEAR PERFORMANCE OF AISI 304 STAINLESS STEEL WITH IN625 ALLOY TIG WELD CLADDING

NEERAJ KAMBOJ* and LALIT THAKUR

Department of Mechanical Engineering, National Institute of Technology, Kurukshetra, Haryana, India

**Corresponding author. E-mail: neerajkamboj29@gmail.com*

ABSTRACT

In the present study, an Inconel 625 alloy clad layer was fabricated on AISI 304 stainless steel by using an automated tungsten inert gas (TIG) welding machine. The clad layer was deposited on stainless steel substrate to improve its hardness and wear resistance properties. SEM imaging was performed to study the morphology of Inconel 625 alloy powder. The effects of welding current and scanning speed on the microstructure and wear behavior of cladding were studied by using the optical microscopy and ASTM G65 dry sand abrasion wear test. Microhardness and porosity of the claddings deposited by varying the process parameters were also determined. It was observed that the welding current was the dominant factor that governs the properties and wear behavior of cladding. High heat input caused the greater dilution and defects in the cladding and improper melting of powder was also observed at low heat input. The results revealed that with adequate heat input a uniform cladding was deposited to achieve the desired properties and wear resistance.

Modern Manufacturing Systems: Trends and Developments. Rajiv Kumar Garg, Ravi Pratap Singh, Rajeev Trehan, & Ramesh Singh (Eds.)

20.1 INTRODUCTION

Stainless steel is the most demanding material used in chemical, oil refinery, and food industries due to its high oxidation resistance and high corrosion resistance properties. However, poor wear resistance and low hardness properties limit its application in mining, drilling, and ore transportation, where the abrasive and adhesive wear modes are dominant.[1] In abrasive wear, material loss occurs due to the sliding action of more rigid material/surface over softer material/surface. Abrasive wear has two modes, mainly the two-body and three-body abrasive wear. The two-body abrasive wear occurs when hard particles/grits remove material from the softer/opposite material by plowing operation, while in three-body wear, hard particles come in between two softer particles, or we can say that there is no constrained of particles. There is a sliding contact between two solid surfaces in an adhesive type of wear, and the bonding forces at the interface are more potent than the material's surface strength.[2] To avoid the material's degradation or increase the life of components, there are different properties like high creep strength, corrosion resistance, wear, hardness, etc. should be present in the material, and these properties the machining and cost factors come into play. With the help of heat, the deposition of a layer of hard ceramics, nickel–chromium alloy, and a composite mixture of both on the upper face of the stainless steel can combat these types of wear.[3]

Limited work has been published to enhance the properties of different stainless steel grade materials with the tungsten inert gas (TIG) cladding process. AISI 4340 steel coated by WC,[4] AISI 304 stainless steel and AISI 4140 stainless steel, coated by SiC,[5,6] to increased the hardness and wear resistance properties of mild steel, Ti, and Al as TIG cladding powder was used[7] and to improve the properties of grade 316 stainless steel, filler wire of NiTi has been used as coating material[8] with the help of TIG cladding process. a layer of WC-10Co-4Cr successfully deposited on stainless steel to improve its hardness and wear resistance properties.[9] The TIG welding successfully deposited all the coating/cladding. The effect of torch travel/scan speed, preplaced powder thickness, and welding current on the microstructure and hardness of coating were analyzed previously. From the literature, it has been observed that more dilution occurs with an increase in the welding current and also results in decreased hardness. However, by controlling the input parameters (current, scan speed, standoff distance), we can prevent the dilution effect.[10,11]

20.2 MATERIALS AND METHODS

20.2.1 NICKEL–CHROMIUM ALLOY

A combination of mechanical and tribological properties such as high wear, high strength, high creep, and corrosion resistance, the nickel–chromium-based alloys are promising coating materials used to minimize abrasive wear.[12] The presence of niobium and molybdenum in the nickel–chromium matrix makes nickel–chromium alloy a high strength alloy. The nickel alloy has an FCC structure at room temperature and retains its structure even at elevated temperatures.[13,14] Different types of nickel–chromium alloys are IN600, IN625, IN718, IN800, etc. Applications and investigations of IN625 on erosion, corrosion, and wear are becoming a developing area in the field of hard coating/cladding. SEM image of IN625 is shown in Figure 20.1 and the elemental composition of IN625 is given in Table 20.1.

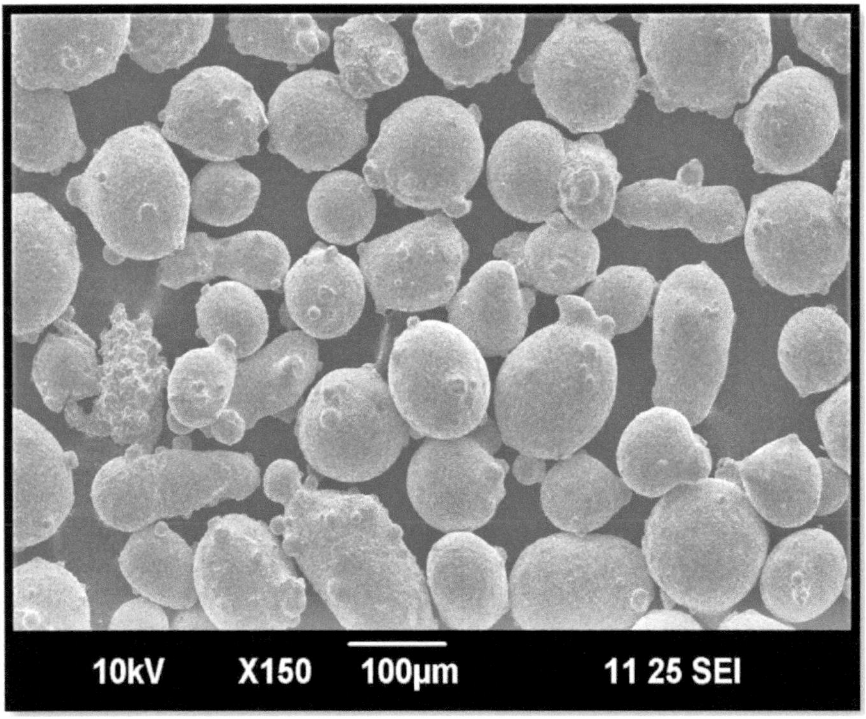

FIGURE 20.1 SEM image of IN625 powder.

TABLE 20.1 Chemical Compositions of IN625 (wt%).

	Ni	Cr	Fe	Mo	Nb	C	Mn	Si	Al	Co
Inconel 625	Balance	21–24	6	7–9	3–5	0.10 max	0.48 max	0.52 max	0.50 max	1.0 max

20.2.2 CLADDING METHOD

Different coating deposition techniques such as high-velocity oxyfuel (HVOF), shielded metal arc welding, atmospheric plasma spraying, laser cladding, and tungsten inert gas (TIG) welding, etc. are used to deposit a layer of superior material over stainless steel substrate. Every coating technique has its own merits and demerits. Out of these techniques, TIG cladding is the most economical, simple, and high deposition rate process for the fabrication of the cladding. Another name of TIG welding is gas tungsten arc welding. Russel Meredith invented TIG welding, and Linde Company developed TIG welding. On February 24th,1942, Mr. Russel Meredith got his patent for TIG welding. In TIG cladding/welding, we produced an electric arc (AC power source) between nonconsumable tungsten electrode (melting point 3380°C) present in the welding torch and workpiece. As a result, this arc generates heat used to melt preplaced powder over a substrate material. A ceramic nozzle is used for the flow of shielding gas (argon) that protects the weld pool from oxidation to ensure a high-quality weld bead.[15–17] The working mechanism of the TIG welding process is presented in Figure 20.2.

AISI 304 stainless steel (SS) material was selected as a substrate for the deposition of claddings. Firstly, the substrate material was burnished on to the emery paper and then cleaned with acetone to increase the surface roughness and remove any impurities. It also helps in achieving a good bonding between the clad layer and the substrate. IN 625 alloy was used as the clad layer powder feedstock material, which shows particles' spherical morphology and particle size varies in the range of 50–100 μm. Before the cladding deposition, a preplaced layer of feedstock powder was deposited on the SS substrate. A liquid solution was prepared by mixing 5 wt.% polyvinyl alcohol (PVA) and distilled water with the help of a mechanical stirrer. Thereafter, PVA solution and IN625 powder were gently mixed until a semisolid paste was obtained. The semisolid paste was dispersed over the SS substrate's surface and then dried for 6 h in an oven maintained at 150°C, as shown in Figure 20.3.

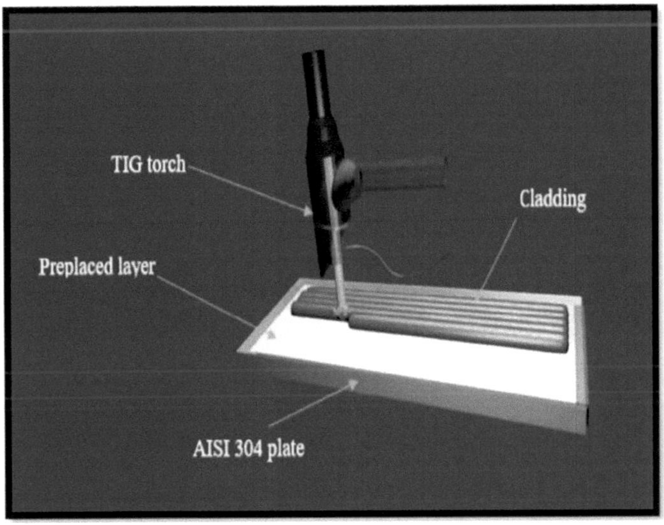

FIGURE 20.2 TIG weld cladding process.

Source: Reprinted from Ref. [15]. © 2020 The Author(s). Published by IOP Publishing Ltd on behalf of the IMMT. https://creativecommons.org/licenses/by/3.0/

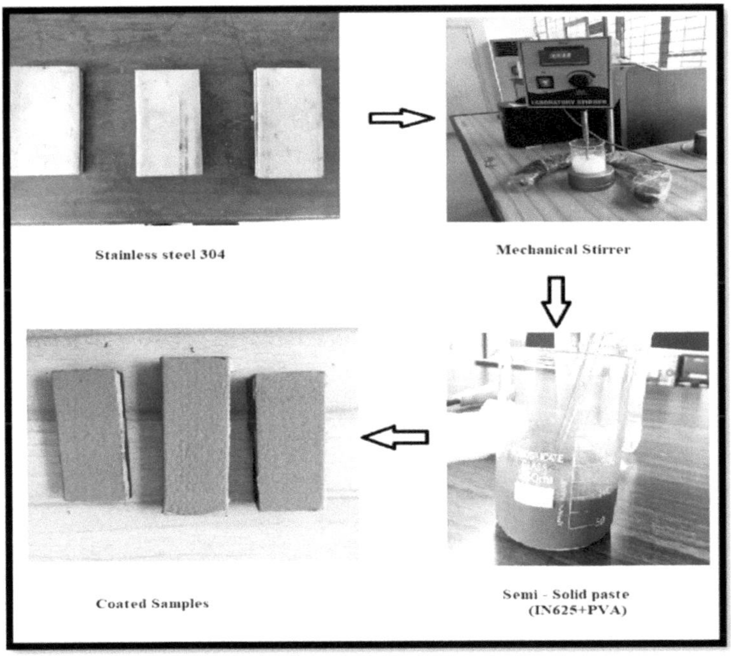

FIGURE 20.3 Preplaced powder layer deposition steps.

After preplaced the IN 625 layer of 0.7 mm thickness on the surface of SS substrate, an automatic TIG welding manipulator was used to deposit the cladding over the substrate, as shown in Figure 20.4. In the TIG weld cladding process, a tungsten electrode of nonconsumable type was used to produce the arc. For proper arc generation, the gap between the workpiece and electrode was maintained as 2.5 mm. The flow rate of shielding gas (argon) was set to 6 L/min. Using equation $H = \eta(V.I/S.1000)$, heat input was calculated. Where η is the heat absorption efficiency, having a value of 48%.[15] Table 20.2 provides the TIG weld cladding parameters used in the present work. The welding current and scanning speeds were varied, as illustrated in Table 20.3.

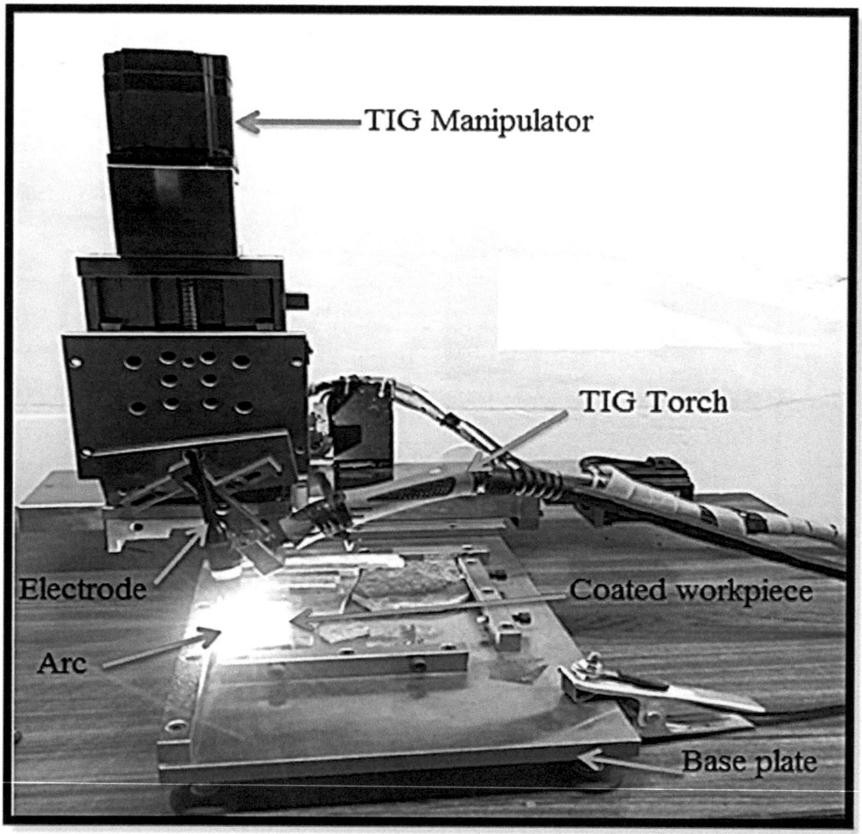

FIGURE 20.4 3-axes TIG welding manipulator.

TABLE 20.2 Operating Parameters for the TIG Cladding Process.

Voltage	20 V
Polarity	DCEN
Shielding gas	Argon
Gas flow rate	6 L/min
Electrode diameter	2.4 mm
Standoff distance	2.5 mm

TABLE 20.3 Details of Experimental Conditions for the TIG Cladding Process.

Exp. no.	Current (A)	Torch scan speed (mm/min)	Heat input (kJ/m)	Hardness ($HV_{0.3}$)	Weight loss (mg)	Observation (physically)
1	60	200	172.97	220.8	56.6	Proper bonding
2	60	250	138.46	255.9	50	Proper bonding
3	60	300	115.2	–	–	Improper bonding
4	70	200	201.80	190	64	Proper bonding
5	70	250	161.53	200	62.7	Proper bonding
6	70	300	134.4	209	58	Proper bonding
7	80	200	230.63	170	70	High heat input
8	80	250	184.61	200	66	Proper bonding
9	80	300	153.6	220	58	Proper bonding

The cladding samples obtained by using different TIG welding process parameters (Table 20.2) are shown in Figures 20.5a–c. Thereafter, the specimens were cut from the clad layer's center for microstructural examination and hardness testing. With the help of disc-polishing machine, cut samples were polished using different grades of emery papers (220–1200 grit size). The IN 625 clad layer morphology and weld beads were analyzed by Vision Measuring Microscope (Sipcon Instruments, Ambala, India). The microstructure of claddings was studied by using a Digital Metallurgical Inverted Microscope (Suxma, Conation Technologies, Pune, India).

20.2.3 ABRASION WEAR TEST

To measure the abrasive wear of samples dry-sand rubber wheel abrasive wear test was done according to ASTM G65 standard. Dry sand of prescribed particle size is passed from the nozzle between the specimen/workpiece and a rotating wheel of diameter 222.8 mm with a layer of chlorobutyl rubber over it, as shown in Figure 20.6. Test specimens were weighed before and after the wear test. The difference in the specimen's initial weight and final weight was calculated; this gives the weight loss. Table 20.4 provides the parameters for the dry sand abrasion test.

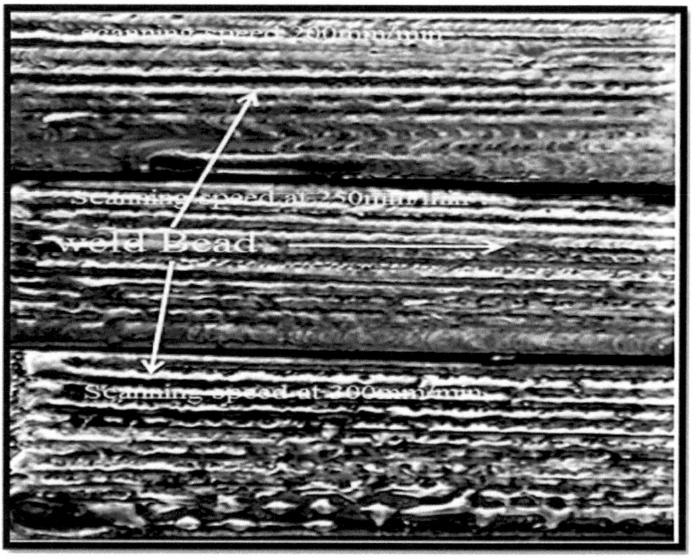

FIGURE 20.5a AISI 304 SS with IN 625 claddings at current 60 A.

FIGURE 20.5b AISI 304 SS with IN 625 coating at current 70 A.

FIGURE 20.5c AISI 304 SS with IN 625 coating at current 80 A.

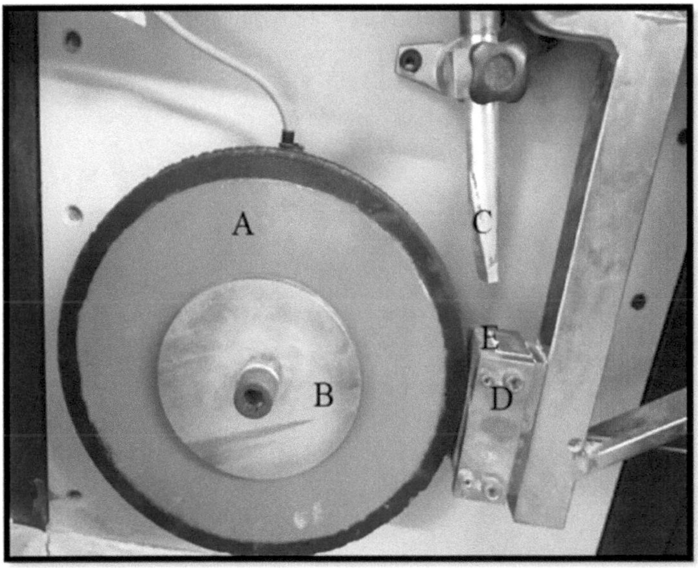

A– Chlorobutyl Rubber Wheel
B– Steel Disc
C– Nozzle
D– Workpiece Holder
E – Workpiece

FIGURE 20.6 Dry sand abrasion test machine.

TABLE 20.4 Test Condition for Dry Sand Abrasion Test.

Abrasive used	Sand
Mean particle size of abrasive	250 m
Rotational speed	200
Load	130 N
Sand flow rate	250 g/min
Duration for test	10 min

20.2.4 MICROHARDNESS TEST

The microhardness of the samples was measured at a normal load of 300 g with a dwell time 10 s by using a Vickers' microindentation hardness tester, as shown in Figure 20.7. The average value of microhardness was calculated by considering eight numbers of indents on the cross-section of claddings and substrate. The maximum value of cladding hardness was obtained as 256 $HV_{0.3}$, and the minimum value of hardness was found to be170 $HV_{0.3}$.

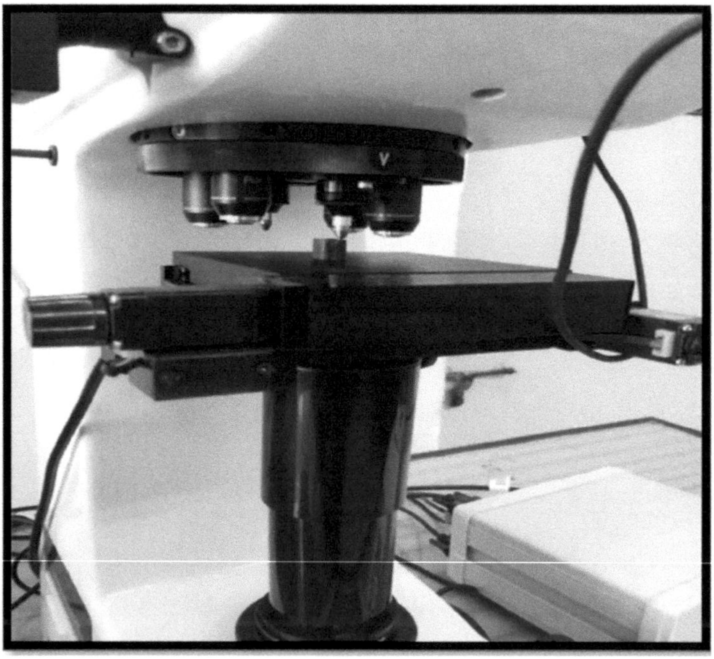

FIGURE 20.7 Microhardness tester.

20.3 RESULT AND DISCUSSION

The microhardness of IN 625 claddings was evaluated at their cross-sections and cladding–substrate interface. The average values of microhardness and weight loss of the substrate were found to be 125 $HV_{0.3}$ and 348 mg, respectively. From the microhardness test, the maximum hardness value was observed in the cladding developed at the current of 60 A and 250 mm/min scanning speed (138.46 kJ/m of heat input). However, the minimum hardness was seen at a current of 80 A and 200 mm/min scanning speed (230.63 kJ/m of heat input). From Table 20.2, it is clear that at low heat input value (60 A, 300 mm/min, 115.2 kJ/m) the hardness value and abrasive wear were undetected. The clad layer was not formed due to the improper melting of powder particles. There was a lack of interparticle bonding between the powder particles and the adhesion of the clad layer with the substrate. The clad layer was not formed properly and exhibited visible porosity and improper bead formation.

Similarly, with high heat input value (80 A, 200 mm/min, 230.63 kJ/m), the microhardness was found to be minimum, and abrasive wear was maximum (Figures 20.8a and b) because the high heat input results in the greater dilution and porosity in the cladding, as shown in Figure 20.9a. High porosity contents lead to decreased hardness in the cladding. However, at adequate heat input (60 A, 250 mm/min, 138.46 kJ/m of heat input), the maximum value of microhardness and minimum value of abrasive wear was observed. The improved hardness of the cladding may be attributed to the lower levels of dilution and porosity in the cladding, as demonstarted in Figure 20.9b. The enhanced wear resistance of the cladding may be attributed to its high microhardness value.

20.4 CONCLUSION

IN 625 claddings were fabricated on AISI 304 stainless steel substrate with the automated TIG welding. TIG welding current and torch scanning speed were varied to study their effects on the microhardness, porosity, and the wear of claddings.

- It was observed that cladding deposited with 60 A current and 250 mm/min of scanning speed exhibited the maximum hardness and wear resistance. In contrast, the cladding deposited with 80 A welding current and 200 mm/min of torch scanning speed showed the minimum hardness and low wear resistance.

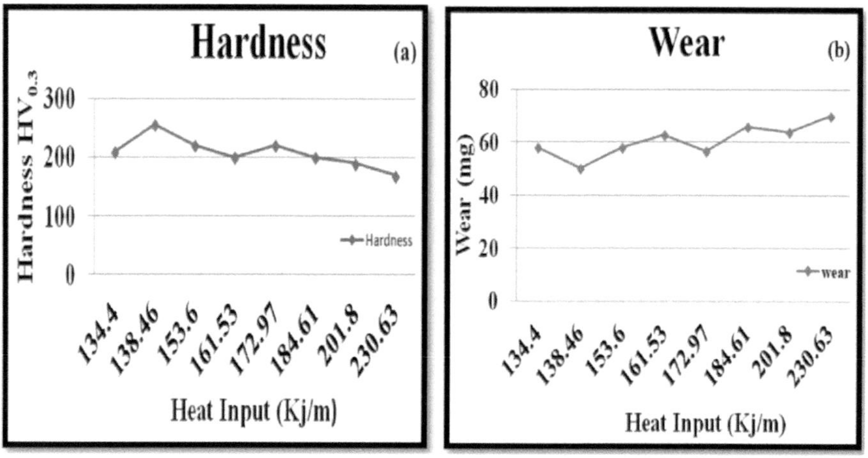

FIGURE 20.8 (a) Cladding hardness vs. heat input, (b) cladding wear vs. heat input.

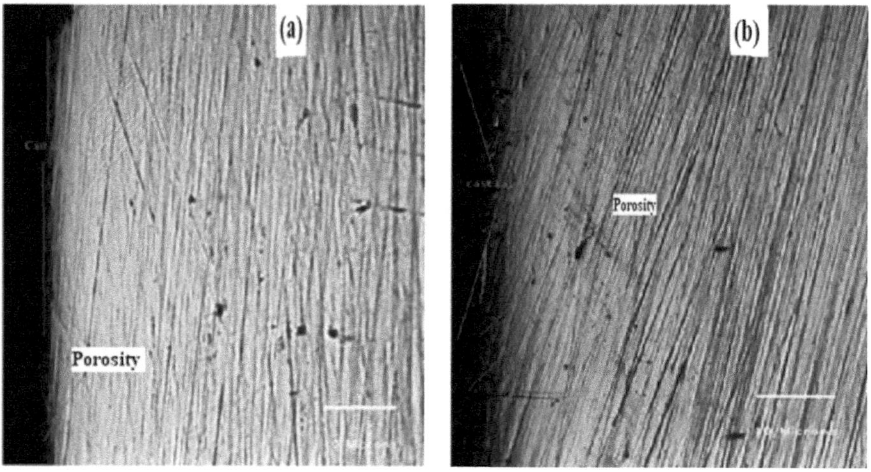

FIGURE 20.9 Cladding deposited with (a) 80 A and 200 mm/min (b) 60 A and 250 mm/min.

- The parameter setting with 60 A current and 250 mm/min of scanning speed exhibited the minimum dilution, heat-affected zone, and porosity. The parameter setting with 80 A current and 200 mm/min of scanning speed resulted in higher dilution, bigger heat-affected zone, and higher porosity contents.
- During the dry sand abrasion test, wear of material was found to be in the form of abrasion and plowing. The low value of current with optimum scanning speed reduces the heat input, which results

in minimum dilution and porosity in the cladding. Hence the wear resistance and hardness properties of the substrate are enhanced with the TIG weld clad.

KEYWORDS

- **TIG**
- **abrasive**
- **wear**
- **inconel**
- **cladding**
- **sand**

REFERENCES

1. Axén, N.; Zum Gahr, K. H. Abrasive Wear of TiC-steel Composite Clad Layers on Tool Steel. *Wear* **1992,** *157* (1), 189–201. DOI: 10.1016/0043-1648(92)90197-G
2. Kato, K.; Adachi, K. Wear Mechanisms. In *Mod. Tribol. Handbook,* 2001; Vol 1, p 28.
3. Jankauskas, V.; Antonov, M.; Varnauskas, V.; Skirkus, R.; Goljandin, D. Effect of WC Grain Size and Content on Low Stress Abrasive Wear of Manual Arc Welded Hardfacings with Low-Carbon or Stainless Steel Matrix. *Wear* **2015,** *328–329,* 378–390. DOI: 10.1016/j.wear.2015.02.063
4. Buytoz, S.; Ulutan, M.; Yildirim, M. M. Dry Sliding Wear Behavior of TIG Welding Clad WC Composite Coatings. *Appl. Surf. Sci.* **2005,** *252* (5), 1313–1323. DOI: 10.1016/j.apsusc.2005.02.088
5. Ulutan, M.; Yildirim, M. M.; Buytoz, S.; Çelik, O. N. Microstructure and Wear Behavior of TIG Surface-Alloyed AISI 4140 Steel. *Tribol. Trans.* **2011,** *54* (1), 67–79. DOI: 10.1080/10402004.2010.519859
6. Buytoz, S.; Ulutan, M. In Situ Synthesis of SiC Reinforced MMC Surface on AISI 304 Stainless Steel by TIG Surface Alloying. *Surf. Coatings Technol.* **2006,** *200* (12–13), 3698–3704. DOI: 10.1016/j.surfcoat.2005.02.178
7. Dyuti, S. et al. Surface Modification of Mild Steel Using Tungsten Inert Gas TorchSurface Cladding. *Am. J. Appl. Sci.* **2010,** *7* (6), 815–822.
8. Cheng, F. T.; Lo, K. H.; Man, H. C. NiTi Cladding on Stainless Steel by TIG Surfacing Process Part I. Cavitation Erosion Behavior. *Surf. Coatings Technol.* **2003,** *172* (2–3), 308–315. DOI: 10.1016/S0257-8972(03)00345-1
9. Singh, J.; Thakur, L.; Angra, S. Abrasive Wear Behavior of WC-10Co-4Cr Cladding-deposited by TIG Welding Process. *Int. J. Refract. Met. Hard Mater.* January **2020,** *88,* 105198. DOI: 10.1016/j.ijrmhm.2020.105198

10. Chakraborty, G. Study on Microstructure and Wear Properties of Different Nickel Base Hardfacing Alloys Deposited on Austenitic Stainless Steel. *Surf. Coatings Technol.* **2014,** *244*, 180–188. DOI: 10.1016/j.surfcoat.2014.02.013
11. Chen, J. H.; Hua, P. H.; Chen, P. N.; Chang, C. M.; Chen, M. C.; Wu, W. Characteristics of Multi-Element Alloy Cladding Produced by TIG Process. *Mater. Lett.* **2008,** *62* (16), 2490–2492. DOI: 10.1016/j.matlet.2007.12.038.
12. Sateesh, N. H.; Kumar, G. C. M.; Prasad, K.; Vinod, A. R. Microstructure and Mechanical Characterization of Laser Sintered Inconel-625 Superalloy. *Procedia Mater. Sci.* **2014,** *5*, 772–779. DOI: 10.1016/j.mspro.2014.07.327
13. Phiri, R. R.; Oladijo, O. P.; Maledi, N.; Akinlabi, E. T. Effect of Coating Thickness on Wear Performance of Inconel 625 Coating. *IOP Conf. Series: Mater. Sci. Eng.* **2018,** *423*, 012159. DOI:10.1088/1757-899X/423/1/012159
14. Abioye, T. E.; McCartney, D. G.; Clare, A. T. Laser Cladding of Inconel 625 Wire for Corrosion Protection. *J. Mater. Process. Technol.* **2015,** *217*, 232–240.
15. Singh, J.; Thakur, L.; Angra, S. Effect of Argon Flow Rate and Standoff Distance on the Microstructure and Wear Behaviour of WC-CoCr TIG Cladding. *J. Phys. Conf. Ser.* **2019,** *1240* (1). DOI: 10.1088/1742-6596/1240/1/012162
16. Sahoo, C. K.; Soni, L.; Masanta, M. Evaluation of Microstructure and Mechanical Properties of TiC/TiC-steel Composite Coating Produced by Gas Tungsten Arc (GTA) Coating Process. *Surf. Coatings Technol.* **2016,** *307*, 17–27. DOI: 10.1016/j.surfcoat. 2016.08.056
17. Welding History: Timeline and Information, http://weldguru.com/ OLDSITE/welding-history.html.

CHAPTER 21

INFLUENCE OF WELDING SPEED ON THE METALLURGICAL AND CORROSION PERFORMANCE OF AISI 316 SS WELDED JOINTS

SUBODH KUMAR[1*], BHUVNESH BHARDWAJ[2], VARUN SHARMA[3], and SHAILENDER SINGH BHADAURIA[3]

[1]*Department of Mechanical Engineering, DIT University, Dehradun 248009, India*

[2]*Department of Mechanical Engineering, Jaipur Engineering College & Research Centre, Jaipur, Rajasthan 302022, India*

[3]*Department of Industrial and Production Engineering, Dr B R Ambedkar National Institute of Technology, Jalandhar, Punjab 144011, India*

[*]*Corresponding author. E-mail: subodh.kumar@dituniversity.edu.in*

ABSTRACT

The influence of welding speed on the microstructure and corrosion properties of tungsten inert gas (TIG) welded AISI 316 SS joints were examined in the present research. Three welding speeds were chosen from the working envelope of the TIG welding and welded joints fabricate using these speeds were exposed to microstructural, microhardness and corrosion evaluations. Microstructural examination showed the lathy and skeletal δ-ferrite was present in the weld zone microstructures, while variable amount of grain coarsening was detected in the heat affect zone microstructures of the weldments. Double loop electrochemical potentiokinetic reactivation practice

Modern Manufacturing Systems: Trends and Developments. Rajiv Kumar Garg, Ravi Pratap Singh, Rajeev Trehan, & Ramesh Singh (Eds.)

was used to assess the corrosion performance of the welds, which showed that low-welding speed specimen revealed relatively higher value of degree of sensitization (indicate higher amount of corrosion rate) as compared to high-welding speed specimen. This was because low-welding speed specimen has δ-ferrite skeletal morphology, which promoted the corrosion rate. However, lathy morphology of δ-ferrite was found in the high-welding speed specimen, which demotes the corrosion rate.

21.1 INTRODUCTION

Steel is an alloy of iron and carbon with less than 2.14 wt.% carbon. By adding the chromium (Cr) more than 10.5 wt.%, steel becomes stainless steel. Stainless steels (SS) are most popular materials used by different industries due to their good corrosion resistance properties. These steels are widely split into four grades, that is, austenitic, ferritic, duplex (austenite and ferrite blend), and martensitic. Austenitic stainless steels (ASS) provide an appealing alternative as nominee materials for use in wide-ranging industrial sectors such as nuclear, automobile, chemical, biomedical and food industries, etc., because of the appropriate combination of strength, ductility, and outstanding corrosion resistance. These steels consist of 200 and 300 series, out of these series AISI 316 grade is broadly used by different industries as it has noble mechanical and corrosion resistance performance.[1–4] Welding is one of the fabrication practice to join the AISI 316L austenitic stainless steels. During the welding of these steels, some definite regions of the welded joints experience temperatures in between 500°C and 900°C, which can make them susceptible to the problem of grain boundary carbide precipitation, which is identified as sensitization. In stainless steels, where chromium carbides precipitates at the grain boundaries, sensitization is a phenomenon that contributes to the creation of depleted chromium areas across the grain boundaries and degrades their mechanical and corrosion efficiency.[5–7] In addition, the literature survey indicates that most research on the corrosion performance of these steels is recorded on wrought materials, including the analysis of the effects of various alloys composition[8–12] and different degrees of cold work.[13,14] Chuaiphan and Srijaroenpramong[15] investigated the corrosion performance of AISI 201 SS welded joints. It was found that the high-welding speed weld exhibited higher pitting corrosion resistance as compared to low- and medium-welding speed welds. Dendritic length and distance between dendrite in the welded metal microstructure were also observed to affect the mechanical and corrosion properties of these welded joints. Few studies on the corrosion performance of these steels by welding conditions have also been published. Welded assemblies generally require the use of welding processes for their fabrication, resulting in variable and confined heating and cooling, due to which separate zones of welded joints such as welding metal (due to microsegregation) and heat affected zone (due to variable cooling rates) may result in different corrosion behaviors in these zones.[16–18] Therefore, the consequences of welding may be so severe that joints can become vulnerable to sensitization during actual service, which can ultimately degrade their corrosion performance, leading to early failures. In the abovementioned literature review, however, a few important gaps were identified, as no research was described on the AISI 316L SS where the influence of welding settings (varied by selecting different welding speeds from the TIG operating envelope) on the corrosion and mechanical performance of welds was examined. Therefore, in view of this research gap, the current work was carried out in which the position of different welding speeds was investigated in order to develop the database of this industrially significant alloy in order to achieve a greater understanding of the corrosion actions of these welds, in addition to facilitating the design of welding processes by manufacturing industries.

21.2 EXPERIMENTATION

21.2.1 MATERIAL AND WELDING METHOD

The parent material taken in the current work was in the form of AISI 316L SS plates of 100 mm length, 100 mm width, and 5 mm thickness. According to the single-V groove design, the edges of these plates were prepared. The welding of these plates was carried out by tungsten inert gas (TIG) welding using the 1.6 and 2.4 mm diameter solid electrode filler AISI 316L SS to provide the root and main weld passes, respectively. The groove volume for each joint was filled with one root pass and two main welding passes to achieve sound quality welded joints. Parent and filler metal chemistry are presented in Table 21.1 for this present work.

The details of the welding conditions are shown in Table 21.2, together with the three different welding speed levels used for welding. Throughout welding, due care and caution was applied where all essential precautions such as precleaning of the groove ends and inter-pass cleaning between successive weld passes were trailed to prevent any kind of contamination; thus confirming defect-free welds. Table 21.3 displays the other details of the current work.

TABLE 21.1 Chemical Composition (wt.%) of Parent and Filler Metal.

Alloy element	C	Si	Mn	P	S	Cr	Ni	Mo	Fe
Parent metal (AISI 316L SS)	0.029	0.413	1.215	0.022	0.014	18.421	9.94	2.14	Balance
Filler metal (AISI 316L SS)	0.027	0.421	1.430	0.027	0.017	19.252	10.22	2.26	Balance

TABLE 21.2 Parameters of Welding Employed in the Current Work.

Welding speed	Kind of passes	Current (A)	Voltage (V)	Welding speed (mm/s)
Low	Root pass	90	10	1.45
	Middle pass	100	12	1.15
	Capping pass	100	12	1.15
Medium	Root pass	90	10	1.45
	Middle pass	140	15	1.35
	Capping pass	140	15	1.35
High	Root pass	90	10	1.45
	Middle pass	180	18	1.55
	Capping pass	180	18	1.55

TABLE 21.3 Supplementary Particulars Allied to the Current Work.

Welding process	Tungsten inert gas (TIG) welding
Electrode	EW-Th-2 thorium tungsten with a diameter of 2.4 mm
Shielding gas and its flow rate	Argon Industrially Pure (15 L/min)
Welding position	Flat
Interpass temperature	150°C
Polarity	(DCEN) DC electrode negative

21.2.2 METALLOGRAPHY AND HARDNESS TESTING

Appropriate dimension of the welded specimens were cut out from each welded plate for acting microstructural and microhardness testing. To view the microstructure of various zones of the welded joints, first of all grind the side face of the specimen then polish the surface with the help of different grades of emery papers. Electrolytic etching was applied after polishing under conditions such as oxalic acid = 10 g, distilled water = 100 mL, cell voltage = 6 V, and etching time = 60 s.[19] The metallurgical microscope was used to display and record microstructures of different regions of welded joints in conjunction with image processing software. The microhardness of the various areas of the welds was estimated with Vickers' microhardness measurement system with a weight of 500 g.

21.2.3 DLEPR CORROSION TESTING

As indicated by previously published studies,[20–22] a double loop electrochemical potentiokinetic reactivation (DLEPR) testing method was used to determine the corrosion performance of the welds. For performing DLEPR tests on the specimens, a software-supported potentiostat (Make: Gamry Instruments, USA) was used. The testing samples contained the combined region of welded joint, that is, weld zone, fusion boundary zone, heat affect zone (HAZ) besides the base metal. Each testing sample (mounted in the epoxy resin) has cross-sectional area of 15×6 mm^2, that is, 90 mm^2, which was specifically exposed to the test solution in the DLEPR test. Various grit size emery papers were used to prepare the specimens. The electrolyte for the present work was a combination of 0.01M KSCN and 0.5M H_2SO_4 solutions formulated with pure reagents and distilled water. The DLEPR experiments were conducted using a 250 mL glass cell with a working electrode (testing sample), reference electrode (saturated calomel electrode), and a counter electrode (graphite). Three stages were used in the DLEPR procedure. The first phase was a 5-min delay at open circuit voltage (V_{oc}). The second step was anodic polarization scanning from 50 mV below V_{oc} to 300 mV with 100 mV/min scan rate in the passive range. The third step was a cathodic reactivation scan from 300 mV to V_{oc} with the same scan rate. The degree of sensitization was calculated by the formula $\{(I_r/I_a) \times 100\}$; where I_r is reactivation current and I_a is activation current.[23] Three DOS readings were taken from each testing sample, and each single reading was taken using a new test solution. The mean value of the DOS thus obtained was recorded for each sample examined.

21.3 RESULTS AND DISCUSSION

21.3.1 OPTICAL MICROSTRUCTURAL STUDIES

After performing the optical microstructural investigation of the welded joints, few pictures of the microstructures were taken and shown in Figure 21.1a–f. The welding metal microstructure of low-, medium- and high-welding speed specimens is seen in Figure 21.1a–c, respectively. Both low- and medium-welding speed specimens (as shown in Figure 21.1a and

b) show skeletal morphology of δ-ferrite, while the length of dendrites in low-welding speed specimen is relatively high as compared to medium-welding speed specimen. However, from Figure 21.1c, it can be seen that high-welding speed weld metal has lathy morphology of delta-ferrite. HAZ microstructures of the low-, medium- and high-welding speed specimen are shown in Figure 21.1d–f. These HAZs microstructure show the variable degree of grain-coarsening. It can be shown from these figures that the grain coarsening of the HAZ reduces as the welding speed increases.

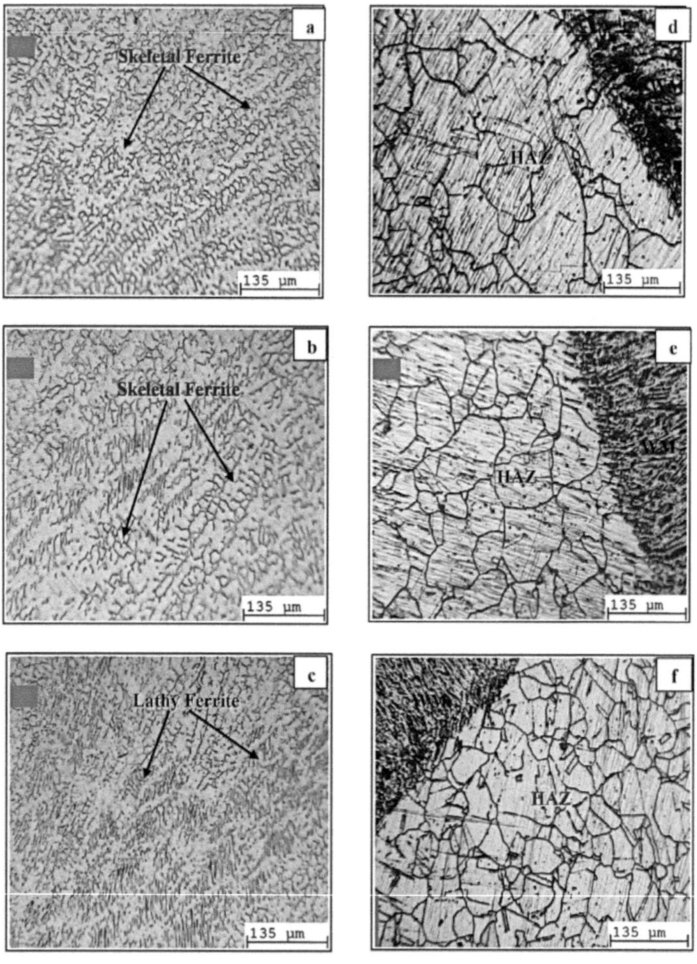

FIGURE 21.1 Optical photomicrographs (at 100×) of the weldments showing: (a), (b) and (c) weld metal corresponding to low-, medium- and high-welding speed respectively; (d), (e) and (f) HAZ corresponding to low-, medium- and high-welding speed respectively.

21.3.2 MICROHARDNESS STUDIES

Figure 21.2 and Table 21.4 display the various areas (welded metal and HAZ) of the weldments for the microhardness of the low-, medium- and high-welding speed samples measured (average of five values) around the cross-section of the welded joints. For weld metals, the average microhardness values of 196.2, 208.6, and 222.1 VHN were measured at low-, medium- and high-welding speed, respectively. The microstructural modifications that occurred attributable to differing weld thermal cycles during welding is liable for such weld metal microhardness deviation. With the low-welding speed specimen containing ferrite skeletal morphology, a comparatively low microhardness value was identified, while with the high-welding speed specimen confining delta-ferrite lathy morphology, a high microhardness value was found. Table 21.4 shows the relatively high microhardness value (190.5 VHN) of high-welding speed specimen as compare to low microhardness value (167.3 VHN) of low-welding speed specimen. This microhardness variation can be explained by the variable grain-coarsening of the HAZ of different welded joints. Low-welding speed welded joint experiences higher degree of grain coarsening and hence relatively low value of microhardness, however high value of microhardness was observed of high-welding speed HAZ, which was due to lower degree of grain-coarsening.

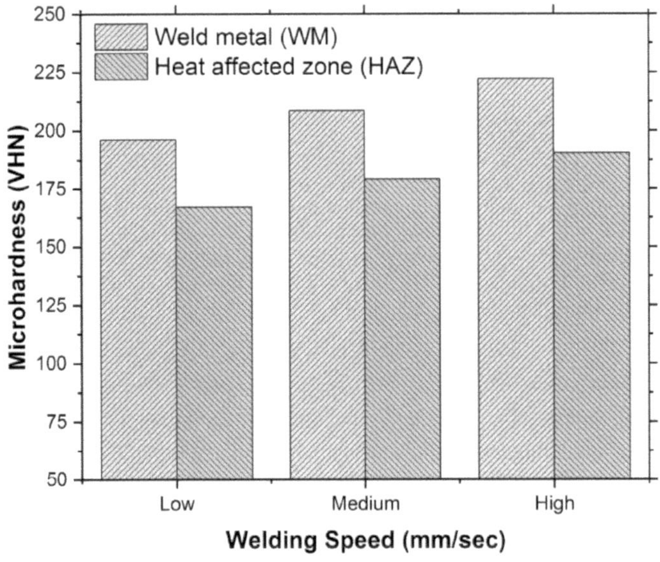

FIGURE 21.2 Microhardness values of weld metal and HAZ of different welding speed welds.

TABLE 21.4 Microhardness Values of Different Welding Speed Welded Joints.

Region	Microhardness value (VHN)		
	Low-welding speed specimen	**Medium-welding speed specimen**	**High-welding speed specimen**
Weld metal (WM)	196.2	208.6	222.1
Heat affected zone (HAZ)	167.3	179.2	190.5

21.3.3 CORROSION EVALUATION

Corrosion evaluation of different welded joints was carried out by DLEPR technique and the curves generated from this test are shown in Figures 21.3–21.5. DOS values of different welds were calculated with the help of DLEPR curves and recorded in Table 21.5. Low-, medium- and high-welding speed welds exhibited the DOS values of 1.18, 0.61, and 0.27 respectively. From this result, it can be seen that low-welding speed weld showed relatively greater DOS value as compared to the high-welding speed weld. This disparity in DOS values was attributed to higher amounts of carbide formation in low-welding speed welds because of the development of high-energy areas allied with skeletal ferrite, which subsequently accounted for greater DOS values for these welds. Furthermore, high-welding speed welded joint showed relatively lower DOS values because in these joints, there was a lesser tendency of carbide formation, which was due to the lathy ferrite morphology owing to the formation of relatively low energy regions.

21.4 CONCLUSION

The main findings of this current work are as follows:

1. Weld metal microstructure of low-welding speed specimen showed skeletal morphology of δ-ferrite while a lathy morphology of δ-ferrite was observed in high-welding speed specimen.
2. HAZ microstructure of the welded joints show the variable degree of grain coarsening; however, as the welding speed increases the degree of grain coarsening decreases.
3. Weld metal and HAZ of low-welding speed specimen showed relatively lower value of hardness whereas high-welding specimen showed relatively higher values of hardness.

4. DOS evaluation of the welded joints showed the relatively higher DOS values (indicate higher corrosion rate) of low-welding speed specimen as compared to the high-welding speed specimen.

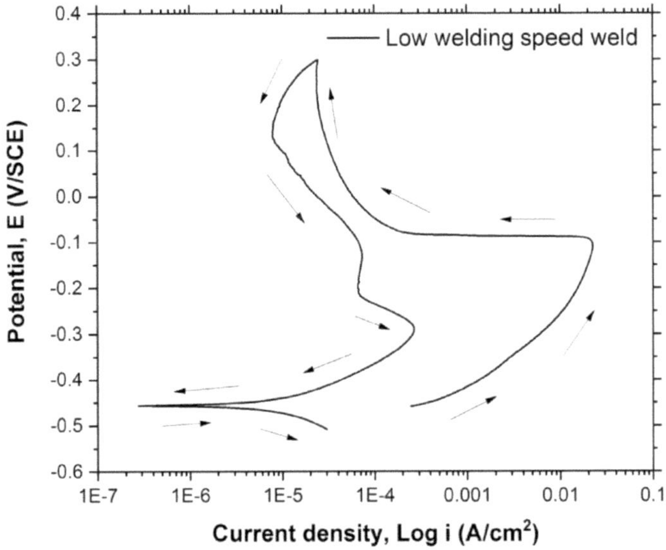

FIGURE 21.3 DLEPR curves of the low-welding speed weld.

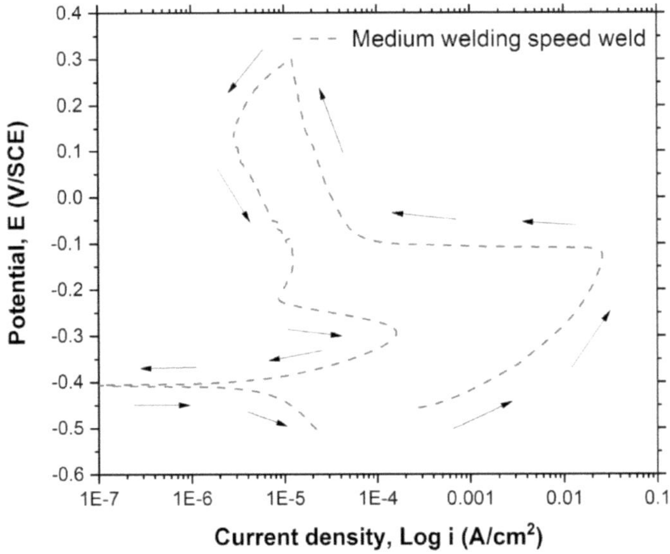

FIGURE 21.4 DLEPR curves of the medium-welding speed weld.

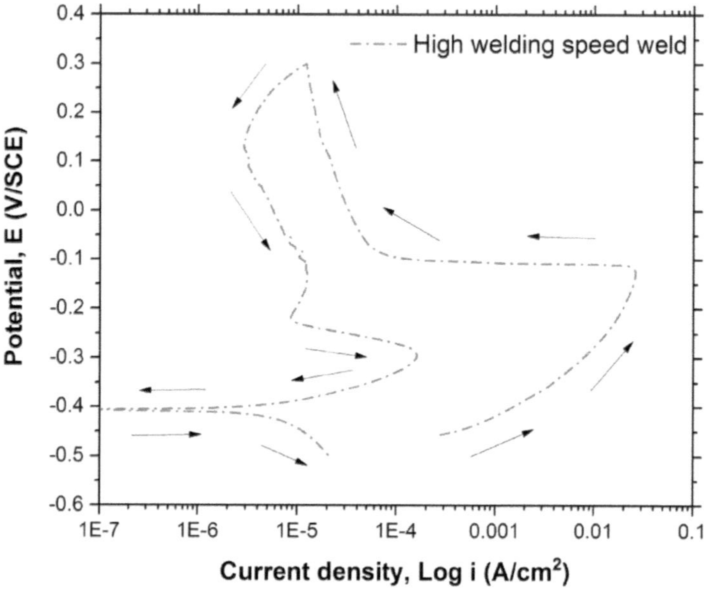

FIGURE 21.5 DLEPR curves of the high-welding speed weld.

TABLE 21.5 DOS (Degree of Sensitization) Values of Various Welding Speed Welds.

Welding speed	Degree of sensitization (DOS)
Low-welding speed welds	1.18
Medium-welding speed welds	0.61
High-welding speed welds	0.27

KEYWORDS

- **welding speed**
- **AISI 316 SS**
- **TIG**
- **microstructure**
- **microhardness**
- **DOS**

REFERENCES

1. Sedric, A. J. *Corrosion of Stainless Steels*, 2nd ed.; John Wiley & Sons: New York, 1996.
2. Khatak, H. S.; Raj, B. *Corrosion of Austenitic Stainless Steels*; Narosa Publishing House: India, 2002.
3. Finsgar, M.; Milosev, I. Corrosion Behavior of Stainless Steels in Aqueous Solutions of Methanesulfonic Acid. *Corros. Sci.* **2010,** *52*, 2430–2438.
4. Kuang, W.; Han, E. H.; Wu, X.; Rao, J. Microstructural Characteristics of the Oxide Scale Formed on 304 Stainless Steel in Oxygenated High Temperature Water. *Corros. Sci.* **2010,** *52*, 3654–3660.
5. Bruemmer, S. M.; Charlot, L. A. Development of Grain Boundary Chromium Depletion in Type 304 and 316 Stainless Steels. *Scripta Metallurgica* **1986,** *20*, 1019–1024.
6. Povich, M. J. Low Temperature Sensitization of Type 304 Stainless Steel. *Corrosion* **1978,** *34*, 60–65.
7. Bruemmer, S. M.; Charlot, L. A.; Bagchi, A.; Atteridge, D. G. Influence of Grain Boundary Carbides and Phosphorus Segregation on the Low Temperature Intergranular Embrittlement of Type 316 Stainless Steel. *Scripta Metallurgica* **1989,** *23*, 1549–1554.
8. Folkhard, E. *Welding Metallurgy of Stainless Steels*; Springer-Verlag: Vienna, 1988.
9. Mudali, U. K.; Dayal, R. K.; Gnanamoorthy, J. B.; Rodriguez, P. Influence of Thermal Aging on the Intergranular Corrosion Resistance of Types 304 LN and 316 LN Stainless Steels. *Metal. Mater. Trans. A* **1996,** *27A*, 2881–2887.
10. Beneke, R.; Sandenbergh, R. F. The Influence of Nitrogen and Molybdenum on the Sensitization Properties of Low-Carbon Austenitic Stainless Steels. *Corros. Sci.* **1989,** *29*, 543–555.
11. Pardo, A.; Merino, M. C.; Coy, A. E.; Viejo, F.; Arrabal, R.; Matykina, E. Pitting Corrosion Behaviour of Austenitic Stainless Steels- Combining Effects of Mn and Mo Additions. *Corros. Sci.* **2008,** *50*, 1796–1806.
12. Mudali, U. K.; Dayal, R. K. Pitting Corrosion Resistance of as Welded and Thermally Aged Nitrogen Containing Type 316 Stainless Steel Weld Metal. *Mater. Sci. Technol.* **2000,** *16*, 393–398.
13. Oh, Y. J.; Hong, J. H. Nitrogen Effect on Precipitation and Sensitization in Cold-Worked Type 316 L (N) Stainless Steels. *J. Nucl. Mater.* **2000,** *278*, 242–250.
14. Mannepalli, S.; Gupta, R. K.; Kumar, A. V.; Parvathavarthini, N.; Mudali, U. K. Influence of Prior Deformation on Sensitization Kinetics of Nitrogen Alloyed 316L Stainless Steels. *J. Mater. Eng. Perform.* **2015,** *24*, 1848–1855.
15. Chuaiphan, W.; Srijaroenpramong, L. Effect of Welding Speed on Microstructures, Mechanical Properties and Corrosion Behavior of GTA-welded AISI 201 Stainless Steel Sheets. *J. Mater. Proces. Technol.* **2014,** *214*, 402–408.
16. Cui, Y.; Lunding, C. D. Evaluation of Initial Corrosion Location in E316 Austenitic Stainless Steel Weld Metals. *Mater. Lett.* **2005,** *59*, 1542–1546.
17. Cui, Y.; Lunding, C. D. Austenite-Preferential Corrosion Attack in 316 Austenitic Stainless Steel Weld Metals. *Mater. Des.* **2007,** *28*, 324–328.
18. Dadfar, M.; Fathi, M. H.; Karimzadeh, F.; Dadfar, M. R.; Saatchi, A. Effect of TIG Welding on Corrosion Behaviour of 316L Stainless Steel. *Mater. Lett.* **2007,** *61*, 2343–2346.
19. ASTM International E407-07. *Standard Practice for Microetching Metals and Alloys.*

20. Lopez, N.; Cid, M.; Puiggali, M.; Azkarate, I.; Pelayo, A. Application of Double Loop Electrochemical Potentiodynamic Reactivation Test to Austenitic and Duplex Stainless Steels. *Mater. Sci. Eng. A* **1997,** *229*, 123–128.
21. Cihal, V.; Stefec, R. On the Development of the Electrochemical Potentiokinetic Method. *Electrochimica Acta* **2001,** *46*, 3867–3877.
22. Momeni, M.; Moayed, M. H.; Davoodi, A. Tuning DOS Measuring Parameters Based on Double-Loop EPR in H_2SO_4 Containing KSCN by Taguchi Method. *Corros. Sci.* **2010,** *52*, 2653–2660.
23. Kain, V.; Chandra, K.; Adhe, K. N.; De, P. K. Effect of Cold Work on Low-Temperature Sensitization Behaviour of Austenitic Stainless Steels. *J. Nucl. Mater.* **2004,** *334*, 115–132.

CHAPTER 22

EFFECT OF FRICTION WELDING CONDITIONS ON THE MECHANICAL PROPERTIES OF WELDED JOINTS DURING WELDING OF DISSIMILAR MATERIALS

BHUVNESH BHARDWAJ[1], SUBODH KUMAR[2*], VARUN SHARMA[3], and SHAILENDER SINGH BHADAURIA[3]

[1]*Department of Mechanical Engineering, Jaipur Engineering College & Research Centre, Jaipur, Rajasthan 302022, India*

[2]*Department of Mechanical Engineering, DIT University, Dehradun 248009, India*

[3]*Department of Industrial and Production Engineering, Dr B R Ambedkar National Institute of Technology, Jalandhar, Punjab 144011, India*

[*]*Corresponding author. E-mail: subodh.kumar@dituniversity.edu.in*

ABSTRACT

The friction welding process is useful for the joining of those materials that are tough to join using conventional joining processes. The present research explores the influence of friction welding parameters on the tensile properties of the fabricated weldments during the welding of mild steel with aluminum alloy AA7075. The results show that the rotational speed of the welding tool was found to be the most significant welding parameter that affects the strength (yield and ultimate tensile strength) of welded joints followed

Modern Manufacturing Systems: Trends and Developments. Rajiv Kumar Garg, Ravi Pratap Singh, Rajeev Trehan, & Ramesh Singh (Eds.)

by the diameter of the workpiece and welding time. The yield and ultimate tensile strength initially increase up to a certain value with an increase in the rotational speed of tool and the joining time. Further increase in the revolving velocity and joining time decreases the yield strength and tensile strength.

22.1 INTRODUCTION

It always remains a challenge to join similar and dissimilar metallic materials using conventional welding processes due to the degradation of tensile properties of the weldments.[1] The tensile properties of the substrate material degraded near the weld interface due to the melting of base material.[2] To overcome this problem, researchers introduced friction welding. Friction stir welding is a solid-state welding, mostly used in defense and aerospace applications. Moreover, the process is used for joining of those materials that are difficult to join using conventional joining processes like high-strength aluminum alloys, etc. In this process, heat at the interfaces of the materials generates due to mechanical action.[3] The quality of the weldments depends on the selected friction welding parameters, and the range of the parameters. Commonly used friction welding parameters are pressure, welding time, and welding speed.[4] The mechanical and corrosion resistance of the weldments depends on the combination of the selected parameters and the selected range of the parameters. Hence proper selection of optimal welding parameters becomes necessary. In the past, enormous research has been reported related to the parametric effect of friction welding parameters on the mechanical (tensile, hardness, impact, etc.) properties of the weldments. A brief discussion in this regard has been presented in this section.

Bassu and Irving[5] performed stir welding process on the 2024-T351aluminum alloy to examine the residual stress, heat-affected zone cracks propagation in the weldments due to fatigue. Mustafa and Adem[6] optimized the input parameters, namely, dwell time, speed of rotation, and depth of plunge of tool shoulder of friction stir spot welding process for maximizing joint strength during the welding of high-density polypropylene using Taguchi methodology. Uzun et al.[7] joined two dissimilar materials using friction stir welding process and studied the welded joints for microstructure and fatigue properties, and characterized the joints using EDX analysis. Cavaliere et al.[8] examined the impact of input parameters of friction stir welding (FSW) on the tensile properties of AA6056 aluminum alloy welded joints developed using friction stir welding process. Single-factor effect of the revolving speed of welding tool on the average grain size of the nugget region of the AA2024–T351 aluminum alloy joint was studied by Jariyaboon et al.[9]. Hwang et al.[10] developed a weldments temperature prediction model for the friction stir welded joint of AA6061-T6 aluminum alloy. The influence of tool movement speed on the metallurgical and tensile properties of weldments of aluminum alloy developed using friction stir welding was investigated by Sakthivel et al.[11]. Juan et al.[12] compared the efficiency of the Al-Mg-Sc alloy welded joints obtained using gas tungsten arc welding process and friction stir welding process. Palanivel et al.[13] developed a mathematical correlation among the process parameters of friction stir welding process and the tensile properties of the weldments of AA6351 aluminum alloy. Koilraj et al.[14] used the friction stir welding process for the joining of dissimilar AA2219 and AA5083 materials. Asif et al.[15] employed Grey relational-based Taguchi approach for optimizing the ultimate tensile strength, hardness, Charpy V notch impact toughness, and corrosion rate of the weldments. Elfar et al.[16] used friction stir welding for welding two different metallic materials for the study of the ultimate tensile strength of weldments using Taguchi methodology. Devaiah et al.[17] effectively joined the two grades of aluminum, namely, AA5083 and AA6061 using the friction stir welding process. The joint efficiency of the similar alloys was further improved by Prasad and Namala.[18]

From the review of past published literature, it has been revealed that a significant amount of research has been made in the past related to the joining of similar materials. However, most of the research was focused on investigating the impact of FSW process parameters over the properties (mechanical) of the weldments. On the other hand, fewer efforts have been made in the past related to the joining of dissimilar metallic materials using friction welding process. In addition to this, fewer efforts have been made to develop prediction models. Therefore, the foremost objective of the current experimental study is to study the influence of friction welding parameters on the yield strength and ultimate tensile strength of the joints fabricated from the dissimilar welding of mild steel and aluminum alloy AA7075.

22.2 EXPERIMENTATION AND MEASUREMENT

In the present work, AA 7075 alloy and mild steel (each having 6 cm lengths) have been selected as dissimilar materials for friction welding. The aluminum alloy AA7075 has exceptional tensile properties and better corrosion resistance. It is widely employed for highly stressed structural applications like the fabrication of aircraft structural parts. On the other hand, mild steel is the most common grade of plain carbon steel being used extensively for the number of engineering applications owing to its low price and good mechanical properties.

The friction welding on the samples has been carried out on a conventional lathe machine. Initially, the bars of AA7075 alloy and mild steel in desire length have been kept in 10% NaOH solution for 20–30 min followed by in the solution of nitric acid and water to remove the flit to remove the oil, dirt, grease, etc. After that these clean pieces of bars have been dried at room temperature. After that turning and facing operations have been carried out on the pieces to get the desire length and diameter. Later than one piece of mild steel and another piece of AA7075 alloy have been fitted on the friction welding setup. For the welding on a lathe, one piece of the sample has been fitted in chuck while another piece has been fixed in tool post under pressure. The pressure has been applied using tailstock. The rotation has been provided to one sample while axial pressure is applied on other samples.

The tensile tests on the fabricated samples have been conducted to estimate the yield and ultimate tensile strength of the weldments. The tensile tests on the specimens have been carried out using universal testing machine at room temperature at the loading speed of 1 mm/min.

22.3 DEVELOPMENT OF PREDICTION MODELS FOR TENSILE PROPERTIES

In this work, the prediction models of yield strength and tensile strength have been developed using response surface methodology based on face-centered design using design expert software (Trial version). The levels of friction welding parameters and the experimentation plan according to FCD have been depicted in Tables 22.1 and 22.2, respectively.

TABLE 22.1 Levels of Process Parameters of Friction Welding.

Friction welding parameters	Levels		
	Minimum value	Center value	Maximum value
Rotational speed (rpm)	300	600	900
Welding time (min)	20	25	30
Diameter (mm)	5	7.5	10

TABLE 22.2 Experimentation Plan and Measured Value of Tensile Properties.

Std	Run	A: Rotational speed (rpm)	B: Welding time (min)	C: Diameter (mm)	Yield strength (MPa)	Tensile strength (MPa)
1	7	300	20	5	119.6	197.34
2	8	900	20	5	96.3	154.08
3	10	300	30	5	95.7	153.12
4	5	900	30	5	82.5	128.7
5	3	300	20	10	96.6	154.56
6	14	900	20	10	82.4	128.54
7	17	300	30	10	85.1	132.76
8	2	900	30	10	78.2	122.00
9	18	300	25	7.5	119.6	194.95
10	12	900	25	7.5	105.95	169.52
11	6	600	20	7.5	119.6	200.00
12	1	600	30	7.5	109.7	181.00
13	15	600	25	5	136.3	231.71
14	9	600	25	10	121.9	203.57
15	11	600	25	7.5	129.4	219.98
16	4	600	25	7.5	134	227.8
17	13	600	25	7.5	130.2	221.34
18	16	600	25	7.5	127.8	217.26

To develop the mathematical model, the first step is to apply the analysis of variance test (ANOVA). The ANOVA is employed to check the (1) significance of the developed model, (2) significance of the individual model terms, and (3) to check the adequacy of the lack of fit. In this work, ANOVA is carried out at the confidence level of 95%. The ANOVA tables for yield strength and tensile strength have been presented in Tables 22.3 and 22.4, respectively.

From Table 22.3, it has been cleared that "Prob. > *F*" value for yield strength prediction equation is less than 0.0001, which is less 0.05; therefore, yield strength prediction model is significant at 95% confidence level. This shows the positive influence of model terms on the yield strength. In addition to this, Prob. > *F* value of main effect of rotational speed, welding time, and diameter are less than 0.05, therefore these terms significant. Also, Prob. > *F* values of different interactions, namely, A&B, A&C, and B&C, and the quadratic terms of welding time and rotational speed are less than 5%, which depicts that the terms have significant influence on variability of the developed mathematical model. On the other hand, the quadratic term of diameter has been found insignificant model term. The "Prob. > *F*" for the lack of fit is greater than 0.05, that is, 0.8538, which indicates that the lack of fit is nonsignificant and is desirable quantity for fitting the model. The coefficient of determination (R^2) value is equal to 0.995 and adjusted R^2 value is equal to 0.989. It is further inferred from the table that the adjusted value of R^2 is in the proximity of ordinary value of R^2, which indicates that the variation in the yield strength is due to parameters not due to uncontrollable parameters. The signal to noise is referred as adequate precision value. This value in the present model was found to be 91.33, which is more than 4 as per the requirement to adequately fit the empirical model. The yield strength prediction model in terms of parameters is shown in eq 22.1.

TABLE 22.3 Variance Analysis for the Yield Strength.

Terms	SS: sum of squares	DOF: degrees of freedom	MS: mean square value	F-value	p-value Prob > F
Model	6486.88	9	720.8	174.81	<0.0001
A—Rotational speed	507.66	1	507.7	123.12	<0.0001
B–Welding time	400.69	1	400.7	97.18	<0.0001
C—Diameter	438.24	1	438.2	106.29	<0.0001
AB	37.85	1	37.85	9.18	0.0163
AC	29.65	1	29.65	7.19	0.0279
BC	60.50	1	60.5	14.67	0.0050
A^2	979.54	1	979.5	237.57	<0.0001
B^2	795.87	1	795.87	193.02	<0.0001
C^2	19.58	1	19.58	4.75	0.0609
Residual	32.99	8	4.1232		
Lack of Fit	12.24	5	2.4472	0.35	0.8538
Pure Error	20.75	3	6.9167		
Cor Total	6519.86	17			
Std. Dev.	2.03		R^2		0.995
Adjusted R^2	0.989		Predicted R^2		0.982

Table 22.4 shows that "Prob. > F" for tensile strength prediction model which is less than 0.0001, therefore tensile strength prediction model is significant. Also, the main effect terms (namely, diameter, tool revolving speed, and welding time), the interaction terms (namely, rotational speed—welding time, rotational speed–diameter, welding time–diameter), and quadratic terms of

revolving speed of tool and welding time have been found to be the significant model terms. The lack of fit for the prediction model of tensile strength was found to be nonsignificant owing to the higher value of Prob. > F than 0.05. The ordinary R^2 value and adjusted R^2 values have been found as 0.997 and 0.993, respectively.

TABLE 22.4 Variance Analysis for the Tensile Strength.

Terms	SS: sum of squares	DOF: degrees of freedom	MS: mean square value	F-value	p-value Prob > F
Model	24069.56	9	2674.40	259.09	<0.0001
A—Rotational speed	1687.09	1	1687.09	163.44	<0.0001
B—Welding time	1361.49	1	1361.49	131.90	<0.0001
C—Diameter	1525.84	1	1525.84	147.82	<0.0001
AB	145.28	1	145.28	14.07	0.0056
AC	119.35	1	119.35	11.56	0.0094
BC	212.63	1	212.63	20.60	0.0019
A^2	4252.49	1	4252.49	411.98	<0.0001
B^2	2685.40	1	2685.40	260.16	<0.0001
C^2	47.98	1	47.98	4.65	0.0632
Residual	82.58	8	10.32		
Lack of Fit	22.61	5	4.52	0.23	0.9284
Pure Error	59.97	3	19.99		
Cor Total	24,152.14	17			
Std. Dev.	3.21		R^2		0.997
Adjusted R^2	0.993		Predicted R^2		0.988

Eqs 22.1 and 22.2 show the prediction models for the yield strength and tensile strength, respectively, in terms of welding parameters.

$$\text{Yield strength} = -257.39 + 0.174 \times \text{Rotational speed} + 30.49 \times \text{Welding time} - 3.24 \times \text{Diameter} + 1.45000\text{E}{-003} \times \text{Rotational speed} \times \text{Welding time} + 2.56667\text{E}{-003} \times \text{Rotational speed} \times \text{Diameter} + 0.22 \times \text{Welding time} \times \text{Diameter} - 2.11257\text{E}{-004} \times \text{Rotational speed}^2 - 0.68 \times \text{Welding time}^2 - 0.43 \times \text{Diameter}^2 \quad (22.1)$$

$$\text{Tensile strength} = -497.13 + 0.37 \times \text{Rotational speed} + 55.83 \times \text{Welding time} - 8.24 \times \text{Diameter} + 2.84100\text{E}{-003} \times \text{Rotational speed} \times \text{Welding time} + 5.15000\text{E}{-003} \times \text{Rotational speed} \times \text{Diameter} + 0.41244 \times \text{Welding time} \times \text{Diameter} - 4.40170\text{E}{-004} \times \text{Rotational speed}^2 - 1.26 \times \text{Welding time}^2 - 0.67 \times \text{Diameter}^2. \quad (22.2)$$

22.4 EFFECT OF WELDING PROCESS PARAMETERS ON TENSILE PROPERTIES

The effect of welding parameters (namely, workpiece diameter, tool rotational speed, and welding time) on the tensile properties (yield strength and ultimate tensile strength) of the weldments was expressed as main effect plots and was plotted using design expert software (trial version).

22.4.1 INFLUENCE OF THE ROTATIONAL VELOCITY OF TOOL

The main effect plots between the rotational velocity and yield strength, rotational velocity and tensile strength are shown in Figures 22.1 and 22.2, respectively. The curvature of the plots indicates that the variation of yield strength and tensile strength is quadratic with rotational speed. Also, it has been revealed from the figures that the yield strength and tensile strength initially increases up to a certain value with the rise in the revolving velocity of the welding tool. Further rise in rotational speed, decreases the yield strength and tensile strength.

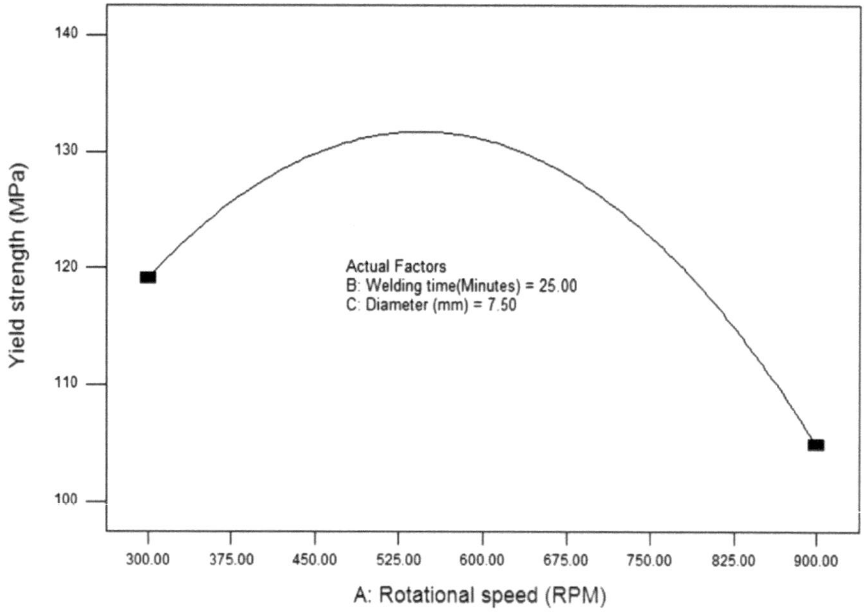

FIGURE 22.1 Influence of the rotational velocity of tool on the yield strength.

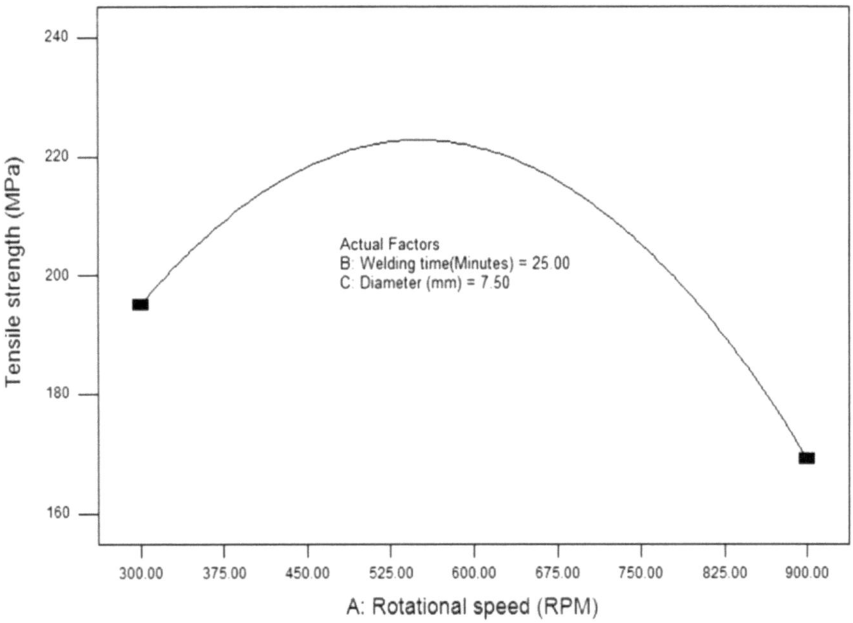

FIGURE 22.2 Influence of the rotational velocity of tool on the ultimate tensile strength.

22.4.2 *INFLUENCE OF WELDING TIME*

The variations in yield strength and tensile strength with welding time have been presented in Figures 22.3 and 22.4, respectively. These plots have been plotted at fixed value of rotational speed (600 rpm) and fixed value of workpiece diameter (7.5 mm). The curvature of the plots indicates that the variation of yield strength and tensile strength is quadratic with welding time. In addition to this, it has been revealed from the figures that yield strength and tensile strength initially increases up to a certain value with the rise in joining time. Further rise in welding time, decreases the yield strength and tensile strength.

22.4.3 *INFLUENCE OF WORKPIECE DIAMETER*

Figures 22.5and 22.6 depict the influence of workpiece diameter on the yield and ultimate tensile strength of the weldments. These plots have been plotted at a constant value of revolving velocity (600 rpm), and duration of welding (25 min). It is well evident from the plot that the yield strength and ultimate tensile strength continuously decreases with an upsurge in workpiece diameter.

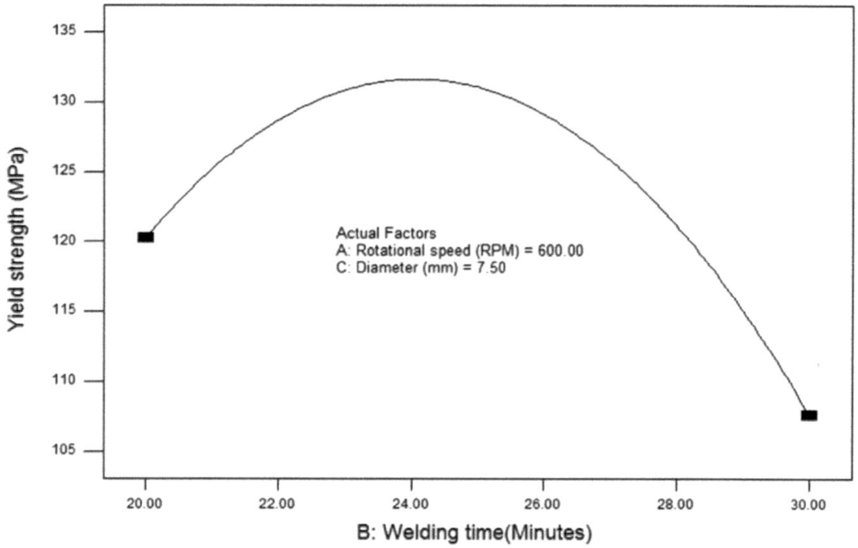

FIGURE 22.3 Impact of friction welding time on yield strength.

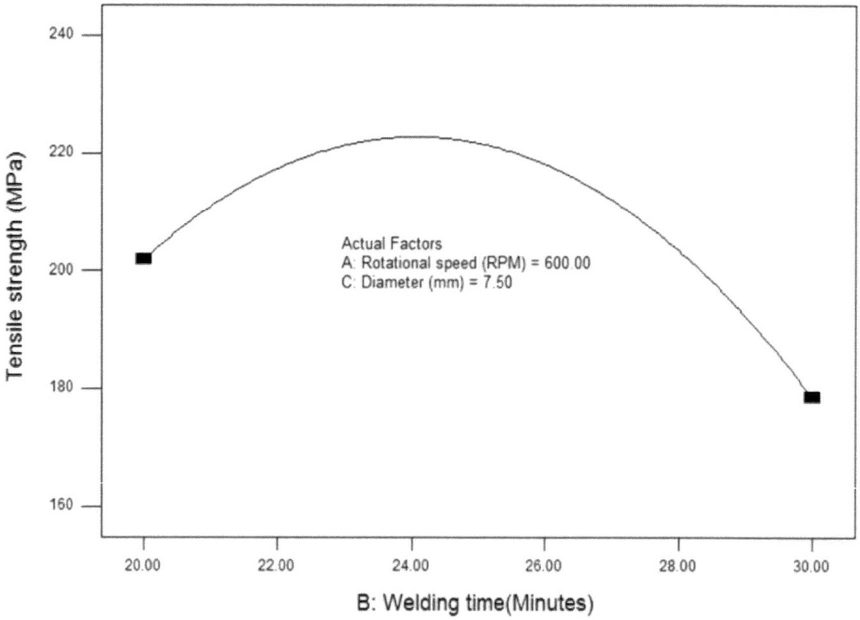

FIGURE 22.4 Impact of frictional welding time on ultimate tensile strength.

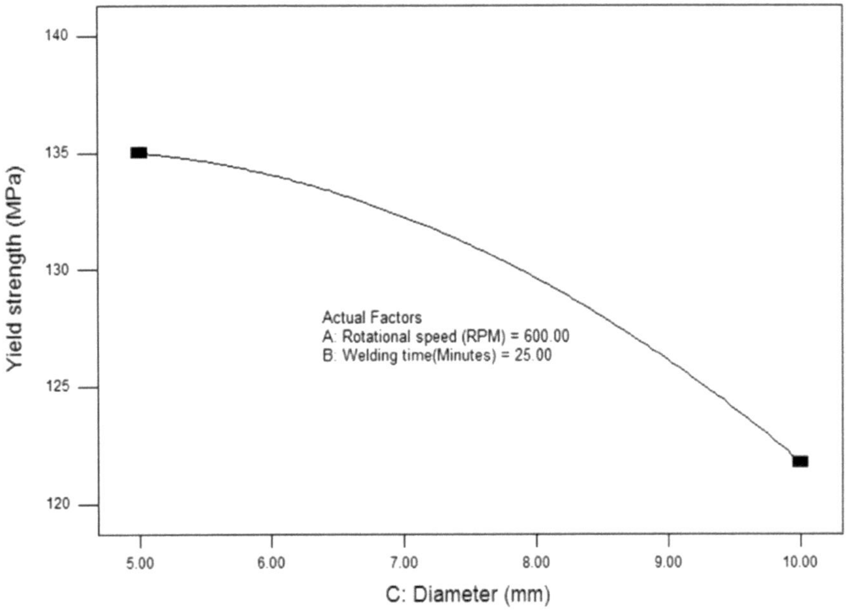

FIGURE 22.5 Influence of workpiece diameter on the yield strength.

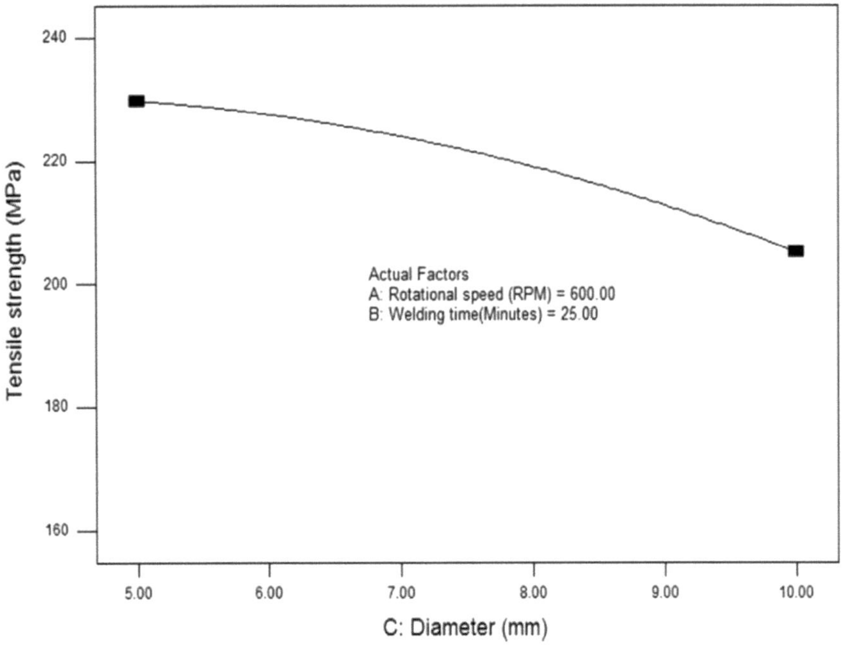

FIGURE 22.6 Influence of workpiece diameter on the ultimate tensile strength.

22.5 CONCLUSION

The current piece of research work aims to explore the impact of friction welding parameters on the tensile properties of the fabricated weldments while joining mild steel with aluminum AA7075 alloy. To achieve the objective, face-centered design has been applied. The key deductions drawn from the current study are as follows:

1. The rotational speed, welding time, and workpiece diameter have been found significant model terms for yield strength and tensile strength.
2. The rotational velocity of the welding tool was observed to be the most significant friction welding parameter that influenced the yield strength and tensile strength of weldments followed by workpiece diameter and welding time.
3. Rise in rotational velocity and welding time showed a significant rise in yield and ultimate tensile strength of the welded joint up to a certain value. However, at higher values of rotational velocity and the duration of welding reduced both the yield as well as the ultimate tensile strength of the welded joint.
4. Both the tensile properties, that is, the yield strength and ultimate tensile strength decreased continuously with the rise in workpiece diameter.

KEYWORDS

- **friction welding**
- **mild steel**
- **AA7075 alloy**
- **process parameters**
- **tensile properties**

REFERENCES

1. Silva, D. A. A. An Investigation on the Structure/Property Relationships of Solid State Welding Processes in a Titanium Matrix Composite Alloy (Ti-6Al-4V). *Adv. Manuf.* **2006,** *1*, 15–22.

2. Palanival, R.; Laubscher, R. F.; Dinaharan, I.; Hattingh, D. G. Microstructure and Mechanical Characterization of Continuous Drive Friction Welded Grade 2 Seamless Titanium Tubes at Different Rotational Speeds. *Int. J. Pressure Vessels Piping* **2017,** *154*, 17–28.
3. Rambabu, G.; Naik, D. B.; Rao, C. H. V.; Rao, K. S.; Reddy, G. M. Optimization of Friction Stir Welding Parameters for Improved Corrosion Resistance of AA2219 Aluminum Alloy Joints. *Defence Technol.* **2015,** *11*, 330–337.
4. Sathiya, P.; Aravindan, S.; Haq, A. N. Optimization of Friction Welding Parameters Using Evolutionary Computational Techniques. *J. Mater. Proces. Technol.* **2009,** *209*, 2576–2584.
5. Bassu, G.; Irving, P. E. Throle of Residual Stress and Heat Affected Zone Properties on Fatigue Crack Propagation in Friction Stir Welded 2024-T351 Aluminum Joints. *Int. J. Fatigue* **2003,** *25*, 77–88.
6. Mustafa, B.; Adem, K. The Influence of Stirrer Geometry on Bonding and Mechanical Properties in Friction Stir Welding Process. *Mater. Des.* **2004,** *25*, 343–347.
7. Uzun, H.; Donne, C. D.; Argagnotto, A.; Ghidini, T.; Gambaro, C. Frictionstir Welding of Dissimilar Al 6013-T4 To X5CrNi18-10 Stainless Steel. *Mater. Des.* **2005,** *26*, 41–46.
8. Cavaliere, P.; Campanile, G.; Panella, F.; Squillace, A. Effect of Welding Parameters on Mechanical and Microstructural Properties of AA6056 Joints Produced by Friction Stir Welding. *J. Mater. Proces. Technol.* **2006,** *180*, 263–270.
9. Jariyaboon, M.; Davenport, A. J.; Ambat, R.; Connolly, B. J.; Williams, S. W.; Price, D. A. The Effect of Welding Parameters on the Corrosion Behavior of Friction Stir Welded AA2024–T351. *Corros. Sci.* **2007,** *29* (2), 877–909.
10. Hwang, Y. M.; Kang, Z. W.; Chiou, Y. C.; Hsu, H. H. Experimental Study on Temperature Distributions within the Work Piece during Friction Stir Welding of Aluminum Alloys. *Int. J. Machine Tools Manuf.* **2008,** *48*, 778–787.
11. Sakthivel, T.; Segar, G. S.; Mukhopadhyay, J. Effect of Welding Speed on Microstructure and Mechanical Properties of Friction-Stir Welded Aluminum. *Int. J. Adv. Manuf. Technol.* **2009,** *43* (5–6), 468–473.
12. Juan, F.; Marian, N.; Thompson, L.; Edgar, A. A. Principal Component Analysis Reveals Age-Related and Muscle-Type-Related Differences in Protein Carbonyl Profiles of Muscle Mitochondria. *J. Gerontol. Series A Biol. Sci. Med. Sci.* **2010,** *63*, 1277–1288.
13. Palanivel, R.; Koshy, M. P.; Murugan, N. Development of Mathematical Model to Predict the Mechanical Properties of Friction Stir Welded AA6351 Aluminum Alloy. *J. Eng. Sci. Technol. Rev.* **2011,** *4* (1), 25–31.
14. Koilraj, M.; Sundareswaran, S.; Vijayan, S.; Koteswara, Rao, S. R. Friction Stir Welding of Dissimilar Aluminum Alloys AA2219 to AA5083-Optimization of Process Parameters Using Taguchi Technique. *Mater. Des.* **2012,** *42*, 1–7.
15. Asif, M. M.; Shrikrishna, K. A.; Sathiya, P. Optimization of Processparameters of Friction Welding of UNS S31803 Duplex Stainless Steels Joints. *Adv. Manuf.* **2015,** *4*, 1–11.
16. Elfar, O. M. R.; Rashad, R. M.; Megahed, H. Process Parameters Optimization for Friction Stir Welding of Pure Aluminium to Brass (CuZn30) Using Taguchi Technique. *MATEC Web Conf.* 2016, *43*, 1–6.

17. Devaiah, D.; Kishore, K.; Laxminarayana, P. Parametric Optimization of Friction Stir Welding Parameters Using Taguchi Technique for Dissimilar Aluminum Alloys (AA5083 and AA6061). *IOSR J. Eng.* **2017,** *07* (09), 44–49.
18. Prasad, M. V. R. D.; Namala, K. K. Process Parameters Optimization in Friction Stir Welding by ANOVA. *Mater. Today: Proc.* **2018,** *5*, 4824–4831.
19. Eslami, N.; Hischer, Y.; Harms, A.; Lauterbach, D.; Bohm, S. Optimization of Process Parameters for Friction Stir Welding of Aluminum and Copper Using the Taguchi Method. *Metals* **2019,** *9* (63), 1–15.

CHAPTER 23

EFFECT OF ELECTRODE GEOMETRY IN RESISTANCE SPOT WELDING OF LOW CARBON STEEL SHEETS

ANAND KUMAR MANDAL[1,2*], RAKESH KUMAR[1],
KAPIL KUMAR GOYAL[1], and R. S. JADOUN[2]

[1]*Department of Industrial and Production Engineering,
Dr B R Ambedkar National Institute of Technology, Jalandhar, India*

[2]*Department of Industrial and Production Engineering,
College of Technology, Pantnagar, Uttrakhand, India*

Corresponding author. E-mail: anandkm.ip.19@nitj.ac.in

ABSTRACT

A finite element method (FEM)-based electrothermal model for analyzing the effect of variation in electrode geometry in the resistance spot welding (RSW) process has been developed. The effect of various nonlinearities in terms of variation in material properties, contact resistance, geometry, and its variation with temperature has been included in the model. The difference in heat generated for different electrode geometry along electrode–sheet and sheet–sheet interfaces have been explained with the help of variation in electric current density (ρ) along the two interfaces. The computed results of one of the geometries of electrode for different electric currents have been compared with the experimental result. The experimental results are in good agreement with the computed result, which validates the computation model developed.

Modern Manufacturing Systems: Trends and Developments. Rajiv Kumar Garg, Ravi Pratap Singh, Rajeev Trehan, & Ramesh Singh (Eds.)

23.1 INTRODUCTION

Resistance spot welding (RSW) is the most popular method for joining sheet metals and finds major application in automotive, aeronautical, and space industries. The process finds high applicability due to its high productivity. However, in the present scenario where new advanced materials are discovered on regular basis, the process needs to be optimized accordingly. Degradation of electrodes is one of the major problems that need to be taken care of in order to keep up with the high productivity requirement. The various phenomena of electrode wear are recovery, recrystallization, surface alloying, pitting, and erosion,[1] and the continuous mechanical and thermal fatigue of the electrodes are the reasons for the above phenomenon. The process of RSW involves multiple physics like electric, thermal, mechanical, and metallurgical making the analysis of the process difficult.[2] FEM-based model can make the analysis simple so to make optimum choice of weld parameters and design of electrode according to the material of sheet metal to be welded.

23.2 RESEARCH BACKGROUND

The effect of contact resistance has been studied by Wyant et al.,[3] which states the value of contact resistance decreases as welding proceeds and the best weld is produced when the value of contact resistance is minimum. The effect of pressure and temperature on contact resistance for uncoated steel sheets using Truncated Cone (TC) electrode have been investigated by Volger et al.[4] The author found that with the increase in pressure from 0 to 30 psi the contact resistance along the sheet–sheet interface decreased from 1200 to 600 μΩ and from 400 to 0 μΩ along the electrode sheet interface. The contact resistance at both the interfaces becomes zero when there is an increase in temperature from 0°C to 400°C. The electrode wear characteristic during RSW of aluminum has been investigated by Kondo et al.[5] The authors found out that high heat requirement for RSW of aluminum sheets also affect the heat generated at electrode sheet interface. Crinon et al.[6] investigated the effect of surface roughness and oxide film thickness on contact resistance in aluminum sheets to find that for the least rough surface, the value of contact resistance is highest, which is unaffected by axial electrode force and for most rough surface the surface roughness falls by many folds as the electrode pressure is increased. The variation of dynamic contact resistance in nickel plate has been studied by Tan et al.[7] It was observed that the stages of RSW process of nickel plate involved asperity heating, surface breakdown, asperity softening, partial surface melting, nugget growth, and expulsion. The first peak in dynamic curve is observed when breakdown of oxide film occurs and the second peak occurs when nugget formation take place. The surface condition of magnesium sheet of thickness 2 mm was evaluated by Liu et al.[8] It was found that for as received surface condition there is high variable contact resistance, in which case Weld Nugget (WN) formation starts earlier but it also inherits defects like porosity and weld spatter. For chemically treated surface; however, contact resistance is low but it is uniform and gives good weld strength for same weld size due to less weld defects. Rashid et al.[9] studied the electrode–sheet contact resistance for three types of surfaces, that is, as-received, abraded, and grounded. The specimen made by as-received sheet metal is found to have the highest value of contact resistance followed by specimen made by abraded sheet metal. It has been found that faster pitting of electrode to be dominant in as received surface due to high heat generation at electrode sheet interface.

Greenwood et al.[10] developed a contact model based on Hertzian theory, which explained the area of contact is based on the number and size of the asperity in contact. However, it is a very cumbersome task to find this asperity contact, which is temperature dependent. A different way of calculating contact resistance assuming contacts are made in the elastic range has been made by Holm et al.[11] The local stresses on these asperities would be much higher to yield it plastically. And the mean contact pressure applied on these asperities would be equal to asperity hardness or a fraction of it. Nied et al.[12] developed an axisymmetric FE model to simulate the RSW process using solid iso-parametric elements. The coupled electrothermal and thermomechanical analysis are done to find the effect of the squeeze and weld cycle. Ninshu et al.[13] investigated the RSW of three pieces of high-strength steel of different thickness both experimentally and analytically for three current magnitudes of 5, 6, and 7 kA, respectively, each for time period of 17 cycles and at weld force application of 250 kg F. The result showed that WN formation takes place in between the sheets and its size increases with the increase in weld current and weld time weld. There is significant increase in WN diameter but has less effect on WN depth. The effect of Curie temperature and magnetic field due to high current intensity on the transport phenomenon in the RSW process has been studied by Wei et al.[14] These parameters control solute mixing inside the WN and if optimized would give a good quality weld.

Optimizing the RSW process requires controlling multiple parameters like weld current, weld time, electrode pressure, electrode material, electrode geometry, and much more. Experimentally optimizing these RSW variables for different sheet materials would require large-scale lab experimentation. Developing a robust computational model can reduce the scale of experimentation. Whereas the different FEM models were developed to characterize the process very well in terms of electrothermal and thermomechanical analysis. These models have found out the effect of welding variables like weld current, weld time, weld pressure, contact resistance, and material properties used on the quality of weld produced. The effect of change in electrode geometry on weld characteristics needs more analysis.

Since the TC electrode is the most versatile electrode shape that is used to weld uncoated steel sheets in any industry. Further, it has been reviewed earlier that pressure has little effect in terms of change in the contact area and contact resistance for TC electrode.[4] Thus, an electrothermal analysis is sufficient to model the RSW process using this electrode. The inclusion of temperature-dependent material properties, contact resistance, and phase change during melting and solidification of the process make the model nonlinear. The FEM model is able to solve the heat transfer and electric conduction problem of the RSW process with the help of software like Ansys (APDL). The effect of change in shape of electrode with respect to change in TC angle and radius of the electrode on WN quality and electrode heating is investigated. The results obtained by FEM are very well compatible with the results obtained from lab experiments to validate the FEM model developed.

23.3 THEORETICAL BACKGROUND

Since in the process of RSW, localization of heat due to Joule heating occurs for a very short duration on a small area of contact interfaces of the sheet–electrode system, therefore conduction is the only dominant mode of heat transfer for the process. Water flowing inside the hollow electrode is heated due to convection, and the water flow rate is high there to keep the internal wall of the electrode at ambient room temperature. The electric current flowing at the contact zones is determined by the contact area of the sheet electrode interface. Electrical analysis is carried out due to which Joule heating takes place, which is high at high resistant zones. Subsequently, thermal analysis is carried out to find the distribution of heat and melt zone in the sheet–electrode assembly. Therefore, this process can be regarded as an electrothermal phenomenon in which transient heat conduction and steady electric conduction equations are used as governing equations with suitable boundary and initial conditions.

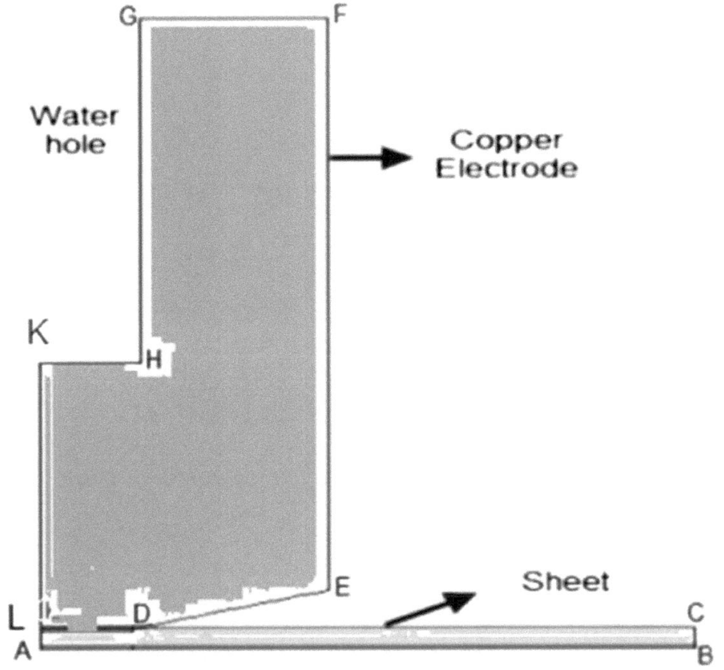

FIGURE 23.1 An axisymmetric quarter model of sheet–electrode system.

An axisymmetric quarter model of the RSW process is used as a solution domain. Since there is no variation of heat transfer along the azimuthal or θ direction, makes the problem axisymmetric. The two-dimensional axisymmetric problem is symmetric along the *x*-axis and *y*-axis, so only quarter of the model is analyzed as shown in Figure 23.1. The cylindrical cross-section of the electrode, due to which the heat transfer, electric current density, and pressure variation occur in cylindrical manner, allows the problem to be solved in polar coordinate system.

The dimensions of electrode are according to the AWS standard D8.6:2005. The outer radius of electrode is having dimension of 11 mm, the radius of hole and its depth is having measurement equivalent to 4 mm and 15 mm, respectively, the measurement of electrode tip radius and height is equivalent to 3 mm and 25 mm, respectively. The sheet metals used in lap joint configuration is having a thickness of 2 mm each.

Table 23.1 shows the temperature-dependent material properties of copper and steel assigned to the electrode and sheet metals, respectively. Contact resistance between the electrode and sheet and in between sheet-to-sheet interface is accounted by using surface elements. The thickness of these surface elements is equivalent to the oxide film thickness on the electrode and sheet surface.

TABLE 23.1 Temperature Dependent Material Properties of Steel and Copper.[15]

Temperature (°C)		**100**	**300**	**500**	**700**	**900**	**1100**	**1300**	**1500**
Steel	Bulk material resistivity(μm) × 10^{-6}	267	495	818	1115	1179	1209	1255	1299
	Thermal conductivity (W/m.K) × 10^{-3}	63.2	49.97	39.95	30.5	27.66	28.5	27	26.5
	Specific-heat(W/m.K)	452	561	661	762	1500	1000	1000	1000
Copper	Bulk material resistivity(μm) × 10^{-6}	30	50.5	69.9	89.8	94.8	99.8	103.5	100
	Thermal conductivity (W/m.K) ×10^{-3}	380	355	334.9	315	310	305	300	300
	Specific-heat(W/m.K)	402	431.2	452.2	477	488	501	495	495

23.3.1 MATHEMATICAL MODELING OF HEAT TRANSFER AND ELECTRIC CONDUCTION PROCESS

The governing equation for heat conduction in the cylindrical coordinate system (generalized form of Fourier equation) with no variation of temperature along the azimuthal angle is as follows.

$$\frac{1}{r}\frac{\partial}{\partial r}\left(rk\frac{\partial T}{\partial r}\right)+\frac{1}{r}\frac{\partial}{\partial z}\left(rk\frac{\partial T}{\partial z}\right)+\mathring{Q}=sC\frac{\partial T}{\partial t} \tag{23.1}$$

where k, s, and C are thermal conductivity, density, and specific heat. The first term on the left side of the equation is the heat transfer in radial direction, the second term is the heat transfer in axial direction, $\mathring{Q}$ refers to the amount of internal heat generation per unit volume, and the right-hand term is the transient variation of stored energy. The internal heat generation is due to bulk material resistance and contact resistance of the sheet electrode system.

Thermal boundary condition with reference to Figure 23.1 can be represented as follows:

a) Since the problem is axisymmetric, the boundary condition along KL and AL will be according to the below equation

$$\frac{\partial T}{\partial r}=0 \quad \text{at axis KL, AL} \tag{23.2}$$

b) The inner surface of the electrode (GH and HK) is assumed to be at ambient temperature.
c) The external surface of sheet–electrode system is assumed to be adiabatic.

$$\frac{\partial T}{\partial r} = 0 \quad \text{as surfaces EF and BC} \tag{23.3}$$

$$\frac{\partial T}{\partial z} = 0 \quad \text{at surfaces GF, DC, and AB (due to symmetry)} \tag{23.4}$$

$$\frac{\partial T}{\partial n} = 0 \quad \text{at surfaces ED} \tag{23.5}$$

In electrical analysis, the electric current is assumed to flow in axial direction due to comparatively large electrode tip radius to the sheet thickness ratio. The governing equation used for electrical analysis is based on Maxwell's first law $\nabla \cdot E = 0$, which if expressed in terms of electric potential is $\nabla(\nabla \cdot \varphi) = 0$and in axisymmetric coordinate system eq. 23.6 is given as

$$\frac{1}{r}\frac{\partial}{\partial r}\left(r\frac{1}{\sigma}\frac{\partial \varphi}{\partial r}\right)+\frac{1}{r}\frac{\partial}{\partial z}\left(r\frac{1}{\sigma}\frac{\partial \varphi}{\partial z}\right)=0 \tag{23.6}$$

Since the electric current is flowing axially in downward direction, the electric flux along surface BC, CD, DE, and EF is zero as demonstrated by

$$\frac{\partial \varphi}{\partial n} = 0 \quad \text{at surface BC, CD, DE, EF} \tag{23.7}$$

Electric current is applied on the surface FG in the negative z-direction, which is expressed in terms of electric current density, J is as follows as given by

$$\frac{1}{\sigma}\frac{\partial \varphi}{\partial n} + J = 0 \quad \text{at surface FG} \tag{23.8}$$

The electric flux lines entering the surface FG must leave along the surface AB as demonstrated by

$$\frac{\partial \varphi}{\partial z} = 0 \qquad \text{on the surface AB} \tag{23.9}$$

The incorporation of Joule heat due to electric current flow is incorporated by using the below equation

$$\overset{o}{Q} = \sigma \tag{23.10}$$

where $\overset{\circ}{Q}$ refers to the amount of internal heat generation per unit volume, J is the current density, σ is the electrical resistivity, E is the electric field, and Φ is electric potential.

23.3.2 ESTIMATION OF CONTACT AREA AND CONTACT RESISTIVITY

The contact area between the electrode–sheet and sheet–sheet interface is important for the conduction of heat and electric current. In the present case of electrothermal analysis using truncated tip electrode, the contact area between the electrode–sheet and sheet–sheet interface is taken equivalent to the electrode tip area. There is limited variation of contact area and contact resistance with the increase in pressure,[16,17] so the effect of pressure on contact area and contact resistance is not that significant in this case.

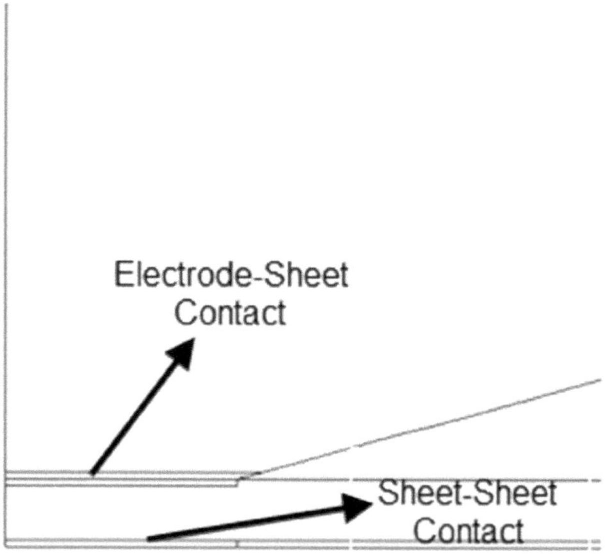

FIGURE 23.2 Sheet–electrode system showing contact at the electrode–sheet and sheet–sheet interface.

The estimation of contact resistance along sheet–sheet interface and electrode–sheet interface as shown in Figure 23.2 is of vital importance for modeling the RSW process. Since contact resistance is temperature dependent, so its evaluation along the two interfaces is very much cumbersome both experimentally and mathematically. Various authors have given different mathematical model to evaluate contact resistance based on number and radius of asperity in contact, which is very difficult to evaluate. So, the present work mainly takes into account an experimentally determined robust relation that has been validated by many researchers.[6] The exponential relationship to determine contact resistance is given by Grietmann and Rothers, which is given as

$$R_K(T) = \alpha \frac{R_0}{\beta^{(T-T_0)/(T_l-T_0)}} \tag{23.11}$$

where $R_K(T)$is the temperature-dependent contact resistance, R_0 is the initial contact resistance, α and β are factors, which depends on the type of sheet metal and their surface coating, T_L, T_o, and T are liquidus, ambient, and variable temperatures, respectively. The value of α and β is taken as 1 and 25, respectively, which has been derived from experimental results. Once contact resistance has been evaluated, its value is used to derive contact resistivity that can be implemented in the electrothermal RSW model. The expression for evaluation of resistivity from given contact resistance is given by

$$(\rho_k)_i = \left(R_k(T) \cdot A / h\right)\left(A / A_i\right) \tag{23.12}$$

where $(\rho_k)_i$, the contact resistivity for the ith contact element along the contact surface. A, A_i, and h are the area of the contacting surfaces, area of ith element, and height of the contact area, respectively.

Net resistivity of ith contact element is given by

$$(\rho)_i = (\rho_k)_i + (\rho_b)_i, \tag{23.13}$$

Where $(\rho_b)_i$ is the temperature-dependent bulk resistivity of ith element. Table 23.2 shows the calculated values of contact resistivity that have been implemented into the electrothermal RSW model.

23.4 FINITE ELEMENT ANALYSIS

Due to a large number of nonlinearities involved in the model developed, FEM is used in finding out the approximate solution to the problem. The nonlinearities involved are mainly due to complex geometry, different materials, and variation of material properties with temperature. The calculation of temperature at each node in an element becomes cumbersome with different material properties in every element at that instant of time. Thus, FEM can provide the best alternative for a good solution to the problem defined.

There are many methods of discretizing the governing equation and boundary condition for transient heat conduction and steady-state electric current analysis. The method chosen here, for performing the discretization procedure is Galerkin method of weighed residuals. This discretization

TABLE 23.2 Calculated Values of Contact Resistivity for Varying Contact Resistance at sheet–sheet and electrode–sheet Interface.

Temperature (°C)		100	300	500	700	900	1100
Sheet–sheet	Contact resistance (μ-ohm)	Contact-resistivity (ohm-mm) × 10^{-6}					
	100	26899.2	10364.2	3993.28	1538.6	592.8	228
	200	53798.4	20728.3	7986.57	3077.2	1185	456.8
	300	80697.5	31092.5	11979.9	4615.8	1778	685
	400	107597	41456.7	15973.1	6154.4	2371	913
	500	134496	51820.9	19966.4	7693	2964	1142
Steel		Bulk-resistivity(ohm-mm) × 10^{-4}					
		186	376	648	1011	1115	1179
		Net contact-resistivity(ohm-mm) × 10^{-6}					
	100	27085	10740.2	4641.28	2549.6	1707	1407
	200	53984	21104.3	8634.57	4088.2	2300	1635
	300	80883	31468.5	12627.9	5626.8	2893	1864
	400	107783	41832.7	16621.1	7165.4	3486	2092
	500	134682	52196.9	20614.4	8704	4079	2321
Copper-steel	0	0	0	0	0	0	0
	70	14144	2541	456	82	14	2.646
		Copper bulk-resistivity (ohm) × 10^{-4}					
		30	50.5	70	90	98	100
		Net contact-resistivity(ohm-mm) × 10^{-6}					
	0	30	50.5	70	90	98	100
	70	14174.1	2591.52	526.499	172.01	112.7	102

procedure mainly involves transformation of governing equation into a set of algebraic equations in matrix form that can be solved for unknown temperature and potential field subsequently.

23.4.1 DISCRETIZATION FOR TRANSIENT HEAT CONDUCTION EQUATION (GALERKIN METHOD)

The governing transient heat equation used for modeling the thermal part of the analysis is discretized by using the Galerkin method of weighted residue technique. The element used to discretize the domain is two-dimensional iso-parametric plane element 223. The element has eight nodes with up to four degrees of freedom per node.[18] The element has capability of simulating thermoelectric effect, which includes Joule heating as required for the RSW process. The element is capable for axisymmetric analysis.

The method of discretization is done in polar coordinate system considering the model to be axisymmetric and cylindrical, which is demonstrated as given below.

- Divide the domain "Ω" into '"*e*" finite elements of "8" nodes each.
- Assume suitable variation of "*T*" in each finite element and express T^e (*r*, *z*, *t*) in element "e" as given by

$$T^e\left(r,z\right) = N_1T_1 + N_2T_2 + N_3T_3 + ...N_8T_8$$

$$= \left[N_1 \quad N_2 \quad N_3 .... \quad N_8\right]\begin{Bmatrix} T_1 \\ T_2 \\ T_3 \\ . \\ . \\ . \\ T_8 \end{Bmatrix} \tag{23.14}$$

Where N_i's are the shape functions and T_i's are the nodal temperatures.

- Apply Galerkin method of weighted residue where the integral of weighted residue over domain of the element is set equivalent to zero by taking weight same as interpolation function, N_i. Since the solution is not exact, substitution it into the diffusion equation would give a nonzero value instead of zero, which is also called residue.
- Diffusion equation with boundary condition to be solved by Galerkin method is shown by eqs. 23.15 and 23.16, respectively.

$$\frac{1}{r}\frac{\partial}{\partial r}\left(r\,k\frac{\partial T}{\partial r}\right)+\frac{1}{r}\frac{\partial}{\partial z}\left(r\,k\frac{\partial T}{\partial z}\right)+\overset{o}{Q}=s\,C\frac{\partial T}{\partial t} \tag{23.15}$$

And the associated boundary condition is as follows:

$$k_n\frac{\partial T}{\partial n}+q+h\left(T-T_\infty\right)+\sigma\varepsilon\left(T^4-T_\infty^4\right)=0 \tag{23.16}$$

where the term k_n refers to the thermal conductivity of the material normal to the surface, h is convection heat transfer coefficient, ε is Stefan–Boltzmann constant, and σ is radiative heat transfer coefficient. The first, second, and third term on the left side of the equation represents heat transfer by conduction, convection, and radiation. Since the process of RSW is a very fast process so the heat transfer by convection is minimal and assumed to be zero and since external surface of electrode and sheet metal is always exposed to atmosphere, it remains at ambient temperature so the heat transfer by radiation is assumed to be zero. So, finally, the boundary condition reduces to the expression as demonstrated by

$$k\frac{\partial T}{\partial n}=0 \tag{23.17}$$

- Using Galerkin method of weighted residual to the governing equation in the axisymmetric domain, the integral becomes as given below

$$\int_\Omega\left[\frac{\partial}{\partial r}\left(r\frac{\partial T}{\partial r}\right)+r\left(\frac{\partial^2 T}{\partial z^2}\right)+\frac{r}{\alpha}\left(\frac{1}{sC}\dot{Q}-\frac{\partial T}{\partial t}\right)\right]N_i\,d\Omega=0 \tag{23.18}$$

Ω is the domain of integration along the r- and z-axis and $d\Omega=2\prod drdz$

- Equations 23.19 and 23.20 dothe partial derivative of the first term and second term on the left side of the equation in order to reduce it from strong form to weaker form

$$N.\frac{\partial}{\partial r}\left(r\frac{\partial T}{\partial r}\right)=\frac{\partial}{\partial r}\left(N.r\frac{\partial T}{\partial r}\right)-r.\frac{\partial N}{\partial r}.\frac{\partial T}{\partial r} \tag{23.19}$$

$$N.\frac{\partial}{\partial z}\left(r\frac{\partial T}{\partial z}\right)=\frac{\partial}{\partial z}\left(N.r\frac{\partial T}{\partial z}\right)-r.\frac{\partial N}{\partial z}.\frac{\partial T}{\partial z} \tag{23.20}$$

Reducing the equation to weaker form and applying the boundary condition, the equation becomes

$$\int_\Omega\left[r\left(\frac{\partial N_i}{\partial r}\frac{\partial T}{\partial r}+\frac{\partial N_i}{\partial z}\frac{\partial T}{\partial z}\right)-\frac{r}{\alpha}\left(\frac{1}{sC}\dot{Q}-\frac{\partial T}{\partial t}\right)N_i\right]d\Omega=0 \tag{23.21}$$

Substituting piecewise integral over the element,

$$\sum_{i=1}^{8} 2\Pi \iint_A \left(\frac{\partial N_i}{\partial r}\frac{\partial N_j}{\partial r}+\frac{\partial N_i}{\partial z}\frac{\partial N_j}{\partial z}\right)\text{Trdrdz}-\frac{2\Pi}{\alpha}\sum_{i=1}^{8}\iint_A \dot{Q}N_i\text{rdrdz}$$
$$+\frac{2\Pi}{\alpha}\sum_{i=1}^{8}\iint_A \frac{\partial T}{\partial t}N_iN_j\text{rdrdz}=0 \quad (23.22)$$

So the equation reduces in the form of stiffness equation

$$\left[h^e\right]\left\{T^e\right\}+\left\{f^e\right\}=0 \quad (23.23)$$

where $$h^e_{ij}=2\Pi\iint_A \left(\frac{\partial N_i}{\partial r}\frac{\partial N_j}{\partial r}+\frac{\partial N_i}{\partial z}\frac{\partial N_j}{\partial z}\right)\text{rdrdz} \quad (23.24)$$

$$f_i^e=\frac{2\Pi}{\alpha}\iint_A \left(\frac{1}{\text{sC}}\dot{Q}-\frac{\partial\left(N_iT_i\right)}{\partial t}\right)N_i\text{rdrdz}$$
$$=\frac{2\Pi}{k}\iint_A \dot{Q}N_i\text{rdrdz}-\frac{2\Pi}{\alpha}\iint_A N_iN_i\text{rdrdz}\frac{\partial T_i}{\partial t} \quad (23.25)$$

Similarly applying Galerkin method of weighted residual steady-state electrical conduction equation can be discretized to be solved for electrical analysis.

23.4.2 BOUNDARY CONDITIONS AND INITIAL SOLUTION

The analysis has been done in FEM software, Ansys version 16.1. Plane eight noded element that has the capability of simulating thermoelectric effect as required for RSW process is used for meshing the domain. Care is taken that each zone is map-meshed and fine meshing is done at critical areas of the domain. The critical zones are the regions where maximum nonlinearity in terms of change in geometry and material properties is present and which are crucial for getting solution are as shown in Figure 23.3.

The potential gradient and current magnitude (7 kA) are applied as boundary conditions for electrical analysis, for a weld time of 250 ms. Constant ambient temperature at the water hole wall and external wall of the electrode are used as boundary conditions for thermal analysis. The FEM model is solved in small substeps and the solution is stored for further analysis. The temperature distribution at different region of sheet electrode system is highlighted by contours of five different colors. The red-colored contour lines represent temperature range of 1250°C–1500°C. In this temperature range mild steel melts to form WN at the sheet–sheet interface, so red colored contour lines represent WN or the fusion zone. The extent of WN in the radial direction signifies the WN radius and in axial direction signifies WN penetration.

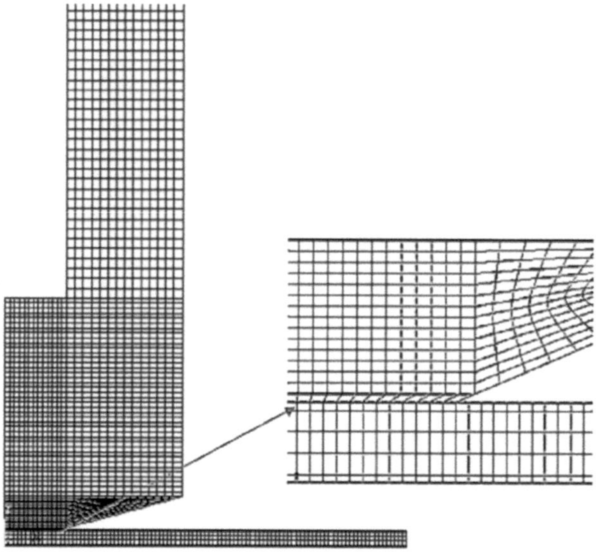

FIGURE 23.3 Discretized RSW model used for simulation and magnified mesh of the contact region.

The solution of this model showing WN development at different time steps is shown in Figure 23.4. The Time-Step of computational model is equivalent to weld time of RSW process. The metal at the sheet–sheet interface started melting to form WN at Time-Step 100 ms. The WN has maximum size at the Time-Step of 200 ms after which expulsion is observed at Time-Step 350 ms.

23.5 RESULTS AND DISCUSSION

After the development of robust electrothermal model that could very well simulate WN growth with a given set of welding conditions, similar FEM models have been developed to incorporate the effect of change in electrode geometry. The electrode geometry change is done by changing electrode tip radius and electrode TC angle.

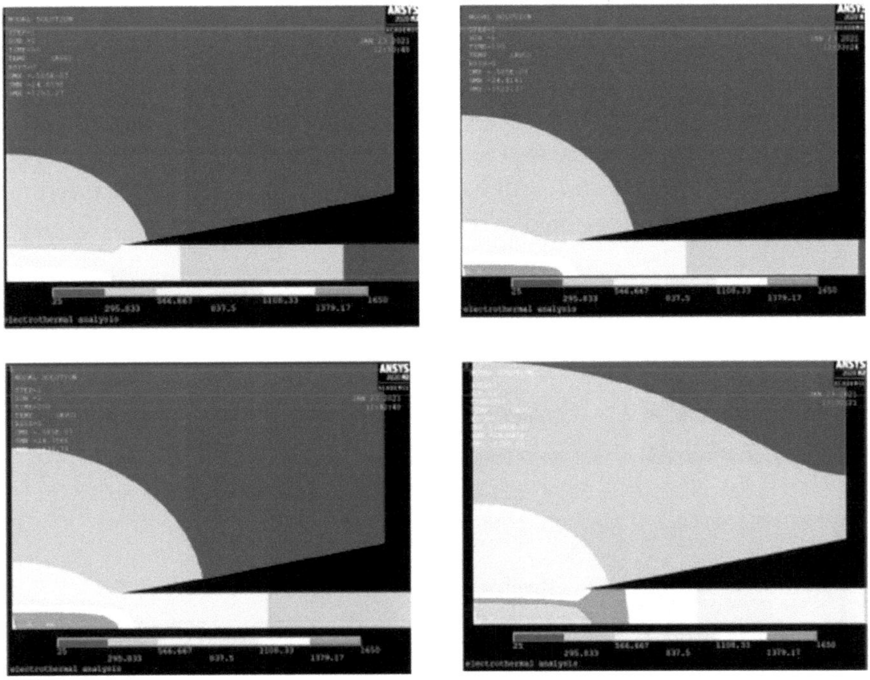

FIGURE 23.4 The computed WN diameters with increasing time-steps from 50 to 350 ms.

23.5.1 EFFECT OF ELECTRODE TRUNCATED ANGLE ON WN DIMENSIONS

The effect of change in TC angle to 15°, 30°, 45°, and 60° on weld joint and electrode heating is demonstrated by Figure 23.5. It is observed from Figure 23.5 that for TC of 15° the WN dimensions (radius =3 mm, depth of penetration = 1.5 m) is minimum, making it as the most weakest weld joint. But as the TC angle increases to 30° the WN dimension (radius =3.25 mm, depth of penetration =2.2 mm) increases, with significant increase in the depth of penetration of the weld.

Further increase of TC angle to 45° follows the same trend of increase in WN dimension but there is significant heating of electrode (660°C) taking place. The TC angle of 60° gives the largest WN, but the electrode is heated the most at the electrode sheet interface. The variation of WN dimensions with TC angle can be best understood by contour plot of ρ diagram as shown in Figure 23.6 and from graphical plot shown in Figure 23.7A, comparing the intensity of ρ accumulated at the edge of electrode tip for various TC angle electrodes. The contour and graphical plots analyzes that electrode with minimum TC angle has maximum ρ accumulated at the edge of the electrode tip, therefore, it waste maximum proportion of electric current to bulk material. It further makes the analysis that with the gradual increase in TC angle there is decrease in ρ accumulated at the edge of the electrode and there is more electric current flowing toward sheet–sheet interface forming larger WN as can be justified from Figure 23.5.

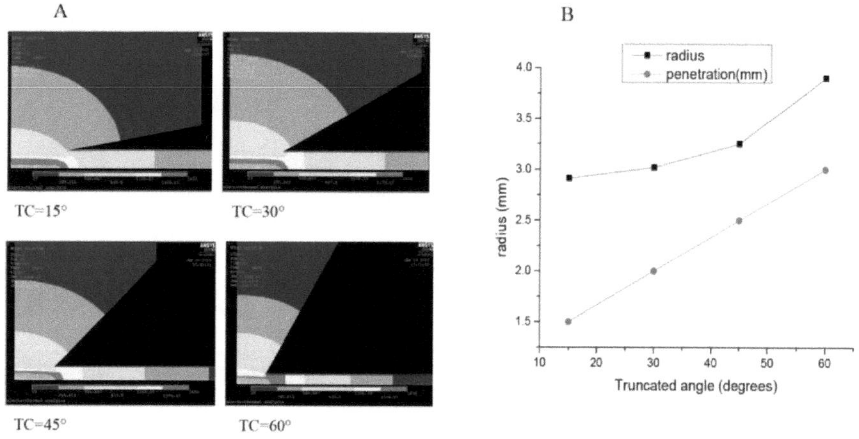

FIGURE 23.5 (A) The temperature contour profile for the variation of electrode truncated angle. The truncated angles are 15°, 30°, 45°and 60°, respectively, for which contour plots have been plotted. (B) Graphical plot comparing WN dimension for given truncated angles.

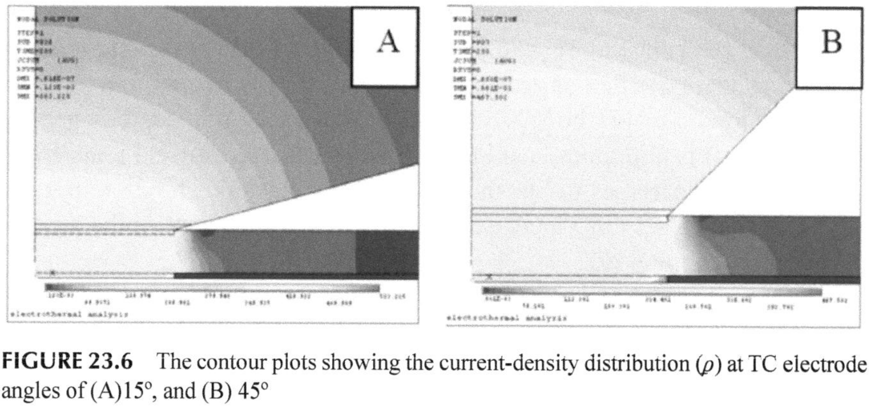

FIGURE 23.6 The contour plots showing the current-density distribution (ρ) at TC electrode angles of (A)15°, and (B) 45°

The ρ distribution uniformity along the electrode tip also decides the volume of melt along the sheet–sheet interface. A more uniform distribution of ρ is observed for electrode of TC angle 45° compared with electrode of TC angle 15° as can be observed from Figure 23.6. Similarly for 60° TC angle electrode, ρ distribution is most uniform at the electrode tip and as TC angle decreases, the ρ distribution uniformity decreases as shown in the graphical plot of Figure 23.7B. From the correlation made between Figures 23.5 and 23.7B, it can be concluded that a more uniform ρ distribution along the electrode tip will increase the WN dimensions. Nonuniform ρ distribution along the electrode tip for low TC angle electrode will lead to uneven wear of the electrode compromising the reliability of the weld joint produced. But more electrode mass present at the electrode tip for low TC angle electrode act as heat sink which keeps the electrode tip cold and resist its wear.[19] For high TC angle electrode, the electrode mass is less at the electrode tip providing less area for heat dissipation, making the electrode hotter and more prone to wear.

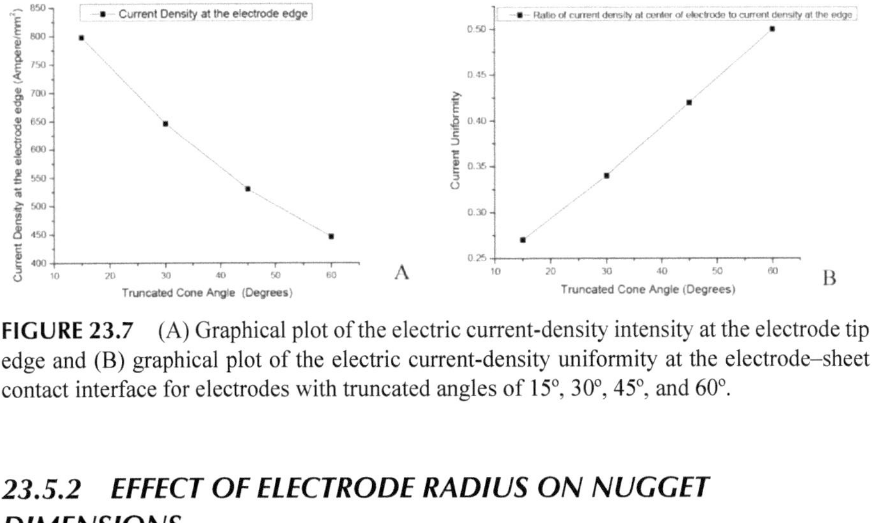

FIGURE 23.7 (A) Graphical plot of the electric current-density intensity at the electrode tip edge and (B) graphical plot of the electric current-density uniformity at the electrode–sheet contact interface for electrodes with truncated angles of 15°, 30°, 45°, and 60°.

23.5.2 EFFECT OF ELECTRODE RADIUS ON NUGGET DIMENSIONS

Keeping all the parameters constant including TC angle constant at 15°, electrode tip radius is varied from 2.5 to 3.5 mm to find its effect on the weld joint. It is observed that the smallest electrode of radius 2.5 mm gives the largest WN(weld radius = 4.5 mm and weld penetration=1.5 mm), but there is also expulsion of weld metal taking place and the electrode is heated to more than 600°C making it unfit to be used. As the electrode radius increases the WN dimension starts to decrease as can be seen from the contour and graphical plots of Figure 23.8. The electrode with radius of

3 mm produces a satisfactory weld with minimum electrode heating and the formation of WN do not initiate with given set of parameters for electrode of radius 3.5 mm.

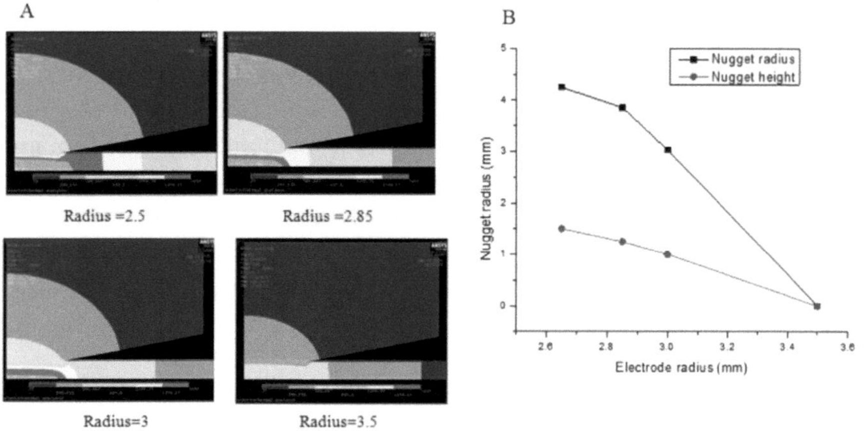

FIGURE 23.8 (A) The temperature profile for the variation electrode radius of 2.5, 2.85, 3 and 3.5 mm at 7 kA current, 240 ms weld time. (B) And a graphical comparison on the weld characteristics has been plotted.

For explanation of this behavior of electrode when the data of ρ distribution for all electrode tip radius is compared as shown in Figure 23.9A, it is found that the intensity of ρ distribution is maximum for electrode tip with least radius. The uniformity of ρ distribution at the electrode–sheet interface remains constant for all the electrodes of same radius. It is the convergence of electric current from a large radius electrode shank to a small radius electrode tip that increases the intensity of ρ at the electrode tip. For a constant electrode shank radius, less is radius of the electrode tip, more is the convergence of electric current and more is the intensity of ρ distribution at the electrode tip, more metal melts at sheet–sheet interface.

23.5.3 VALIDATION OF THE COMPUTED RESULTS

The computed results for its robustness and accuracy need to be validated with the results obtained by performing real experiments. The experiments are performed on a medium frequency direct current pedestal welding machine using 2 mm thick sheets of mild steel. The electrode used is water-cooled copper electrode. The RSW is performed to check the effect of variation in magnitude of electric current on WN diameter and penetration. These results are the optimized values of many experiments that have been conducted in the lab. The computational model has used similar setup that has been used while doing experimentation.

The weld time is kept constant at 240 ms with an electrode force of 3 kN and electrode diameter is 6 mm with a truncated angle of about 15°. The electric current magnitude is taken as 5.5, 6.5, 7.5, and 8 kA to find its effect on WN diameter. The computation model is based on the same data and results are compared as shown in Figure 23.9 (B). The above comparison depicts that for weld current of 5.5 kA there is error of 0.7 mm in the radius of WN but for other weld currents of magnitude 6.5, 7.5, and 8 kA the comparison of radius of WN obtained by computational analysis exactly fits into the results obtained by doing experimentation. Thus the experimental results very well validate the result obtained by computation analysis.

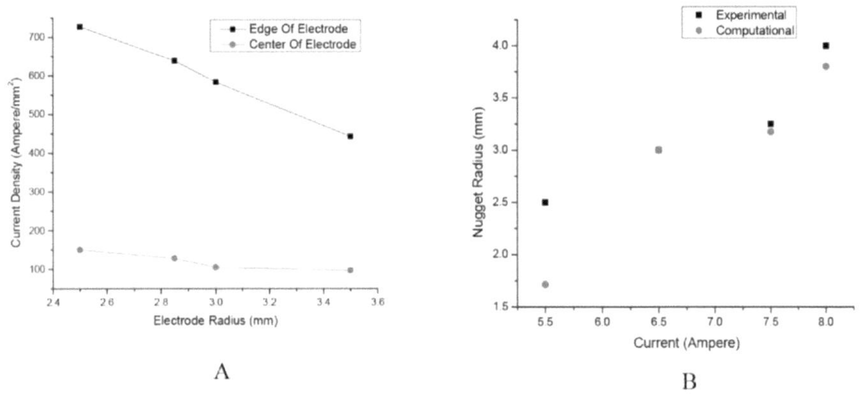

FIGURE 23.9 (A) The graphical plot of intensity of electric-current density at the edge and center of electrode tip for electrode radius of 2.5, 2.90, 3 and 3.5 mm at 7 current, 240 ms weld time. (B) Comparison of nugget radius obtained by experimental and computation method by varying current magnitude from 5.5 to8 kA.

23.6 CONCLUSION

The effect of change in electrode geometry is very well analyzed through the computational method in this research article. The FEM model developed has been validated with the results obtained through lab experiments. The model very well simulates the effect of change in electrode geometry on WN and electrode heating. Various conclusions drawn are given as follows:

1. The increase in TC angle although increases WN dimensions but also causes electrode heating. The electrode being heated above 600°C is undesirable as it will lead to frequent electrode wear leading to productivity loss.
2. The effect of unevenness of ρ distribution at the electrode interface for low angle electrode is neutralized by more electrode mass near the tip of the electrode. For the present set of welding conditions, the TC angle of the electrode should range between 15° and 30° for a compromise between best WN and minimum electrode wear.
3. The increase in radius of electrode causes a decrease in electric current density (ρ) at the electrode tip due to which poor weld joint is obtained. The uniformity of ρ at the electrode sheet interface is not affected by change in radius of the electrode. The ρ is more for low radius electrodes due to convergence of electric current from a large area to a smaller area. Here a 3 mm electrode is the best choice for high productivity.

Further research can be carried out by finding the effect of A.C, D.C, and MFDC on WN both numerically and experimentally. The wear of electrodes of different geometries, materials can be analyzed while welding zinc coated sheet material. New hybrid electrode providing maximum life can be documented. The effective way to reduce electrode heating by using finned electrode or electrode having more water channels without affecting the stiffness of electrode can be tested.

KEYWORDS

- **resistance spot welding**
- **finite element method**
- **electrode geometry**
- **weld nugget**
- **electric current density (ρ)**
- **truncated cone**

REFERENCES

1. Williams, N. T.; Parker, J. D. Review of RSW of Steel Sheets Part 2 Factors Influencing Electrode Life. *Int. Mater. Rev.* **2004,** *49* (2), 77–108.
2. Williams, N. T.; Parker, J. D. Review of RSW of Steel Sheets Part 1 Modeling and Control of WN Formation. *Int. Mater. Rev.* **2004,** *49* (2), 45–75.
3. Wyant, R. A. Measurement and Effect of Contact Resistance in Spot Welding. *Electric. Eng.* **1946,** *65* (1), 26–33.
4. Vogler, M.; Sheppard, S. Electrical Contact Resistance Under High Loads and Elevated Temperatures. *Surfaces* **1993,** *9* (10), 11.
5. Kondo, M.; Nagata, H.; Nishimura, A.; Kokawa, H. Degradation Mechanism of Electrode Tip during RSW of Aluminium Alloy Sheets. *Sci. Technol. Weld. Join.* **2011,** *16* (2), 126–132.
6. Crinon, E.; Evans, J. T. The Effect of Surface Roughness, Oxide Film Thickness and Interfacial Sliding on the Electrical Contact Resistance of Aluminium. *Mater. Sci. Eng. A* **1998,** *242* (1–2), 121–128.
7. Tan, W.; Zhou, Y.; Kerr, H. W.; Lawson, S. A Study of Dynamic Resistance during Small Scale Resistance Spot Welding of Thin Ni Sheets. *J. Phys. D Appl. Phys.* **2004,** *37* (14), 1998.
8. Liu, L.; Zhou, S. Q.; Tian, Y. H.; Feng, J. C.; Jung, J. P.; Zhou, Y. N. Effects of Surface Conditions on Resistance Spot Welding of Mg Alloy AZ31. *Sci. Technol. Weld. Join.* **2009,** *14* (4), 356–361.
9. Rashid, M. Some Tribological Influences on the Electrode-Worksheet Interface during Resistance Spot Welding of Aluminum Alloys. *J. Mater. Eng. Perform.* **2011,** *20* (3), 456–462.
10. Greenwood, J. A.; Williamson, J. B. P. Electrical Conduction in Solid–II: Theory of Temperature Dependent Conductors. *Proc. R. Soc. Lond.* **1958,** 13–31.
11. Holm, R. *Electrical Contacts: Theory and Application*, 4th ed.; Springer-Verlag: Berlin/Heidelberg/New York, 1967.
12. Nied, H. The Finite Element Modeling of the Resistance Spot Welding Process. *Weld. J.* **1984,** *63* (4), 123.
13. Ma, N.; Murakawa, H. Numerical and Experimental Study on Nugget Formation in Resistance Spot Welding for High Strength Steel Sheets in Automobile Bodies. *Trans. JWRI* **2009,** *38* (2), 19–24.
14. Wei, P. S.; Wu, T. H. Magnetic Property Effect on Transport Processes in Resistance Spot Welding. *J. Phys. D Appl. Phys.* **2011,** *44* (32), 325501. DOI: 10.1088/0022-3727/44/32/325501
15. Mills, K. C. *Recommended Values of Thermo Physical Properties for Selected Commercial Alloys*; Woodhead Publishing.
16. Savage, F.; Nippes, E. F.; Wassell, F. a. Static Contact Resistance of Series Spot Welds. *Weld. Res. Suppl.* **1977,** 365–370.
17. Savage, B. Y. W. F. Dynamic Contact Resistance of Series Spot Welds. *Weld. Res. Suppl.* **1978,** 43–50.
18. Manual, ANSYS User'S., and Elements Volume III. Rev. 5.0. 4–425.
19. Bowers, R. J.; Sorensen, C. D.; Eagar, T. W. Electrode Geometry in Resistance Spot Welding. *Welding J.* **1990,** *69* (2), 45S.

PART IV

Studies on Advanced/Hybrid Machining Solutions

CHAPTER 24

PARAMETRIC OPTIMIZATION OF ELECTRODISCHARGE FACE GRINDING: A HYBRID MACHINING PROCESS

AKSHAT SRIVASTAVA KULSHRESTHA[1], ASHOK KUMAR DARGAR[1*], and DEEPAK RAJENDRA UNUNE[1,2]

[1]*Department of Mechanical-Mechatronics Engineering, the LNM Institute of Information Technology, Jaipur 302031, India*

[2]*Department of Materials Science and Engineering, INSIGNEO Institute for in silico Medicine, University of Sheffield, Sheffield, S1 3JD, UK*

**Corresponding author. E-mail: ashok.dargar@lnmiit.ac.in*

ABSTRACT

Nowadays, hybrid machining has become a hot cake for both industries as well as for academics due to its capability to fabricate the products in a more proficient and industrious way. In the present work, the material removal rate (MRR) and average surface roughness (ASR) of D2 steel were measured while performing electrical discharge face grinding (EDFG), a hybrid machining process. Taguchi gray relational analysis technique was applied to determine the best combination of process parameters to optimize the process performance. Total 31 experiments were performed based on response surface methodology, and their machining responses were measured. The gray relational grade for each level of the machining parameter was established. The significance of each machining parameter was achieved by analysis of variance based on gray relational analysis. From the evaluated results of the

Modern Manufacturing Systems: Trends and Developments. Rajiv Kumar Garg, Ravi Pratap Singh, Rajeev Trehan, & Ramesh Singh (Eds.)

confirmatory experiment, an improvement in MRR and ASR by 14.58% and 13.92%, respectively, was perceived.

24.1 INTRODUCTION

Accompanying the development of the mechanical industry, the demand for hard materials with high degree of dimensional accuracy is still a challenging job for the conventional machining processes.[1] The unconventional machining processes, like electrical discharge machining (EDM), can machine such materials easily irrespective of their hardness. However, there are a few problems associated with EDM such as low machining rate and poor surface integrity which limit the performance of the process.

The hybrid machining processes have been developed by the researcher's to improve the demerits of individual machining processes by combining or assisting the individual machining process with some other machining technique or energy assistance[2] Electrodischarge grinding (EDG) is a hybrid machining process that applies a collaborative effect of the grinding and EDM to augment the machining efficiency and enhance the surface integrity of EDM process.[3]

The EDG can be operated in three different configurations, viz., (1) electrodischarge face grinding (EDFG), (2) electrodischarge cut-off grinding (EDCG), and (3) electrodischarge surface grinding (EDSG). In the case of EDFG, the grinding wheel (tool) face side is used to perform the machining action and the feed toward the workpiece in a perpendicular direction. The rotational motion of the grinding wheel assists in the easy removal of the debris particle generated during EDM action. The machining performance of the EDFG depends on several control input machining parameters, viz., discharge current (IP), pulse-on-time (T_{on}), electrode rotational speed (ER), and pulse-off-time (T_{off}). The optimization of these process parameters is crucial to achieve better surface finish, low tool wear rate, and a high MRR.

In various machining processes for achieving the high-quality surface finish, optimization of process parameters is one of the important benchmarks.[4] Optimization of multiple performances characteristics is more intricate than optimization of a single performance characteristic.[5] Taguchi methodology is a single parameter optimization technique that is based on the signal-to-noise ratio. However, gray relational analysis (GRA) is applied for those parameters that are having the multi responses. Theory of gray system makes use of gray relational generating and calculates gray relational coefficient along with gray relational grade. The gray relational analysis optimization for various machining processes has been widely reported in the literature by various researchers for improving the MRR and surface finish as performance characteristics. Jangra et al.[6] optimized multiple machining characteristics in WEDM; they selected D3 steel as work material and applied GRA for optimization of WEDM process parameters. Nipanikar et al.[7] used the Taguchi methods to determine the optimal process parameter setting in EDM during the machining of AISI D3 steel. Hsiao et al.[8] Applied GRA to optimize the plasma arc welding process. Lin et al. showed that GRA is more straightforward than the fuzzy-based Taguchi method for optimizing the EDM process with multiple process responses.[9]

Thus, selecting the best levels of process parameters for optimal machining, which can introduce high MRR and surface quality, is really a tough task. It is found from the literature that no study has been available on AISI D2 steel machining with multi-objective optimization. This paper presents the optimization of EDFG for AISI D2 steel so as to optimize the responses using Taguchi-based gray relational analysis (TGRA). Thirty-one experiments were conducted to determine the best level combination of process parameters, and the experimental runs were planned according to central composite design (CCD) in response surface methodology (RSM).

The prime advantage of employing CCD is to reduce the number of experimental runs required to generate sufficient information for a statistically adequate result. The levels were assessed according to the response variables, namely, MRR and ASR. Their percussion of the control parameters on response variables was studied by assessing the single gray relational grade (GRG) for all performance characteristics.

The outline of the paper is as follows. The paper begins with the experimental plan and procedure followed by the TGRA of the multiple responses as discussed in detail. Analysis of variance (ANOVA) of the GRG has been performed to predict the optimum parameters. In the last of the paper, experimental validation of the results is carried out.

Nomenclature	
TGRA	Taguchi-based gray relational analysis
MRR	Material removal rate
ASR	Average surface roughness
EDFG	Electrodischarge face grinding
CCD	Central composite design

24.2 EXPERIMENTAL SETUP

Experiments were conducted on an EDM machine with a designed grinding attachment in face grinding mode. The setup is shown in Figure 24.1. The electrode is made of copper, a 25 mm diameter circular rod used as a grinding wheel. The chemical composition of D2 steel is given in Table 24.1. The experiments were planned by considering three levels for each of the four input process parameters that are depicted in Table 24.2. Each D2 steel was machined for 30 min before measuring response parameters.

TABLE 24.1 D2 Steel Chemical Composition.

Elements	Weight in%
Carbon	1.55
Chromium	12.0
Molybdenum	0.80
Vanadium	0.90
Iron	Balance
Silicon	0.25
Manganese	0.35

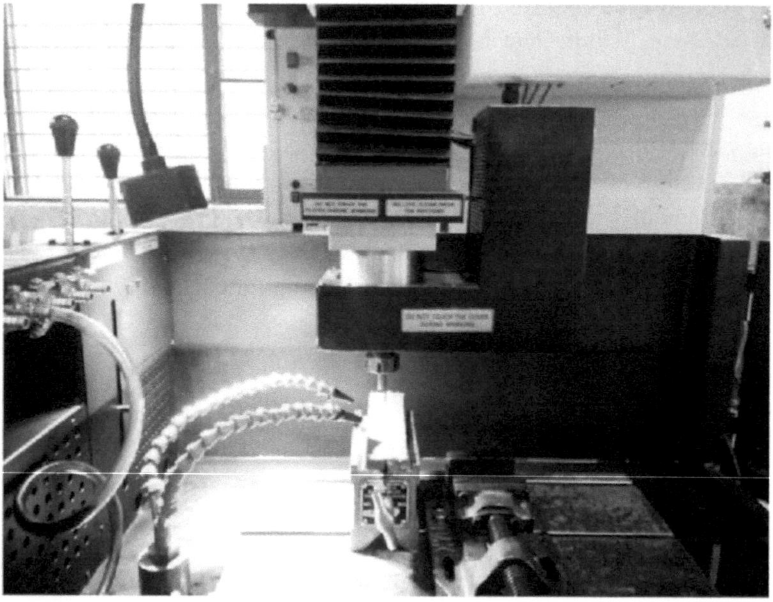

FIGURE 24.1 EDFG setup on ZNCEDM.

TABLE 24.2 Parameters and Their Levels.

Levels	Discharge current (IP, A)	Pulse-on-time (T_{on}, μs)	Pulse-off-time (T_{off}, μs)	Wheel rotation (ER, rpm)
1	5	50	4	200
2	10	75	7	400
3	15	100	10	600

In CCD, the limits of the experimental domain to be explored are defined and are made as wide as possible to obtain a clear response. So, the arrangement to conduct the experiments with four variables includes total of 31 experiments that were performed as displayed in Table 24.3.

TABLE 24.3 CCD and Their Response.

Exp. no.	Process parameters					
	(ER rpm)	(IP, A)	(Ton, μs)	(Toff, μs)	MRR (mm3/min)	ASR (μm)
1	600	15	50	10	12.180	8.993
2	600	15	50	4	12.354	8.987
3	600	15	100	10	11.617	8.949
4	400	15	75	7	12.180	6.537
5	200	5	100	4	7.022	5.613
6	200	15	50	4	11.877	6.898
7	200	5	50	4	6.935	6.121
8	200	15	50	10	12.050	6.899
9	200	15	100	10	12.224	7.999
10	400	10	75	7	9.536	5.991
11	400	10	50	7	9.406	5.509
12	600	5	50	10	6.025	5.472
13	400	10	100	7	10.230	6.789
14	400	10	75	7	9.536	5.992
15	600	5	100	4	6.242	6.244
16	600	5	100	10	5.765	6.168
17	600	5	50	4	5.808	5.545
18	400	10	75	10	8.799	5.626
19	200	5	50	10	6.285	6.611
20	600	10	75	7	10.403	7.458
21	200	10	75	7	8.062	6.691

TABLE 24.3 *(Continued)*

Exp. no.	Process parameters					
	(ER, rpm)	(IP, A)	(Ton, μs)	(Toff, μs)	MRR (mm3/min)	ASR (μm)
22	400	10	75	7	9.536	5.898
23	400	10	75	7	9.536	5.888
24	400	10	75	7	9.536	5.873
25	400	5	75	7	6.979	5.843
26	200	5	100	10	6.805	6.655
27	400	10	75	4	9.970	5.714
28	600	15	100	4	10.360	8.756
29	400	10	75	7	9.536	5.897
30	400	10	75	7	9.536	5.882
31	200	15	100	4	12.094	8.321

On completing the experiments, MRR response values were calculated by weight loss formula as shown below.

$$\text{MRR} = \frac{(\text{Wa} - \text{Wb}) \times 1000}{\rho \times t} \text{ mm}^3/\text{min}$$

Where Wa = weight of work material before machining, Wb = weight of work material and after machining, ρ is the density of work material, and t is the machining time. ASR of the machined surface was measured by using a surface roughness tester (Make: MITUTOYO). Three observations were carried out and their average value was taken as an experimental outcome for further analysis.

24.3 RESULTS AND DISCUSSION

24.3.1 GRAY RELATIONAL ANALYSIS FOR EXPERIMENTAL RESULTS

For EDFG process, MRR is a higher-the-better performances characteristic, whereas, ASR is lower-the-better performances characteristic. Hence, optimization of such conflicting performance characteristics is more complicated than optimization of a single performance characteristic. In this regard, GRA has found wide application areas for determining optimal parameters for different machining processes. So, in this investigation, GRA is carried out to obtain the optimal set of process parameters. The overall quality characteristics of the multi-response process depend on the calculated GRG.

24.3.1.1 DATA PRE-PROCESSING

Data pre-processing is a procedure where the original sequence is converted to a comparable sequence that is in the range between 0 and 1, and this is called gray relational generating. In this investigation, normalization of the experimental results was attained for MRR and ASR. For larger-the-better type responses such as MRR, the original sequence is normalized based on eq (24.1).

$$x_i^*(k) = \frac{x_i(k) \min x_i(k)}{\max x_i(k) \min x_i(k)} \tag{24.1}$$

Where for smaller-the-better type responses such as ASR, minimum values of each are fit into the reference sequences by normalizing their original sequences using eq (24.2).

$$x_i^*(k) = \frac{\max x_i(k) - x_i(k)}{\max x_i(k) - \min x_i(k)} \tag{24.2}$$

In eqs (24.1) and (24.2), $x_i^*(k)$ are the responses after normalizing, respectively, for $i = 1, 2, 3...31$, and $x_i(k)$ are the original values that are obtained for MRR and ASR.

After normalizing, the deviation sequences for each parameter are obtained by calculating the amount of each response that deviates from the reference sequences using eq (24.3).

$$\Delta_{0i}(k) = | x_0^*(k) - x_i^*(k) |_{\,I} \tag{24.3}$$

$\Delta_{0i}(k)$ is denoted as deviation sequences and $x_0^*(k) = 1$, the references sequences. The sequences after normalizing and the deviation sequence are shown in Table 24.4.

24.3.1.2 COMPUTING GRAY RELATION COEFFICIENT AND GRADE

Gray relation coefficient (GRC) ξ_i gives the degree of closeness of each of the actual normalized responses, and it is calculated by the below mentioned eq (24.4).[10]

$$\xi_i(k) = \frac{\Delta_{min}(k) + \zeta\Delta_{max}(k)}{\Delta_{0i}(k) + \zeta\Delta_{max}(k)} \tag{24.4}$$

Where ζ is the distinguishing coefficient and taken as 0.5 to give equal importance to both the responses. GRC is concerned with each individual performances characteristic, so for the evaluation of multi-performance characteristics, GRG, γ, is calculated by averaging individual GRC for each experiment using eq (24.5).

$$\gamma_i(k) = \frac{1}{2}\sum_{k=1}^{2}\xi_i(k) \tag{24.5}$$

From Table 24.4, it can be seen that experiment 4 has the best multi-performance characteristics as it has higher GRG value.

TABLE 24.4 Normalizing, Deviation Sequence Value, Gray Relation Coefficient, and Grades.

Exp. no.	Normalized sequence values		Deviation sequence values		Gray relation coefficient		Gray relation grade	Rank
	MRR	ASR	MRR	ASR	MRR $\xi i(1)$	ASR $\xi i(2)$	γ	
1	0.974	0.000	0.026	1.000	0.950	0.333	0.642	20
2	1.000	0.002	0.000	0.998	1.000	0.334	0.667	14
3	0.888	0.012	0.112	0.988	0.817	0.336	0.577	23
4	**0.974**	**0.698**	**0.026**	**0.302**	**0.950**	**0.623**	**0.787**	**1**
5	0.191	0.960	0.809	0.040	0.382	0.926	0.654	18
6	0.928	0.595	0.072	0.405	0.874	0.552	0.713	5
7	0.178	0.816	0.822	0.184	0.378	0.731	0.554	24
8	0.954	0.595	0.046	0.405	0.916	0.552	0.734	3
9	0.980	0.282	0.020	0.718	0.962	0.411	0.686	7
10	0.572	0.853	0.428	0.147	0.539	0.772	0.656	15
11	0.553	0.989	0.447	0.011	0.528	0.979	0.754	2
12	0.039	1.000	0.961	0.000	0.342	1.000	0.671	13
13	0.678	0.626	0.322	0.374	0.608	0.572	0.590	22
14	0.572	0.852	0.428	0.148	0.539	0.772	0.655	16
15	0.072	0.781	0.928	0.219	0.350	0.695	0.523	27
16	0.000	0.802	1.000	0.198	0.333	0.717	0.525	26
17	0.007	0.979	0.993	0.021	0.335	0.960	0.647	19
18	0.461	0.956	0.539	0.044	0.481	0.920	0.700	6
19	0.079	0.677	0.921	0.323	0.352	0.607	0.480	31
20	0.704	0.436	0.296	0.564	0.628	0.470	0.549	25
21	0.349	0.654	0.651	0.346	0.434	0.591	0.513	28
22	0.572	0.879	0.428	0.121	0.539	0.805	0.672	12

TABLE 24.4 *(Continued)*

Exp. no.	Normalized sequence values		Deviation sequence values		Gray relation coefficient		Gray relation grade	Rank
	MRR	ASR	MRR	ASR	MRR ξi(1)	ASR ξi(2)	γ	
23	0.572	0.882	0.428	0.118	0.539	0.809	0.674	10
24	0.572	0.886	0.428	0.114	0.539	0.814	0.677	8
25	0.184	0.895	0.816	0.105	0.380	0.826	0.603	21
26	0.158	0.664	0.842	0.336	0.373	0.598	0.485	30
27	0.638	0.931	0.362	0.069	0.580	0.878	0.730	4
28	0.697	0.067	0.303	0.933	0.623	0.349	0.486	29
29	0.572	0.879	0.428	0.121	0.539	0.806	0.672	11
30	0.572	0.884	0.428	0.116	0.539	0.811	0.675	9
31	0.961	0.191	0.039	0.809	0.927	0.382	0.654	17

Since the purpose of this work is to find an optimal level of machining parameters with maximum MRR and minimum ASR, the optimization of EDFG process parameters for the D2 steel is converted to optimization of GRG. So mean GRG for each level of the input parameters and their total mean are calculated as shown in Table 24.5. The influence of each parameter can be more clearly presented by the GRG graph in Figure 24.2. It can be observed that an increase in IP results in increase in GRG (i.e., increase in MRR and decrease in ASR). As the IP increases the spark energy also increases and also the numbers of ions that are attacking the workpiece and causes the temperature of the workpiece to rise, thereby melting and evaporating of material takes place from the workpiece surface and leads to an increase in MRR.

TABLE 24.5 Mean GRG for Each Level.

Machining parameters	GRG			Main effect	Rank
	Level 1	Level 2	Level 3	(Max-min)	
(IP)	0.5713	0.6435	0.6607*	0.0893	2
(Ton)	0.6513*	0.6501	0.5756	0.0758	3
(Toff)	0.6253	0.6378*	0.6111	0.0267	4
(ER)	0.6081	0.6904*	0.5874	0.1029	1

Total mean of GRG = 0.626.

*Optimal level of GRG for each parameter.

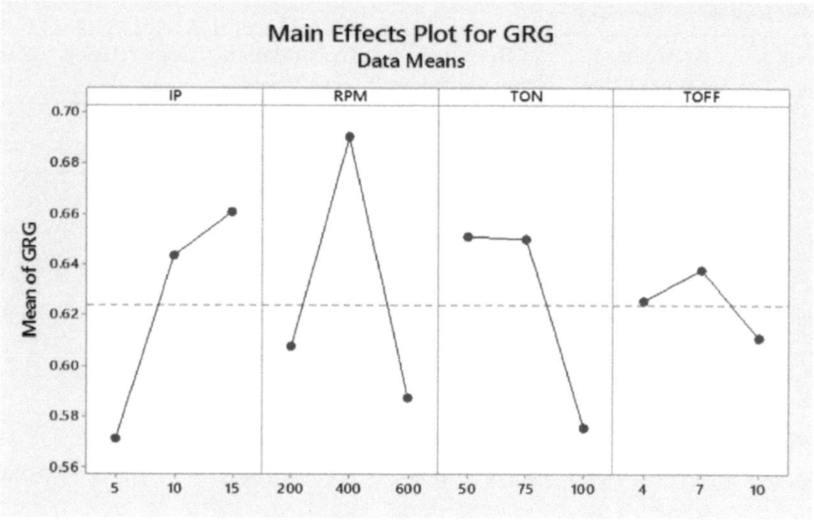

FIGURE 24.2 Effect of process parameters on GRG.

The higher the GRG, the better performance of the process is. From Figure 24.2, it is observed that **IP3 Ton1Toff 2 ER 2** (where IP3 = level 3 of factor IP, similarly Ton1 = level 1 of factor Ton, Toff2 = level 2 of factor Toff and ER2 = level 2 of factor ER) are the optimum condition for maximum GRG, that is, the optimal setting for maximum MRR and minimum ASR.

24.3.2 ANALYSIS OF VARIANCE (ANOVA) FOR EXPERIMENTAL RESULTS

ANOVA assists in an extensive understanding of the obtained data. In statistics, it is generally used to test for the hypothesis that the confidence interval for the data is 95%. But in this case, ANOVA is mainly used to determine the percentage contribution of each of the input parameters for the variation caused in the output during experimentation.

The variation caused in the output can be caused by the chance causes or assignable causes Misra.[11] Hence, it is necessary in this case to find the percentage contribution of each input process parameter and also the error toward the variation of the responses as shown in Table 24.6. The Fisher's ratio (*F*-ratio) determines the significances of a particular parameters. For a particular input parameter, the *F*-ratio has to be more than 1 to be considered as a significant parameter. As we are considering 95% confidence interval, it means if the P is lesser than 0.05 then it indicates stronger evidences against the null hypothesis (i.e., that if P value < 0.05, then the source is significant).

TABLE 24.6 Analysis of Variance of GRG.

Source	Degree of freedom	Sum of squares	Adjusted mean square	F-ratio	P	Remark	Percentage contribution%
Ton	2	0.025865	0.012932	3.60	0.044	Significant	12
IP	2	0.037677	0.018839	5.24	0.014	Significant	18
Toff	2	0.006421	0.003211	0.89	0.424	Insignificant	3.07
ER	2	0.051285	0.025642	7.13	0.004	Significant	24.5
Error	22	0.079073					
Total	30	0.209118					

24.4 CONFIRMATION TEST

After identification of the optimal levels of the machining parameters, the last step is to predict and confirm the improvement of performance characteristic using the optimal level of the machining parameters. The estimated GRG $\hat{\gamma}$ using the optimal level of the machining parameters can be calculated as

$$\hat{\gamma} = \gamma_m + \sum_{i=1}^{q} (\gamma_i - \gamma_{\text{im}}) \tag{24.6}$$

Where = is the total mean of the GRG, γ_i is the mean of the GRG at the optimal level, and q is the number of the machining parameters that significantly affect the multiple performance characteristics. Using eq (24.6), predicted MRR, ASR, and GRG for optimal machining parameters are obtained as shown in Table 24.7.,which shows the comparison of experimental results using initial machining parameters.

TABLE 24.7 Improvement in GRG with Optimized Parameters.

Machining parameters	Initial machining parameters	Optimal machining parameters	
	$IP_3\,Ton_2\,Toff_3\,ER_1$	Predicted $IP_3\,Ton_1\,Toff_2\,ER_2$	Experimental $IP_3\,Ton_1\,Toff_2\,ER_2$
MRR	12.180	12.283	13.957
ASR	6.537	10.029	5.627
GRG	0.787	0.796	0.796

Improvement in GRG = (0.009).

24.5 CONCLUSIONS

The application of the CCD in RSM with GRA to optimize the EDFG of D2 steel with the multiple performance characteristics has been testified in this study. GRA of the experimental results of MRR and ASR can convert optimization of the multiple performance characteristics into the optimization of a single performance characteristic called GRG. The ER is recognized to be the most significant factor. ER of 400 rpm, IP of 15 A, Ton 50 μs and Toff 7 μs are the recommended levels for the four parameters for a controllable process to obtain better MRR and lower ASR simultaneously. ANOVA of GRG for multi-performance characteristics. The investigation result for the optimal setting shows that there is a considerable improvement in the machining efficiency of the process, and GRG is also improved. This analysis shows that the RSM and GRA analysis can be successfully implemented to find the best parametric combination for EDFG.

ACKNOWLEDGMENTS

The authors would like to express their deep gratitude to the Department of Mechanical Mechatronics Engineering of LNMIIT, Jaipur, India for providing the laboratory facilities and financial support.

KEYWORDS

- **electrical discharge machining**
- **response surface methodology**
- **gray relational analysis**
- **material removal rate**
- **average surface roughness**

REFERENCES

1. Benedict, G. F. *Non-Traditional Manufacturing Processes*; Marcel Dekker: New York, 1987.

2. Yadava, V.; Jain, V. K.; Dixit, P. M. Theoretical Analysis of Thermal Stresses in Electro-discharge Diamond Grinding, Machining Science and Technology. *Int. J. Taylor & Francis* **2004,** *8* (1), 119–140.
3. Unune, D. R.; Mali, H. S. A Study of Multiobjective Parametric Optimisation of Electric Discharge Diamond Cut-off Grinding of Inconel 718. *Int. J. Abrasive Technol.* **2016,** *7* (3), 187–199.
4. Ranganathan, S.; Senthilvelan, T. Multiresponse Optimization of Machining Parameters in Hot Turning Using Grey Analysis. *Int. J. Adv. Manuf. Technol.* **2011,** *56*, 455–462.
5. Ross, P. J. *Taguchi Technique for Quality Engineering*, 2nd ed.; McGraw-Hill: NewYork, 1996.
6. Jangra, K.; Jain, A.; Groover, S. Optimization of Multiple-Machining Characteristics in Wire Electrical Discharge Machining of Punching Die Using Grey Relational Analysis. *J. Sci. Indust. Res.* **2010,** *67*.
7. Nipanikar, S. R. Parameter Optimization of Electro Discharge Machining of AISI D3 Steel Material by Using Taguchi Method. *J. Eng. Res. Studies* **2012,** *3*, 7–16.
8. Hsiao, Y. F.; Tarng, Y. S.; Huang, W. J. Optimization of Plasma Arc Welding Parameters by Using the Taguchi Method with the Grey Relational Analysis. *J. Mater. Manuf. Process.* **2008,** *23*, 51–58.
9. Fung, C. P. Manufacturing Process Optimization for Wear Property of Fiber-Reinforced Poly Butylene Terephthalate Composites with Grey Relational Analysis. *Wear* **2003,** *254*, 298–306.
10. Natarajan, N.; Arunachalam, R. M. Optimization of Micro–EDM with Multiple Performances Characteristics Using Taguch Method and Grey Relational Analysis. *J. Sci. Indust. Res.* **2011,** *70* (07), 500–505.
11. Misra, K. B. *The Hand Book of Performability Engineering*; Springer, 2008.

CHAPTER 25

A STUDY ON THE MACHINING BEHAVIOR AND OPTIMIZATION OF PROCESS PARAMETER OF AA5051 USING ELECTRODISCHARGE MACHINE

RICHA MISHRA, NEERAJ KUMAR BHOI, and HARPREET SINGH*

Department of Mechanical Engineering, PDPM Indian Institute of Information Technology Design and Manufacturing, Jabalpur 482005, Madhya Pradesh, India

Corresponding author. E-mail: hps.dme@gmail.com

ABSTRACT

The requirement of high productivity and minimal tooling cost is a prime consideration in the competitive world. The optimum value of the process parameter plays a critical role in determining the overall machining performance as well as product quality. In the present case of aerospace-grade, AA5051 was machined with the help of an electrodischarge machine (EDM) to discover the consequences of machining factors on the material removal rate and tool wear rate of the material. For the present case, the copper tool and EDM 30 oil as the dielectric medium was utilized to find out the process parameter effect. In the current study, Taguchi orthogonal array (L9) was utilized for the selection of process parameter and for the optimum process parameter value. The obtained results from the optimum set of input parameter are in close agreement with the confirmation experiments. The data

Modern Manufacturing Systems: Trends and Developments. Rajiv Kumar Garg, Ravi Pratap Singh, Rajeev Trehan, & Ramesh Singh (Eds.)

set derived from the present set of the experiment will help to find out the potential process parameters for the machining.

25.1 INTRODUCTION

The use of aluminum and its alloy materials is in growing demand due to their lightweight and higher stiffness in the field of medical, aerospace, and automotive. The excellent lightweight characteristics, damping properties, excellent ductility lead to numerous uses in the industry, as well as household applications.[1–8] Electric-discharge machining (EDM) is one of the most broadly used nonconventional machining processes, with a lot of applications in the modern industrial environment for desired complex geometry development. The unique capabilities of the intricate shape generation, complex product development with minimal work surface damage lead to the deterministic place of EDM in sectors, such as biomedical, aerospace, automotive, and micromanufacturing compoenents.[9,10] The removal of material takes place due to the generation of higher temperature and melting of the work material. The dielectric medium during sparking between the tool and work material. In the similar context, Surekha et al.[11] tried to optimize the EN-19 work material using powder mixed EDM process. Their studies narrate the response surface modeling and fuzzy mixed with grey relational approach optimizations process. It was noticed that current is having maximum influence 22.23% toward removal of material. Further, the wear rate in the tool was in increasing trends with enhancement in the input process conditions to higher operating levels. On a similar examination, Rouniyar and Shandilya[12] develop a new approach for the machining of Al6061 alloy material. They combined the magnetic-assisted powder mixed with electrodischarge machining action to enhance the process capability of the conventional machining operation with EDM. The authors examined the full factorial experimental design with five input level of conditions. It was perceived that the increase in the magnetic field improves the material removal rate (MRR) with decrement in the wear of tool material. Uthayakumar et al.[13] describe the machining of Al-SiC functionally graded composite material with die-sinking EDM. The authors utilized full factorial experimental design with a grey relational approach to meet the standard and optimized process conditions. The surface obtained after the machining was observed with microscopy method to judge the quality of the surface produced. They observed that the machining with hard carbide zone was moderately difficult as compared to that of monolithic phase. The major contributing factor for machining was found to be influenced by pulse current (39.4%). Sidhu et al.[14] explored the machining of metal-based composite material with EDM. They observed that a significant amount of hard ceramics and carbides get deposited on the tool surface with the machining with monolithic copper material. In a similar context, Paswan et al.[15] utilized the steam-assisted surface for the machining of composite material under normal working mode with EDM. They observed that the utilization of steam during machining reduces the recast layer formation. However, the metal removal process is reduced (5.6 times) compared to the kerosene dielectric medium. The overall sustainability is greatly improved with the assistance of steam with significant reduction in the cost. Kumar et al.[1] explored EDM with Al (6351)/5SiC/5B_4C hybrid composite with copper electrode material. They model the machining characteristics with Taguchi experimental design (L27) for material removal and electrode wear behavior. The analysis of variance results for the conducted experimental trails exhibit pulse current that have greater influence on the machining productivity. Similarly, for Al (6351)/5SiC/5B_4C hybrid composite surface integrity was measured by Kumar et al.[16] with EDM. They noticed that the heat-affected zone was higher in the case of composite machining compared to pure metallic phase. Also, the surface roughness for the developed machined zone was tend to increase with increasing current. The presence of carbide phase in the matrix leads to the formation of higher crater in the machined surface.

From the extensive literature work, it was found that a little effort was given for the machining of Al5051 machining with EDM. Accordingly, the present aim of the study was to conduct experimental observation on the MRR and tool wear rate (TWR) for the Al-5051 material with the assistance of Taguchi experimental design approach.

25.2 WORKING PRINCIPLE

EDM is an advanced machining process that works on high voltage and low current. In the EDM process, two conducting materials, one is electrode (cathode) and other is workpiece (anode), are submerged in a dielectric fluid. When the two conducting materials are brought in contact, arc or spark is produced due to which heat is generated which in term removes the material from the workpiece through fusion and vaporization. Schematic representation of the EDM setup used in the experiment is shown in Figures 25.1 and 25.2.

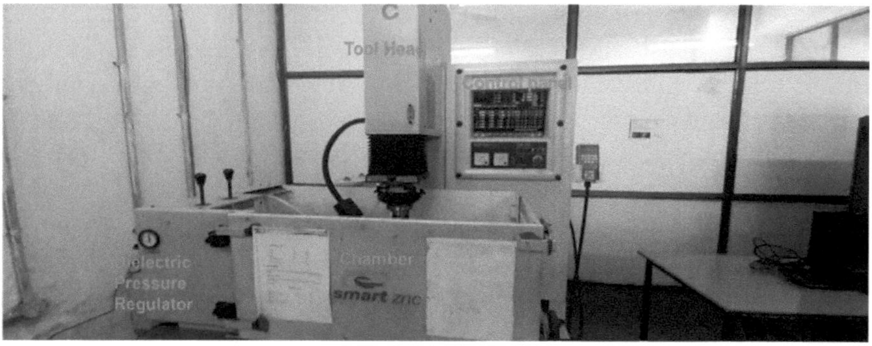

FIGURE 25.1 Pictorial view of the electrodischarge machining process.

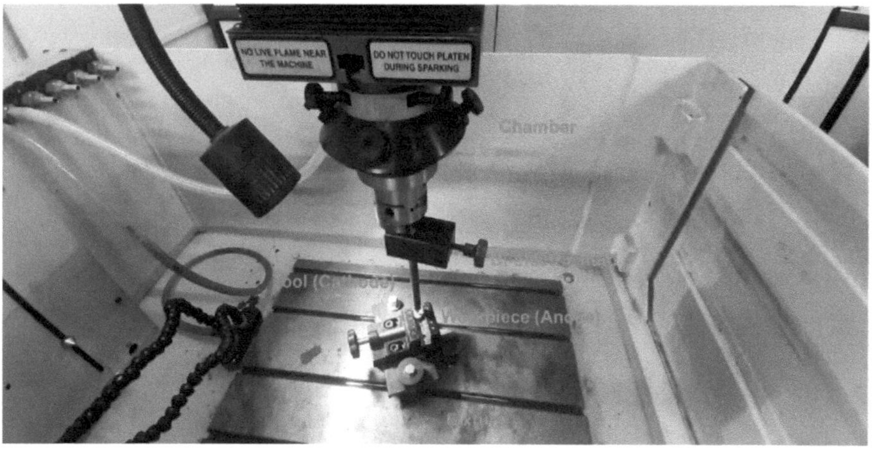

FIGURE 25.2 Sectional view of the electrodischarge machine.

The EDM process is based on the RC-circuit in which the capacitor is fist allowed to charged (V_C) to threshold voltage value, once the threshold value is reached, charging is stopped and discharging (V_d) of capacitor begins. Since discharging of capacitor is a spontaneous process, so the spark is produced resulting in large amount of heat generation between the tool and the workpiece leading to removal of material from workpiece by spark erosion as a result of fusion and vaporization. The discharge voltage and charge voltage are related as shown in eq 25.1:

$$V_d = V_0\left(1 - e^{\frac{-t_1}{RC}}\right). \tag{25.1}$$

Energy required for charging is given by

$$E = \frac{1}{2} C V_d^{\,2}. \tag{25.2}$$

Power equation is given by

$$P = \frac{\frac{1}{2} C V_d^{\,2}}{t_1}. \tag{25.3}$$

For maximum power

$$V_d = .718 V_0. \tag{25.4}$$

25.3 EXPERIMENTAL PROCEDURES

The workpiece material chosen for the present work is AA5051 with the specimen dimension of 150 × 35 × 2 mm^3 for all the experiments. The physical and mechanical properties of AA5051 material are shown in Table 25.1.

TABLE 25.1 Physical and Mechanical Properties of AA5051 Material.

Density	2.7 g/cm^3
Specific capacity	900 J/kg k
Thermal conductivity	150 w/m k
Poisson's ratio	0.33
Modulus of elasticity	68 GPa
Melting point	640
Modulus of resilience	23 kJ/m^3
Thermal diffusivity	63 m^2/s
Tensile strength	170 MPa
Elongation	21%
Shear strength	110 MPa

The electronica ZNC EDM machine was employed for the present study. The process parameters used for the current study are shown in Table 25.2. For the present research work, three parameters, namely, (1) peak current (Ip), (2) pulse duration (Ton), and (3) gap voltage (Vg), are employed to investigate the MRR and EWR under different experimental conditions. Machining of 1 mm depth of diameter 4 mm hole is fixed for experimentation. MRR and

TWR after machining are calculated using eqs 25.5 and 25.6, respectively. Acetone solution was to clean the prepared samples and then an electronic weighing machine was used to measure the weight of samples. An electronic timer with an accuracy of 0.1 s was used to calculate the machining time.

$$\text{MRR}\left(\text{g}/\text{min}\right)=\frac{\text{Difference in weight loss for work material in gram}}{\left(\text{Overall machining time}\right)\text{min}} \quad (25.5)$$

$$\text{TWR}\left(\text{g}/\text{min}\right)=\frac{\text{Diffrence in weight loss for tool material in gram}}{\left(\text{Overall machining time}\right)\text{min}} \quad (25.6)$$

TABLE 25.2 Working Range of the Machining Process Parameters.

S. No.	Process variable	Range/value of the variable
1	Ip	5, 10, and 15 amp
2	Vg	40, 60, and 80
3	Ton	50, 75, and100 μs
4	t	6

TABLE 25.3 Experimental Settings.

Working condition	Description
Workpiece	AA5051 (150 × 65 × 2 mm^3)
Electrode	Copper
Dielectric	Commercial EDM Oil
Flushing	Side flushing with pressure 0.5 MPa
Polarity	Positive
Supply voltage	240 V
Machining depth	1 mm

25.4 RESULTS AND DISCUSSION

For optimization of (1) process parameters and (2) quality improvement of the manufactured components, a statistical approach namely Taguchi method is employed.[10] The obtained results were then analyzed using the S/N ratio and were found to be in strict accordance with the experimental results, as shown in Table 25.4. Optimality setting for the machining parameters (1) MRR and (2) TWR was calculated and then successfully verified. For the experimental data analysis, MINITAB version 16 software was employed.

TABLE 25.4 MRR and TWR at Different Working Conditions.

S no.	Ip (A)	Ton (μs)	Vg (V)	T (min)	MRR (g/min)	TWR (g/min)
1	5	50	40	6	0.00145	0.000197
2	5	75	60	6	0.00262	0.000380
3	5	100	80	6	0.00240	0.000188
4	10	50	60	6	0.01030	0.000706
5	10	75	80	6	0.00809	0.000449
6	10	100	40	6	0.00667	0.000571
7	15	50	80	6	0.00965	0.000706
8	15	75	40	6	0.00904	0.000822
9	15	100	60	6	0.00974	0.000256

25.4.1 DETERMINATION OF S/N RATIO FOR MRR

The objective function here is to perform the maximization of MRR, "large the better" approach has been used as it is desired that MRR should be maximum. The ratio of signal-to-noise for "larger the better" approach is defined by eq. 25.7, as follows:

$$S / N \text{ ratio} = -\log_{10} 1 / n \sum_{i=1}^{i=n} 1 / y_i^{2} \tag{25.7}$$

where y_i = ith measured value in a row and n = total number of trails.

From Figure 25.3 and the response table of *S/N* ratios for MRR, we can tell that the most significant parameter which affects MRR is peak current. With the increase in the pulse-on-time up to 75 μs MRR increases and then gradually decreases when the peak current is increased. The MRR is higher at peak current of 15 amp and is lower at 5 amp. The effect of the duty cycle is very minimal on MRR so it is taken constant with value of 6. The gap voltage indicates that MRR is higher at 60 V and is lower at 40 V.

The confirmation test was conducted with optimum input parameters, (1) peak current of 15 A, (2) pulse-on-time of 75 μs, and (3) voltage of 60 V for MRR. The parameter which has the highest impact on MRR is peak current as per the response table found from experimental values.[8–10] From the ANOVA analysis of MRR, it was noticed that the current value having most influencing process conditions followed voltage and pulse-on-time for MRR. This fact and obtained results are in-line with the previously reported value for machining with experimental design at different levels.[1,17,18]

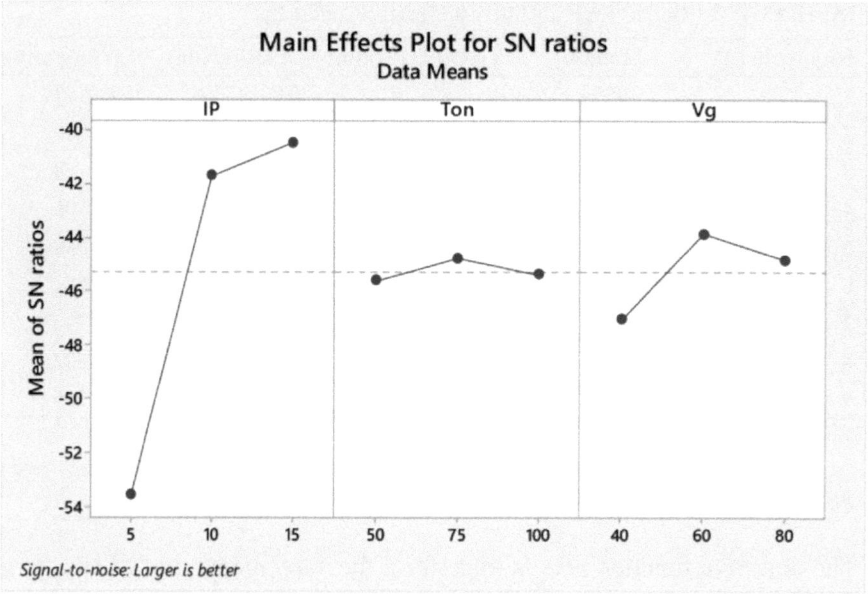

FIGURE 25.3 Main effect plot of *S/N* ratio for MRR.

Analysis of Variance for SN Ratios for MRR

Source	DF	Seq SS	Adj SS	Adj MS	*F*	*p*
Ip	2	315.500	315.500	157.750	55.28	0.018
Ton	2	1.088	1.088	0.544	0.19	0.840
Vg	2	15.987	15.987	7.993	2.80	0.263
Residual error	2	5.707	5.707	2.854		
Total	8	338.282				

Response Table for Signal-to-Noise Ratios for MRR

Larger is better

Level	Ip	Ton	Vg
1	−53.60	−45.61	−47.06
2	−41.70	−44.78	−43.87
3	−40.47	−45.38	−44.85
Delta	13.13	0.82	3.19
Rank	1	3	2

25.4.2 S/N RATIO DETERMINATION FOR TWR

The key objective for the existing work is the minimization of TWR. "Smaller the better" approach has been used as it is desired that TWR should be minimum. The ratio of signal-to-noise for "Smaller the better" approach is defined by eq 25.8, as follows:

$$S/N \text{ ratio} = -\log_{10} 1/n\sum_{i=1}^{i=n} 1/y_i^{\,2} \quad (25.8)$$

where y_i = ith measured value in a row and n = total number of trail.

From Figure 25.4 and Table 25.4, we can tell that the tool process parameter affects the TWR. We are using copper toll material apart from other tool materials.

With the increase in peak current value up to 10 A, TWR decreases, and then it shows a gradually increasing behavior. Duty cycle has very little effect on TWR, so keep it constant. When pulse-on-time increases up to 75 μs, TWR decreases and then starts increasing gradually. As we increase the gap voltage, TWR also increases.

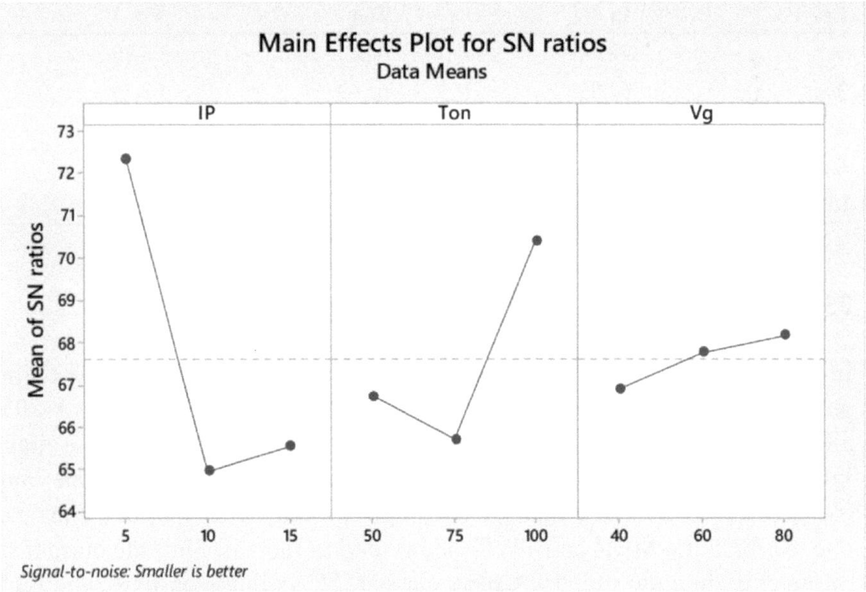

FIGURE 25.4 Main effect plot of S/N ratio for TWR.

The confirmation test is conducted with optimum input parameters, namely, (1) peak current of 10 A, (2) pulse-on-time of 75 μs, and (3) voltage of 40 V for TWR. The ANOVA analysis of the TWR reports confirms that the peak current having most affecting parameter for wear progression. The rank order for input conditions considering wear rate of tool material can be given as peak current, voltage, and on time for machining. The similar reports were given in the literature for EDM.[11,19]

Analysis of Variance for Signal-to-Noise Ratios for TWR

Source	DF	Seq SS	Adj SS	Adj MS	*F*	*p*
Ip	2	101.575	101.575	50.787	1.94	0.340
Ton	2	36.929	36.929	18.465	0.71	0.586
Vg	2	2.527	2.527	1.264	0.05	0.954
Residual error	2	52.329	52.329	26.165		
Total	8	193.360				

Response Table for Signal-to-Noise Ratios for TWR

Smaller is better

Level	Ip	Ton	Vg
1	72.34	66.72	66.89
2	64.95	65.69	67.75
3	65.52	70.41	68.17
Delta	7.40	4.72	1.27
Rank	1	2	3

25.5 CONCLUSION AND FUTURE PROSPECTS

In this current work, the response of input process parameters on the output performance measures in terms of MRR, and TWR in EDM of AA-5051 alloy was investigated. It was noticed that the peak current is the major influencing factor responsible for the MRR and TWR. As the current value is increased, the surface damage and high melting of the surface enhances due to which the MRR and TWR values tend to increase. For the current set of the experimental run, 15 A peak current, 75 μs pulse-on-time, and 60 V voltage were found to be optimal setting for MRR. Similarly, for TWR peak current as 10 A, pulse-on-time as 75 μs, and voltage as 40 V found to optimal for minimum wear during machining. The ANOVA analysis of the MRR and TWR suggests that the peak current is having more affecting process conditions for machining with EDM of Al and its kind of alloy material. The process can be extended for metaheuristic algorithms along with Taguchi experimental design to obtain more precise and significant conditions for machining and process optimization.

KEYWORDS

- **AA5051**
- **material removal rate**
- **tool wear rate**
- **process parameter optimization**

REFERENCES

1. Kumar, S. S.; Uthayakumar, M.; Kumaran, S. T.; Parameswaran, P. Electrical Discharge Machining of Al(6351)-SiC-B4C Hybrid Composite. *Mater. Manuf. Process*. **2014,** *29*, 1395–1400. https://doi.org/10.1080/10426914.2014.952024.
2. Rouniyar, A. K.; Shandilya, P. Experimental Investigation on Recast Layer and Surface Roughness on Aluminum 6061 Alloy During Magnetic Field Assisted Powder Mixed Electrical Discharge Machining. *J. Mater. Eng. Perform*. **2020,** *29*, 7981–7992. https://doi.org/10.1007/s11665-020-05244-4.
3. Bhoi, N. K.; Singh, H.; Pratap, S. Developments in the Aluminum Metal Matrix Composites Reinforced by Micro/Nano Particles—A Review. *J. Compos. Mater*. **2020,** *54*, 813–833. https://doi.org/10.1177/0021998319865307.
4. Pattnaik, S. K.; Bhoi, N. K.; Padhi, S.; Sarangi, S. K. Dry Machining of Aluminum for Proper Selection of Cutting Tool: Tool Performance and Tool Wear. *Int. J. Adv. Manuf. Technol*. **2018,** *98*, 55–65. https://doi.org/10.1007/s00170-017-0307-0.
5. Bhoi, N. K.; Singh, H.; Pratap, S. Synthesis and Characterization of Alumina Nanoparticles: A Case Study. *J. Inst. Eng. India Ser. C*. **2019,** *101*, 407–413. https://doi.org/10.1007/s40032-019-00542-w.
6. Singh, H.; Jain, P. K.; Bhoi, N.; Pratap, S. Experimental Study Pertaining to Microwave Sintering (MWS) of Al-Metal Matrix Composite—A Review. *Mater. Sci. Forum*. **2018,** *928*, 150–155. https://doi.org/10.4028/www.scientific.net/MSF.928.150.
7. Bhoi, N. K.; Singh, H.; Saurabh, P. Promise of Self-lubricating Aluminum-Based Composite Material. In *Functional Smart Materials*; Prakash, C.; Singh, S.; Davim, J. P., Eds.; CRC Press Taylor & Francis Group, 2020; pp 65–82.
8. Bhoi, N. K.; Singh, H.; Pratap, S. Synthesis and Characterization of Zinc Oxide Reinforced Aluminum Metal Matrix Composite Produced by Microwave Sintering. *J. Compos. Mater*. **2020,** *54*, 3625–3636. https://doi.org/10.1177/0021998320918646.

9. Li, L.; Niu, Z. W.; Zheng, G. M. Ultrasonic Electrodeposition of Cu-SiC Electrodes for EDM. *Mater. Manuf. Process.* **2016,** *31*, 37–41. https://doi.org/10.1080/10426914.2015.1025968.
10. Patowari, P. K.; Mishra, U. K.; Saha, P.; Mishra, P. K. Surface Integrity of C-40 Steel Processed with WC-Cu Powder Metallurgy Green Compact Tools in EDM. *Mater. Manuf. Process.* **2011,** *26*, 668–676. https://doi.org/10.1080/10426914.2010.512652.
11. Surekha, B.; Lakshmi, T. S.; Jena, H.; Samal, P. Response Surface Modelling and Application of Fuzzy Grey Relational Analysis to Optimise the Multi Response Characteristics of EN-19 Machined Using Powder Mixed EDM. *Aust. J. Mech. Eng.* **2019,** *00*, 1–11. https://doi.org/10.1080/14484846.2018.1564527.
12. Rouniyar, A. K.; Shandilya, P. Fabrication and Experimental Investigation of Magnetic Field Assisted Powder Mixed Electrical Discharge Machining on Machining of Aluminum 6061 Alloy. *J. Eng. Manuf.* **2019,** *233*, 2283–2291. https://doi.org/10.1177/0954405419838954.
13. Uthayakumar, M.; Babu, K. V.; Kumaran, S. T.; Kumar, S. S.; Jappes, J. T. W.; Rajan, T. P. D. Study on the Machining of Al–SiC Functionally Graded Metal Matrix Composite Using Die-Sinking EDM. *Part. Sci. Technol.* **2019,** *37*, 103–109. https://doi.org/10.1080/02726351.2017.1346020.
14. Sidhu, S. S.; Batish, A.; Kumar, S. Study of Surface Properties in Particulate-Reinforced Metal Matrix Composites (MMCs) Using Powder-Mixed Electrical Discharge Machining (EDM). *Mater. Manuf. Process.* **2014,** *29*, 46–52. https://doi.org/10.1080/10426914.2013.852211.
15. Paswan, K.; Pramanik, A.; Chattopadhyaya, S.; Basak, A. K. A Novel Approach towards Sustainable Electrical Discharge Machining of Metal Matrix Composites (MMCs). *Int. J. Adv. Manuf. Technol.* **2020,** *106*, 1477–1486. https://doi.org/10.1007/s00170-019-04816-6.
16. Suresh Kumar, S.; Uthayakumar, M.; Thirumalai Kumaran, S.; Varol, T.; Canakci, A. Investigating the Surface Integrity of Aluminium Based Composites Machined by EDM. *Def. Technol.* **2019,** *15*, 338–343. https://doi.org/10.1016/j.dt.2018.08.011.
17. Wang, C. C.; Chow, H. M.; Yang, L. D.; Te Lu, C. Recast Layer Removal after Electrical Discharge Machining via Taguchi Analysis: A Feasibility Study. *J. Mater. Process. Technol.* **2009,** *209*, 4134–4140. https://doi.org/10.1016/j.jmatprotec.2008.10.012.
18. Kumar, Y.; Singh, H. Multi-response Optimization in Dry Turning Process Using Taguchi's Approach and Utility Concept. *Procedia Mater. Sci.* **2014,** *5*, 2142–2151. https://doi.org/10.1016/j.mspro.2014.07.417.
19. Singh, S.; Yeh, M. F. Optimization of Abrasive Powder Mixed EDM of Aluminum Matrix Composites with Multiple Responses Using Gray Relational Analysis. *J. Mater. Eng. Perform.* **2012,** *21*, 481–491. https://doi.org/10.1007/s11665-011-9861-z.

CHAPTER 26

DISQUISITION OF IMPACT OF KEY ELECTRIC DISCHARGE MACHINING INPUT PARAMETERS ON VARIOUS PROCESS CHARACTERISTICS: A REVIEW

RAHUL SONI, RAVI PRATAP SINGH*, and
SHAILENDER SINGH BHADAURIA

Department of Industrial and Production Engineering, Dr. B. R. Ambedkar National Institute of Technology Jalandhar, Punjab, India

**Corresponding author. E-mail: singhrp@nitj.ac.in*

ABSTRACT

Different alloys and superalloys are prominently known for their flexile and exceptional mechanical properties. Possessing such characteristics, they are extensively used in countless engineering and industrial applications, specifically; automobile, aeronautics, marine, power, etc. In the available literature, different alloys and superalloys have been investigated and analyzed on electric discharge machining in order to achieve the best performance, that is, enhanced rate of material erosion, diminished rate of tool wear, better quality of eroded surface, etc. In this chapter, an effort has been made to censoriously examine the impact of several process inputs such as discharge current, pulse-on time, pulse-off time, gap voltage, flushing pressure of dielectric, electrode diameter, electrode material type, coating of electrode, etc. on the output responses. Most significant factors for different outputs have also been discussed. The input variable (i.e., discharge current) has

Modern Manufacturing Systems: Trends and Developments. Rajiv Kumar Garg, Ravi Pratap Singh, Rajeev Trehan, & Ramesh Singh (Eds.)

been publicized as the major influential variable for most of the responses such as MRR, SR, and TWR trailed by pulse-on time. In the last section of this review chapter, the research gaps have also been discussed.

26.1 MACHINING

Removal of material to get a required final structure and dimensions is called machining process. Material removal takes place in a controlled manner and today collectively named as subtractive manufacturing processes. There are two kinds of machining processes: conventional (traditional) and nonconventional (nontraditional) machining processes.

26.1.1 CONVENTIONAL MACHINING

In conventional machining processes, a sharp cutting tool making direct contact with the work material is used for removing the material to obtain the desired work geometry. Due to the direct contact between tool and work material, there are some problems in conventional machining:

- Poor machining characteristics is found
- The geometries of cutting tool get worn-out
- Exactitude of dimensions is not achieved
- There is a lack of surface rectitude

When these processes are scaled down from macro levels to micro levels, then these problems get much severe. So, to overcome these limitations, non-traditional machining came into existence.

26.1.2 NONCONVENTIONAL MACHINING

These are special type of machining processes in which a form of energy such as mechanical, chemical, electrochemical, thermoelectrical is applied to erode the material to get the desired final profile and dimensions. The material is eroded without the direct contact of tool with the work material. It has the capability to machine the complex geometries and the materials with excessive strength and hardness which is very problematic to machine by traditional machining processes can be machined in an easy way. Electric discharge machining is a type of nonconventional machining.

26.1.2.1 ELECTRIC DISCHARGE MACHINING

In electric discharge machining, material is removed from a conductive workpiece due to the electric spark when the high pulsed voltage is employed between the tool electrode and the conductive work material. Localized melting and vaporization take place because of high temperature generated which helps in material erosion from the workpiece. The flow of dielectric fluid carries away the most of the eroded material from the interelectrode gap. The materials that are electrically conductive in nature can be machined by the electric discharge machining process. Typically, in oil die-sinking electric discharge machining, pulsed direct current (DC) power supply is applied where generally the tool electrode is attached to the (+) ve terminal (tool electrode as anode) and the work material is attached to the (−) ve terminal (workpiece as cathode) (Figure 26.1).

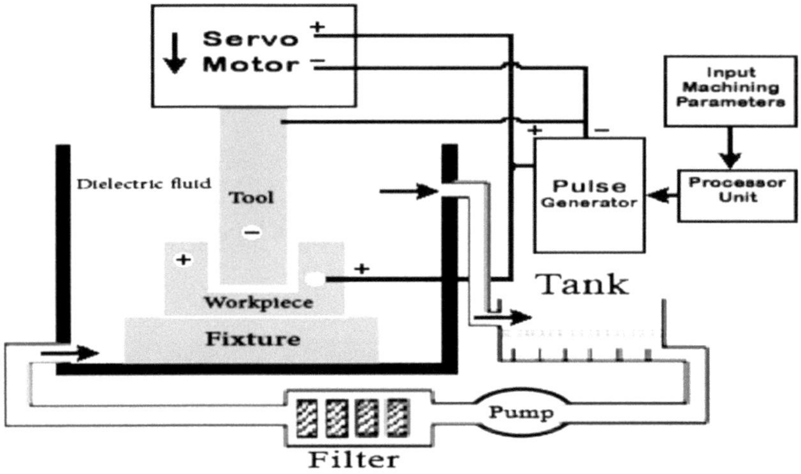

FIGURE 26.1 Schematic setup diagram of electric discharge machining.

26.2 WORK MATERIAL: AL7075 T651

Alloy Al 7075 is an alloy of aluminum in which the principal debasing ingredient is Zn. It possesses superior mechanical properties, good ductility, high strength, durability, and good fatigue resistance. While it is more impressionable because of microsegregation to embrittlement than many other aluminum alloys, it has considerably finer corrosion resistance than the 2000 series (Tables 26.1 and 26.2).

TABLE 26.1 Composition of Al7075 T651 Alloy.

Element	Percentage
Aluminum	90%
Zinc	5.6%
Magnesium	2.5%
Copper	1.6%
Chromium	2.3%

TABLE 26.2 Properties of Al7075 T651 Alloy.

Property	Value (magnitude)
Density	2.81 g/cm^3
Ultimate tensile strength	570 MPa
Yield strength	500 MPa
Elongation	(3–9) %

26.3 APPLICATIONS

- First mass production worldwide, usage—MITSUBISHI A6M Zero Fighter.
- Transport applications due to their enhanced specific strength including:
    - Marine
    - Automotive
    - Aviation
- Mold tool manufacturing
- M-16 Rifles, AR-15 Style Rifles, Desert Tactic Arms
- Aerospace application
- Space shuttle solid rocket booster (SRB) nozzles
- External tank SRB in the portion of the inner tank
- Rock climbing tools and equipment
- Bicycle parts and components
- Hang glider air structure and many more.

26.4 LITERATURE REVIEW

Khan et al.[1] conducted an experiment on stainless steel D3 of 10 mm diameter and 5 mm thickness with the help of electric discharge machine. The influence of input variables; pulse-on time (T_{on}), current (A), and voltage (V) on the several response parameters were analyzed. Machine off-set and copper wire electrode of 2 mm diameter were kept fixed. The concerned parameters, that is, MRR, SR, and TWR have been investigated by varying the machining parameters. The experiments were designed employing Taguchi L9 OA technique. The calculated optimum parameters for MRR (current—7 A, pulse on-time—20 μs, voltage—125 V), SR (current—7 A, pulse-on time—20 μs, voltage—125 V) and TWR (current— 7 A, pulse on-time—20 μs, voltage—125 V) with the help of S/N ratio and ANOVA. R^2 values and regression analysis provide the significant results for MRR, SR, and TWR as 92.5%, 89.9%, and 80.4%, respectively. Sharma et al.[2] scrutinized the multioptimization of output responses MRR and TWR for Inconel 718 (specimen of 14 mm diameter) with Cu–Cd as electrode. Taguchi L9 OA was applied for designing the experiments and the further analysis of results was made with the help of two multicriteria decision-making techniques named TOPSIS and PROMITHEE. The experimentation was done on a usual CNC electric discharge machine with positive polarity.[3–6] The observed optimum values of the parameters were current—7 A, gap voltage—50 V, pulse-on time—200 μs and pulse-off time—6 μs.

Gudipudi et al.[7] fabricated AA6061-B_4C composite by stir casing process. An experimental investigation was conducted for the optimization of the output parameters. MRR and ARLT were taken as the responses to be studied. Taguchi L9 OA was employed as the technique for designing the experiments. SEM micrographs were used to analyze the ARLT and topography of the surface of machined zone. The statistical software Minitab 18 was used to develop the linear regression model to the optimize the MRR (maximum) and ARLT (minimum).[8,9] ANOVA was conducted to find the significant parameters of the responses. Discharge current (I) was found as the most significant input parameter affecting the MRR with percentage contribution of 85.81% followed by the discharge duration (T_{on}) with contribution of 13.187%. For ARLT, T_{on} was the most significant machining parameter with percentage contribution of 68.50%.

Nallusamy[10] analyzed the MRR and TWR on OHNS die steel with Cu and Cu–Zn electrodes using electric discharge machine (ELECTRONICA—M2S) with electromechanical servo control and positive polarity. The size of the work plate used was 50 mm × 50 mm with 24 mm thickness. Discharge current (I) and electrode material were chosen as machining variables. Sparking voltage (V), pulse-on time (T_{on}), and pulse-off time (T_{off}) were selected as input machining parameters. Kerosene Commercial of grade oil was used as dielectric fluid. SEM was used for analyzing the heat affected regions.[11,12] It was concluded that the increment in discharge current leads in improving the MRR but decrease in TWR. It was also stated that at 4 A discharge current, the MRRs for both the electrodes were equal. On visual inspection, it was found that the surface finish with both the electrodes were almost same.

Khanna et al.[13] conducted an investigational study on electric discharge drilling machine to study and achieve the optimized performance for Al7075. A brass rod of 2 mm diameter was adopted as a tool electrode for drilling. Peak current, pulse-on (T_{on}), pulse-off (T_{off}), and speed of electrode rotation (ω) were taken as controllable variables for the experimentation. Other input variables such as air pressure and voltage were kept fixed at 3 kg/cm^2 and 65 V, respectively, all over the experiment. TWR and MRR were taken as output responses. From this experimental study, T_{on} was found as the most significant input parameter having maximum effect on MRR with 94.63% contribution tracked by T_{off} with contribution of 4.55%. The percentage contributions of T_{on} and T_{off} for TWR were 75.27% and 4.25% respectively. Multiquality characteristics for the same material were given by TGRA. The optimum results given by GRG were observed to be 0.0359 mm^3/min for MRR and 0.252 g/s for TWR.

El-Taweel[14] inspected the relationship of input variables with Al–Cu–Si–TiC composite tool electrode formed by powder metallurgy (P/M) technique in electric discharge machining of CK45 steel. In this study, titanium carbide percentage (TiC%), peak current (I_p), pulse-on time (T_{on}), and flushing pressure of dielectric fluid (P) were considered as the input process variables whereas MRR and TWR were studied as output responses. The rotatable design of the central composite second order was utilized for planning the experiments and experimental models were developed by employing response surface methodology (RSM) technique. The appropriacy of developed regression models was checked by carrying out the ANOVA test. In the above experimental study, peak current was appeared as the key significant input process variable influencing both MRR and TWR whereas dielectric flushing pressure had a little impact on both the responses.[15–17] It was also addressed that the reactivity of the formed tool electrodes to I_p and T_{on} were more than the other traditional electrodes. The optimal combination of input process variable settings showed by the multiresponse optimization was TiC%—18%, I_p—6 A, P—1.2 MPa and T_{on}—182 µs to achieve the required MRR and TWR.

Kao et al.[18] conducted an experimental examination on alloy Ti-6Al-4V for optimizing the machining parameters with multiple quality characteristics.

Electrolytic copper of 10 mm diameter and 50 mm height was chosen as the tool electrode. The discharge current (amp), gap voltage (V), duration of pulses (μs), and the duty factor (%) were chosen as the input process variables whereas TWR, MRR, and SR were taken as the responses to be analyzed. The experiments were governed by introducing the Taguchi L9 OA. The experimental results of the responses were normalized by GRA.[19,20] The contribution of each parameter was calculated by using ANOVA results. According to results provided by ANOVA, discharge current was identified as the most dominating input variable with percentage contribution of almost 70% followed by the duty factor of 18.64%. Discharge current of 5 A, gap voltage of 150 V, pulse duration of 200 μs, and the duty factor of 70% were the optimal values for the EWR. According to the experimental results of MRR and S/N ratio, it was noted that MRR is largely affected by the discharge current with percentage contribution of 74% followed by the duty factor with 22.78%. For MRR, discharge current of 20 A, gap voltage of 100 V, pulse duration of 100 μs, and the duty factor of 30% were the optimal results. According to S/N ratio and the corresponding ANOVA results, surface roughness was largely affected by the discharge current with contribution of 67.06% followed by the duty factor of 17.32% and the pulse duration of 14.01%. The optimal values for surface roughness were discharge current of 5 A, open voltage of 200 V, pulse duration of 200 μs, and the duty factor of 70%.

Viswanth et al.[21] experimentally studied the MRR, EWR, and SR on AISI 2507 super duplex stainless steel on ARD EDM ZNC die-sinker EDM. Pure grade electrolytic copper rod (99.71% Cu) was selected as the tool electrode for the experimentation. Pongamia Green Dielectric was used as the dielectric medium. T_{on}, T_{off}, I_p and (IEG) were taken as the process parameters for multiresponse optimization. The duty factor (75–90%) was kept constant throughout the experimental trials. MRR, EWR, and SR were the responses to be calculated and analyzed for the multiresponse optimization. Taguchi L_{27} OA was introduced as a technique for designing the experiments. GRA was conducted to identify the optimal setting of the machining variables. In this experimental study, it was found that on increasing the value of T_{on}, MRR tends to increase and EWR tends to decrease. The minimum value of the EWR was 0.221 mm^3/min at 400 μs of T_{on}, 90 μs of T_{off}, 15 A of I_p, 50 V of V and 120 μm of the IEG. The minimum value of the SR was 6.711 μm at 300 μs of T_{on}, 90 μs of T_{off}, 9 A of I_p, 60 V of V and 120 μm of the IEG. The optimal settings for multi-optimization given by GRA were 400 μs of T_{on}, 60 μs of T_{off}, 21 A of I_p, 50 V of V and 60 μm of the IEG. According to the ANOVA test, peak current (I_p) was found as the key influential factor trailed by T_{on}.

Chauhan et al.[22] experimentally investigated the impact of different input variables on MRR and TWR for M2 high speed steel using Taguchi L9 OA. The experimentation was carried out on EDM (model C-3822 with PSR-220). Cylindrical shaped Cu was used as the tool electrode for the study. I_p, T_{on}, and diameter of the electrode were selected as the process variables. The optimum values were obtained by using the MINITAB software. In this experimental study, it was discussed that the MRR improves with increment in peak current but with increment in pulse-on time, the MRR enhances sharply up to a maximum value after that there is a fall in the slope. MRR is directly proportional to the electrode diameter whereas TWR is inversely proportional to the same. With the elevation in peak current, TWR first increases up to a certain point then decreases. It was also concluded that MRR is highly affected by T_{on} and least influenced by the electrode diameter whereas the I_p has the moderate influence on MRR.

Kumar et al.[23] conducted an experimental investigation on Ti-6Al-7Nb to achieve the best microdrilling efficiency and performance in EDM. The various parameters such as electrode material, rotation of the electrode, pulse-on time (T_{on}), pulse-off time (T_{off}), work material thickness, and drill angle were taken as the input variables for the microdrilling whereas DR and EWR were chosen as the output parameters to be studied. The input variables and their ranges were selected for the further experimentation using Taguchi OA. The optimal setting of variables was attained using NSGA-II approach. In this experimental examination, it was concluded that the DR and EWR both have increasing trend with increase in peak current and pulse-on whereas the rise in T_{off} leads to improvement in DR but decrement in EWR. The high aspect ratio microholes were drilled very easily but at higher hole depth, front side of electrode becomes conical. So, tapered electrode was introduced for drilling.

Jagadeeswararao et al.[24] performed experiments on Incoloy-800 super-alloy steel plate to optimize the outputs; MRR, EWR, and SR by varying input parameters such as current, pulse-on, pulse-off, and flushing pressure. During the experiments, duty cycle and voltage (50 V) were kept constant. Statistical technique was used for limiting the number of experiments and GRA was used to optimize the responses for reaching the better performance. The maximum value of MRR was found to be 0.027423 mm^3/min. The minimum value of EWR and SR was 0.0001 mm^3/min and 2.82667 μm, respectively. The optimal setting for achieving the best performance was found $I_p = 16$ A, $T_{on} = 150$ μs, $T_{off} = 20$ μs and $F_p = 3.3$ kg/cm^2.

Rouniyar and Shandilya[25] conducted an experimental study on MRR of Al 6061 alloy machined with magnetic-field-assisted powder mixed EDM.

Peak current (I_p), concentration of powder (P_c), pulse-on (T_{on}), magnetic field (M_F), and pulse-off (T_{off}) were considered as the input variables. RSM-based Box-Behnken design approach was employed to perform the experimentation and to investigate the effect of input parameters on MRR. A quadratic model was developed for predicting the MRR based on the analysis. Optimum set of parameters were determined using the desirability function approach. In this study, it was concluded that peak current is the most significant input variables influencing MRR trailed by pulse-on, powder concentration, and magnetic field.

Bejaxhin et al.[26] experimentally studied the impact of TiN/AlCrN nano-coated EN steel tool electrode on SF and MRR of EDM with optimized outcomes. TiN/AlCrN nanocoatings, gap current (I), T_{on}, and T_{off} were selected as the machining parameters. The experiments were designed using Taguchi method and results were optimized through Regression Analysis. In this investigation, it was concluded that TiN/AlCrN coated electrodes result in achieving the eroded surface of better quality. It was also stated that the higher MRR was dominated by reverse polarity tracked by gap current.

Singaravel et al.[27] conducted an optimization study of EDM responses on AISI D2 steel. Cryogenically treated and tempered copper was taken as the tool electrode for experimentation. Dielectric fluid type, current, and pulse-on was considered as the machining variables whereas MRR and TWR were taken as the responses to be optimized. Taguchi method was combined with VIKOR to solve the multiple objectives simultaneously. From this study, current = 8 A, T_{on} = 400 μs with cottonseed oil dielectric were the optimized process parameters. It was also concluded that cryogenic treatment on electrodes enhanced the properties of electrode. Hence reduction in tool wear was observed.

Hanif[28] conducted an experimental study for optimizing the machining variables in EDM of AISI D2 steel using graphene-based dielectric. T_{off}, I_p, T_{on} and concentration of nanoparticles were the input machining parameters whereas MRR, SR, and white layer thickness (WLT) were considered as the responses to be optimized. Graphene nanoplatelets mixed with kerosene oil were utilized as the dielectric fluid. Regression models for the output responses were developed using RSM. The most significant factors were identified by carrying out the ANOVA. In this investigation, it was observed that MRR = 21.93 mm^3/min, SR = 3.9 μm, and WLT = 19.13 μm were the optimum performance measures. In this research, it was also stated that WLT was significantly affected by T_{on}. Maximum MRR achieved at higher level of I_p and concentration of nanoparticles. Minimum SR and WLT achieved at lower level of I_p and higher level of concentration of nanoparticles (Table 26.3).

TABLE 26.3 Literature Review in Tabular Form.

S. No.	Authors	Work material	Tool electrode	Process/machine/ technique used	Parameters/ variants	Response studied	Remarks
1	Khan et al. [1]	Stainless Steel D3	Copper	ELECTRONICA S-70 ZNC, Taguchi L9 Design, ANOVA	T_{on} I V	MRR SR TWR	Optimum parameters calculated by ANOVA and S/N Ratio for MRR: Current—7 A, Pulse on-time—20 µs, Voltage—125 V TWR: Current—1 A, Pulse on-time- 10 µs, Voltage—100V SR: Current— 1 A, Pulse on—time—10 µs, Voltage—150 V Regression analysis and values of R^2 for MRR, SR, and TWR were 92.5%, 89.9% & 80.4%, respectively, provides significant results.
2	Sharma et al. [2]	Inconel 718	Copper–cadmium	ZNC 50A Sparkonix, TOPSIS, PROMITHEE	I_p T_{on} T_{off} V	MRR TWR	Optimal values: Current = 7 A, Gap voltage = 50 V, Pulse-on time = 200 µs, Pulse-off time = 6 µs
3	Gudipudi et al. [7]	AA6061-B4C Composite	Copper	Die Sinker EDM, Taguchi L9 OA, ANOVA, SEM Micrographs	I T_{on} T_{off}	MRR ARLT	ANOVA results: For MRR—% contribution of I = 85.81% % contribution of T_{on} = 13.187% For ARLT—% contribution of T_{on} = 68.50%
4	Nallusamy [10]	OHNS Die Steel	Brass and copper	ELECTRONICA-M2S EDM, SEM	I_p T_{on} T_{off}	MRR TWR	With increase in I_p: MRR increases and TWR decreases At I_p = 4 A, MRRs for both electrodes are equal.

TABLE 26.3 *(Continued)*

S. No.	Authors	Work material	Tool electrode	Process/machine/ technique used	Parameters/ variants	Response studied	Remarks
5	Khanna et al. [13]	Al 7075	Brass	Electronica Make ED 300 EDDM, Taguchi L27 OA, ANOVA, Grey Relational Analysis	T_{on} T_{off} F_p	MRR TWR	MRR: Range—0.0148–0.0359 mm^3/min TWR: Range—0.1459–0.2590 g/s For MRR: % contribution of T_{on} = 94.63% % contribution of T_{off} = 4.55% For TWR: % contribution of T_{on} = 75.27% % contribution of T_{off} = 4.25%
6	El Taweel [14]	CK45 Steel	Al–Cu–Si–TiC P/M composite	AGETRON Advance Die Sinking EDM, RSM, ANOVA	TiC % I_p T_{on} F_p	MRR TWR	Optimal values: TiC % = 18% Peak current = 6 A DFP = 1.2 MPa T_{on} = 182 µs Peak current is most dominating parameter affecting both MRR and TWR
7	Kao et al. [18]	Ti-6Al-4V alloy	Electrolytic copper	YAWJET, M35B CNC EDM, Taguchi Technique, Grey Relational Analysis, ANOVA	I V PD DF	EWR MRR SF	For electrode wear ratio: % contribution of discharge current = 70% % contribution of the duty factor = 18.64% For material removal rate: % contribution of discharge current = 74% % contribution of the duty factor = 22.78% For surface roughness: % contribution of discharge current = 67.06% % contribution of the duty factor = 17.32% % contribution of pulse duration = 14.01%

TABLE 26.3 (*Continued*)

S. No.	Authors	Work material	Tool electrode	Process/machine/ technique used	Parameters/ variants	Response studied	Remarks
8	Viswanth et al. [21]	AISI 2507 Super Duplex Stainless Steel	Electrolytic copper	ARD EDM ZNC Die-Sinker EDM, Taguchi L27 OA, Grey Relational Analysis	I_p T_{on} T_{off} IEG	EWR MRR SF	With increase in T_{on}: • MRR increases • EWR decreases Least EWR = 0.221 mm^3/ min Minimum SR = 6.711 μm Multiresponse optimization: T_{on} = 400 μs, T_{off} = 60 μs, I_p = 21 A, IEG = 60 μm
9	Chauhan et al. [22]	M2 High Speed Steel	Copper	Taguchi L9 OA, MINITAB, ANOVA	T_{on} I_p ED	TWR MRR	1. Effect of electrode diameter: • MRR is directly proportional to electrode diameter. • TWR is inversely proportional to electrode diameter. 2. MRR and TWR both increases with pulse-on time (T_{on}) but after attaining the peak value there is a decrease in the slope of the curve. 3. MRR is moderately affected by the peak current.
10	Kumar et al. [23]	Ti-6Al-7Nb	Copper	Fast Drill EDM, Taguchi Method, NSGA-II	ERR EM T_{on} T_{off} WMT Drilling angle	DR EWR	1. Optimum parameter for high DR and low EWR I_p = 4 A, T_{on} = 60 μs, T_{off} = 50 μs 2. DR and EWR both increase with the rise in I_p and T_{on}. With the increase in T_{off}, DR increases but EWR decreases.

TABLE 26.3 (*Continued*)

S. No.	Authors	Work material	Tool electrode	Process/machine/ technique used	Parameters/ variants	Response studied	Remarks
11	Jagadeeswara Rao et al. [24]	Incoloy-800	Copper	Die-sinking EDM, Taguchi L9 OA, Grey Relational Analysis	I_p T_{on} T_{off} IEG F_p	MRR EWR SR	Optimum input variables for best performance: $I_p = 16$ A, $T_{on} = 150$ µs, $T_{off} = 20$ µs, $F_p = 3.3$ kg/cm^2
12	Rouniyar and Shandilya [25]	Al 6061 alloy		Die-sinking EDM, Box-Behnken design approach, Desirability function approach	I_p P_c T_{on} M_F T_{off}	MRR	1. Peak current is the most significant parameter. 2. Optimum set of parameters for the best performance $I_p = 6.995$ A, $P_c = 9.998$ g/L, $T_{on} = 68.771$ µs, $T_{off} = 36.287$ µs, $M_F = 0.308$ T
13	Bejaxhin et al. [26]	Al 6061 alloy	EN steel electrode	SPARKONIX MOS-15A, Taguchi method, Regression models, SEM analysis	TiN/AlCrN nanocoating, I T_{on} T_{off} Polarity	SR MRR	1. Good quality of eroded surface was achieved in case of nanocoating. 2. Coated electrode type and pulse duration was found as the most significant parameter for surface roughness 3. Higher MRR was dominated by negative polarity followed by current.
14	Singaravel et al. [27]	AISI D2 steel	cryogenically treated and tempered copper	Die-sinking EDM, Taguchi+VIKOR hybrid	Dielectric fluid type, I T_{on}	MRR TWR	Optimum settings: Current = 8 A $T_{on} = 400$ µs With cottonseed oil

TABLE 26.3 (*Continued*)

S. No.	Authors	Work material	Tool electrode	Process/machine/ technique used	Parameters/ variants	Response studied	Remarks
15	Hanif et al. [28]	AISI D2 steel	Copper, Aluminum, Brass, Graphite	CM 655C CNC EDM, ANOVA, RSM, SEM	T_{on} T_{off} I_p C	MRR SR WLT	1. T_{on} significantly affects WLT. 2. Maximum MRR achieved at higher level of I_p and conc. of nanoparticles. 3. Minimum SR and WLT achieved at lower level of conc. of nanoparticles.

26.5 GAPS OBSERVED

- In the available literatures, very few studies have been reported on aluminum alloy 7000 and 8000 series like Al 7075, Al 7175, Al 8009, etc.
- Very few researchers have employed more than one optimization technique to compare the optimal settings.
- Development of a novel tool electrode can be attempted by different machining operations.
- None of the researchers have compared the Material removal rates after the starting of machining and before the finishing of machining.

26.6 CONCLUSIONS

The major interpretations that can be made as per the data available in previous research papers on EDM of different materials are as below-

- In the available literature, it is concluded that the most of the studies have conducted on Inconel-718, 625, 600, 825, Incoloy-800, AISI D2 steel, Al6061, Ti-6Al-4V alloy, etc. However, various other categories of alloys and superalloys are still needed to study with EDM process.
- The process variable, that is, discharge current is the key influential factor for most of the output responses (such as MRR, TWR, and SR, etc.) followed by the pulse-on time.
- As there is a rise in the value of discharge current, the MRR and SR keep up increasing but TWR first increases then after achieving a maximum value it starts falling.
- With the enhanced pulse-on times, the MRR first increases sharply then after attaining a certain value, the slope of increasing trend diminishes but the TWR decreases.
- As the pulse-off time increases, the MRR improves and the TWR reduces. Also, the SR gets decreased.
- The MRR is directly proportional to the electrode diameter whereas the TWR is inversely proportional to the electrode diameter.
- MRR is dominated by negative polarity as compared to the positive polarity.
- Cryogenic treatment enhances the properties of tool electrode. Hence, reduction in tool wear is observed.

- Nanocoating on the tool electrode results in achieving the better quality of eroded surface.

KEYWORDS

- **EDM**
- **machining parameters**
- **Taguchi**
- **optimization**
- **design of experiments**

REFERENCES

1. Khan, F.; Kumar, J.; Soota, T. Optimization of EDM Process Parameter for Stainless Steel D3. *Mater. Today Proc.* **2019,** *25*, 635–638. DOI: 10.1016/j.matpr.2019.07.529.
2. Sharma, V.; Prakash Misra, J.; Singhal, P. Multi-Optimization of Process Parameters for Inconel 718 While Die-Sink EDM Using Multi-Criterion Decision Making Methods. *J. Phys. Conf. Ser.* **2019,** *1240* (1). DOI: 10.1088/1742-6596/1240/1/012166.
3. Singh, R. P.; Singhal, S. Rotary Ultrasonic Machining: A Review. *Mater. Manuf. Process.* **2016,** *31*, 1795–1824.
4. Singh, R. P.; Singhal, S. Investigation of Machining Characteristics in Rotary Ultrasonic Machining of Alumina Ceramic. *Mater. Manuf. Process.* **2017a,** *32*, 309–326.
5. Singh, R. P.; Singhal, S. Experimental Investigation of Machining Characteristics in Rotary Ultrasonic Machining of Quartz Ceramic. *J. Mater. Des. Appl.* **2018a,** *232,* 870–889.
6. Singh, R. P.; Singhal, S. Rotary Ultrasonic Machining of Alumina Ceramic: Experimental Study and Optimization of Machining Responses. *J. Eng. Res.* **2018c,** *6*, 01–24.
7. Gudipudi, S.; Nagamuthu, S.; Subbian, K. S.; Chilakalapalli, S. P. R. Fabrication and Experimental Study to Optimize the Recast Layer and the Material Removal in Electric Discharge Machining (EDM) of AA6061-B4C Composite. *Mater. Today Proc.* **2019,** Volume 19, Part 2, 448–454. DOI: 10.1016/j.matpr.2019.07.634.
8. Singh, R. P.; Singhal, S. Rotary Ultrasonic Machining of Macor Ceramic: An Experimental Investigation and Microstructure Analysis. *Mater. Manuf. Process.* **2017b,** *32*, 927–939.
9. Singh, R. P.; Singhal, S. An Experimental Study on Rotary Ultrasonic Machining of Macor Ceramic. *J. Eng. Manuf.* **2018b,** *232,* 1221–1234.
10. Nallusamy, S. Analysis of MRR and TWR on OHNS Die Steel with Different Electrodes Using Electrical Discharge Machining. *Int. J. Eng. Res. Africa* **2016,** 22, 112–120. DOI: 10.4028/www.scientific.net/JERA.22.112.

11. Singh, R. P.; Tyagi, M.; Kataria, R. Selection of the Optimum Hole Quality Conditions in Manufacturing Environment Using MCDM Approach: A Case Study. In *Operations Management and Systems Engineering*; Springer: Singapore, 2019a; pp 133–152.
12. Singh, R. P.; Kataria, R.; Singhal, S. Decision-Making in Real-Life Industrial Environment through Graph Theory Approach. In *Computer Architecture in Industrial, Biomechanical and Biomedical Engineering*; IntechOpen, 2019b. DOI: 10.5772/intechopen.82011
13. Khanna, R.; Kumar, A.; Garg, M. P.; Singh, A.; Sharma, N. Multiple Performance Characteristics Optimization for Al 7075 on Electric Discharge Drilling by Taguchi Grey Relational Theory. *J. Ind. Eng. Int.* **2015,** *11* (4), 459–472. DOI: 10.1007/s40092-015-0112-z.
14. El-Taweel, T. A. Multi-response Optimization of EDM with Al-Cu-Si-TiC P/M Composite Electrode. *Int. J. Adv. Manuf. Technol.* **2009,** *44* (1–2), 100–113. DOI: 10.1007/s00170-008-1825-6.
15. Singh, R.; Singh, R. P.; Tyagi, M.; Kataria, R. Investigation of Dimensional Deviation in Wire EDM of M42 HSS Using Cryogenically Treated Brass Wire. *Mater. Today: Proc.* **2019c**.
16. Singh, R.; Singh, R. P.; Trehan, R. State of the Art in Processing of Shape Memory Alloys with Electrical Discharge Machining: A Review. In *Proceedings of the Institution of Mechanical Engineers, Part B: Journal of Engineering Manufacture*, 2020. 0954405420958771.
17. Tyagi, M.; Panchal, D.; Singh, R. P.; Sachdeva, A. Modeling and Analysis of Critical Success Factors for Implementing the IT-Based Supply-Chain Performance System. In *Operations Management and Systems Engineering*; Springer: Singapore, 2019; pp 51–67.
18. Kao, J. Y.; Tsao, C. C.; Wang, S. S.; Hsu, C. Y. Optimization of the EDM Parameters on Machining Ti-6Al-4V with Multiple Quality Characteristics. *Int. J. Adv. Manuf. Technol.* **2010,** *47* (1–4), 395–402. DOI: 10.1007/s00170-009-2208-3.
19. Singh, R. P.; Kumar, J.; Kataria, R.; Singhal, S. Investigation of the Machinability of Commercially Pure Titanium in Ultrasonic Machining Using Graph Theory and Matrix Method. *J. Eng. Res.* **2015,** *3*, 75–94.
20. Singh, R. P.; Kataria, R.; Kumar, J.; Verma, J. Multi-response Optimization of Machining Characteristics in Ultrasonic Machining of WC-Co Composite through Taguchi Method and Grey-fuzzy Logic. *AIMS Mater. Sci.* **2018,** *5*, 75–92.
21. Srinivas Viswanth, V.; Ramanujam, R.; Rajyalakshmi, G. Performance Study of Eco-Friendly Dielectric in EDM of AISI 2507 Super Duplex Steel Using Taguchi-fuzzy TOPSIS Approach. *Int. J. Product. Qual. Manag.* **2020,** *29* (4), 518–541. DOI: 10.1504/IJPQM.2020.106425.
22. Chauhan, V. S.; Ahmad, F. Experimental Investigation of Mrr and Twr on Edm for M2 High Speed Steel Using Taguchi Method. **2019,** *14* (9), 194–198.
23. Dewangan, S.; Deepak Kumar, S.; Kumar Jha, S.; Kumar Biswas, C. Optimization of Micro-EDM Drilling Parameters of Ti–6Al–4V Alloy. *Mater. Today Proc.* **2020,** Proceedings 33, 2–6. DOI: 10.1016/j.matpr.2020.03.307.
24. Jagadeeswararao, M.; Shaik, R. U.; Buschaiah, K. *Grey Relational Analysis of EDM Process Parameters for Incoloy-800*; Springer International Publishing, 2020.
25. Rouniyar, A. K.; Shandilya, P. Experimental Study on Material Removal Rate of Al6061 Machined with Magnetic Field Assisted Powder Mixed Electrical Discharge Machining. *J. Phys. Conf. Ser.* **2019,** *1240* (1). DOI: 10.1088/1742-6596/1240/1/012018.

26. Bovas Herbert Bejaxhin, A.; Paulraj, G.; Aravind, S. Influence of TiN/AlCrN Electrode Coatings on Surface Integrity, Removal Rates and Machining Time of EDM with Optimized Outcomes. *Mater. Today Proc.* **2020,** *21*, Proceedings 21(4 Ver. VI) 340–345. DOI: 10.1016/j.matpr.2019.05.459.
27. Singaravel, B.; Prasad, S. D.; Shekar, K. C.; Rao, K. M.; Reddy, G. G. *Optimization of Process Parameters Using Hybrid Taguchi and VIKOR Method in Electrical Discharge Machining Process*; Springer: Singapore, 2020.
28. Hanif, M. Optimization of Process Parameters Using Graphene-Based Dielectric in Electric Discharge Machining of AISI D2 Steel, 2019; pp 3735–3749.

CHAPTER 27

POWDER-MIXED MICRO-ELECTRICAL DISCHARGE MACHINING OF BIODEGRADABLE AZ91D MAGNESIUM ALLOY: AN EXPERIMENTAL INVESTIGATION

SANTOSH KUMAR YADAV[1*], ABHISHEK SINGH[1], and KISHORE DEBNATH[2]

[1]*Department of Mechanical Engineering, National Institute of Technology Patna, Patna 800005, India*

[2]*Department of Mechanical Engineering, National Institute of Technology Meghalaya, Shillong 793003, India*

[*]*Corresponding author. E-mail: santosh.me18@nitp.ac.in*

ABSTRACT

Magnesium is one of the important elements found in human bone. It is widely used as biocompatible material due to its density close to the density of human bone. AZ91D magnesium alloy is used as an implant material due to light-in-weight and favorable mechanical characteristics. The nontraditional machining method namely microelectrical discharge machining (μEDM) is widely used for producing intricate shapes with a tight tolerance in a variety of conductive materials. In this chapter, powder-mixed μEDM (PM-μEDM) of AZ91D magnesium alloy was conducted considering different input parameters namely (1) tool speed, (2) feed rate, (3) voltage, and (4) powder concentration. The values of capacitance (100 μF), depth

Modern Manufacturing Systems: Trends and Developments. Rajiv Kumar Garg, Ravi Pratap Singh, Rajeev Trehan, & Ramesh Singh (Eds.)

of cut (0.25 mm), and pulse on time (10 μs) were kept constant during the PM-μEDM. The tungsten carbide (WC) tool of 0.97 mm in diameter was used to perform the experiments. Zinc particles (average diameter = 80 nm) at different concentrations (2–6 g/L) were mixed in the dielectric to perform PM-μEDM. The major objective was to measure the machining time and overcut of the micro-hole. The overcut of the machined micro-hole was studied through microscopic images obtained using optical microscopy. The minimum machining time and the overcut were 04:25:18 min and 301 μm, respectively.

27.1 INTRODUCTION

The drilling of micro holes in various substrates has become important for the fabrication of the miniature product, microfluidic devices, biomedical application, printer, automobile, and heat exchanger. Microelectric discharge drilling (μEDD) is a micromanufacturing process basically used for machining complex material irrespective of its hardness. In μEDM, the initiation of spark occurs at the narrowest gap between tool and workpiece that causes the breakdown of the dielectric, which promotes the growth of plasma channel. Due to the expansion of the plasma channel, temperature rises so melting and vaporization start.[1] Nowadays, the performance of micromanufacturing process is increased by additive mixed dielectric. Generally, SiC, Zn, graphite, and aluminum nanopowder are mixed in dielectric for enhancing the performance of μEDM. μEDM is a thermo-electric machining process where there is no physical interaction between the shaped tool and workpiece. Therefore, there is no generation of machining force and mechanical stress that cause the deformation of tool and workpiece. In μEDM, the material is removed from the workpiece through a series of recurring sparks produced in the gap between the tool and workpiece immersed in the dielectric. Magnesium alloy AZ91D used in biomedical implants due to its nontoxicity, biocompatibility, and easily dissolve characteristics in the human body.[2–4] The density of magnesium alloy is in the range of 1.75–1.85 g/cm^3 which is close to the human bone, that is, 1.75 g/cm^3. Thus, magnesium alloy is quite suitable for a bone fixation device.[5] Magnesium alloy is stable only below its melting temperature that is 650°C. The machining of magnesium alloy by means of conventional machining techniques is not preferable due to the formation of built-up edges, cracks, and chatter.[6] The fine chips and dust formed during the machining of magnesium alloy are also highly flammable.

However, the latest research promoted PM-μEDM method due to its better performance than μEDM. The spark distribution is relatively more uneven during μEDM than PM-μEDM.[7] The sparking efficiency increases due to added particles in the PM-μEDM. The added particles can be mixed in the dielectric in the same or different working tanks. A stirrer and a circulating pump are used for uniform distribution of conductive particles. Figure 27.1 shows the working principle of the PM-μEDM process. The electrodes (cathode and anode) are connected to a proper pulsed DC power supply and a constant gap is maintained between them throughout the machining through the servo feed mechanism. The workpiece and tool are submerged in a dielectric or forced to submerge in dielectric through an external pump. Deionized water, kerosene, or paraffine oils are generally used as a dielectric. The insulating effect of the dielectric decreases due to the chain-like formation of conductive particles. Hence early discharge in the gap occurs due to an increase in the chance for short circuit. So shallow craters on the workpiece surfaces produced due to effective distribution of discharge energy among the dielectric, powder particles, and electrodes take place.[8] It is a very difficult task to explain the microelectric spark phenomenon as it occurs in microseconds or nanoseconds as well as due to small size of tool electrode. Mechanism of electric spark in μEDM is explained by two models namely electrothermal and electromechanical.

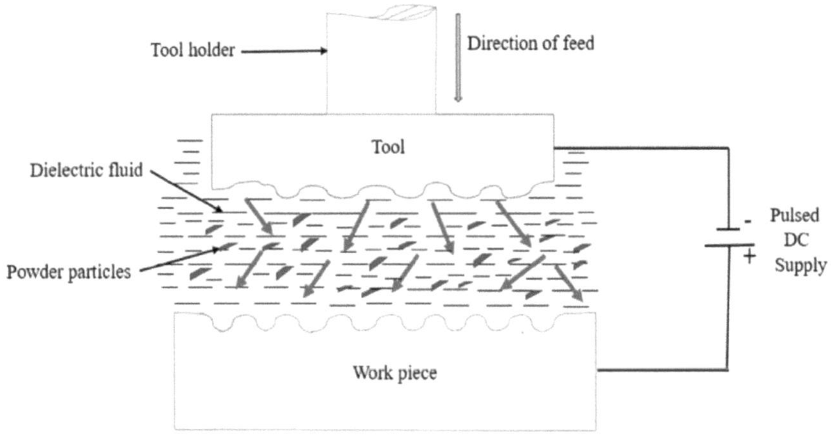

FIGURE 27.1 Principle of PM-μEDM.

Yielding is the main mechanism of erosion in the electromechanical model. Out of these two theories, the electrothermal theory is broadly accepted as it can explain maximum practical phenomenon associated with EDM/μEDM/PM-μEDM material removal.[9] It was found that the gap size increases due to the addition of conductive particles in PM-μEDM, which subsequently decreases the discharge power density. It was also seen that the added conductive particles decrease the dielectric strength that enables early and uniform discharges at low energy.[10,11] In conventional EDM, generally, two types of pulse generators are used: RC type pulse generator and transistor-type pulse generator shown in Figures 27.2 and 27.3. The fabrication of smaller parts in the range of micrometer requires minimization of pulse energy in the gap between tool and workpiece. So finishing operation in μEDM requires dozen of nanoseconds of pulse duration. The RC type pulse generator can produce minimum discharge energy by reducing the capacitance but have some demerits namely (1) very less material removal rate due to low discharge energy, (2) obtain poor surface finish due to variation of discharge energy, and (3) thermal damage on the workpiece surface due to continuous flow of current without charging the capacitor.

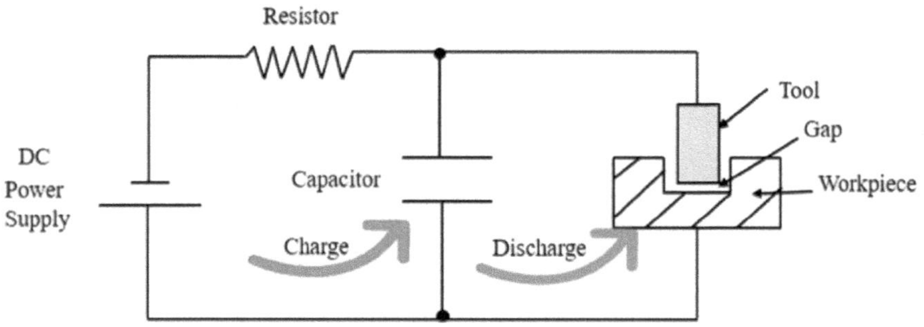

FIGURE 27.2 Relaxation-type pulse generator.

Therefore, transistor-type pulse generator is generally used in conventional μEDM. It gives high material removal rate compared to RC pulse generator due to high discharge frequency and no need of charging capacitor.[12] The pulse duration and discharge current can be changed arbitrarily according to the requirement of machining condition. Therefore, transistor-type pulse generator is mostly used in μEDM due to better removal rate at high discharge frequency. On the other hand, a uniform surface finish is obtained due to equal interval of pulse frequency. This chapter investigates the effect of different input parameters on the performance of PM-μEDM of AZ91D magnesium alloy. The PM-μEDM performance of chosen materials was evaluated in terms of machining time and overcut of the microhole.

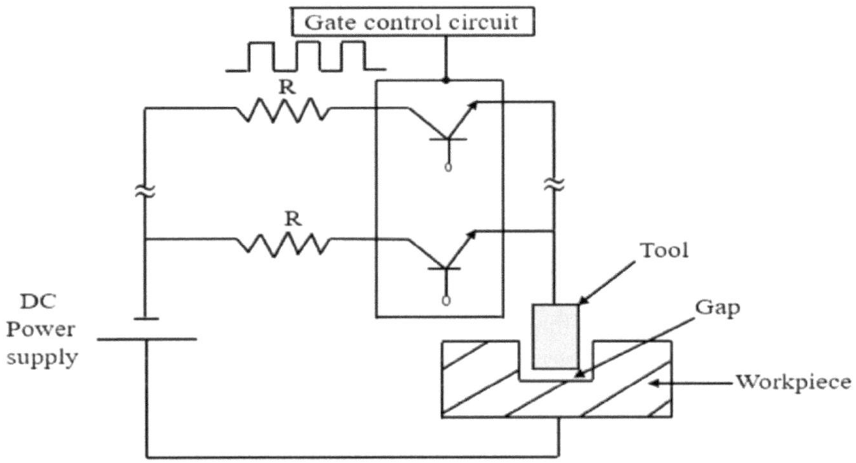

FIGURE 27.3 Transistor-type pulse generator.

27.2 LITERATURE REVIEW

The material removal in the μEDM is based on the spark erosion principle. The μEDM process requires low discharge energy and a micron-size tool. The electrodes (tool and workpiece) are separated by an interelectrode gap (IEG) and are connected to direct current power supply. The IEG is controlled using the servo mechanism. Generally, the specific energy of 10^{-7} J is required for a single discharge. The material removed in each spark is in the range of microgram (μg) as low voltage and current is used for a few microseconds or nanoseconds in μEDM. A good surface quality (roughness of around 0.1 μm) is achieved at high frequency (greater than 200 Hz) and voltage range of 40–100 V. A duty factor of 0.5 is preferred in μEDM.[13] Generally, deionized water is used as the dielectric fluid. Jahan et al.[14] experimentally investigated the performance of microelectric-discharge drilling of cemented carbide. The authors investigated the effect of several parameters namely (i) polarity, (ii) resistance, (iii) voltage, (iv) pulse duration, (v) current, (vi) pulse interval, (vii) duty ratio, (viii) speed, and (ix) EDM speed on the (a) machining time, (b) spark gap, (c) material removal rate (MRR), (d) surface finish, (e) relative electrode wear ratio (RWR), and (f) dimensional accuracy. Jahan et al.[15] further evaluated the performance of μEDM of tungsten carbide using both transistor and RC-type generators. The effect of gap voltage and capacitance on MRR and RWR was also studied. It was reported that MRR increased with the capacitance due to an increase in the discharge energy.

It was also found that deeper craters formed as the capacitance increased during machining.

Wang et al.[16] compared the stability of the plasma channel in aluminum powder-mixed kerosene and pure kerosene. It was found that the plasma channel is more stable in aluminum powder-mixed kerosene and stability increased further with the current. Tiwary et al.[17] found that there was a significant improvement in the performance of μEDM in terms of MRR, tool wear rate (TWR), taper, and overcut (OC) when copper powder was mixed in deionized water as compared to the pure deionized water and DEF-92 EDM oil. Jahan et al.[18] showed that the machining performance was improved when graphite nanopowders were mixed in the dielectric during μEDM of tungsten carbide. Prakash et al.[19] investigated the effect of cryogenic treatment of the tungsten carbide tool on the performance of μEDM of biocompatible AZ31B magnesium alloy. It was concluded that MRR increased due to high spark intensity during the machining of magnesium alloy using the cryogenically treated tool. Wang et al.[20] compared the performance of μEDM of Inconel 706 between aluminum powder-mixed EDM oil and pure EDM oil. The authors investigated the effect of several parameters namely (i) peak current (I), (ii) Voltage (V), (iii) Pulse-on time (T_{on}), and (iv) Pulse-off time (T_{off}) on the (a) surface roughness, (b) material removal rate (MRR), (c) tool removal rate (TWR), and (d) microstructure. It was found that higher MRR and better surface finish were obtained in the case of aluminum powder-mixed μEDM as compared to pure EDM oil. Prakash et al.[21] reviewed different types of dielectrics used in μEDM in place of normal hydrocarbon oil. It was reported that some of the organic compound such as ethylene glycol 400 mixed water-based dielectrics showed better result than hydrocarbon oil.

Tan et al.[22] developed a new numerical model to explain the powder mixed-dielectric micro-EDM process. It was reported that due to the presence of powder particles, there was a large expansion in plasma channel and less heat flux distributed into the workpiece. It was also found that at low powder concentration, the expansion of plasma channel was more but reduction in heat flux to the workpiece was observed at higher powder concentration. Prihandana et al.[23] found that the surface finish of micro-EDM of Inconel 718 improved due to MoS2 nanopowder (50 nm)-suspended dielectric. It was concluded that 5 g/L of powder concentration can achieve better micro-holes and maximum material removal. The effect of carbon nanofiber mixed dielectric in μEDM was reported by Liew et al.[24] They found that MRR, machinability, discharge gap, and frequency increase but TWR and Taper angle (TA) decrease due to addition of powder-mixed dielectric. The effect of SiC concentration in the dielectric and microelectric discharge energy of Ti-6Al-4V using WC tool was presented by Ali et al.[25] SiC concentration and electric discharge energy were taken as input variables. Surface roughness was reduced by adding SiC upto 20 g/L concentration in dielectric. From ANOVA result, powder concentration was the most influential parameter for surface roughness, whereas electric discharge energy mostly affects MRR and TWR.

Yeo et al.[26] studied the effect of powder-mixed dielectric on crater characteristics for micro-EDM using single RC discharge experiments at different low discharge energies at 2.5, 5, and 25 μJ. The formed craters having small diameter and depth were more consistent and in circular shape due to presence of powder particles. The machinability of titanium alloy on micro-EDM at different electrolyte namely kerosene, aluminum mixed kerosene, and SiC mixed kerosene was investigated by Chow et al.[27] The material removal depth, discharge gap, and surface roughness were improved by adding both powders. Material removal depth was more by adding SiC powder whereas an improvement in surface roughness by adding aluminum powder. The largest gap between tool and workpiece was found by mixing aluminum powder in dielectric. Whenever the adherence of carbon nuclides attached to the surface of electrode disturbed by adding powder particles in dielectric, the TWR increased. In this paper, machining time and overcut of the microhole were evaluated during PM-μEDM AZ91D magnesium alloy.

27.3 METHODOLOGY

Generally, conventional design experimental procedures are more complex so usually it is not adopted in experiment. It becomes laborious and more rigid with an increase in process parameters. The Taguchi method solves these problems by adopting an orthogonal array method. The study of all process parameters can be done with a minimum number of experiments.[28,29] The Taguchi method (orthogonal array) is mostly useful in engineering analysis. It consists of a plan of experiments with the objective of collection of data in a controlled way for getting information about the behavior of the given process. The important advantages of this method are to reduce effort for conducting experiment, reduce time for conducting experiment, and finding significant factor in a faster way.[30] In this work, magnesium alloy AZ91D was used as the workpiece and tungsten carbide was used as the tool. The melting point of tungsten carbide is 2870°C and have high value of toughness and hardness so it is mostly used as a tool material in μEDM process. Tables 27.1 and 27.2 show the mechanical properties of the tungsten carbide and magnesium alloy AZ91D.

TABLE 27.1 Mechanical Properties of Tungsten Carbide (WC).

Property	Unit	Value
Density	gm/cm	15.8
Melting point	°C	2870
Thermal conductivity	W/m/K	84.2
Specific heat capacity	J/g/K/	0.044
Electrical resistivity	Ohm mm^2/m	0.2
Modulus of elasticity	GPa	700
Hardness	HRC	70

TABLE 27.2 Mechanical Properties of Magnesium Alloy AZ91D.

Property	Unit	Value
Tensile strength	MPa	230
Yield strength	MPa	150
Elastic modulus	GPa	44.8
Shear modulus	GPa	17
Poisson's ratio	–	0.35
Hardness, Brinell	HRC	63

The cutting depth was kept constant to 0.25 mm during the experiments. Figure 27.4 shows a photographic view of the experimental setup. μEDM experiments were conducted at three levels of input parameters namely (1) tool speed, (2) feed rate, (3) voltage, and (4) powder concentration, as shown in Table 27.3. Each experimental run was replicated at least three times to avoid the erroneous result. The optimum input parameters for obtaining minimum machining time were obtained using the Taguchi approach. The PM-μEDM experiments were carried out using a machining tank, which was equipped with a stirrer and a pump. Zinc particles (average diameter = 80 nm) at different concentrations (2, 4, and 6 g/L) were mixed in the dielectric. It is generally seen from literature that efficiency of PM-μEDM process increases by increasing powder concentration upto a certain limit after that machining efficiency decreases by increasing the powder concentration. Zinc particles were used due to its conductive behavior. Zinc is also one of the major elements of the AZ91D. The performance of the PM-μEDM of chosen material was evaluated in terms of machining time and overcut of the microhole.

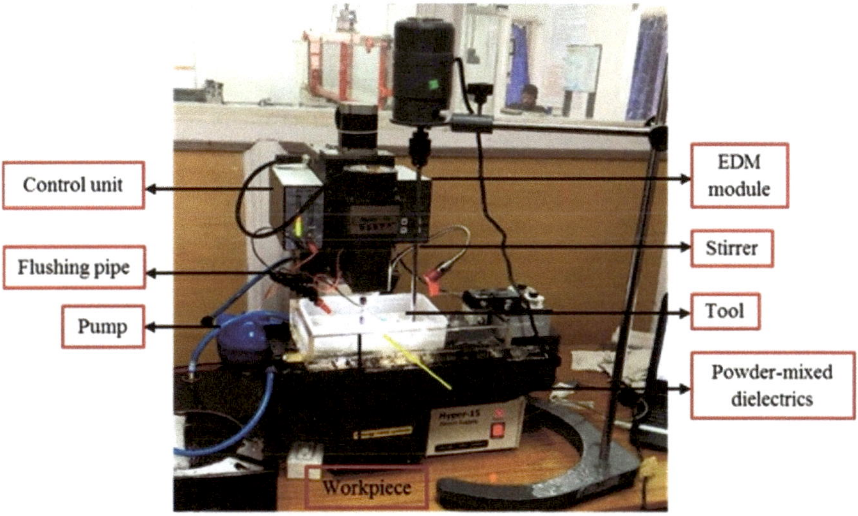

FIGURE 27.4 Photograph of the experimental setup.

TABLE 27.3 Variable Process Parameters.

Parameter	Level 1	Level 2	Level 3
Speed	300 rpm	350 rpm	400 rpm
Feed rate	2 μm/s	3 μm/s	4 μm/s
Voltage	100 V	130 V	160 V
Powder concentration	2 g/L	4 g/L	6 g/L

27.4 RESULTS AND DISCUSSION

In μEDM, the generation of an electrical spark is a critical phenomenon as it occurs for a fraction of microseconds or nanoseconds to remove the material. The electrothermal and electromechanical theories were proposed to explain the principle of material removal during EDM. The electrothermal theory proposed that material removed due to localized melting and vaporization of the material as a superheated plasma channel established in the IEG during machining. The electromechanical theory explained that yielding is the main mechanism of material removal during EDM. All the electrons coming from the tool (cathode) accelerated toward the workpiece (anode) and ions accelerated toward the cathode. Thereby, a plasma channel is established in the gap between the cathode and anode. The bombardment of the plasma channel due to the development of high temperature and pressure results in the removal of material from the work surface. A comparison of spark phenomenon in μEDM and PMμEDM is shown in Figure 27.5.

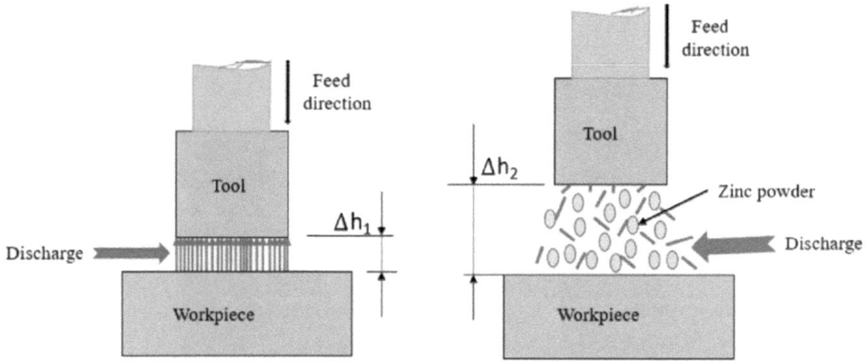

FIGURE 27.5 Comparison of sparking phenomena in μEDM and PMμEDM.

From Figure 27.5, it is clear that when powder particles are mixed in dielectric, the additional gap between tool and workpiece increases and primary sparking takes place between tool and zinc particles instead of tool and workpiece. These conductive or semiconductive energized powder particles decrease the effective gap (gap between tool and powder particles) and reduces the breakdown voltage under the influence of electric field established between IEG. So early discharges of capacitance take place before fully charge and secondary discharge may take place between powder particles and workpiece hence total effective gap increases as compared to pure dielectric. The value of IEG mainly depends on the type of material and electrical properties of powder particles.[9]

The results of machining time obtained under different experimental conditions are shown in Table 27.4. The maximum machining time of 12:20:22 min was obtained at tool speed of 300 rpm, feed rate of 2 μm/s, voltage of 100 V, and powder concentration of 2 g/L. The minimum machining time of 04:25:18 min was obtained at the input condition of 300 rpm, 4 μm/s, 160 V, and 6 g/L. From the analysis, it was found that the powder concentration has a significant effect on the machining time.

TABLE 27.4 Machining Time Obtained during PM-μEDM of AZ91D Magnesium Alloy.

Experiment no.	Speed (rpm)	Feed rate (μm/s)	Voltage (V)	Powder concentration (g/l)	Machining time (minutes)
1	300	2	100	2	12:20:22
2	300	3	130	4	06:12:12
3	300	4	160	6	04:25:18
4	350	2	130	6	06:13:68
5	350	3	160	2	04:47:73
6	350	4	100	4	09:09:88
7	400	2	160	4	04:53:13
8	400	3	100	6	09:34:06
9	400	4	130	2	07:13:91

Figure 27.6 shows that machining time decreases with an increase in powder concentration from 2 to 6 g/L. The increasing number of zinc particles in dielectric resulted in more conductivity of the dielectric. Thus, machining time decreased or material removal increased with powder concentration. When pure dielectric with small flow rate is used in narrow gap in μEDM process, there is a chance of adhesion of debris in the IEG thereby bridging and short-circuit takes place. If short-circuiting is detected by the servo feed mechanism, tool direction reversed, and searches for new optimum IEG. Hence machining time increases and machining efficiency decreases. Prihandana et al.[31] presented a study about suspension of graphite powder in the dielectric. It was found that when the concentration of powder particles increases from 0 to 10%, the machining time reduce by 14–20% in a faster way. It was also reported that at 10 g/L powder concentration, a proper bridging in the IEG takes place due to uniform pulse frequency.

Figure 27.7 shows the effect of feed rate on machining time. It was observed that a higher feed rate resulted in more material removal due to the decrease in the machining time. Figure 27.8 indicates that machining time decreased with an increase in the voltage from 100 to 160 V. The spark intensity increases with the voltage which resulted in more material removal.

Figure 27.7 indicates that when feed rate increases from 2 to 3 μm/s, the machining time rapidly decreases but when feed rate increases from 3 to 4 μm/s, the machining time decreases in slower way.

Gap voltage indicates the spark gap and intensity of dielectric strength. When the value of voltage increases, the dielectric strength reduces so early discharge taking place and machining time reduces. From Figure 27.8, it is clear that when the value of voltage increases from 100 to 120 V, the machining time decreases rapidly. With further increase in voltage from 120 to 160 V, the machining time further decreases but the decrease rate is lower than previous.

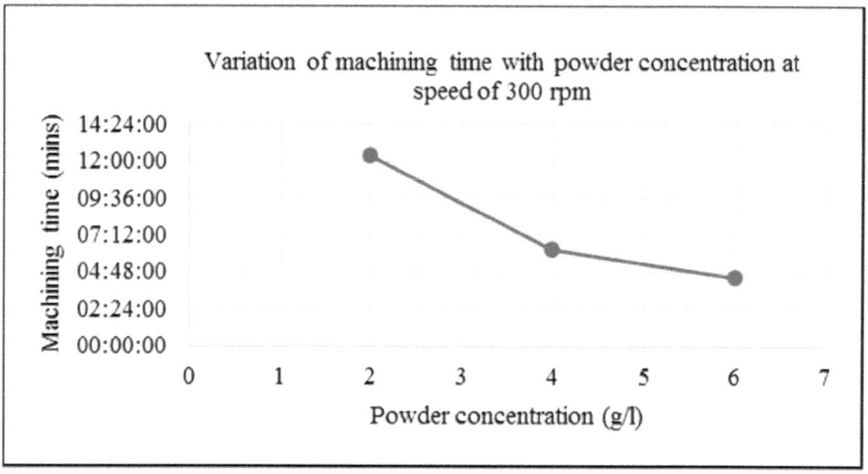

FIGURE 27.6 Variation of machining time with powder concentration.

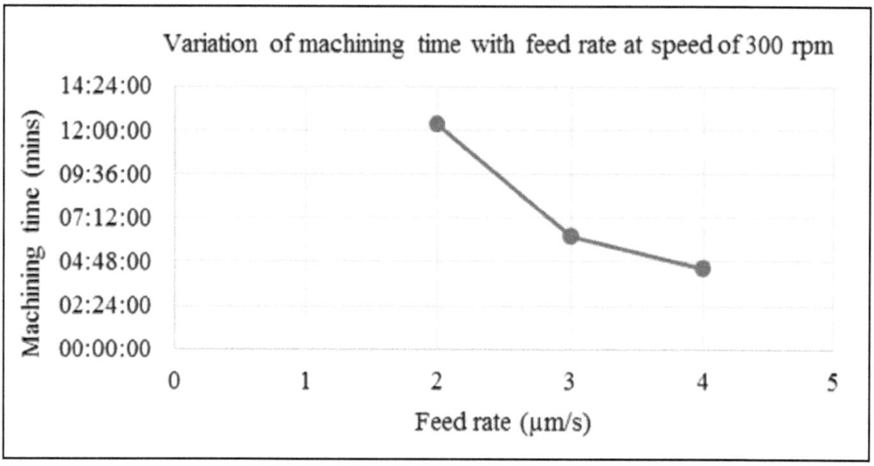

FIGURE 27.7 Variation of machining time with the feed rate.

From the experimental investigation, it was observed that the diameter of the machined microhole is greater than the diameter of the microtool. The difference in the diameter of the microhole and the diameter of the microtool is known as overcut (OC) of the microhole. The overcut of the microhole can be expressed as follows:

$$OC = (D_I - D) \tag{27.1}$$

where, D_I = diameter of the microhole and D = diameter of the microtool, respectively.

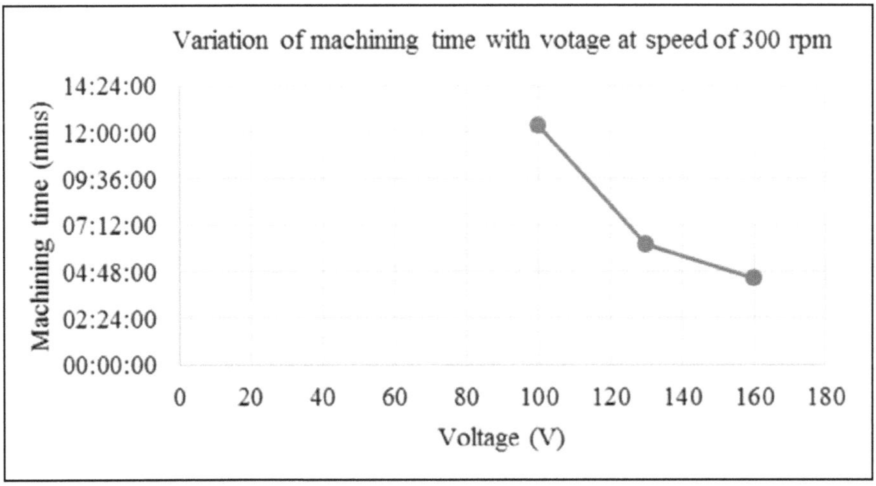

FIGURE 27.8 Variation of machining time with voltage.

The numerical value of overcut of the microhole for all nine experiments is shown in Table 27.5. It was observed that the maximum value of overcut (0.326 mm) was obtained at tool speed of 300 rpm, feed rate of 2 μm/s, voltage of 100 V, and powder concentration of 2 g/L. Similarly, the minimum value of overcut (0.301 mm) of the microhole was obtained at 400 rpm, 2 μm/s, 160 V, and 4 g/L. The optical microscopic images of nine machined microhole are shown in Figure 27.9. Some white spots are seen in some microscopic images, it indicates some powder particles absorbed on the machined surface.

Figure 27.9a machined microhole took maximum machining time due to less powder concentration (2 g/L) while Figure 27.9c machined microhole took minimum machining time due to high powder concentration (6 g/L). Figure 27.9a have maximum overcut (OC) due to long machining time and Figure 27.9g have minimum overcut due to proper machining.

Figure 27.10 shows the SEM image of the zinc powder-mixed μEDM machined surface.

TABLE 27.5 The Numerical Value of Overcut of Nine Experiments.

Experiment no.	D_I (mm)	D (mm)	OC (μm)
1	1.296	0.97	326
2	1.279	0.97	309
3	1.285	0.97	315
4	1.282	0.97	312
5	1.284	0.97	314
6	1.271	0.97	305
7	1.273	0.97	301
8	1.278	0.97	303
9	1.275	0.97	308

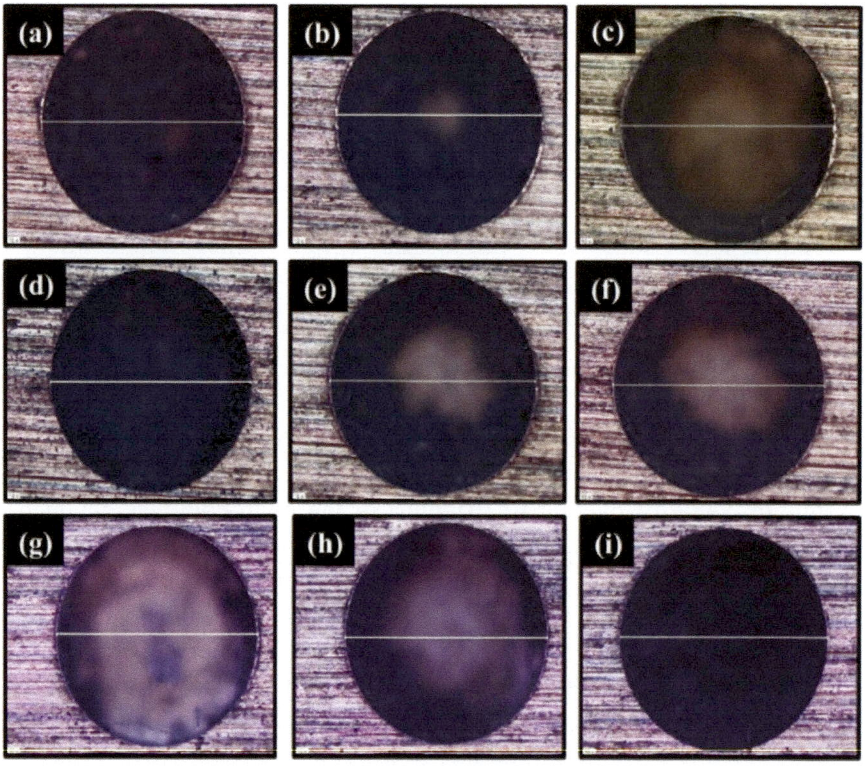

FIGURE 27.9 Optical microscopic images of machined microhole.

It is earlier discussed that by increasing the concentration of powder particles, more number of powder particles present in the IEG. So primary discharge takes place between tool and powder particles and secondary discharge takes place between powder particles and workpiece. Therefore, effective discharge energy decreases and surface roughness improves hence uniform craters are formed on the machined surface in place of high intensive shallow craters in pure dielectric.

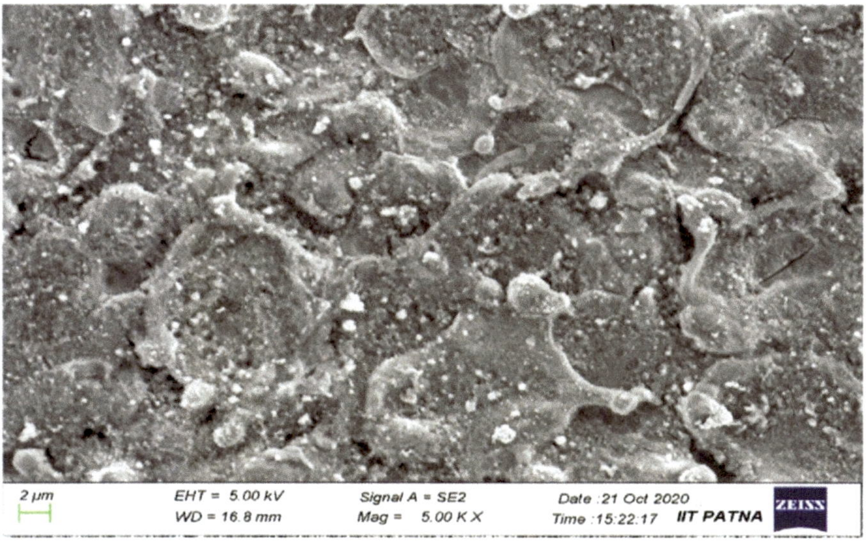

FIGURE 27.10 SEM image of the zinc PMµED machined surface.

27.5 CONCLUSIONS

In this paper, PM-µEDM experiments were conducted on AZ91D magnesium alloy at three different levels of input parameters (tool speed, feed rate, voltage, and powder concentration). It was found that the minimum machining time of 04:25:18 min was obtained at input parameter condition of tool speed of 300 rpm, feed rate of 4 µm/s, voltage of 160 V, and powder concentration 6 g/L. The maximum machining time of 12:20:22 min was obtained at 300 rpm, 2 µm/s, 100 V, and 2 g/L. It was also found that the machining time was reduced with an increase in the zinc particles in the dielectric. Similarly, the machining time was found to be reduced with the feed rate and voltage. The minimum overcut of 0.301 mm was obtained at 400 rpm, 2 µm/s, 160, and 4 g/L. In this study, principle of powder-mixed micro EDM in detail was studied. RC type pulse generator and transistor type pulse generator were studied and it was concluded that transistor type pulse generator produced uniform pulse frequency for the fabrication of microproducts in the range of nanometer. The effect of powder particles in dielectric in terms of IEG was studied. It was concluded that the IEG increases by mixing powder.

KEYWORDS

- **magnesium alloy**
- **electrical-discharge machining**
- **powder-mixed**
- **zinc particles**
- **machining time**
- **overcut**

REFERENCES

1. Cyril, M. P. J.; Asokan, D. P.; Jerald, D. J.; Kumar, M. S.; Kanagaraj, D. G. Experimental Investigation on Performance of Additive Mixed Dielectric during Micro-Electric Discharge Drilling on 316L Stainless Steel. *Mater. Manuf. Process.* 2016. http://dx.doi.org/10.1080/10426914.2016.1221107
2. Cha, P. R.; Han, H. S.; Yang, G. F.; Kim, Y. C.; Hong, K. H.; Lee, S. C.; Jung, J. Y.; Ahn, J. P.; Kim, Y. Y.; Cho, S. Y.; Byun, J. Y.; Lee, K. S.; Yang, S. J.; Seok, H. K. Biodegradability Engineering of Biodegradable Mg Alloys: Tailoring the Electrochemical Properties and Microstructure of Constituent Phases. *Sci. Rep.* **2013,** *3* (1), 1–6.
3. Klocke, F.; Schwade, M.; Klink, A.; Kopp, A. EDM Machining Capabilities of Magnesium (Mg) Alloy WE43 for Medical Applications. *Procedia Eng.* **2011,** *19*, 190–195.
4. Andani, M. T.; Moghaddam, N. S.; Haberland, C.; Dean, D.; Miller, M. J.; Elahinia, M. Metals for Bone Implants. Part 1. Powder Metallurgy and Implant Rendering. *Acta Biomaterialia* **2014,** *10* (10), 4058–4070.
5. Li, L.; Gao, J.; Wang, Y. Evaluation of Cyto-Toxicity and Corrosion Behavior of Alkali-Heat-Treated Magnesium in Simulated Body Fluid. *Surface Coatings Technol.* **2004,** *185* (1), 92–98.
6. Razak, M. A.; Abdul-Rani, A. M.; Rao, T. V. V. L. N.; Pedapati, S. R.; Kamal, S. Electrical Discharge Machining on Biodegradable AZ31 Magnesium Alloy Using Taguchi Method. *Procedia Eng.* **2016,** *148*, 916–922.
7. Singh, A.; Singh, R. Effect of Powder Mixed Electric Discharge Machining (PMEDM) on Various Materials with Different Powders: A Review. *Int. J. Innov. Res. Sci. Technol.* **2015,** *2* (3), 164–169.

8. Kuriachen, B.; Mathew, J. Effect of Powder Mixed Dielectric on Material Removal and Surface Modification in Microelectric Discharge Machining of Ti-6Al-4V. *Mater. Manuf. Process.* **2016,** *31* (4), 439–446.
9. Kuriachen, B. Powder-mixed Microelectric Discharge Machining. In *Micro-electrical Discharge Machining Processes, Materials Forming, Machining and Tribology*, 2019. https://doi.org/10.1007/978-981-13-3074-2_7
10. Kansal, H. K.; Singh, S.; Kumar, P. Technology and Research Developments in Powder Mixed Electric Discharge Machining (PMEDM). *J. Mater. Process. Technol.* **2007,** *184* (1–3), 32–41.
11. Mohal, S.; Kumar, H.; Kansal, S. K. Nano-finishing of Materials by Powder Mixed Electric Discharge Machining (PMEDM): A Review. *Sci. Adv. Mater.* **2015,** *7* (10), 2234–2255.
12. Han, F.; Wachi, S.; Kunieda, M. Improvement of Machining Characteristics of Micro-EDM Using Transistor Type Isopulse Generator and Servo Feed Control. *Precis. Eng.* **2004,** *28*, 378–385.
13. Klocke, F.; Lung, D.; Antonoglou, G.; Thomaidis, D. The Effects of Powder Suspended Dielectrics on the Thermal Influenced Zone by Electrodischarge Machining with Small Discharge Energies. *J. Mater. Process. Technol.* **2004,** *149* (1–3), 191–197.
14. Jahan, M. P.; Wong, Y. S.; Rahman, M. Experimental Investigations Into the Influence of Major Operating Parameters during Micro-Electro Discharge Drilling of Cemented Carbide. *Mach. Sci. Technol.* **2012,** *16* (1), 131–156.
15. Jahan, M. P.; Wong, Y. S.; Rahman, M. A Study on the Quality Micro-Hole Machining of Tungsten Carbide by Micro-EDM Process Using Transistor and RC-type Pulse Generator. *J. Mater. Process. Technol.* **2009,** *209* (4), 1706–1716.
16. Wang, X.; Liu, Y.; Zhang, Y.; Sun, Q.; Li, Z.; Shen, Y. Characteristics of Plasma Channel in Powder-Mixed EDM Based on Monopulse Discharge. *Int. J. Adv. Manuf. Technol.* **2016,** *82* (5–8), 1063–1069.
17. Tiwary, A. P.; Pradhan, B. B.; Bhattacharyya, B. Investigation on the Effect of Dielectrics during Micro-Electro-Discharge Machining of Ti-6Al-4V. *Int. J. Adv. Manuf. Technol.* **2018,** *95* (1–4), 861–874.
18. Jahan, M. P.; Rahman, M.; Wong, Y. S. Study on the Nano-Powder-Mixed Sinking and Milling Micro-EDM of WC-Co. *Int. J. Adv. Manuf. Technol.* **2011,** *53* (1–4), 167–180.
19. Prakash, D.; Tariq, M.; Davis, R.; Singh, A.; Debnath, K. Influence of Cryogenic Treatment on the Performance of Micro-EDM Tool Electrode in Machining of Magnesium Alloy AZ31B. *Mater. Today: Proc.* **2020**. DOI: 10.1016/j.matpr.2020.03.589
20. Wang, C.; Qiang, Z. Comparison of Micro-EDM Characteristics of Inconel 706 between EDM Oil and an Al Powder-Mixed Dielectric. *Adv. Mater. Sci. Eng.* **2019**. https://doi.org/10.1155/2019/5625360
21. Prakash, V.; Das, A. K.; Chattopadhyay, S. Micro-EDM Performance Using Different Dielectrics. *Micro-electrical Discharge Machining Processes, Materials Forming, Machining and Tribology*, 2019. https://doi.org/10.1007/978-981-13-3074-2_6
22. Tan, P. C.; Yeo, S. H. Simulation of Surface Integrity for Nanopowder-Mixed Dielectric in Micro Electrical Discharge Machining. *The Minerals, Metals & Materials Society and ASM International*, 2013. DOI: 10.1007/s11663-013-9819-7
23. Prihandana, G. S.; Sriani, T.; Mahardika, M.; Hamdi, M.; Miki, N.; Wong, Y. S.; Mitsui, K. Application of Powder Suspended in Dielectric Fluid for Fine Finish Micro-EDM of Inconel 718. *Int. J. Adv. Manuf. Technol.* **2014,** *75* (1–4), 599–613.

24. Liew, P. J.; Yan, J.; Kuriyagawa, T. Carbon Nanofiber Assisted Micro Electro Discharge Machining of Reaction-Bonded Silicon Carbide. *J. Mater. Process. Technol.* **2013,** *213* (7), 1076–1087. DOI: 10.1016/j.jmatprotec.2013.02.004.
25. Ali, M. Y.; Atiqah, N.; Erniyati. Silicon Carbide Powder Mixed Micro Electro Discharge Milling of Titanium Alloy. *Int. J. Mech. Mater. Eng.* **2011,** *6* (3), 338–342.
26. Yeo, S. H.; Tan, P. C.; Kurnia, W. Effects of Powder Additives Suspended in Dielectric on Crater Characteristics for Micro Electrical Discharge Machining. *J. Micromech. Microeng.* **2007,** *17*, N91.
27. Chow, H. M.; Yang, L. D.; Lin, C. T.; Chen, Y. F. The Use of SiC Powder in Water as Dielectric for Micro-Slit EDM Machining. *J. Mater. Process. Technol.* **2008,** *195*, 160–170.
28. Aravindan, S.; Naveen, S. A.; Noorul, H. A. A Machinability Study of GFRP Pipes Using Statistical Techniques. *Int. J. Adv. Manuf. Technol.* **2008,** *37* (11–12), 1069.
29. Montgomery, D. C. *Design and Analysis of Experiments*, 5th edn.; Wiley: New York, 2006.
30. Ponappa, K.; Aravindan, S.; Rao, P. V.; Ramkumar, J.; Gupta, M. The Effect of Process Parameters on Machining of Magnesium Nano Alumina Composites through EDM. *Int. J. Adv. Manuf. Technol.* **2009,** *46*, 1035–1042.
31. Prihandana, G. S.; Mahardika, M.; Hamdi, M.; Wong, Y. S.; Miki, N.; Mitsui, K. Study of Workpiece Vibration in Powder-Suspended Dielectric Fluid in Micro-EDM Processes. *Int. J. Precis. Eng. Manuf.* **2013,** *14* (10), 1817–1822.

CHAPTER 28

ASSESSING THE EFFECT OF NONELECTRICAL PROCESS PARAMETERS DURING THE STED OF HOLES ON INCONEL 718 ALLOY

ANUJ VATS*, RAMVER, AKSHAY DVIVEDI, and PRADEEP KUMAR

Department of Mechanical and Industrial Engineering, IIT Roorkee, Roorkee, India

**Corresponding author. E-mail: vatsanuj26@gmail.com*

ABSTRACT

Drilling of high aspect ratio holes on Inconel 718 is difficult owing to its high hot hardness and corrosion resistance, and low thermal conductivity. Shaped tube electrolytic drilling (STED) is a variant of the electrochemical machining process, wherein a tubular tool is used as the cathode supplying electrolyte through it to the machining zone. Given the nature of underlying material removal phenomenon, the performance of the STED process is influenced by a wide range of process parameters. These parameters can broadly be categorized into two categories: the electrical and nonelectrical parameters. This research work aims to assess the effect of nonelectrical process parameters like electrolyte type, tool insulation conditions, and tool feed rate on the performance characteristics such as average diametral overcut and material removal rate. The experimental results revealed that the material removal rate (MRR) was maximum in the case of aqueous NaCl solution, while the average diametral overcut (ADOC) was minimum in the case of aqueous $NaNO_3$ solution. The STED in the mixed NaCl + $NaNO_3$

Modern Manufacturing Systems: Trends and Developments. Rajiv Kumar Garg, Ravi Pratap Singh, Rajeev Trehan, & Ramesh Singh (Eds.)

solution resulted higher MRR and lower ADOC compared to the constituents aqueous $NaNO_3$ and aqueous NaCl solutions, respectively. A tool electrode, fully insulated around side face, rendered lower and uniform ADOC than that with the tool electrode having an exposed length of 1 mm, for all types of electrolyte solutions. Moreover, MRR increased with an increase in the tool feed rate, irrespective of the type of electrolyte solutions and tool insulation conditions. The FESEM micrographs of the hole surfaces used to study the surface morphology also highlighted the benefits of using fully insulated tool.

28.1 INTRODUCTION

Nickel-based superalloys have superior properties, such as excellent specific strength, resistance to corrosion, and the capability to sustain loading at elevated temperatures. Rahman et al.[1] stated that these desired characteristics make the Inconel alloys a default choice for material of construction in various engineering applications, for example, the rear casing of gas turbines. However, such specific combinations of engineering properties make the machining of high aspect ratio (AR $\geq$ 5) features using conventional techniques a challenging task.[2] Often nonconventional machining (NCM) techniques, such as electric discharge machining (EDM) and electrochemical machining (ECM), have been employed for machining of such features.[3] The ECM is one of the NCM techniques especially designed to drill high AR holes on materials possessing good electrical conductivity regardless of their thermomechanical characteristics, for example, thermal conductivity, hardness, and toughness.[4] The shaped tube electrolytic drilling (STED) is a process variant of the ECM, wherein a tubular shape tool electrode is used instead of a solid cylindrical tool. The use of a tubular tool electrode facilitates effective replenishment of fresh electrolyte into and removal of the ECM by-products from the machining zone of slender holes. The machined surface does not possess thermal damages as the material removal takes place through electrochemical reactions governed by Faraday's law.[5]

Given the nature of the underlying material removal phenomenon, the performance of the STED process is influenced by a number of factors. These factors can be broadly categorized into two groups, that is, the electrical and nonelectrical parameters. The process parameters that control the characteristics of the electrical signal applied across the electrodes are termed as electrical parameters. Applied voltage, pulse-on/off durations, limiting pulse current, and type of pulse waveform are examples of electrical parameters. The nonelectrical parameters that influence the process performance are the characteristics of the electrolytic solution, the relative motion of tool and work material, and the tool conditioning. The characteristics of electrolytic solution depend upon the type and concentration of the solute and solvent used.

Sharma et al.[6] presented experimental investigations into STED of Inconel using acidified NaCl solution, and recommended the use of low voltage and high feed rate to obtain enhanced quality of the machined holes. Bilgi et al.[7] experimentally investigated the effect of voltage, bare tip length, electrolyte concentration, and feed rate while performing STED with macro tool. Rathod et al.[8] investigated the enhancement in the machining accuracy of the electrochemical drilling process by insulating the sides of the tool electrode. They reported that about 33.78% decrease in the radial hole overcut can be obtained while drilling microholes on sheets of stainless steel with the insulated tool. Ali et al.[9] investigated the effect of feed rate, metal ion concentration, voltage while using nitric acid on the hole size obtained in STED. Wang et al.[10] investigated the effect of electrolyte flow conditions on achieving higher feed rates while drilling Inconel with $NaNO_3$. De Silva et al.[11] assessed the environmental impact of using acidic electrolyte during STED. Sathiyanarayanan et al.[12] assessed the effect of feed rate and electrolyte concentration on the Material Removal Rate (MRR) and overcut. Geethapriyan et al.[13] noted the effect of microtool feed rate and type of electrolytes on process outputs.

The afore-cited literature on the STED process makes it clear that these existing investigations recognize the significance of nonelectrical process parameters during the drilling of high AR holes, however, a dedicated experimental investigation assessing their influence on the quality characteristics of the drilled holes is still not present. Therefore, the current investigation attempts to assess the effect of nonelectrical process parameters, such as the type of electrolyte solution used, tool feed rate, and tool insulations conditions, on the response outputs such as the MRR and average diametral overcut (ADOC) during the STED of high AR holes on Inconel 718 alloy.

28.2 MATERIALS AND METHODS

Figure 28.1 illustrates a graphic representation of the in-house developed STED facility used to perform the current experimental investigation. The machining facility consisted of three major systems, namely tool-work relative motion control system, electrolyte circulation system, and DC power supply unit. The relative positioning of a base table with respect to the tool electrode in the horizontal plane was realized through two linear axes (i.e., x- and y-axes) drive. A machining tank along with a fixture to hold work material was attached to the base table. The relative positioning of the tool electrode with respect to the base table in the vertical plane was realized through one linear axis (i.e., z-axis) drive. A tool holder was fixed to the z-axis drive, and the electrical connection was made to the tool holder to minimize tool misalignment during mounting of the tools for the subsequent experiments. The electrolyte was circulated into the machining zone through the tubular tool at a constant flow rate using a metering pump.

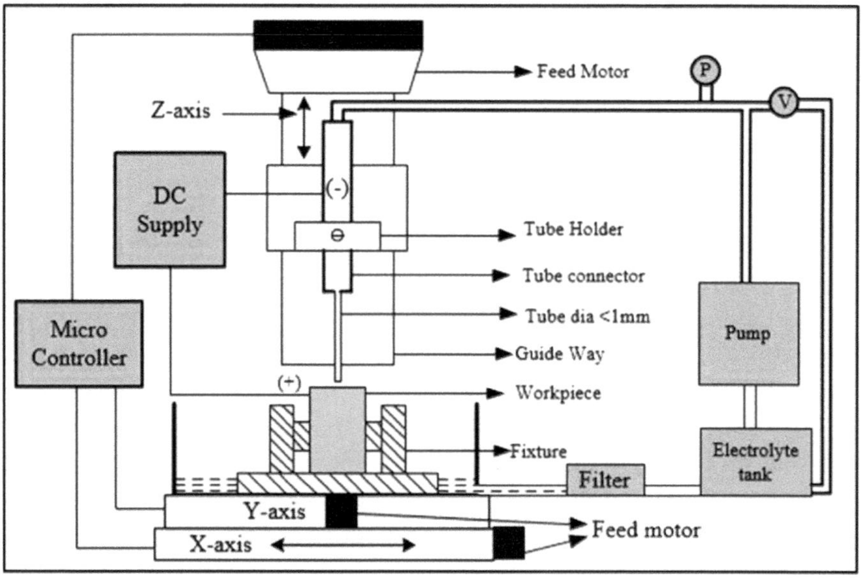

FIGURE 28.1 Schematic of electrochemical drilling facility.

Stainless steel was selected as a tool material based on the existing literature due to its ability to sustain in aggressive electrolytic solutions frequently used in the ECM process.[10,14] A tubular tool having an outer and inner diameter of 0.72 and 0.415 mm, respectively, was chosen to drill high AR holes on the 8-mm thick Inconel 718 work samples. Two sets of tool electrodes having different exposed lengths (ETL) were prepared. The first set of tools was completely insulated at the periphery, that is, having zero ETL. Whereas the second set was having 1 mm of the ETL, and the remaining length of the tool was insulated. The ETL was achieved by covering the front end of the tip during coating, and subsequently, it was cross-examined using the stereo zoom microscope (Nikon SMZ745 T). For both sets, the thickness of insulating coating was approximately 60 μm. To protect the insulation coating of the tool electrode from scratching with the sidewalls of the machined hole, a stand-off or initial interelectrode gap (IEG) of 150 μm was maintained. Drilling a hole using the STED is illustrated schematically in Figure 28.2. Figure 28.2a shows an initial position of the tubular tool above the work surface and just prior to start the machining action. As the tool approaches the surface of work material, the IEG shrinks and the electric field intensity in the electrolyte in the machining zone starts increasing. The corresponding electrochemical reactions start evolving gases increasingly at the surfaces of both the electrode, and consequently, the work material starts dissolving as per the Faraday's law of electrolysis. The continuous evacuation of the reaction by-products from the IEG and the tool electrode feed continues the material dissolution beneath the tool electrode essential for the drilling operation to proceed (Figure 28.2b–c).

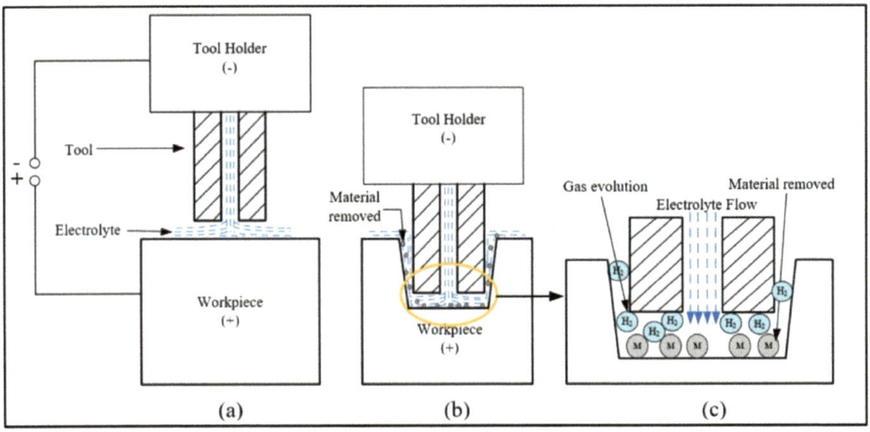

FIGURE 28.2 Electrochemical drilling progression (a) initial IEG at beginning of the drilling process (b) equilibrium IEG during the process (c) an enlarged view of IEG illustrating phenomena material removal and gas evolution in the machining zone.

The performance of ECM processes strongly depends upon the selection of electrolyte solution.[15] Most of the investigations reported on drilling of high AR holes on Inconel 718 used either concentrated acidic or acidified electrolyte solutions.[7,11] However, the use of neutral electrolytes is gaining popularity due to their environment-friendly characteristics.[16] Therefore, freshly prepared aqueous solutions of NaCl, $NaNO_3$, and a mixture of $NaNO_3$ and NaCl were chosen as electrolytes for the present investigation. Ali et al. reported that the tool feed rate influenced the hole's size and taper significantly.[9] Thus, the tool feed rate was chosen as a parameter of investigation. During the preliminary trials, the feed rate was varied to obtain its feasible range for a constant value of applied voltage (12 V). Initially, a higher value of feed rate (1.4 mm/min) was tested, which resulted in a frequent short-circuiting and retraction of the tool electrode in the first minute of the drilling. Frequent short-circuiting deteriorates the integrity of machined surface as well as damages the insulation of the tool electrode. Moreover, frequent tool retraction diminishes the machining rate. Subsequently, a lower value of feed rate (1.2 mm/min) was explored, which exhibited the few instances of short-circuiting and frequent tool retractions toward the latter half of the drilling depth. At feed rate of 1 mm/min, the hole was drilled successfully with a few instances of tool retractions. Thus, the tool was fed at maximum 1 mm/min for the selected concentration of electrolytes and tool diameter. The experimental parameters enlisted in Table 28.1 were selected for the experiments.

TABLE 28.1 Experimental Parameters Used during Electrochemical Drilling.

Input Parameters	Parameter values
Variable parameters	
Electrolytes	100 g/L $NaNO_3$, 100 g/L NaCl, 50 g/L $NaNO_3$ + 50 g/L NaCl
Feed rate	0.6, 0.8, 1 mm/min
Exposed tool length (ETL)	0 mm (side face completely insulated), 1 mm
Constant parameters	
Voltage	12 V
Tool material	Stainless steel
Tool diameter outer and inner	0.72, 0.415 mm
Initial interelectrode gap (IEG)	0.15 mm
Electrolyte flow rate	120 mL/min

The material removed from the machined sample was obtained by weighing the work sample before and after each drilling operation with the help of a precision digital weighing balance (Shimadzu: AUW220D; least count: 0.01 mg). A digital stopwatch was used to find out the drilling duration for completing each hole. The values of MRR were obtained as per eq 28.1. The machined samples were cleaned thoroughly in acetone before measuring the diameter of machined holes using the microscope (Nikon:

SMZ745 T). The diameter of holes was measured at five different depths, including the entry and exit plane of the hole and three equally distanced planes revealed after grinding the work samples to the corresponding depths as shown in Figure 28.3. Subsequently, these values of diameters were used to calculate the values of depth-averaged diameter of the hole (D_{dav}) using eq 28.2. Further, the ADOC was calculated by subtracting the outer diameter of the tool electrode from the D_{dav} as per eq 28.3.

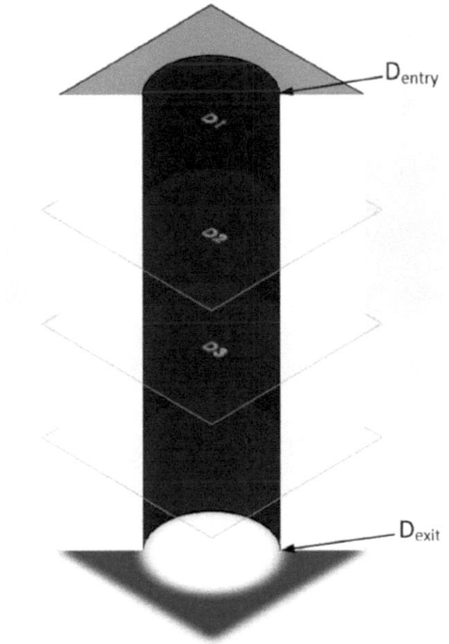

FIGURE 28.3 Measurement scheme to obtain depth averaged diameter of the hole fabricated through STED.

$$\text{Material removal rate}\left(\text{MRR}\right) = \frac{\text{Initial weight}\left(\text{mg}\right) - \text{Final weight}\left(\text{mg}\right)}{\text{Machining time}\left(\text{minutes}\right)} \quad (28.1)$$

$$\text{Depth averaged diameter}\left(D_{\text{dav}}\right) = \frac{d_{\text{entry}} + d_1 + d_2 + d_3 + d_{\text{exit}}}{5} \quad (28.2)$$

$$\text{Average diametral overcut (ADOC)} = (D_{\text{dav}} - \text{Outer diameter of the tool}) \quad (28.3)$$

Each experiment was conducted thrice for the same parametric settings and their average values were considered in the analysis (refer to Table 28.2).

TABLE 28.2 Design of Experiment Table along with Obtained Response Characteristics.

S. no.	Type of electrolyte used	Tool insulation condition	Feed rate (mm/min)	Avg. MRR (mg/min)	ADOC (μm)
1	Aqueous $NaNO_3$ (10% wt./v)	Exposed tool (i.e., ETL 1 mm)	0.6	6.25	353
2			0.8	6.83	321
3			1.0	7.73	281
4		Fully insulated tool (i.e., ETL 0 mm)	0.6	4.05	249
5			0.8	4.94	231
6			1.0	5.92	211
7	Aqueous NaCl (10% wt./v)	Exposed tool (i.e., ETL 1 mm)	0.6	6.96	421
8			0.8	8.32	349
9			1.0	10.06	319
10		Fully insulated tool (i.e., ETL 0 mm)	0.6	4.55	303
11			0.8	5.86	278
12			1.0	7.23	252
13	Aqueous $NaNO_3$ and NaCl (5+5% wt./v)	Exposed tool (i.e., ETL 1 mm)	0.6	6.64	378
14			0.8	7.31	337
15			1.0	9.12	308
16		Fully insulated tool (i.e., ETL 0 mm)	0.6	4.41	275
17			0.8	5.46	251
18			1.0	6.81	225

28.3 RESULTS AND DISCUSSION

Effect on Material Removal Rate

The effects of electrolyte type and feed rate on MRR for both the types of tool electrode are shown in Figure 28.4. It was observed that the amount of material dissolved was high when the aqueous solution of NaCl was used as the electrolyte, for corresponding tool feed rates and tool insulation conditions. The MRR was higher for higher feed rates irrespective of the type of electrolyte solution and tool insulation used. Moreover, the MRR was higher in the case of exposed tool electrode than that in the case of fully insulated tool, for all values of feed rate and type of electrolyte solution used. For the tool feed rate of 1 mm/min, about 30% and 18% higher MRR were obtained in the case of exposed tool when electrolyte was changed from $NaNO_3$ to NaCl and to a mixture of $NaNO_3$ and NaCl solutions, respectively.

Whereas, approximately 22% and 15% increase in corresponding MRR was obtained in the case of fully insulated tool. Besides, when the value of tool feed rate was 0.6 mm/min, the increase in MRR was relatively smaller for corresponding cases of electrolyte solutions and tool electrodes compared to higher values of tool feed rate.

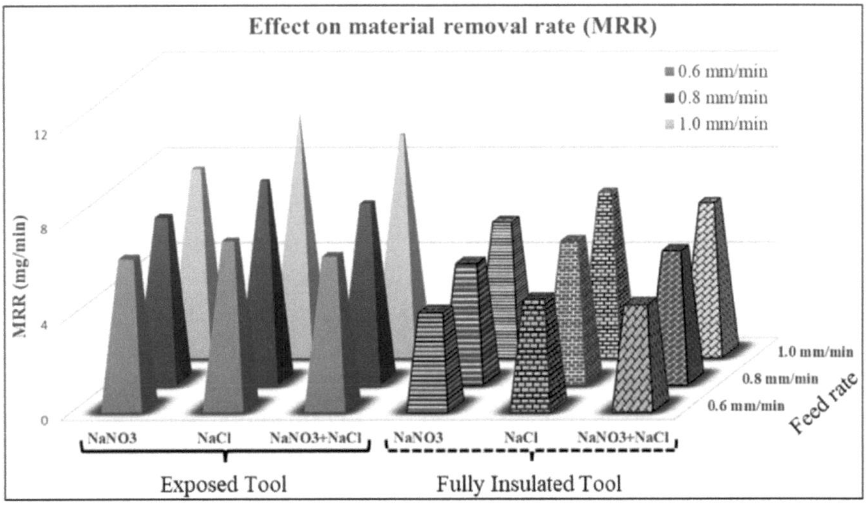

FIGURE 28.4 Effect of the type of electrolyte used and the tool feed rate on MRR during the STED with the exposed tool (having ETL of 1 mm) as well as fully insulated tool (FIT) electrode (having ETL of 0 mm).

The higher material dissolution obtained using an aqueous solution of NaCl as the electrolyte medium can be attributed to the presence of aggressive chemical reactivity of halide ions. The bulk work material is exposed for subsequent anodic dissolution as the chloride ion (Cl^-) can aggressively break the passivation of the work surface. Therefore, the persistent breakdown of the passivation layer on the work surface results in higher amount of anodic dissolution. Besides, the exposed tool length ETL of 1 mm facilitates an additional cathodic surface area for charge transfer during the chemical reaction, and thus, it increases the amount of material dissolved during the machining action. Consequently, the tool electrode can proceed forward at higher feed rates without frequent tool retractions. A higher tool feed rate results in smaller equilibrium IEG, which causes faster dissolution of material beneath the tool face owing to higher current density in the small IEG. Thus, the maximum value of MRR was obtained in the case of NaCl solution when the exposed tool was fed at 1 mm/min. The $NaNO_3$ solution hinders an active dissolution of the material by forming a passivation layer onto the work surface during the machining action. The formed passivation layer reduces the passage of charge, and so, the anodic dissolution rate. Consequently, frequent incidents of short-circuiting and tool retractions occur as the tool advances, which also prolongs the machining duration. Thus, the increase in the MRR was even lesser than the mixed NaCl + $NaNO_3$ electrolyte solution, which contains moderate concentration of the Cl^- ions.

Effect on Average Diametral Overcut

Figure 28.5 shows the effect of electrolytes' type and tool feed rate on the ADOC for both types of tool insulation conditions. The values of ADOC for the exposed tool electrode case were considerably higher than the values of ADOC obtained with fully insulated tool electrode, irrespective of electrolyte type and the values of tool feed rate. Smaller values of the ADOC were obtained when the $NaNO_3$ solution was used as the electrolyte medium instead of the NaCl solution, for the given value of tool feed rate and the type of tool electrode used. From the experimental results enlisted in Table 28.2, it becomes clear the values of ADOC were smaller when the tool electrode was fed at higher rates for given type of the electrolyte solution and the tool insulation. The decrease of about 12% and 16% in the values of ADOC was observed on using the $NaNO_3$ solution in place of the NaCl solution, when the exposed tool electrode fed at rates of 1 and 0.6 mm/min, respectively. Whereas, the decrease of about 16% and 18% in the ADOC values was obtained when using the $NaNO_3$ solution instead of the NaCl solution, in the case of fully insulated tool electrode fed at rates of 1 and 0.6 mm/min, respectively. The values of ADOC when using mixed electrolyte solution were between that of the NaCl and the $NaNO_3$ solutions for all feed rates. Besides, the STED in the mixed electrolyte solution performs well with fully insulated tool as no short-circuiting and tool retraction instances were occured due to hole choking.

Figure 28.6 shows the representative FE-SEM micrographs illustrating the profiles of machined hole obtained at a depth of 6 mm and a feed rate of 0.8 mm/min for different electrolytes and tool insulation conditions. It was observed that the hole profiles obtained with exposed tool were having more circularity error compared to that in the case of fully insulated tool. Due to the absence of an insulating coating, the exposed surface area near the tool tip actively participates in the material removal phenomenon. Further, the current density in the frontal IEG is low with exposed tool, and thus, the material dissolution in the desired direction (i.e., in the direction of tool electrode feed) reduces. Furthermore, the stray current in the side gap may deteriorate uniformity of the hole overcut as compared with the fully insulated tool. It was observed that the nonuniformity in overcut was more with the NaCl solution as compared to the $NaNO_3$ solution. It may be due to the aggressive material removing ability of the chloride ions even in the region of lower values of applied voltages.[13] Due to the passivating nature of the $NaNO_3$ solution, relatively lesser material is removed from the hole wall, and which causes smaller overcut. However, for a high value of feed rate (1.0 mm/min), the reduction in hole overcut values shrinks the side gap that causes the IEG choking owing to difficulties encountered during sludge evacuation in the narrow, slender side gap. Whereas, at the low feed rate of 0.6 mm/min, the issue of IEG choking was not encountered.

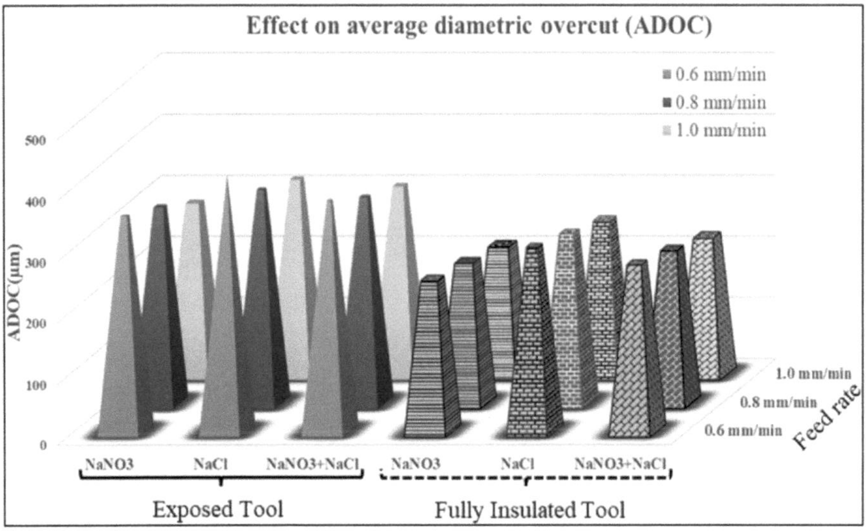

FIGURE 28.5 Effect of the type of electrolyte used and the tool feed rate on the ADOC during the STED with the exposed tool (having ETL of 1 mm) as well as fully insulated tool electrode (having ETL of 0 mm).

A higher feed rate with the exposed tool reduces the machining duration, and so, the total amount of stray dissolution from the sidewalls of the machined hole. Thus, the reduction in machining duration diminished the value of overcut. Further, the reduced overcut with $NaNO_3$ at higher feed rate (1 mm/min) poses sludge evacuation issues occasionally, leading to rubbing of tool wall that deteriorates the efficacy of tool insulation. The exposed tool surface causes gas evolution at the tool circumference, which contributes to nonuniformity in dissolution leading to nonuniformity in the overcut. The out of roundness holes may also be due to dregs sticking to the exposed tool surface, which in turn causes differential dissolution at the hole circumference.[17] However, in the case of the fully insulted tool, relatively uniform localized dissolution was observed because the electrochemical reaction majorly progresses beneath the annular tool face with reduced stray dissolution in the side gap.

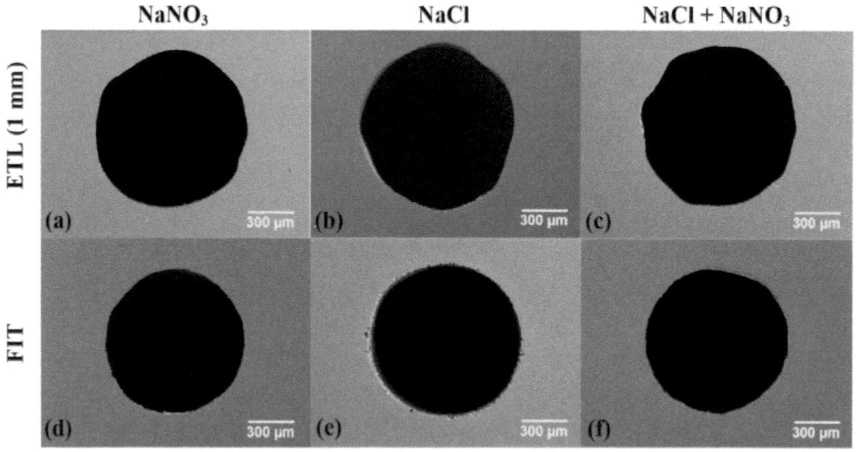

FIGURE 28.6 Representative FE-SEM images of the holes machined captured at a depth of 6 mm for a feed rate of 0.8 mm/min with the exposed tool (ETL, 1 mm) using the aqueous solution of (a) $NaNO_3$, (b) NaCl, and (c) Mixed NaCl+$NaNO_3$; and with the fully insulated tool (FIT) using the aqueous solution of (d) $NaNO_3$, (e) NaCl, and (f) Mixed NaCl+$NaNO_3$.

Surface Morphology of Holes

The type of electrolyte used has a role in the work material's dissolution behavior as the material removal rate differs for different electrolytes used for machining the same work material. Figure 28.7 shows the effect of tool insulation conditions and the type of electrolyte used on the surface morphology of the holes drilled. The hole surfaces obtained with the exposed tool length of 1 mm were found to be uneven than the surfaces obtained with the fully insulated tool. It can be attributed to the fact that in the case of exposed tool, the exposure of the hole surface to the active cathode surface is for a longer time duration, than in the case of insulated tool. The dregs produced as a result of the dissolution also stick to the exposed tool surface more readily as compared to the fully insulated tool with nonsticking insulation coating. Thus, the dregs masking the tool surface also contribute to hinder the uniform dissolution of the work material. It is evident from the FESEM micrograph shown in Figure 28.7b that dregs also stick to the hole surface and alter the dissolution in the locale, promoting the formation of larger pits in the vicinity.

The FESEM micrographs presented in Figure 28.7 depict the extent of electrochemical dissolution on the hole surface in the case of an exposed tool irrespective of the electrolyte type. On comparing Figure 28.7a, d for the $NaNO_3$, the severity of the intergranular attack becomes evident for the exposed tool condition along with the bigger micropits when compared to the surface obtained corresponding to the fully insulated tool. The hole surface obtained with aqueous $NaNO_3$ as a working electrolyte is dark grey with black areas. Zhu et al. also noted the formation of black areas due to the presence of elemental niobium in the Inconel while performing ECM.[18] It contrasts with the shiny grey hole surface obtained with the aqueous NaCl during STED. The hole surfaces obtained with aqueous NaCl as shown in Figure 28.7b, e highlight the pitting action of chloride ion during the electrochemical dissolution. The formation of large micropits in case of exposed tool condition is a result of the continued dissolution of work material even when the annular tooltip has moved downward. Further, the hole surface obtained from STED with a fully insulated tool demonstrates a uniform distribution of smaller micropits pointing toward better dissolution behavior. However, the micropits obtained with a fully insulated tool seem deeper than the micropits obtained with an exposed tool. The comparison of hole surfaces obtained in STED with aq. $NaNO_3$ and NaCl highlights the difference in surface morphology in terms of the extent of intergranular attack and the size of micropits. Hence, a mixture of aq. $NaNO_3$ and aq. NaCl was employed to take advantage of the properties of both electrolytes. The detrimental effects of using an exposed tool length are visible in Figure 28.7c, the hole surface is uneven and the micropits are larger than the ones obtained with individual electrolyte solutions. The uneven surface resulted from the STED with exposed tool and mixed electrolyte is prone to promote scaling in the hole during service application. Whereas, the hole surface obtained with a fully insulated tool during STED with the mixed electrolyte as shown in Figure 28.7f depicted lesser micropits as compared to aq. NaCl and mild intergranular attack. The mixed electrolyte produced lesser dregs

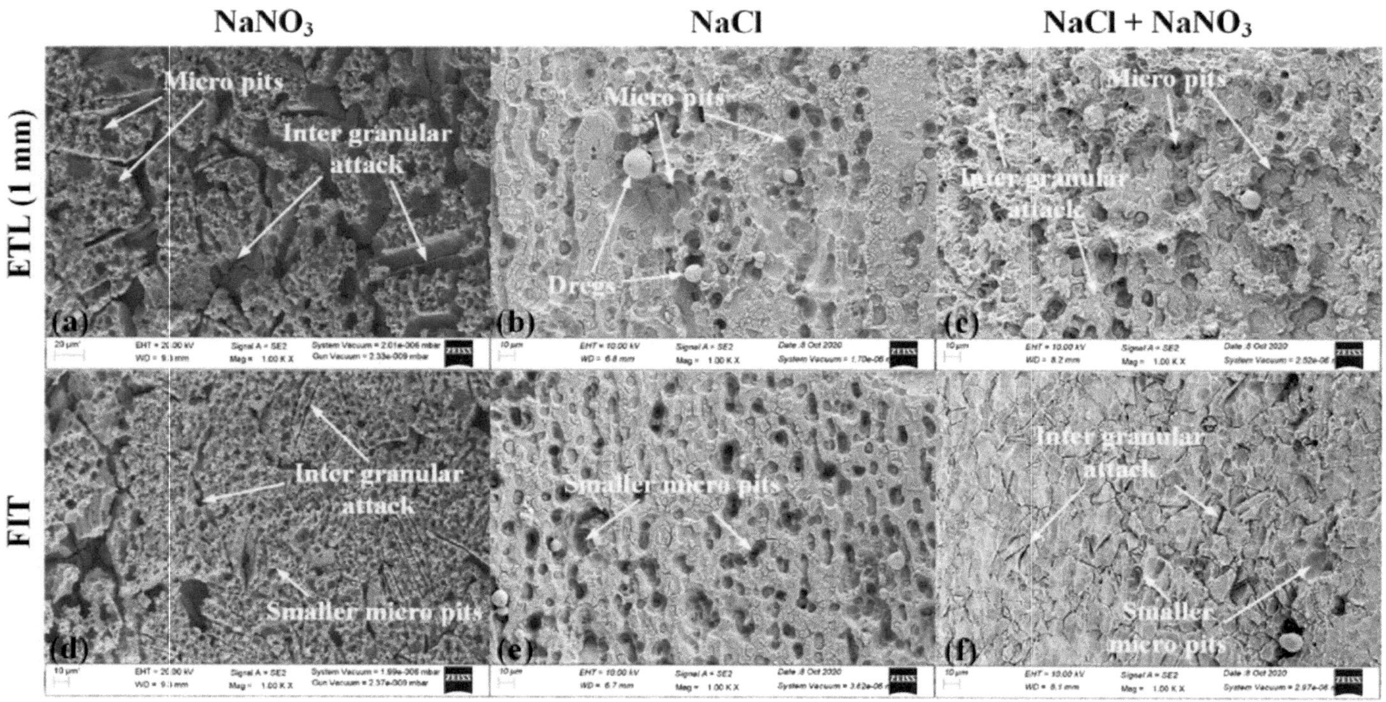

FIGURE 28.7 Representative FE-SEM micrographs depicting the hole surfaces obtained at the feed rate of 0.8 mm/min with the exposed tool (ETL, 1 mm) using the aqueous solution of (a) $NaNO_3$, (b) NaCl, and (c) Mixed NaCl+$NaNO_3$; and with the fully insulated tool (FIT) using the aqueous solution of (d) $NaNO_3$, (e) NaCl, and (f) Mixed NaCl+$NaNO_3$.

as compared to aq. NaCl while producing shiny hole surfaces comparable to the ones obtained with aq. NaCl.

28.4 CONCLUSIONS

In this work, the effects of nonelectrical process parameters like type of electrolyte, tool insulation conditions, and tool feed rate on the response characteristics such as average diametral overcut, and material removal rate has been assessed during the STED of high AR holes on Inconel 718. The general inferences that can be drawn are as follows.

i. The MRR was maximum in the case of aqueous NaCl solution, while the ADOC was minimum in the case of aqueous $NaNO_3$ solution. The STED in the mixed NaCl + $NaNO_3$ solution resulted higher MRR and lower ADOC compared to the constituents aqueous $NaNO_3$ and aqueous NaCl solutions, respectively.
ii. A tool electrode, fully insulated around side face, rendered lower and uniform ADOC than that with the tool electrode having an exposed length of 1 mm, for all types of electrolyte solutions. The values of ADOC obtained with aqueous $NaNO_3$ was least but posed sludge evacuation issue while employing high feed rates at times. However, ADOC obtained with the mixed electrolyte solution was more uniform indicating better control over the material dissolution behavior as compared to that with aqueous NaCl solution.
iii. MRR increased with an increase in the tool feed rate, irrespective of the type of electrolyte solutions and tool insulation conditions. However, frequent instances of tool retractions were observed in the case of aqueous $NaNO_3$ solution owing to the formation passivation layer onto the work surface, which hinders the active dissolution of the anode material during the machining action.
iv. The surface morphology of the holes fabricated during the STED with fully insulated tool comes out more uniform as compared to the holes drilled using a tool with exposed tool length of 1 mm. The FESEM analysis of the hole surfaces obtained with aq. $NaNO_3$ shows visible intergranular attack. Whereas, the hole surface from STED with aq. NaCl was severely pitted. The benefits of using a mixed electrolyte solution show up in the form of smaller micropits and mild intergranular attack visible in the FESEM micrographs.

KEYWORDS

- **electrochemical machining**
- **nickel superalloy**
- **STED**
- **high aspect ratio holes**

REFERENCES

1. Rahman, M.; Seah, W. K. H.; Teo, T. T. The Machinability of Inconel 718 Materials Processing Technology. *J. Mater. Process. Technol.* **1997,** *63* (96), 199–204. DOI: 10.1016/S0924-0136(96)02624-6
2. Yadav, V. K. et al. Investigation on the Effect of Input Parameters on Surface Quality during Rotary Tool Near-Dry EDM. In *Recent Advances in Mechanical Infrastructure, Lecture Notes in Intelligent Transportation and Infrastructure*; Parwani, A. K., Ramkumar, P., Eds.; 2020. DOI: 10.1007/978- 981-32-9971-9_5
3. Ramver et al. Experimental Investigation on Surface Morphology of Micro-EDMed Ti-6Al- 4V Alloy. In *Recent Advances in Mechanical Infrastructure, Lecture Notes in Intelligent Transportation and Infrastructure*; Parwani, A. K., Ramkumar, P., Eds.; Springer Nature Singapore Pte Ltd, 2020. DOI: 10.1007/978-981-32-9971-9_8
4. Singh, R.; Dvivedi, A.; Kumar, P. EDM of High Aspect Ratio Micro-Holes on Ti-6Al-4V Alloy by Synchronizing Energy Interactions. *Mater. Manuf. Process.* **2020,** *35* (11), 1188–1203. DOI: 10.1080/10426914.2020.1762207
5. Ramver; Dvivedi, A.; Kumar, P. On Improvement in Surface Integrity of μ-EDMed Ti-6Al-4V Alloy by μ-ECM Process. In *TMS 2019 148th Annual Meeting & Exhibition Supplemental Proceedings*; Springer International Publishing: Cham, 2019; pp 745–753. DOI: 10.1007/978-3-030-05861-6_73
6. Sharma, S.; Jain, V. K.; Shekhar, R. Electrochemical Drilling of Inconel Superalloy with Acidified Sodium Chloride Electrolyte. *Int. J. Adv. Manuf. Technol.* **2002,** *19* (7), 492–500. DOI: 10.1007/s001700200052
7. Bilgi, D. S. et al. Electrochemical Deep Hole Drilling in Super Alloy for Turbine Application. *J. Mater. Process. Technol.* **2004,** *149* (1–3), 445–452. DOI: 10.1016/j.jmatprotec.2003.09.008
8. Rathod, V.; Doloi, B.; Bhattacharyya, B. Sidewall Insulation of Microtool for Electrochemical Micromachining to Enhance the Machining Accuracy. *Mater. Manuf. Process.* **2014,** *29* (3), 305–313. DOI: 10.1080/10426914.2013.864407
9. Ali, S. et al. Shaped Tube Electrochemical Drilling of Good Quality Holes. *CIRP Ann.—Manuf. Technol.* **2009,** *58* (1), 185–188. DOI: 10.1016/j.cirp.2009.03.070
10. Wang, X. et al. Electrochemical Drilling with Constant Electrolyte Flow. *J. Mater. Process. Technol.* **2016,** *238*, 1–7. DOI: 10.1016/j.jmatprotec.2016.06.033

11. De Silva, A. K. M.; Gamage, J. R.; Harrison, C. S. Assessment of Environmental Performance of Shaped Tube Electrolytic Machining (STEM) and Capillary Drilling (CD) of Superalloys. *CIRP Ann.—Manuf. Technol.* **2017,** *66* (1), 57–60. DOI: 10.1016/j.cirp.2017.04.110
12. Sathiyanarayanan, C.; Manikandan, N.; Kumanan, S. Optimisation of Electrochemical Drilling Process Using Taguchi Method and Regression Analysis. *Int. J. Mach. Machinab. Mater.* **2017,** *19* (2), 136. DOI: 10.1504/ijmmm.2017.10003778
13. Geethapriyan, T.; Muthuramalingam, T.; Kalaichelvan, K. Influence of Process Parameters on Machinability of Inconel 718 by Electrochemical Micromachining Process Using TOPSIS Technique. *Arab. J. Sci. Eng.* **2019,** *44* (9), 7945–7955. DOI: 10.1007/s13369-019-03978-5
14. Chryssolouris, G.; Wollowitz, M.; Sun, N. P. Electrochemical Hole Making. *CIRP Ann.—Manuf. Technol.* **1984,** *33* (1), 99–104. DOI: 10.1016/S0007-8506(07)61388-2
15. Rahman, Z.; Das, A. K.; Chattopadhyaya, S. Microhole Drilling through Electrochemical Processes: A Review. *Mater. Manuf. Process.* **2018,** *33* (13), 1379–1405. DOI: 10.1080/10426914.2017.1401721
16. Li, Z. Y. et al. Process Capability and Effect Size of Vacuum Extraction Shaped Tube Electrolytic Drilling of Inconel Alloy for High-Performance Cooling Hole. *Int. J. Adv. Manuf. Technol.* **2016,** *85* (9–12), 2557–2566. DOI: 10.1007/s00170-015-8085-z
17. Vats, A.; Dvivedi, A.; Kumar, P. Assessing the Performance of STED Process for Fabricating High Aspect Ratio Holes on Inconel 718 Alloy. *Mater. Manuf. Process.* **2020,** *00* (00), 1–16. DOI: 10.1080/10426914.2020.1866189. (In press)
18. Zhu, D.; Yu, L.; Zhang, R. Dissolution Effects with Different Microstructures of Inconel 718 on Surface Integrity in Electrochemical Machining. *J. Electrochem. Soc.* **2018,** *165* (16), E872–E878. DOI: 10.1149/2.0761816jes

CHAPTER 29

STUDY ON ELECTRIC DISCHARGE MACHINING OF AL7075 T651 ALLOY: AN EXPERIMENTAL INVESTIGATION

RAHUL SONI, RAVI PRATAP SINGH*, and
SHAILENDER SINGH BHADAURIA

*Department of Industrial and Production Engineering,
Dr. B. R. Ambedkar National Institute of Technology Jalandhar,
Punjab 144011, India*

**Corresponding author. E-mail: singhrp@nitj.ac.in*

ABSTRACT

Various industries such as aerospace, chemical, power, etc. are now using these types of advanced alloys which also lead in difficulty in the machining of these alloys. Al7075 alloy is an aluminum alloy, in which Zn is the principal alloying element. The present study focuses on the electrical discharge machining of Al7075 T651 to investigate the pre-eminent optimal setting to give best results in the terms of machining responses. The process parameters selected on the basis of past literature and preliminary experiments are: discharge current, gap voltage, pulse-on time and pulse-off time. The experiments are designed by Taguchi L_{25} orthogonal array. In single response optimization of material removal rate, the optimized setting obtained is; I_p—36 A, V—65.54 V, T_{on}—400 μs and T_{off}—60 μs.

Modern Manufacturing Systems: Trends and Developments. Rajiv Kumar Garg, Ravi Pratap Singh, Rajeev Trehan, & Ramesh Singh (Eds.)

29.1 INTRODUCTION

Removal of material to get a required final structure and dimensions is called machining process. Material removal takes place in a controlled manner and today collectively named as subtractive manufacturing processes. There are two kinds of machining processes: conventional (traditional) and nonconventional (nontraditional) machining processes.

In conventional machining processes, a sharp cutting tool making direct contact with the work material is used for removing the material to obtain the desired work geometry. Due to the direct contact between tool and work material, there are some problems in conventional machining:

- Poor machining characteristics is found
- The geometries of cutting tool get worn-out
- Exactitude of dimensions is not achieved
- There is a lack of surface rectitude.

When these processes are scaled down from macro levels to micro levels, then these problems get much severe. So, to overcome these limitations, nontraditional machining processes came into existence. Nontraditional machining processes are special type of machining processes in which a form of energy such as mechanical, chemical, electrochemical, thermoelectrical is applied to erode the material to get the desired final profile and dimensions. The material is eroded without the direct contact of tool with the work material. It has the capability to machine the complex geometries and the materials with excessive strength and hardness which is very problematic to machine by traditional machining processes can be machined in an easy way. Electric discharge machining is a type of nonconventional machining. The optimization-based studies have also been attempted while processing advanced materials with the newest machining processes.[1–4]

In electric discharge machining, material is removed from a conductive workpiece due to the electric spark when the high pulsed voltage is employed between the tool electrode and the conductive work material. Localized melting and vaporization take place because of high temperature generated which helps in material erosion from the workpiece. The flow of dielectric fluid carries away the most of the eroded material from the inter-electrode gap. The materials that are electrically conductive in nature can be machined by the electric discharge machining process. Typically, in oil die-sinking electric discharge machining, pulsed direct current (DC) power supply is applied where generally the tool electrode is attached to the (+) ve terminal (tool electrode as anode) and the work material is attached to the (−) ve terminal (workpiece as cathode).

Khan et al.[5] conducted an experiment on stainless steel D3 of 10 mm diameter and 5 mm thickness with the help of electric discharge machine. The influence of input variables; pulse-on time (T_{on}), current (A), and voltage (V) on the several response parameters were analyzed. Machine off-set and copper wire electrode of 2 mm diameter were kept fixed. The concerned parameters, that is, Material Removal Rate (MRR), Surface Roughness (SR), and Tool Wear Rate (TWR) have been investigated by varying the machining parameters. The experiments were designed employing the Taguchi L9 OA technique. The calculated optimum parameters for MRR (current—7 A, pulse on-time—20 µs, voltage—125 V), SR (current—7 A, pulse-on time—20 µs, voltage—125 V) and TWR (current—7 A, pulse on-time—20 µs, voltage—125 V) with the help of S/N ratio and analysis of variance (ANOVA). R^2 values and regression analysis provide the significant results for MRR, SR, and TWR as 92.5%, 89.9%, and 80.4%, respectively. Sharma et al.[6] scrutinized the multioptimization of output responses MRR and TWR for Inconel 718 (specimen of 14 mm diameter) with Cu–Cd as electrode. Taguchi L9 OA was applied for designing the experiments and the further analysis of results was made with the help of two multicriteria decision-making techniques named TOPSIS and PROMITHEE.[7,8] The experimentation was done on a usual CNC electric discharge machine with positive polarity. The observed optimum values of the parameters were current—7 A, gap voltage—50 V, pulse-on time—200 µs, and pulse-off time—6 µs.

Gudipudi et al.[9] fabricated AA6061-B_4C composite by stir-casing process. An experimental investigation was conducted for the optimization of the output parameters. MRR and Average Recast Layer Thickness (ARLT) were taken as the responses to be studied. Taguchi L9 OA was employed as the technique for designing the experiments. SEM micrographs were used to analyze the ARLT and topography of the surface of machined zone. The statistical software Minitab 18 was used to develop the linear regression model to optimize the MRR (maximum) and ARLT (minimum). ANOVA was conducted to find the significant parameters of the responses. Discharge current (I) was found as the most significant input parameter affecting the MRR with percentage contribution of 85.81% followed by the discharge duration (T_{on}) with contribution of 13.187%. For ARLT, T_{on} was the most significant machining parameter with percentage contribution of 68.50%. Nallusamy10 analyzed the MRR and TWR on OHNS die steel with Cu and Cu–Zn electrodes using electric-discharge machine (ELECTRONICA—M2S) with electromechanical servo control and positive polarity. The size of the work plate used was 50 × 50 mm with 24 mm thickness. Discharge current (I) and electrode material were chosen as machining variables. Sparking voltage (V), pulse-on time (T_{on}), and pulse-off time (T_{off}) were selected as input machining parameters. Kerosene Commercial of grade oil was used as dielectric fluid. SEM was used for analyzing the heat affected regions. It was concluded that the increment in discharge current leads in improving the MRR but decrease in TWR. It was also stated that at 4 A discharge current, the MRRs for both the electrodes were equal. On visual inspection, it was found that the surface finish with both the electrodes were almost same.

Khanna et al.[11] conducted an investigational study on electric discharge drilling machine to study and achieve the optimized performance for Al 7075. A brass rod of 2 mm diameter was adopted as a tool electrode for drilling. Peak current, pulse-on (T_{on}), pulse-off (T_{off}) and speed of electrode rotation (ω) were taken as controllable variables for the experimentation. Other input variables such as air pressure and voltage were kept fixed at 3 kg/cm^2 and 65 V, respectively, all over the experiment. TWR and MRR were taken as output responses. From this experimental study, T_{on} was found as the most significant input parameter having maximum effect on MRR with 94.63% contribution tracked by T_{off} with contribution of 4.55%. The percentage contributions of T_{on} and T_{off} for TWR were 75.27% and 4.25%, respectively. Multiquality characteristics for the same material were given by Traditional Grey Relational Analysis (TGRA). The optimum results given by GRG were observed to be 0.0359 mm^3/min for MRR and 0.252 g/s for TWR. El-Taweel[12] inspected the relationship of input variables with Al–Cu–Si–TiC composite tool electrode formed by powder metallurgy (P/M) technique in electric discharge machining of CK45 steel. In this study, titanium carbide percentage (TiC%), peak current (I_p), pulse-on time (T_{on}), and flushing pressure of dielectric fluid (P) were considered as the input process variables whereas MRR and TWR were studied as output responses. The rotatable design of the central composite second order was utilized for planning the experiments and experimental models were developed by employing response surface methodology (RSM) technique. The appropriacy of developed regression models was checked by carrying out the ANOVA test. In the above experimental study, peak current was appeared as the key significant input process variable influencing both MRR and TWR whereas dielectric flushing pressure had a little impact on both the responses. It was also addressed that the reactivity of the formed tool electrodes to I_p and T_{on} were more than the other traditional electrodes. The optimal combination of input process variable settings showed by the multiresponse optimization was TiC%—18%, I_p—6 A, P—1.2 MPa, and T_{on}—182 μs to achieve the required MRR and TWR.

Kao et al.[13] conducted an experimental examination on alloy Ti-6Al-4V for optimizing the machining parameters with multiple quality characteristics. Electrolytic copper of 10 mm diameter and 50 mm height was chosen as the tool electrode. The discharge current (A), gap voltage (V), duration of pulses (μs), and duty factor (%) were chosen as the input process variables whereas TWR, MRR and SR were taken as the responses to be analyzed. The experiments were governed by introducing the Taguchi L9 OA. The experimental results of the responses were normalized by GRA. The contribution of each parameter was calculated by using ANOVA results. According to results provided by ANOVA, discharge current was identified as the most dominating input variable with percentage contribution of almost 70% followed by the duty factor—18.64%. Discharge current—5 A, gap voltage—150 V, pulse duration—200 μs, and duty factor—70% were the optimal values for the EWR. According to the experimental results of MRR and S/N ratio, it was noted that MRR is largely affected by the discharge current with percentage contribution of 74% followed by the duty factor with 22.78%. For MRR, discharge current—20 A, gap voltage—100 V, pulse duration—100 μs, and duty factor—30% were the optimal results. According to S/N ratio and the corresponding ANOVA results, surface roughness was largely affected by the discharge current with contribution of 67.06% followed by the duty factor—17.32% and the pulse duration—14.01%. The optimal values for surface roughness were discharge current—5 A, open voltage—200 V, pulse duration—200 μs, and duty factor—70%.

Viswanth et al.[14] experimentally studied the MRR, EWR, and SR on AISI 2507 super duplex stainless steel on ARD EDM ZNC die-sinker EDM. Pure grade electrolytic copper rod (99.71% Cu) was selected as the tool electrode for the experimentation. Pongamia Green Dielectric was used as the dielectric medium. T_{ON}, T_{OFF}, I_P and (IEG) were taken as the process parameters for multi-response optimization. Duty factor (75–90%) was kept constant throughout the experimental trials. MRR, EWR, and SR were the responses to be calculated and analyzed for the multiresponse optimization. Taguchi L_{27} OA was introduced as technique for designing the experiments. GRA was conducted to identify the optimal setting of the machining variables.[15,16] In this experimental study, it was found that on increasing the value of T_{ON}, MRR tends to increase and EWR tends to decrease. The minimum value of the EWR was 0.221 mm^3/min at 400 μs of T_{ON}, 90 μs of T_{OFF}, 15 A of I_P, 50

V of V and 120 μm of the IEG. The minimum value of the SR was 6.711 μm at 300 μs of T_{ON}, 90 μs of T_{OFF}, 9 A of I_p, 60 V of V and 120 μm of the IEG. The optimal settings for multioptimization given by GRA were 400 μs of T_{ON}, 60 μs of T_{OFF}, 21 A of I_p, 50 V of V and 60 μm of the IEG. According to the ANOVA test, peak current (I_p) was found as the key influential factor trailed by T_{ON}.

Chauhan et al.[17] experimentally investigated the impact of different input variables on MRR and TWR for M2 high speed steel using Taguchi L9 OA. The experimentation was carried out on EDM (model C-3822 with PSR-220). Cylindrical-shaped Cu was used as the tool electrode for the study. I_p, T_{on}, and diameter of the electrode were selected as the process variables. The optimum values were obtained by using the MINITAB software. In this experimental study, it was discussed that the MRR improves with increment in peak current but with increment in pulse-on time, the MRR enhances sharply up to a maximum value after that there is a fall in the slope. MRR is directly proportional to the electrode diameter whereas TWR is inversely proportional to the same. With the elevation in peak current, TWR first increases up to a certain point then decreases. It was also concluded that MRR is highly affected by T_{on} and least influenced by the electrode diameter whereas the I_p has the moderate influence on MRR. Kumar et al.[18] conducted an experimental investigation on Ti-6Al-7Nb to achieve the best micro-drilling efficiency and performance in EDM. The various parameters such as electrode material, rotation of the electrode, pulse-on time (T_{on}), pulse-off time (T_{off}), work material thickness, and drill angle were taken as the input variables for the micro-drilling whereas DR and EWR were chosen as the output parameters to be studied. The input variables and their ranges were selected for further experimentation using Taguchi OA. The optimal setting of variables was attained using NSGA-II approach.[19–21] In this experimental examination, it was concluded that the DR and EWR both have increasing trend with increase in peak current and pulse-on whereas the rise in T_{off} leads to improvement in DR but decrement in EWR. The high aspect ratio micro-holes were drilled very easily but at higher hole depth, front side of electrode becomes conical. So, tapered electrode was introduced for drilling.

29.2 MATERIALS AND METHODS

The sample material for the research is Al7075 T651 alloy in the form of a flat plate with 150 mm length, 150 mm width, and 20 mm thickness. The chemical composition and properties of the Al7075 T651 alloy have been shown in Tables 29.1 and 29.2. Experiments have been carried out on a CNC electric discharge machine (make: Oscarmax, model: S645 CMAX) available at CIHT Jalandhar. A copper rod of 10 mm diameter is selected as tool electrode and EDM oil is used as dielectric fluid for the experimentation. The holes of 10 mm diameter and 5 mm depth are drilled for the study in this chapter (Figure 29.1).

TABLE 29.1 Chemical Composition of Work Material: Al7075 T651.

Element	Percentage by weight (%)
Aluminum	90
Zinc	5.6
Magnesium	2.5
Copper	1.6
Chromium	2.3

TABLE 29.2 Properties of Work Material: Al7075 T651.

Properties	Magnitude
Tensile strength	570 MPa
Yield strength	500 MPa
Elongation at break	3–9%
Elastic modulus	72 GPa
Shear strength	330 MPa
Hardness, Rockwell	87 HRB
Density	2.81 g/cm^3
Melting point	477°C

The machining setup and machined workpiece after the pilot experiments have been shown in Figure 29.2. Based on pilot experiments conducted and previous researches, five levels for each input variable (i.e., discharge current, gap voltage, pulse-on time, and pulse-off time) have been decided to design the main experiment matrix. The controllable variables and their levels have been shown in Table 29.3. Some of the variables such as working time, jump travel, dielectric flushing pressure, type of dielectric, etc. are kept constant throughout the experimentation.

FIGURE 29.1 CNC EDM machine at the workshop of CIHT Jalandhar.

TABLE 29.3 Specification of CNC EDM at CIHT Jalandhar.

Travel	
Longitudinal Travel (*X*-axis)	600 mm
Latitudinal Travel (*Y*-axis)	450 mm
Vertical Travel (*Z*-axis)	370 mm
Servo travel	400 mm
Dimensions	
Work Tank (L × W × H)	1500 mm × 940 mm × 520 mm
Work Table (L × W)	1000 mm × 600 mm
Dimensions	2830 mm × 2810 mm × 2610 mm
Others	
Maximum electrode weight	250 kg
Maximum workpiece weight	3000 kg
Weight	3900 kg

29.3 RESULTS AND DISCUSSIONS

Taguchi L_{25} orthogonal array has been employed for designing the experiments which reduced the number of experiments from 5^4 to 25.[22,23] The number of experiments and the studied response value corresponding to each experiment number have been shown in Table 29.4.

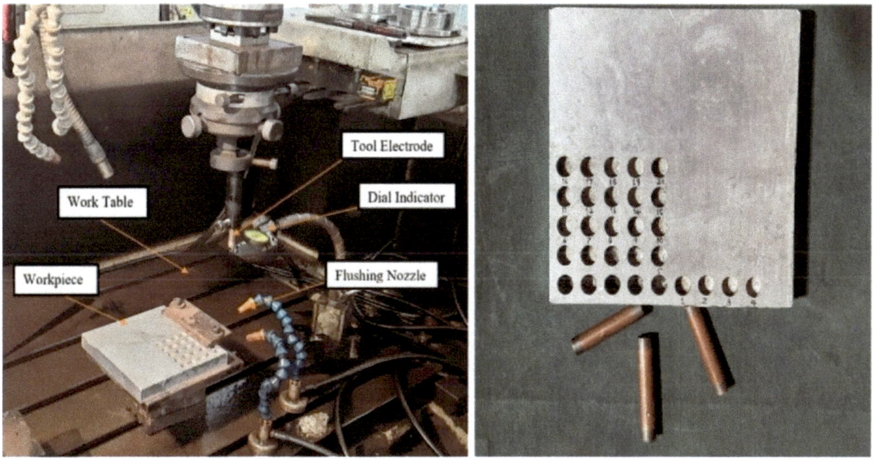

FIGURE 29.2 Machining setup and machined workpiece with Cu electrodes.

TABLE 29.4 Controllable Variables and Their Levels.

Variables (unit)	Symbol	L—I	L—II	L—III	L—IV	L—V
Discharge Current (A)	I	12	18	24	30	36
Gap Voltage (V)	V	50	60	70	80	90
Pulse-on Time (μs)	T_{on}	120	200	300	400	500
Pulse-off Time (μs)	T_{off}	60	90	120	200	300

TABLE 29.5 Taguchi L_{25} (5^4) OA.

Exp. no.	Orthogonal array				MRR$_1$ (mm^3/s)	MRR$_2$ (mm^3/s)	Mean MRR (mm^3/s)
	I (A)	V (V)	T_{on} (μs)	T_{off} (μs)			
1	12	50	120	60	0.1823	0.1611	0.1717
2	12	60	200	90	0.2866	0.2594	0.2730
3	12	70	300	120	0.3426	0.3316	0.3371
4	12	80	400	200	0.2534	0.2652	0.2593
5	12	90	500	300	0.2424	0.2386	0.2405
6	18	50	200	120	0.471	0.4606	0.4658
7	18	60	300	200	0.6008	0.6058	0.6033
8	18	70	400	300	0.5288	0.5144	0.5216
9	18	80	500	60	0.6032	0.601	0.6021

TABLE 29.5 *(Continued)*

Exp. no.	Orthogonal array				MRR_1 (mm^3/s)	MRR_2 (mm^3/s)	Mean MRR (mm^3/s)
	I (A)	*V* (V)	T_{on} (μs)	T_{off} (μs)			
10	18	90	120	90	0.5348	0.5348	0.5348
11	24	50	300	300	0.7348	0.7428	0.7388
12	24	60	400	60	0.7604	0.741	0.7507
13	24	70	500	90	0.8232	0.8084	0.8158
14	24	80	120	120	0.4492	0.4406	0.4449
15	24	90	200	200	0.3896	0.3912	0.3904
16	30	50	400	90	0.8848	0.891	0.8879
17	30	60	500	120	0.8466	0.8382	0.8424
18	30	70	120	200	0.5224	0.5098	0.5161
19	30	80	200	300	0.641	0.6262	0.6336
20	30	90	300	60	0.5785	0.5749	0.5767
21	36	50	500	200	1.1502	1.138	1.1441
22	36	60	120	300	0.6616	0.6736	0.6676
23	36	70	200	60	0.7204	0.7142	0.7173
24	36	80	300	90	0.9989	1.0835	1.0412
25	36	90	400	120	1.2402	1.2316	1.2359

The S/N ratio is used to evaluate the quality characteristics of input variables. Higher value of S/N ratio shows the ideal level of the machining variables. In the meantime, higher value of MRR is desirable, the bigger is better S/N quality characteristics is used. In this experimental study, the S/N ratio values for MRR (bigger is better) is calculated by using eq 29.1 given below and is shown in Table 29.5.

$$S/N = -10\left[\frac{1}{n}\left(\sum_{i=1}^{n} y_i^2\right)\right] \quad (29.1)$$

The optimal setting for MRR can be seen from Table 29.5 and also the same optimal setting can be spotted in Figure 29.3 easily. The optimal setting for the high value of MRR is the highest points (peaks) of the parameters. So, ideal process variable setting for the MRR is A5B2C5D2. The best combination values for maximizing MRR; 36 A of discharge current, 60 V of gap voltage, 500 μs of pulse-on time, and 90 μs of pulse-off time.

ANOVA is employed to determine the statistically significant machining variables that affect the MRR in electric discharge machining of Al7075

T651 and to calculate the contribution (in %) of each input variable on MRR and significance of input variables.

TABLE 29.6 Calculated S/N Ratios of MRR.

Level of process variables	Discharge current	Gap voltage	Pulse-on time	Pulse-off time
1	−12.0256	−4.8867	−7.4061	−5.9738
2	−5.3043	−4.6312	−6.5864	−3.8327
3	−4.4186	−5.0995	−4.1788	−4.5531
4	−3.4025	−5.3557	−3.8122	−5.7718
5	−0.6072	−5.7850	−3.7746	−5.6268
Delta	11.4184	1.1538	3.6315	2.1411
Rank	1	4	2	3

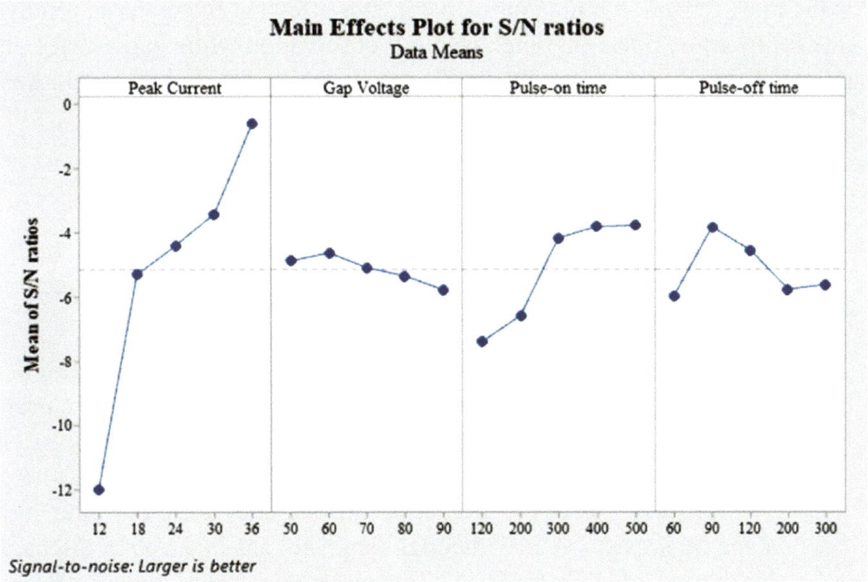

FIGURE 29.3 Main effect plots for S/N ratios.

The results from ANOVA for MRR have been given in Table 29.6. About 95% level of confidence (the level significance is 0.05% or 5%) is made on Minitab 19 software for this analysis (Table 29.7).

TABLE 29.7 ANOVA for MRR.

Source	DOF	Adj-SS	Adj-MS	*f*-Value	*p*-Value
Discharge current	4	1.29668	0.324170	24.94	0.000
Gap voltage	4	0.03215	0.008037	0.62	0.662
Pulse-on time	4	0.32241	0.080603	6.20	0.014
Pulse-off time	4	0.09148	0.022871	1.76	0.230
Error	8	0.10399	0.012998		
Total	24	1.84671			

The *p*-value decides whether a parameter is significant or not. According to ANOVA results, discharge current being the key influential input variable trailed by pulse-on time has the greater physical and statistical significance but voltage between the tool and work electrode and pulse-off time have less significance on MRR because *p*-value for gap voltage and pulse-off time is more than 0.05. In the previous study, it was found that the discharge current is the most remarkable machining parameter affecting the material removal rate followed by pulse-on time[9] as in our observation while gap voltage and pulse-off time have minimum significance level on material removal rate in electric discharge machining of Al7075 T651 alloy.

29.4 CONCLUSIONS

The conclusions can be drawn from the present experimental investigations that are given below:

- Ideal process variable setting for the MRR is A5B2C5D2. The best combination values for maximizing MRR; 36 A of discharge current, 60 V of gap voltage, 500 μs of pulse-on time, and 90 μs of pulse-off time.
- The process variables discharge current and pulse-on time have a major impact on the material removal rate in electric discharge machining of Al7075 T651. The material removal rate is affected majorly by discharge current followed by on time of pulses.
- With the advancement in discharge current, material removal rate increases sharply because of high spark intensity which melts more work material to increase material removal rate and after attaining a fixed value; material removal rate keeps on increasing but at a slower rate.

- The material removal rate is increased with increase in pulse-on time with greater slope up to a definite value because of melting more of work material but after that its slope of increasing trend gets diminished because of no proper flushing of further melted material. Gap voltage and pulse-off time have very minor effects on material removal rate.

KEYWORDS

- **EDM**
- **aluminum alloys**
- **taguchi**
- **Al7075**
- **optimization**

REFERENCES

1. Singh, R. P.; Singhal, S. Rotary Ultrasonic Machining: A Review. *Mater. Manuf. Process*. **2016,** *31*, 1795–1824.
2. Singh, R. P.; Singhal, S. Investigation of Machining Characteristics in Rotary Ultrasonic Machining of Alumina Ceramic. *Mater. Manuf. Process*. **2017a,** *32*, 309–326.
3. Singh, R. P.; Singhal, S. Experimental Investigation of Machining Characteristics in Rotary Ultrasonic Machining of Quartz Ceramic. *J. Mater. Des. Appl*. **2018a,** *232,* 870–889.
4. Singh, R. P.; Singhal, S. Rotary Ultrasonic Machining of Alumina Ceramic: Experimental Study and Optimization of Machining Responses. *J. Eng. Res.* **2018c,** *6*, 01–24.
5. Khan, F.; Kumar, J.; Soota, T. Optimization of EDM Process Parameter for Stainless Steel D3. *Mater. Today Proc.* **2019,** Proceedings 25, 635–638. doi: 10.1016/j.matpr.2019.07.529.
6. Sharma, V.; Prakash Misra, J.; Singhal, P. Multi-Optimization of Process Parameters for Inconel 718 While Die-Sink EDM Using Multi-Criterion Decision Making Methods. *J. Phys. Conf. Ser.* **2019,** *1240* (1). doi: 10.1088/1742-6596/1240/1/012166.
7. Singh, R. P.; Singhal, S. Rotary Ultrasonic Machining: A Review. *Mater. Manuf. Process*. **2016,** *31*, 1795–1824.
8. Singh, R. P.; Singhal, S. An Experimental Study on Rotary Ultrasonic Machining of Macor Ceramic. *J. Eng. Manuf.* **2018b,** *232,* 1221–1234.
9. Gudipudi, S.; Nagamuthu, S.; Subbian, K. S.; Chilakalapalli, S. P. R. Fabrication and Experimental Study to Optimize the Recast Layer and the Material Removal in Electric Discharge Machining (EDM) of AA6061-B4C Composite. *Mater. Today Proc.* **2019,** Volume 19, Part 2, 448–454. doi: 10.1016/j.matpr.2019.07.634.

10. Nallusamy, S. Analysis of MRR and TWR on OHNS Die Steel with Different Electrodes Using Electrical Discharge Machining. *Int. J. Eng. Res. Africa* **2016,** 22, 112–120. doi: 10.4028/www.scientific.net/JERA.22.112.
11. Khanna, R.; Kumar, A.; Garg, M. P.; Singh, A.; Sharma, N. Multiple Performance Characteristics Optimization for Al 7075 on Electric Discharge Drilling by Taguchi Grey Relational Theory. *J. Ind. Eng. Int.* **2015,** *11* (4), 459–472. doi: 10.1007/s40092-015-0112-z.
12. El-Taweel, T. A. Multi-response Optimization of EDM with Al-Cu-Si-TiC P/M Composite Electrode. *Int. J. Adv. Manuf. Technol.* **2009,** *44* (1–2), 100–113. doi: 10.1007/s00170-008-1825-6.
13. Kao, J. Y.; Tsao, C. C.; Wang, S. S.; Hsu, C. Y. Optimization of the EDM Parameters on Machining Ti-6Al-4V with Multiple Quality Characteristics. *Int. J. Adv. Manuf. Technol.* **2010,** *47* (1–4), 395–402. doi: 10.1007/s00170-009-2208-3.
14. Srinivas Viswanth, V.; Ramanujam, R.; Rajyalakshmi, G. Performance Study of Eco-Friendly Dielectric in EDM of AISI 2507 Super Duplex Steel Using Taguchi-fuzzy TOPSIS Approach. *Int. J. Product. Qual. Manag.* **2020,** *29* (4), 518–541. doi: 10.1504/IJPQM.2020.106425.
15. Singh, R. P.; Tyagi, M.; Kataria, R. Selection of the Optimum Hole Quality Conditions in Manufacturing Environment Using MCDM Approach: A Case Study. In *Operations Management and Systems Engineering*; Springer: Singapore, 2019a; pp 133–152.
16. Singh, R. P.; Kataria, R.; Singhal, S. Decision-Making in Real-Life Industrial Environment through Graph Theory Approach. In *Computer Architecture in Industrial, Biomechanical and Biomedical Engineering*; IntechOpen, 2019b. DOI: 10.5772/intechopen.82011
17. Chauhan, V. S.; Ahmad, F. Experimental Investigation of Mrr and Twr on Edm for M2 High Speed Steel Using Taguchi Method. **2019,** *14* (9), 194–198.
18. Dewangan, S.; Deepak Kumar, S.; Kumar Jha, S.; Kumar Biswas, C. Optimization of Micro-EDM Drilling Parameters of Ti–6Al–4V Alloy. *Mater. Today Proc.* **2020,** Proceedings 33, 2–6. doi: 10.1016/j.matpr.2020.03.307.
19. Singh, R.; Singh, R. P.; Tyagi, M.; Kataria, R. Investigation of Dimensional Deviation in Wire EDM of M42 HSS Using Cryogenically Treated Brass Wire. *Mater. Today: Proc.* **2019c**.
20. Singh, R.; Singh, R. P.; Trehan, R. State of the Art in Processing of Shape Memory Alloys with Electrical Discharge Machining: A Review. In *Proceedings of the Institution of Mechanical Engineers, Part B: Journal of Engineering Manufacture*, 2020. 0954405420958771.
21. Tyagi, M.; Panchal, D.; Singh, R. P.; Sachdeva, A. Modeling and Analysis of Critical Success Factors for Implementing the IT-Based Supply-Chain Performance System. In *Operations Management and Systems Engineering*; Springer: Singapore, 2019; pp 51–67.
22. Singh, R. P.; Kumar, J.; Kataria, R.; Singhal, S. Investigation of the Machinability of Commercially Pure Titanium in Ultrasonic Machining Using Graph Theory and Matrix Method. *J. Eng. Res.* **2015,** *3*, 75–94.
23. Singh, R. P.; Kataria, R.; Kumar, J.; Verma, J. Multi-response Optimization of Machining Characteristics in Ultrasonic Machining of WC-Co Composite through Taguchi Method and Grey-fuzzy Logic. *AIMS Mater. Sci.* **2018,** *5*, 75–92.

CHAPTER 30

PARAMETRIC ANALYSIS AND NUMERICAL INVESTIGATION OF HOT ABRASIVE JET MICROMACHINING USING COMPUTATIONAL FLUID DYNAMICS

SUBHADIP PRADHAN*, SOURAV SANKET TRIPATHY, SUSHREESUDHA SAHU, SUDHANSU RANJAN DAS, and DEBABRATA DHUPAL

Production Engineering Department, Veer Surendra Sai University of Technology, Burla, Sambalpur, Odisha 768018, India

Corresponding author. E-mail: subha.mid94@gmail.com

ABSTRACT

The intense hardness and high strength of sintered ceramics create it tough for machining them economically. Hot abrasive jet micromachining (HAJMM) in which target material is eroded by the impingement of a microjet of the heated air-abrasive stream containing fine abrasives, which customs a comparatively high-velocity stream of heated fine abrasive jet for machining precisely with controlled-depth microfeatures. The present surface development models are established for air-abrasive impelled erosion progressions that cannot account for the result of heated abrasive movement on the target channel sidewall erosion that indicates the advanced passage broadening detected in the HAJMM of brittle materials. This paper focused on the ideal numerical–experimental model to forecast the surface contours of microholes

Modern Manufacturing Systems: Trends and Developments. Rajiv Kumar Garg, Ravi Pratap Singh, Rajeev Trehan, & Ramesh Singh (Eds.)

in brittle material as named zirconia using HAJMM. The specific erosion rates of this material are calculated using different abrasive grain sizes as a result of the air-abrasive jet angle. To explain jet impingement angle and trajectories velocities of the abrasive particles, the jet impingement angle-material erosion relations are modified using computational fluid dynamics (CFD) models. In three-dimensional CFD models, the relationships between erosion rate and impingement angle are then implemented for finding the velocity trajectories and jet impingement angles of the abrasive grain path on a shallow eroded target contour, and thus calculate the erosion for deeper target profiles. Compared to experiments showing the previously observed extension of the machined channels, the CFD model is tested as the depth increases due to secondary erosion by abrasive particles impinging the sidewalls of the target channel. Because of its greater erosion resistance, the widening effect is found to be extensive on the zirconia surface but less important in other ceramics. The mathematical–experimental approach can reliably measure the widening and predict the cross-sections of the target channel up to a precise ratio of almost one with a maximum error of less than 5%.

30.1 INTRODUCTION

Hot abrasive jet micromachining (HAJMM) is a nontraditional machining method in which heated abrasive particles are propelled by high velocity along with a pressurized gas stream to erode material from the workpiece. The energized abrasive particles produce stresses on the workpiece surface because of the stresses, the material will get fractured and because of the subsequent impact on the work material surface, the material erodes from the target zone. The significance for the usefulness of hot abrasive particles is very important as the heated abrasive particles reduce the moisture content of carrier gas, improve impingement quality of abrasives, avoid the embedding of abrasive particles on the workpiece surface, and help to improve the abrasion rate and productivity. It has a wide range of applications in the metal-working industry for the machining of sheet and hardened metal, flaws, anodizing marks, cleaning, and dressing of the grinding wheels that are used for glass and diamond cutting of optical fibers without changing its wavelength and machining of critical sections in aircraft and surgical equipment making industries. This paper is focused on computational fluid dynamic (CFD) simulation to measure the erosion per unit area by taking different input factors, that is, sizes of abrasive particles, impact angle of air-abrasive jet, and temperature of abrasive particles along with constant velocity and stand-off distance and comparison between the simulated erosion values with the experimental result.

30.2 RESEARCH BACKGROUND

Jafar et al.[1] investigated the erosion rate and the turbulent kinetic energy on borosilicate glass at the inner surface of the nozzle and at the nozzle exit. Work-piece material erosion per unit area was calculated on aluminum material using CFD tool fluent[2] by varying the nozzle outlet orientation in abrasive water jet micromachining (AWJMM).[3] The flow characteristics, velocity, pressure trajectories, and erosion at the nozzle surface and inside the mixing chamber were analyzed in abrasive water jet cutting (AWJC) by changing the nozzle length, stand-off-distance (SOD), and velocity of the jet.[4,5] Skin friction coefficient, turbulent kinetic energy, velocity, and pressure affects the erosion rate that was analyzed in abrasive water suspension jet machining (AWSJM)[6,7] on different brittle material at the nozzle exit. Improvement of mass flow rate was analyzed in AJM[8] at the nozzle exit by using elliptical nozzle instead of using circular nozzle by which the material removal rate was increased. Jagannatha et al.[9,10] conducted an experiment on soda-lime glass with abrasive hot air jet machining with silica granules and SiC as abrasive particles to explore the result of hot air with different machining constraints, that is, SOD, grain size, and air pressure to calculate the material removal rate and surface texture at the machined zone. K-60 alumina ceramic and quartz were used[11,12] in newly developed FB-HAJM to determine the Material Removal Rate (MRR), Depth of Cut (DOC), and Tool Abrasion (TA) by taking SOD, pressure of air-jet, variable grain size of silicon carbide (SiC), and temperature of abrasive particles by using grey relational analysis (GRA) for optimization. Quartz ceramic was also used as workpiece material by Pradhan et al.[13–15] to investigate the cutting parameters using a genetic algorithm. The past research strategy which was performed by various researchers under different parametric conditions has been presented in Table 30.1.

Although, there is very less number of research on the field of simulation design, experimentation, optimization of the HAJMM, so it is a fairly novel idea of nontraditional machining using heated air-abrasive jet with different machining factors, that is, abrasive temperature, jet impact angle, and abrasive particle sizes on to the workpiece surface.

TABLE 30.1 Past Research Strategy.

Sl. no.	Process	Abrasive material & workpiece	Input & outcomes	Method/tool used	Major findings	References
1	ASJMM	Al_2O_3 & borosilicate glass	SOD, pressure, grain size & erosion rate, turbulent kinetic energy	CFD (FLUENT)	Erosion rate and turbulent kinetic energy is deeply influenced by the influencing parameters	Jafar et al. [1]
2	HAJM	Si_3N_4 & SiC	*P*, *T*, NTD and Rz, OC, FD	CFD	Surface roughness and over cut higher at extreme air pressure	Pradhan et al. [2]
3	AWJMM	Al_2O_3, SiC, SiO_2, & steel material	pressure, grain size, jet velocity, & erosion rate	CFD (FLUENT)	Variation in erosion per unit area occurs on different grades of steel.	Haghbin et al. [3]
4	AWJC	SiC, sand particle, & different grades of alumina	length, inner and outer diameter of nozzle, mixing chamber length, & erosion rate	CFD (FLUENT)	Jet trajectories and erosion at the nozzle surface and inside the mixing chamber is studied	Prisco and D'Onofrio [4]
5	AWJC	SiC, sand particle, Al_2O_3, & carbon steel	Length of nozzle and mixing chamber, nozzle diameter, SOD, & erosion rate	CFD (LS-DYNA)	Erosion occurs at the nozzle and workpiece surface, inside the mixing chamber due to jet trajectories. At different operating conditions	Feng et al. [5]
6	WJM	Water jet	Nozzle diameter, converging angle, focus tube length, & erosion rate	CFD (FLUENT)	Trajectories of jet are studied at inside and exit of nozzle and erosion on the workpiece occurs	Gupta and Sakharwade [6]
7	AWSJM	Garnet	Volume fraction, fluid density, & skin friction coefficient	CFD (FLUENT)	Skin friction coefficient and turbulent kinetic energy are determined at the nozzle exit	Candra and Diwedi [7]
8	AJM	Al_2O_3, SiC, & glass, ceramic	Velocity, jet pressure, abrasive size, SOD, & mass flow rate	CFD (FLUENT)	Mass flow rate is increased in case of using elliptical nozzle	Baranitharan and Zeelan [8]
9	A- HAJM	SiC grain & soda-lime glass	SOD, feed rate, air temperature & MRR, R_a	Taguchi method	Plastic deformation occurs at high temperature	Jagannatha et al. [9]

TABLE 30.1 *(Continued)*

Sl. no.	Process	Abrasive material & workpiece	Input & outcomes	Method/tool used	Major findings	References
10	A-HAJM	SiC, silica sand, & soda-lime glass	Air temperature, SOD, feed rate, & MRR, R_a	One factor approach	MRR and roughness are directly proportional to T_{air}	Jagannatha et al. [10]
11	FB-HAJM	SiC & K-60	Pressure, SOD, GS, T_{abr} & MRR, TA	PSO	The optimal limit is achieved at high P_{air}, GS, and average SOD	Pradhan et al. [11]
12	FB-HAJM	SiC & quartz	P_{air}, T_{abr}, NTD, & MRR, DOC	GRA	P_{air}, and T_{abr} proportional to MRR and DOC	Pradhan et al. [12]
13	FB-HAJM	SiC & quartz	P, T, SOD	GA	GA gives better optimization result	Pradhan et al. [13]
14	FB-HAJM	SiC & quartz	P, T, SOD	RSM, GA	Extreme abrasive temperature gives superior performance	Pradhan et al. [14]
15	FB-HAJM	SiC & quartz	P, T, SOD	GA	High air pressure gives best cutting quality	Pradhan et al. [15]

30.3 EXPERIMENTS

30.3.1 APPARATUS AND TARGET MATERIAL

The experiments were carried out using SiC abrasives with a 1.2 mm internal diameter tungsten carbide nozzle that grasped a nozzle-tip distance of 8 mm from the workpiece surface. The target workpiece surface was zirconia ceramic material, dimension of 20 × 20 × 4 mm. The mechanical and thermal properties of the zirconia as specified by: density: 5680 kg/m^3: specific heat: 540 J/kg k: thermal conductivity: 2 w/mk. Unmasked and straight passages were prepared by fastening the zirconia samples to a computer-operated phase and scanning it under the stationary nozzle at an inspecting velocity of 4 mm/s at different jet impact angles, θ. Figure 30.1 signifies the accurate configuration of HAJMMing of zirconia ceramic with different impingement angles.

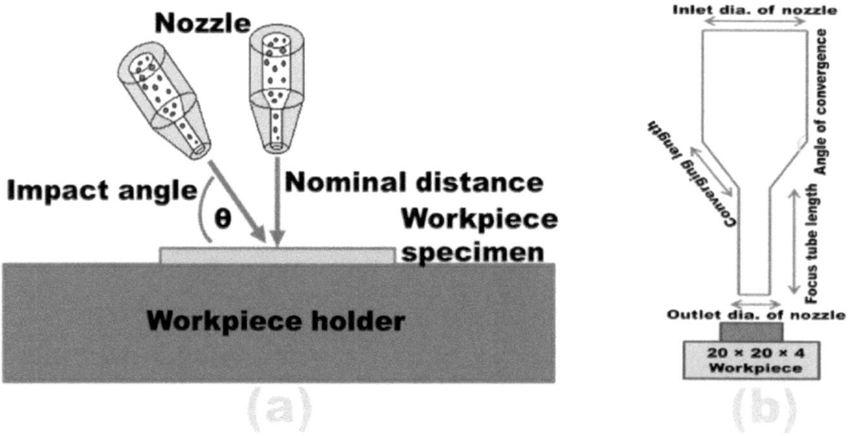

FIGURE 30.1 Schematic diagram of HAJMMing configuration.

30.3.2 CHARACTERIZATION OF ABRASIVE PARTICLE

SiC particles of several particle sizes (250, 550, 745 μm) were used as machining media. The SiC grains were experimented with according to the producer to lessen sampling bias. The circulations of the hot abrasive particles approximately equivalent to spherical diameters, $d = (A / \Pi)^{1/2}$ were determined according to the dimensions of the hot abrasive particle zones, A, acquired by an automated visual analysis system. Abrasive grains with $d < 10$

μm were eliminated, and the size circulations of 25 and 50 μm SiC abrasive particles were close-suited with a log-normal circulation. The machining procedure was executed at a constant airspeed of 550 m/s, and the connected hot abrasive particle velocity circulations within the air-abrasive jet at 8 mm from the outlet zone of nozzle were determined using very high-resolution shadowgraphy. The hot abrasive mass flow rates were between 0.05 kg/s (depending upon the abrasive grain sizes), as determined by weighing the mass of abrasive particles blasted for 10 s into an airtight machining chamber that allows the compressed air to escape from the nozzle entrance zone to the jet impingement zone at the surface of work material.

30.3.3 GEOMETRY

The novel design of nozzle is made on tungsten carbide (WC) and the general dimension of nozzle is displayed in Figure 30.1b. Three geometrical models of nozzle along with workpiece at different impact angles (90°, 115°, and 135°) and mesh model have been constructed in Figure 30.2 with constant stand-off distance and velocity trajectories using the frozen and Boolean operation of ANSYS design modeler. The geometrical parameters of nozzle as well as zirconia workpiece along with the flow parameters and workpiece properties are discussed in Tables 30.2a and 30.2b.

TABLE 30.2a Geometric Parameter of Nozzle.

Geometrical parameter	Value
Inlet diameter of nozzle	10.08 mm
Exit diameter of nozzle	4 mm
Focus tube length	20 mm
Angle of convergence (θ)	144°
Converging length	7.07 mm

30.3.4 PARTICLE EROSION IN DPM MODEL

For solving the transport equations for the continuous fluent phase, it helps us for simulating a discrete phase in a reference Lagrangian frame. Phase 2 consists of spherical particles that are dispersed during the continuous phase. Fluent finishes those phase entities' trajectories.

TABLE 30.2b Flow Parameters and Specifications of Zirconia Workpiece.

Flow parameter and specification of workpiece	Value
Mass flow rate	0.01 kg/s
Density of Zirconia	5680 kg/m^3
Specific heat of Zirconia	540 J/kg k
Thermal conductivity of workpiece	2 w/mk
Density of abrasive (Silicon carbide)	3210 kg/m^3
Specific heat of abrasive	750 J/kg k
Abrasive particle size	250, 550, and 745 μm
Operating temperature	80°C, 100°C, and 120°C
Stand-off distance	8 mm (constant)
Velocity of jet	550 m/s
Turbulence model	k–epsilon model
State	Transient
Dimension of workpiece	20 × 20 × 4 mm

Material erosion at different surfaces of the specimen can be tracked. The material erosion is designated by the following equation:

$$R_{\text{erosion}} = \sum_{P=1}^{N\text{particle}} \frac{\dot{m}_p C(dp) f(\alpha) v^{b(v)}}{A_{\text{face}}} \tag{30.1}$$

where C (dp) is the function of particle diameter, α is the impingement angle of the particle trajectories with the wall face, f (α) is impingement angle function, v is the relative particle velocity, and $b(v)$ is a function of relative particle velocity. Default values are $C = 1, f = 1, b = 0$. As C, f, and b are boundary conditions rather than properties of the material. The density of the wall material that results in the material erosion can be defined as functions C and f. When using the default values for C and f, the mass of material removed will be the erosion rates.

The force balance equates the inertia of the particles with the forces that act on the particles and can be defined as (for x, y, and z-direction in Cartesian coordinates):

$$\frac{du_p}{dt} = F_D\left(u - u_p\right) + \frac{g_z}{\rho_p}\left(\rho_p - \rho\right) + F_x \tag{30.2}$$

where F_D (u–u_p) is the drag force generated by particles

$$F_D = \frac{18\mu}{\left(\rho_p d_p^{\,2}\right)} \times \frac{C_D}{24} \times Re \tag{30.3}$$

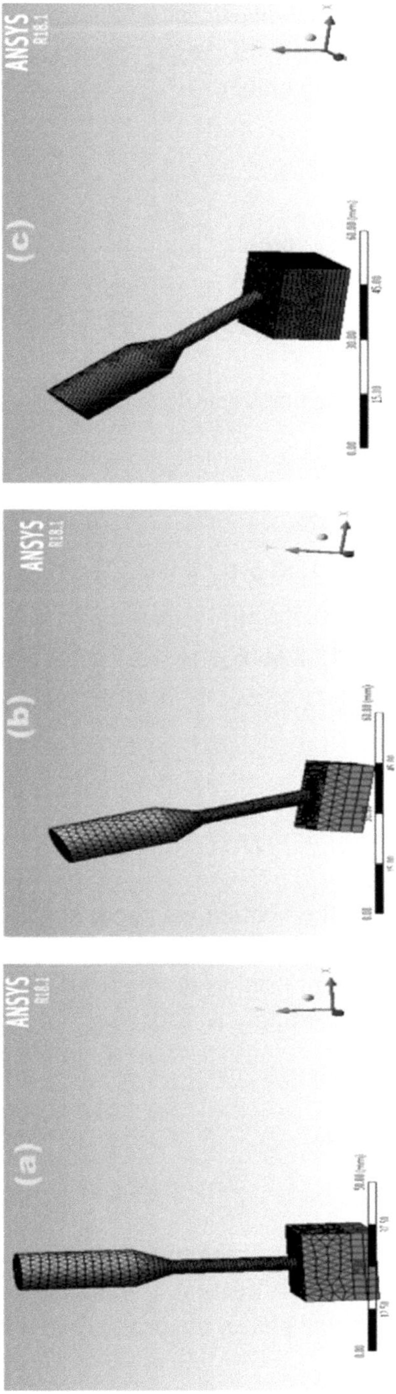

FIGURE 30.2 Meshing at jet impact angle of (a) 90°, (b) 115°, and (c) 135°.

where u = velocity of fluid phase, u_p = velocity of particles, μ = fluid molecular viscosity, ρ = density of the, ρ_p = particle density, d_p = diameter of the particle, Re = Reynolds number.

Here,

$$Re = \frac{\rho}{\mu} \times d_p \times \left| u_p - u \right| \tag{30.4}$$

Drag coefficient, C_D is expressed as:

$$C_D = a_1 + \frac{a_2}{R_e} + \frac{a_3}{R_e^{\,2}} \tag{30.5}$$

where

a_1, a_2, a_3 = constant for the particles over several frames of Re given by the following equation:

$$C_D = \frac{24}{Re} \times \left(1 + b_1 Re^{b_2}\right) + \frac{b_3\, Re}{b_4 + Re} \tag{30.6}$$

where $b_1 = \exp(2.328 - 6.4581\phi + 2.4486\,\phi 2)$

$b_2 = \exp(0.0964 + 0.5565\phi)$

$b_3 = \exp(4.905 + 13.8944\phi + 18.4222\,\phi 2 - 10.2599\,\phi 3)$

$b_4 = \exp(1.4681 + 12.2584\phi + 20.73222\,\phi 2 - 15.8855\,\phi 3)$

ϕ = Shape factor

30.3.4.1 CONTINUOUS PHASE MODEL

In hot abrasive jet flow, the continuous phase is abrasive particles and the Reynolds-averaged Navier–Stokes as incompressible flow equations based on the Eulerian methods are used to define the continuity, momentum, and energy of that process accordingly as follows:

$$\frac{\partial u_i}{\partial x_i} = 0 \tag{30.7}$$

$$\frac{\partial u_i}{\partial t} + \frac{\partial u_i u_j}{\partial x_j} = -\frac{1}{\rho}\frac{\partial p}{\partial x_i} + \mu \nabla 2\, u_i - \frac{\partial u_i' u_j'}{\partial x_j} + g_i \tag{30.8}$$

where u_i' = velocity component due to fluctuation in i-direction, u_i = instantaneous velocity component in i-direction, x_i = space coordinates in i-direction, μ = dynamic viscosity of the phase, p = pressure of the phase, ρ = density of the phase.

Here Realizable k-ε model for turbulence flow-id defined by the following equations:

$$\frac{\partial(\rho k)}{\partial t}+\frac{\partial(\rho k v_i)}{\partial x_i}=\frac{\partial}{\partial x_j}\left[\left(\mu+\frac{\mu_t}{\sigma_k}\right)\frac{\partial k}{\partial x_j}\right]+G_k+G_b-\rho_\varepsilon-Y_M \tag{30.9}$$

$$\frac{\partial(\rho \varepsilon)}{\partial t}+\frac{\partial(\rho \varepsilon v_i)}{\partial x_i}=\mathrm{S}\frac{\partial}{\partial x_j}\left[\left(\mu+\frac{\mu_t}{\sigma_\varepsilon}\right)\frac{\partial \varepsilon}{\partial x_j}\right]$$

$$+\rho C_1 S_\varepsilon-\rho C_2\frac{\varepsilon^2}{k+\sqrt{v\varepsilon}}+C_{1\varepsilon}\frac{\varepsilon}{k}C_{3\varepsilon}G_b \tag{30.10}$$

where i, j = coordinate axis, v = fluid velocity, μ_t = turbulent viscosity, G_k = turbulent kinetic energy generated because of fluid flow, G_b = generation of turbulent kinetic energy due to lift, Y_M = impingement of compressible turbulence inflation on the total dissipation rate, σ_k and σ_ε = Prandtl number due to turbulent flow for k and ε, C_1, C_2, $C_{3\varepsilon}$ = empirical constant.

30.3.4.2 ENERGY EQUATION

Energy equation which is solved by the CFD tool ANYS Fluent is given by

$$\frac{\partial(\rho e)}{\partial t}+\nabla\cdot(\rho U e)=-\nabla\cdot\mathrm{Q}+S_e \tag{30.11}$$

where $e = C_pT$

So eq 30.11 can be written as

$$\frac{\partial(\rho C_p \mathrm{T})}{\partial t}+\nabla\cdot(\rho \mathrm{U} C_p \mathrm{T})=-\nabla\cdot q+S_e \tag{30.12}$$

Use Fourier's law for the heat flux q, $q = -k\nabla T$ and putting q value in the above equation:

$$\frac{\partial(\rho C_p T)}{\partial t}+\nabla\cdot(\rho \mathrm{U} C_p T)=-\nabla\cdot(-\mathrm{k}\nabla T)+S_e \tag{30.13}$$

30.4 RESULTS AND DISCUSSION

Square specimens of zirconia ceramic (ZrO_2) were used as work material to measure the erosion rate using simulation technique and experimental method by taking different processes constraints, that is, temperature of abrasive grains, impact angle of jet, and grain size of abrasive with a constant velocity of jet and stand-off distance.

30.4.1 SIMULATION METHOD FOR TARGET MATERIAL EROSION

The collision model of the hot abrasive particle also exposed that these hot particles' fluidities were adequately low to create the result of interparticle impacts insignificant. Material erosion related target surface damage initiated by individual hot abrasive grains impacts. Succeeding this methodology, single hot abrasive impact places were produced by revealing the zirconia ceramic specimen to the high-velocity stream of hot SiC abrasive particles for 15 s using a heating-cum-mixing chamber. Hot abrasive particles with 250, 550, and 745 μm were blasted with the air jet of 550 m/s on the target surface were measured by the help of optical microscope profilometer with a depth resolution of 5 nm. The equation of motion of HAJMM is solved by CFD tool ANSYS Fluent. The HAJMM setup, nozzle, and workpiece were discretized into 87,655 elements after frozen operation in the design modeler and also the governing equation of HAJMM was solved by pressure–velocity-based transient model and K-epsilon turbulence model with PISO scheme. The governing equation (Navier–Stoke equation) was solved by ANSYS Fluent 18.1. Therefore transient analysis was implemented in HAJMM using abrasive temperature, jet impact angle, and abrasive particle sizes as flow parameter. The governing mass, momentum conservation, and continuity equations are discussed below.

$$\text{Continuity,}\ \frac{\partial P}{\partial x} + \operatorname{div}(\rho u) = 0 \tag{30.14}$$

$$X\text{-momentum,}\ \frac{\partial(\rho u)}{\partial x} + \operatorname{div}(\rho u u) = -\frac{\partial P}{\partial x} + \operatorname{div}(\mu \operatorname{grad} u) + S_{\text{mx}} \tag{30.15}$$

$$Y\text{-momentum,}\ \frac{\partial(\rho v)}{\partial x} + \operatorname{div}(\rho \text{uv}) = -\frac{\partial P}{\partial x} + \operatorname{div}(\mu \operatorname{grad} v) + S_{\text{mx}} \tag{30.16}$$

$$Z\text{-momentum,}\ \frac{\partial(\rho z)}{\partial x} + \operatorname{div}(\rho \text{uz}) = -\frac{\partial P}{\partial x} + \operatorname{div}(\mu \operatorname{grad} z) + S_{\text{mx}} \tag{30.17}$$

$$\text{Energy,}\ \frac{\partial(\rho i)}{\partial x} + \operatorname{div}(\rho \text{iu}) = -P\operatorname{div}(u) + \operatorname{div}(K \operatorname{grad} T) + \phi + S_i \tag{30.18}$$

30.4.2 DESIGN OF EXPERIMENT

Design of experiment (DOE) is a strategy for defining and developing knowledge scientifically so that it can be used in a well-structured manner to improve the machining process. According to the definition, DOE can be applied to the regions where it is necessary to acquire and apply the information efficiently and quickly in any field of requirements. Here, Table 30.3 represents the Taguchi matrix for simulation as well as experiment. According to this, the simulation result of velocity profile and erosion rate and experimental erosion rate have been discussed.

TABLE 30.3 Taguchi Matrix for Simulation and Experiment.

Run no.	Abrasive temperature (°C)	Jet impact angle (°)	Grain size (μm)	Simulation value of erosion (kg/m^2)	Experimental value of erosion (kg/m^2)
1	80	135	250	2.94E–09	2.5613E–09
2	80	115	550	3.67E–08	3.0682E–08
3	80	90	745	7.07E–08	5.9688E–08
4	100	135	550	1.33E–07	1.0266E–07
5	100	115	745	2.5E–08	2.1023E–08
6	100	90	250	1.9E–07	1.4988E–06
7	120	135	745	3.3E–09	1.999E–09
8	120	115	250	4.61E–08	4.0057E–08
9	120	90	550	7.43E–08	5.0325E–08

30.4.3 EFFECT OF MACHINING FACTORS ON MATERIAL EROSION

Figure 30.3 displays the velocity profile of an air-abrasive jet in the machining zone where hot SiC particles deeply influence the zirconia workpiece surface. Abrasive temperature having proportional effect with material erosion and in the case of increasing grain size, also erosion rate is increased. In the case of normal jet impact angle, material erosion rate is maximum because of huge kinetic energy at impingement zone. Thus, all the machining factors proportionally effect the erosion rate of the workpiece surface due to which a machining surface has been generated.

30.4.4 COMPARISON OF EROSION RATE BETWEEN SIMULATION AND EXPERIMENTAL VALUE

Maintaining the abrasive particle temperature varies from 80°C to 120°C, maximum erosion rate of 1.16E–07 kg/m^2, and minimum erosion of 3.68E–08 kg/m^2 were obtained by simulation at abrasive temperature of 100°C and

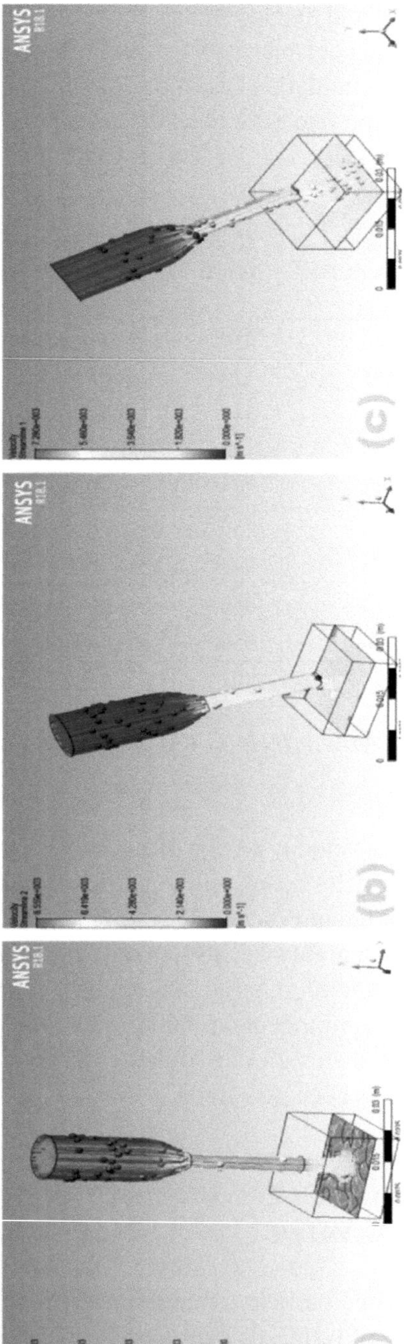

FIGURE 30.3 Velocity profile at (a) 90°, (b) 115°, and (c) 135°.

80°C, respectively, and the corresponding maximum and minimum erosion, that is, 1.02E–07 and 3.09E–08 kg/m^2 were obtained for the same value of abrasive temperature in the case of an experiment, as shown in Figure 30.4a. By illustrating Figure 30.4b, maximum erosion rate of 1.16E–07 kg/m^2 and minimum erosion rate of 3.59E–08 kg/m^2 were obtained at 90° and 115° of jet impact angle in the case of simulation and for this same value of jet impact angle, the experimental value of maximum and minimum erosion rates were 9.60E–08 and 3.08E–08 kg/m^2, respectively, and in the case of comparative study, this results of maximum erosion rate was obtained by simulation, respectively. While considering the grain size of abrasive particles varies from 250 to 745 μm, maximum erosion rate of 8.12E–08 kg/m^2 occurred at 550 μm grain size of abrasive and minimum erosion rate of 3.3E–08 kg/m^2 was obtained at an abrasive grain size of 745 μm using simulation and maximum erosion rate of 7.76E–08 kg/m^2 and minimum erosion rate of 2.80E–08 kg/m^2 were obtained in Figure 30.4c using experimental investigation.

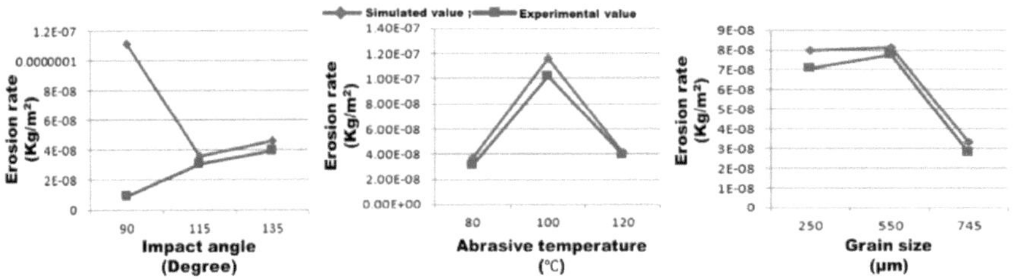

FIGURE 30.4 Effect of machining parameters on erosion rate.

30.4.5 DESIRABILITY FUNCTION APPROACH FOR EROSION PERFORMANCE

For all technical responses, the optimization strategy focused on the desirability feature approach was presented, providing several optimal defined production conditions for heated abrasive jet micromachining of zirconia ceramic with an abrasive temperature of 89.15°C, a jet impact angle of 90.52°, and an abrasive grain size of 618.83 μm. The predicted optimal value of technological response, that is, erosion rate is 2.09273E–07 kg/m^2. The point on the diagram depicts the outcome response value, that is, erosion rate and the actual elevation are achieved for all distinct point that indicates how often the optimal values are desirable from the graph. All the desirable values have been checked by validation testing. The optimal constraint values are obtained from the ramp function points graphs and the near-optimal value, which shows the closeness of response, as found to be almost 1. Figure 30.5 reveals the contour plot for material erosion and desirability, whereas Figure 30.6 reveals the bar graph and ramp function graph for desirability approach. The value for the response's complete desirability function ranges from 0 to 1. It detects the complete desirability value of the response values. The values are determined to be close to 1. It clarifies the accurateness of the values are achieved for machining parameters on increasing erosion rate. For an abrasive temperature, jet impact angle and abrasive grain size, the desirability value obtained is almost equal to 1, which designates that the process parameters met the machining constraints.

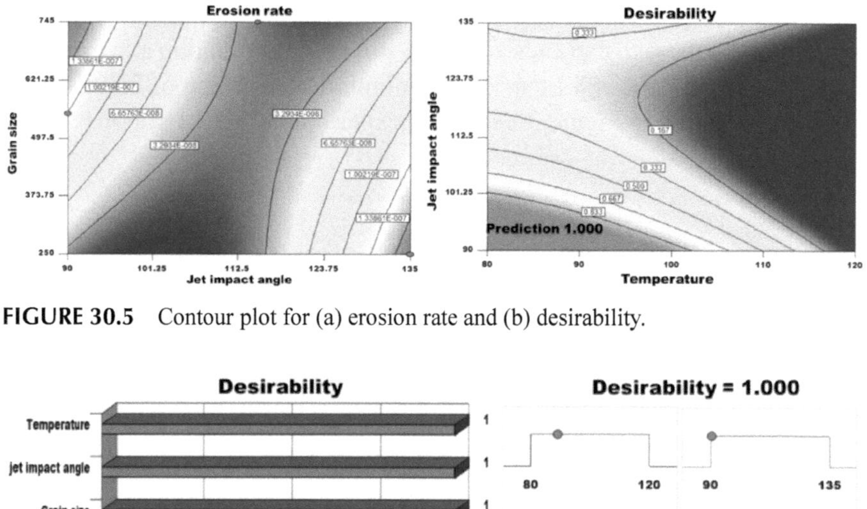

FIGURE 30.5 Contour plot for (a) erosion rate and (b) desirability.

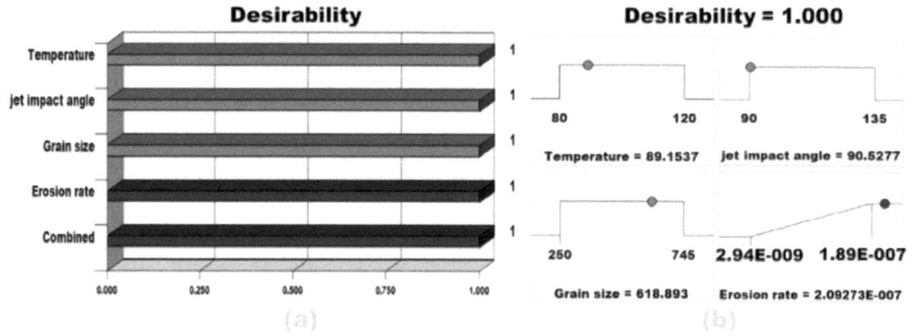

FIGURE 30.6 Plot for (a) desirability and (b) ramp function.

30.5 CONCLUSIONS AND FUTURE TRENDS

The erosion rate resulting from HAJMM of zirconia ceramic was performed systematically both experimentally and modeled. The results of two approaches (simulation and experimental approach) display minor variations if any. The following points are determined in this research.

- It was presented that target material erosion per hot abrasive particle was directly proportional to the kinetic energy of the hot abrasive particle.
- Reduction in jet impingement force at the target zone of work material took place by reducing of hot abrasive jet impingement angle at the striking zone of work material, which flatter passage. The comprehensible analytical explanation of air-abrasive stream in the initial jet region was suggested to attend this purpose.
- The optimal production conditions for hot abrasive jet micromachining on zirconia are obtained at an abrasive temperature of 89.15°C, a jet impact angle of 90.52°, and a grain size of 618 μm by solving the single response optimization problem of erosion efficiency with RSM's desirability feature study. The predicted optimal result of technological responses is 0.000000209273 kg/m^2 for material erosion.
- The result achieved in this investigation is more valuable for additional describing the HAJMM method, namely for decreasing error in prediction of material erosion rate, HAJMMed zirconia profile, and target surface damage.
- Therefore, as a future tendency, we designate the improvement of analytical machine tools capable to identify modifications in abrasive grain dimensions after their breakage, and measurement of extents of this progression in their geometries.

ACKNOWLEDGMENTS

The authors would sincerely thank RIFD Director and NDF Chairman of AICTE, India, for financial support for project n\o. 8-32/RIFD/ RPS—NDF/ Policy-1/2019-2020. The authors also acknowledge the practical support of Production Engineering Department of VSSUT, Burla.

KEYWORDS

- **zirconia ceramic**
- **HAJMM**
- **CFD**
- **material erosion**

REFERENCES

1. Jafar, M. H. R.; Nouraei, H.; Emamifar, M.; Papini, M.; Spelt, J. K. Erosion Modeling in Abrasive Slurry Jet Micro-Machining of Brittle Materials. *J. Manuf. Process.* **2015,** *17*, 127–140.
2. Pradhan, S.; Dhupal, D.; Das, S. R.; Jena, P. C. Experimental Investigation and Optimization on Machined Surface of Si_3N_4 Ceramic Using Hot SiC Abrasive in HAJM. *Mater. Today Proc.* **2021**. https://doi.org/10.1016/j.matpr.2020.12.066.
3. Haghbin, N.; Spelt, J. K.; Papini, M. Numerical Simulation of Single Particle Acceleration Process by SPH Coupled FEM for Abrasive Water Jet Cutting. *J. Mater. Process. Technol.* **2015,** *222*, 399–409.
4. Prisco, U.; D' Onofrio. Three Dimensional CFD Simulation of Two- Phase Flow inside the Abrasive Water Jet Cutting Head. *Int. J. Comput. Methods Eng. Sci. Mech.* **2008,** *9* (5), 300–309.
5. Feng, Y. U.; Jianming, W.; Feihong, L. Abrasive Water Jet Micro Machining of Channel in Metal: Model to Predict High Aspect Ratio Channel Profiles for Submerged and Unsubmerged Machining. *Int. J. Adv. Manuf. Technol.* **2011,** *59*, 193–200.
6. Gupta, M. K.; Sakharwade, S. Computational Optimization of Water Jet Machining: Effect of Nozzle Jet Diameter Ratio. *Int. Res. J. Eng. Technol. Value* **2017,** *04* (Issue), 05.
7. Chandra, R.; Dwivedi, A. Computational Analysis of Abrasive Water Suspension Jet with Impact of Inlet Operating Pressure, Kinetic Energy and Volume Fraction. *Int. J. Latest Res. Eng. Technol.* **2016**, 74–82
8. Baranitharan, P.; Zeelan, B. K. Design and Analysis of Elliptical Nozzle in AJM Process Using Computational Fluid Dynamics. *Sci. Technol. Arts Res. J.* **2015,** *4* (1), 171–179.
9. Jagannath, N.; Hiremath, S.; Sadashivappa, K. Analysis and Parametric Optimization of Abrasive Hot Air Jet Machining for Glass Using Taguchi Method and Utility Concept. *Int. J. Mech. Mater. Eng.* **2012,** *7* (1), 9–15.
10. Jagannatha, N.; Hiremath, S.; Sadashivappa, K.; Arun, K. V. Machining of Soda Lime Glass Using Abrasive Hot Air Jet. *Mach. Sci. Technol.* **2012,** *16*, 459–472.
11. Pradhan, S.; Behera R. K.; Nanda, B. K.; Das, S. R.; Dhupal, D. Drilling of K-60 Alumina Ceramic with Different Grades Abrasives at Various Temperatures using Fluidized BED-Hot Abrasive Jet Machining (FB-HAJM) Process. *IOP Conf. Series: Mater. Sci. Eng.* **2018,** *455*, 012076.
12. Pradhan, S.; Tripathy, S. S.; Sahu, S.; Das, S. R.; Jena, P. C.; Dhupal, D. Investigation on MRR and DOC of the Micro-Holes Generated on Quartz Using Silicon Carbide by FB-HAJM. *Mater. Today: Proc.* **2020.** https://doi.org/10.1016/j.matpr.2020.02.437.
13. Pradhan, S.; Das, S. R.; Nanda, B. K.; Jena, P. C.; Dhupal, D. Experimental Investigation on Machining of Hardstone Quartz with Modified AJM Using Hot Silicon Carbide Abrasives. *J. Braz. Soc. Mech. Sci. Eng.* **2020**. https://doi.org/10.1007/s40430-020-02644-4.
14. Pradhan, S.; Das, S. R.; Nanda, B. K.; Jena, P. C.; Dhupal, D. Machining of Hardstone Quartz with Modified AJM Process Using Hot SiC Abrasives: Analysis, Modelling, Optimization, and Cost Analysis. *Surf. Rev. Lett.* **2020**. https://doi.org/10.1142/S0218625X20500493.
15. Pradhan, S.; Das, S. R.; Dhupal, D. Performance Evaluation of Recently Developed New Process HAJM during Machining Hardstone Quartz Using Hot Silicon Carbide Abrasives: an Experimental Investigation and Sustainability Assessment. *Silicon* **2020**. https://doi.org/10.1007/s12633-020-00641-9.

CHAPTER 31

MATERIAL REMOVAL RATE ANALYSIS FOR ECONOMICAL FEASIBILITY STUDY IN ABRASIVE JET MACHINING WITH DIFFERENT TEMPERATURES USING K-92 CERAMIC

SUBHADIP PRADHAN*, SUSHREESUDHA SAHU, SOURAV SANKET TRIPATHY, SUDHANSU RANJAN DAS, and DEBABRATA DHUPAL

Production Engineering Department, Veer Surendra Sai University of Technology, Burla, Sambalpur 768018, Odisha, India

Corresponding author. E-mail: subha.mid94@gmail.com

ABSTRACT

This research work is concerned with micromachining of high strength ceramic named K-92 which is carried out through abrasive jet machining with different temperatures. As smart and brittle materials are next-generation materials with an intrinsic intelligence to adopt modification in the exterior environment, it becomes a challenge to machine these ceramic materials by the means of traditional methods of machining. The fluidized bed–hot abrasive jet machining (FB–HAJMing) has the potential to machine such smart and high strength materials. However, in abrasive jet assisted drilling and deburring applications, intricate shaped microholes are one of the important features and are often fabricated by FB-HAJM process. This study investigates the possibility of using K-92 ceramic with silicon carbide

Modern Manufacturing Systems: Trends and Developments. Rajiv Kumar Garg, Ravi Pratap Singh, Rajeev Trehan, & Ramesh Singh (Eds.)

abrasive in fluidized bed–hot abrasive jet micromachining. Experimental studies are assessed to execute cost analysis for the economical explanation and viability of K-92 ceramic using the FB-HAJM process. The influence of process constraints such as air pressure, abrasive temperature, stand-off-distance, and grain size on one of the major performance variables, that is, the material removal rate is also investigated and discussed. Sixteen sets of longitudinal microgrooving trials based on Taguchi L16 orthogonal design of experiments (DOEs) are implemented. For the purpose of improving the machining performance, a technique called response surface methodology (RSM) followed by particle swarm optimization (PSO) is employed to progress the mathematical relationships between the influencing input parameters/factors and the response characteristics. Analysis of variance (ANOVA)is applied for the verification of the adequacies of the models. Response surface plot for each response and condition of microholes is analyzed through SEM and optical microscope.

31.1 INTRODUCTION

In search of novel alloys and materials of higher strength and lower weight that are difficult to machine and to achieve the necessary accuracy and precision, research and technology are advancing to an infinite scale. Also many of the recent developments in the aerospace and nuclear engineering are partly related with the increasing use of difficult-to-machine materials such as nimonics, heat-resisting steels, hastalloy, waspalloy, nitralloy, monolithic, stainless steel, carbides, composite ceramics, metal–matrix composites, aluminides, etc. Here, such enormous claim for the fine finished products of these constituents with complex intricate designs and high accuracy affect the machining process and its economy. In today's high-tech precision industries, the necessities for the components are inflexible, such as, unusual properties of materials i.e. high hardness, high strength, corrosion resistant, high heat resistant, intricate 3D modules (say, turbine blade), nanolevel surface finish on intricate geometries, miniature features (filters for food handling and textile industries having a limited tens of micrometer as hole diameter and thousands in numbers like thousands of tubulated cooling holes in a turbine blade, making and finishing of micro fluidic channels in the electrically conducting an nonconducting materials (say, ceramics, quartz, glass) are impossible to achieve by any traditional method. The conventional machining processes are inadequate to machine these materials because of the following: (1) dimensional accuracy and precision, (2) low machinability, (3) low production rate, (4) higher economy, and (5) undesirable deformations and residual stresses on the workpiece. Some structures on a component can be accomplished only through innovative manufacturing processes. Considering the seriousness of these problems, Merchant first emphasized the need for the development of a newer technique in machining process by studying the steady growth in strength of material with yearly development of material for the aerospace industry. This manufacturing revolution was the introduction to nontraditional manufacturing (NTM) processes (or nonconventional machining process or modern machining process), a new manufacturing process that can be applied for material removal, forming, and joining processes. They are unconventional machining processes because the conventional machine tools are not employed for metal cutting, there is no shear stress at the interface of the tool-workpiece, and they directly utilize some form of energy. More than 20 various unconventional manufacturing processes have been designed and effectively executed into precision and production engineering during the last 50 years.

31.2 RESEARCH BACKGROUND

Jagannatha et al.[1,2] investigated the compressed air temperature effect on material removal rate (MRR) and target substrate roughness of soda-lime glass plate and found that better result attained using abrasive hot-air jet by using the improved Taguchi robust design method. The novel idea of application of hot abrasives was positively shown while drilling of alumina ceramic and borosilicate glass material using fluidized bed-hot abrasive jet machining[3,4] and simultaneously the effect of abrasive temperature, bed pressure, stand-off-distance (SOD), different grades of grains on machining performance were investigated. The effects of a set of machined parameters pressure, abrasive temperature, and SOD on displacement of cut and MRR were investigated and optimized by using grey relational analysis (GRA) in FB-HAJM of quartz by Pradhan et al.[5] The effect of air pressure, nozzle-tip distance, abrasive grit size, stand-off distance[6–8] on MRR, surface roughness, taper angle, and depth of cut was studied and optimized and air pressure was found to be more significant followed by SOD. The effect of simulated machining parameters had been considered[9,10] by the use of ANOVA (GRA) to find out the displacement of cut, kerf width (both top and bottom), surface roughness, and machinability using machining parameters, that is, jet pressure, traverse speed, feed rate, SOD on Ti6AL4V, and other brittle materials using garnet abrasive in AWJM. A theoretical model was presented[11,12] to analyze the advancement of AJM, AJP, and its use in different manufacturing industry in which quartz was utilized as work material for making different types of products. Quartz ceramic and Si_3N_4[13] were also used as workpiece material by Pradhan et al.[14–16] to investigate the cutting parameters using genetic algorithm. The past research strategy selection done by various researchers under different criteria in different conditions is presented in Table 31.1.

Although, there is negligible research in the field of design, fabrication, experimentation, and optimization of the hot abrasive jet machining (FB-HAJM), so it is a relatively new concept of nontraditional machining employing high temperature air-abrasive stream with different machining parameters, that is, different materials and grit sizes on to the workpiece.

31.3 RESEARCH METHODS

31.3.1 EXPERIMENTAL SETUP

The novel idea of heating-cum-mixing chamber is fabricated according to the principle of fluidized bed mixing so that the homogenous mixing of air and abrasive is possible at more turbulent flow of the mixture with minor damage to the hot abrasive cutting edges on the workpiece surface. The abrasive feed chamber is fabricated on the basis of pressurized powder feed system in which abrasive is forced into the heating-cum-mixing chamber with the help of pressurized air which ensures an efficient use of the abrasives with lesser wastage. Thereafter, abrasive particles are heated within the requisite temperature by the help of heating coil which is surrounded to mixing chamber followed by a suitable regulator. The horizontal plane motion (X-Y table), vertical (Z-axis) motion, fluidized bed heating-cum-mixing chamber, nozzle, nozzle holder, pressurized power feed abrasive chamber, etc., are appropriately designed and, associated with the help of a CAD software. Finally, the indigenously developed fluidized bed HAJMing is set up and installed. Figure 31.1 represents the developed HAJM setup for cutting operation of ceramic materials.

31.3.2 MATERIALS AND METHODS

Materials used are K-92 alumina ceramic with 99.8% Al_2O_3 and less percentage of SiO_2, Fe_2O, Na_2O, MgO, and CaO. Different physical properties like porosity, melting point, and higher density of 3.90 g/cc with

TABLE 31.1 Criteria for Past Research Strategy.

Sl. no.	Process	Abrasive and work-piece material used	Machining parameters and outcomes	Methods studied	Major findings	References
1.	A-HAJM	SiC grain and Soda-lime glass plate	SOD, feed rate, air temperature, and Ra, MRR	Taguchi method	Plastic deformation occurred at high temperature.	Jagannatha et al. [1]
2	A-HAJM	SiC, Silica sand, and Soda-lime glass	Air temperature, SOD, feed rate, and MRR, Ra	One factor approach	MRR and roughness are directly proportional to T_{air}.	Jagannatha et al. [2]
3	FB-HAJM	SiC K-60 alumina ceramics	Pressure, SOD, GS, T_{abr}, and MRR, TA	PSO	The optimal limit achieved at high P_{air}, GS and average SOD.	Pradhan et al. [3]
4	FB-HAJM	Zircon sand borosilicate glass	P_{air}, T_{abr}, NTD, grain size, and MRR, FD, Ra	ANOVA	All the inputs at their higher level cause better MRR.	Nanda et al. [4]
5	FB-HAJM	SiC abrasive and quartz	P_{air}, T_{abr}, NTD, and MRR, depth of cut (DOC)	GRA	Pressure has high significance on both MRR and DOC.	Pradhan et al. [5]
6	µ-AJM	SiC abrasive and quartz	P_{air}, SOD, abrasive grit size, and MRR, ROC, TA	GRA, ANOVA	At high temperatures more plastic deformation occurred following brittle fracture.	Tomy and Hiremath [6]
7	P-FB AJD	SiC and K-60 Alumina ceramic	Pressure, NTD, Grain size, and MRR, Ra, DOC	PCA	Grain size was found to be potential factor.	Nanda et al. [7]
8	FB-AJM	SiC and GFRP composite	P_{bed}, NTD, GS, and MRR, Ra, DOC, TA	Desirability based PSO	The shearing and rupture causing due to pull out effect.	Mogul et al. [8]
9	AWJC	Garnet and titanium alloy Ti-6Al-4V	P_{water}, mass flow rate, orifice size, and DOC, Ra	GRA	Water pressure found to be the most effective factor.	Mogul et al. [8]
10	AWJM	Angular shaped abrasives and CFRP	Jet pressure, feed rate, SOD, and kerf taper, Ra	ANOVA	Minimum Ra is achieved at high P_{air} with low feed rate.	El-Hofy et al. [9]
11	AWJM	Theoretical study				Natarajan et al. [10]

TABLE 31.1 *(Continued)*

Sl. no.	Process	Abrasive and work-piece material used	Machining parameters and outcomes	Methods studied	Major findings	References
12	AAJP	Al_2O_3 and Quartz glass	Jet pressure, abrasive size, and Ra	Predictive model	Elastic spring-back nature studied after abrasive impact.	Wang et al. [11]
13	HAJM	Si_3N_4 and SiC	P, T, NTD, and Rz, OC, FD	CFD	Surface roughness and over cut higher at extreme air pressure.	Pradhan et al. [12]
14	FB-HAJM	SiC and Quartz	P, T, and SOD	GA	GA gives better optimization result.	Pradhan et al. [13]
15	FB-HAJM	SiC and quartz	P, T, and SOD	RSM, GA	Extreme abrasive temperature gives superior performance.	Pradhan et al. [14]
16	FB-HAJM	SiC and quartz	P, T, and SOD	GA	High air pressure gives best cutting quality.	Pradhan et al. [15]

airtight arrangement indicates its difficult-to-machine characteristics. Square samples of K-92 alumina ceramic material has been chosen as the workpiece for investigation with specification of 25 mm × 25 mm × 3 mm (with ± 2% dimensional accurateness). The scope, nature, and shape of the workpiece and the preferred finish will affect the selection of the abrasive particle used. The engineering materials above feature tremendous hardness for machining in developed HAJMing setup. Hence black-colored silicon carbide (SiC) abrasive with 97% purity is used here for conducting the experiments with four grain sizes of 109, 260, 525, and 745 μm are taken to conduct experiments. Figure 31.2 represents the scanning electron microscopic image of SiC abrasive particles and CAD model of K-92 ceramic. The physical properties and chemical composition of the SiC abrasives as well as K-92 ceramic are given below in Tables 31.2a and 31.2b.

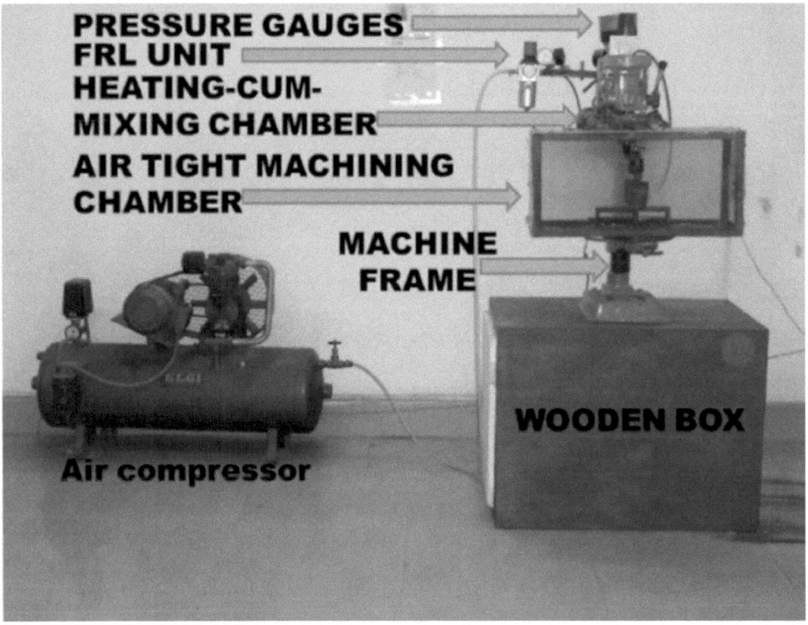

FIGURE 31.1 Developed FB-HAJMing setup.

31.4 RESULTS AND DISCUSSION

K-92 alumina ceramic samples were arranged sensibly by the help of various conventional techniques to measure the material removal rate. To determine the results of this study, for the measurement the MRR, only easy-to-control process constraints, that is, different abrasive particle size (GS_{ap}) of SiC, compressed air pressure (P_{Cair}), stand-off-distance (Z), and abrasive particle temperature (T_{ap}) were considered.

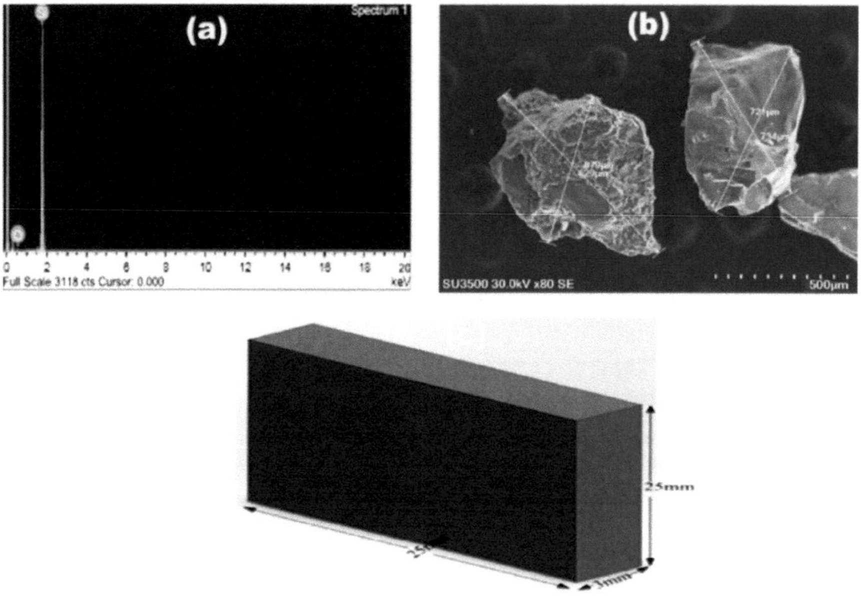

FIGURE 31.2 (a) EDS of SiC, (b) SEM of SiC and (c) CAD view of K-92.

TABLE 31.2a Physical Properties and Chemical Composition of SiC Abrasive.

Physical properties	Hardness (MOH)		Bulk density:		Specific gravity
	9.4		**1.4–1.53 g/cc**		**3.2.**
Chemical composition	SiC	Si+SiO_2	Fe_2O_3	Al_2O_3	Free C
	97.04	1.4	0.8	0.32	0.4

TABLE 31.2b Physical and Chemical Properties of K-92 Workpiece.

Physical properties	Melting Point	Limit of use	Porosity	Density	Structure	
	2000°C	**1900°C**	**0.1%**	**3.90 g/cc**	**Gas tight**	
Chemical composition	Al_2O_3	SiO_2	Fe_2O	Na_2O	MgO	CaO
	99.8%	0.03%	0.02%	0.07%	0.05%	2.02%

31.4.1 DESIGN OF EXPERIMENTS

Design of Experiments (DOEs) is a technique of scientifically finding and establishing information so that it can be used to improve the processes in the best well-organized manner possible. According to the definition, DOE can be applied to any area in which it is important to obtain and apply knowledge quickly and efficiently in any field of human endeavor. So, DOE should be seen as an opportunity for providing almost limitless benefits to the user. If the user concentrates on the knowledge acquired by DOE and the application of that knowledge to the organization, then instead of the statistical process of obtaining and manipulating the results, better benefits will be available. The most important fact for the DOE practitioner is to remember that the ultimate capability and responsibility for improving the process, design, or other operation, lies with the person whose job assignment requires it. DOE and the statistical methods associated with it can generate no knowledge, they can only help to extract and organize the knowledge already there. Table 31.3 specifies the MRR value obtained with the newly designed FB-HAJM setup with coded and actual value.

TABLE 31.3 The BBDOE Matrix for Experimentation.

Run no.	P_{Cair} (kgf/cm^2)	Z (mm)	T_{ap} (°C)	GS_{ap} (μm)	MRR (g/s)
1	4	4	20	109	0.00837
2	4	6	50	260	0.00581
3	4	8	80	525	0.00808
4	4	10	110	745	0.01158
5	5	4	50	525	0.0071
6	5	6	20	745	0.01594
7	5	8	110	109	0.01352
8	5	10	80	260	0.01271
9	6	4	80	745	0.01276
10	6	6	110	525	0.003504
11	6	8	20	260	0.02554
12	6	10	50	109	0.0174
13	7	4	110	260	0.0175
14	7	6	80	109	0.00787
15	7	8	50	745	0.02349
16	7	10	20	525	0.05123

31.4.2 DEVELOPMENT OF PREDICTIVE MODEL AND ANOVA

The process models for the outcome (MRR) found according to the Box-Behken's design (BBD) of response surface methodology (RSM) were investigated with the given machining constraints (P_{Cair}, Z, T_{ap} and GS_{ap}) through regression analysis and are given below.

$$\begin{aligned}\text{MRR} &= +0.18444 - 0.076099 \times \text{P}_{\text{Cair}} + 1.08180 \times 10^{-3} \times \text{SOD} - 4.28614 \times 10^{-4} \\ &\times T_{\text{ap}} + 1.04657 \times 10^{-4} \times \text{GS}_{\text{ap}} - 1.16278 \times 10^{-3} \times P_{\text{Cair}} \times \text{SOD} - 1.55623 \times 10^{-4} \\ &\times P_{\text{Cair}} \times T_{\text{ap}} + 9.67056 \times 10^{-6} \times P_{Cair} \times \text{GS}_{\text{ap}} + 2.90904 \times 10^{-5} \times \text{SOD} \times T_{\text{ap}} \\ &- 1.50983 \times 10^{-5} \times \text{SOD} \times \text{GS}_{\text{ap}} - 2.07635 \times 10^{-7} \times T_{ap} \times \text{GS}_{\text{ap}} + 8.68166 \times 10^{-3} \\ &\times P_{\text{Cair}}^2 + 7.10538 \times 10^{-4} \times \text{SOD}^2 + 7.57273 \times 10^{-6} \times T_{\text{ap}}^2 - 5.04987 \times 10^{-8} \times \text{GS}_{\text{ap}}^2 \quad (31.1)\end{aligned}$$

ANOVA is carried out systematically via statistical analysis, based on MRR results that are achieved through FB-HAJMing experimentation, which denotes a table with degrees of freedom (DoF), mean as well as sum of squares (MS and SS), probability and Fisher's values (*P* and *F*), and is used to check and determine the validity of the established regression mode with significance. Naturally, if the *P*-value is below 0.05, and the *F*-value considered is beyond the standardized Fisher's value, the statistical significance is measured at 95% of the confidence level. It is clearly seen from Figure 31.3a that MRR is initially decreased and thereafter increased from the rising abrasive temperature from 65°C to 110°C due to high kinetic energy propulsion. This shows that the compressed air pressure has proportional effect on MRR. Figure 31.3b reveals that MRR is directly proportional to SOD and abrasive particle size, that means if SOD and GS_{ap} increases from 4to 10 mm and 109to 745 μm then MRR is also increased due to variation of abrasive mesh size as well as nozzle entrance to workpiece surface distance. From Table 31.4, it is detected that the established modeling approach for the outcome MRR in HAHMing is significant. Figure 31.4b provides a suitable approach to find out an appropriate transformation without engaging in a lot of operative test and error. Ideally, the residual values are distributed along the horizontal axis steadily and at random manner to find the regression problems for MRR as shown in Figure 31.4a.

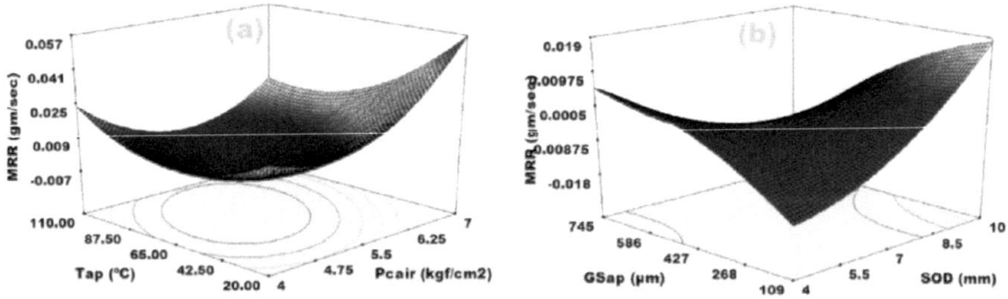

FIGURE 31.3 Surface plots for illustration of machining constraints effect on MRR.

TABLE 31.4 ANOVA Table for MRR.

Source	Sum of square	df	Mean square	F value	P-value Prob > F
Model	1.954E–003	14	1.396E–00400	3.083 + 005	<0.0014 (Significant)
Residual	4.528E–010	1	0.093		
Cor total	1.954E–003	15			
Std deviation	2.128E–005	R-Squared	1.000		
mean	0.015	Adj R-Squared	1.000		
C.V %	0.14	Pred R-Squared	0.9998		
Press	3.699E–003	Adeq Precision	2316.536		

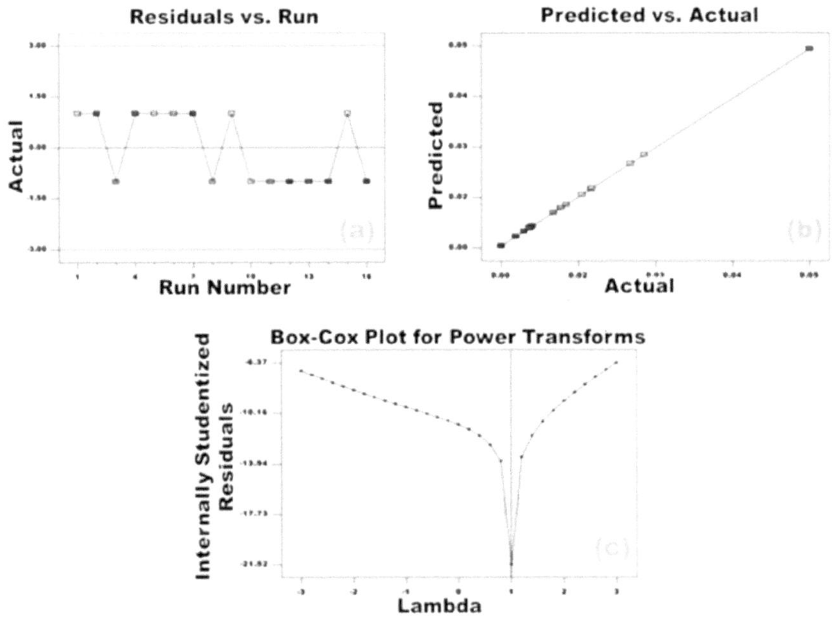

FIGURE 31.4 Plot of (a) Residuals vs. Run, (b) Predicted vs. Actual and (c) Power transformation for MRR.

31.4.3 SEM STUDY

Figure 31.5 shows the surface micrograph of machined K-92 using SiC abrasive particles of 109, 260525and 745 μm at 20°C, 50°C, 80°C, and 110°C of abrasive temperatures. Larger sized radial crater was detected in the micrographs due to larger size of abrasive particles. Deep penetration and microcrack were seen from K-92 machined surface for 745 μm grain size. In this SEM review, it was found that the rate of material removal for abrasive grain size of 745 μm was higher than of 109 μm as large chips supporting a growth in erosion rate. Hence, abrasive grain size and abrasive temperature also has an influence on MRR. The surface quality of FB-HAJMed surface is also explained in terms of crater (due to the existence of various types of interaction between hot abrasives, debris, and continuous heating effects), circular debris modules (due to the cohesion effect of SiC abrasive, compressed air effects, and SOD on the HAJMed surface), microvoids (due to the hot abrasive embedded on the workpiece), microcrack (attributed to crack), and various micropores due to several thermal shock resistance.

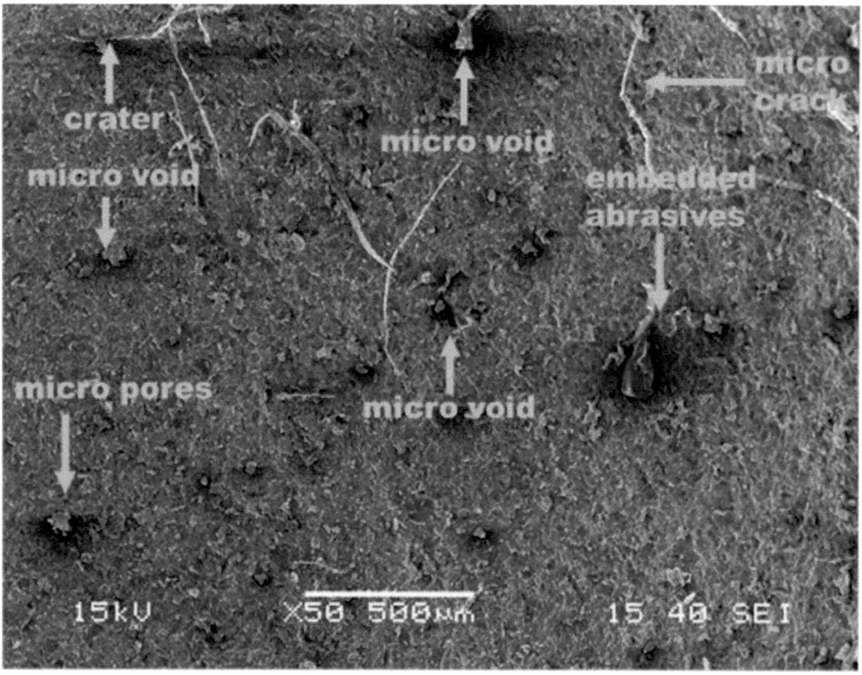

FIGURE 31.5 SEM image of FB-HAJMed K-92.

31.5 DESIRABILITY-BASED PARTICLE SWARM OPTIMIZATION

The major logic of PSO procedure defines the interaction between the group followers to determine the optimization for the group movement in which each subdivision notifies about its duration of time for completion a work.

Then each subdivision updates its path with the other particle having the minimum time period for completion the work and the target is completed in least time duration. A discrete desirability function $d_i(\hat{y}_i)$ for the response $\hat{y}_i(k)$ must be generated by the fitted models and create optimization philosophies. Desirability is continuous, where $d_i(\hat{y}_i) = 0$ for an undesirable answer is between 0 and 1, and $di(\hat{y}_i) = 1$ means a fully desirable value (i.e., an ideal response). Intermediate values of $d_i(\hat{y}_i)$ show less or more desirable results. In a satisfactory range of the result values given by $(U_i–L_i)$ depending upon the accepted optimization standards, several functions may be assembled, where the U_i is the highest acceptable result value and L_i is the lowest. Therefore, if the result is to be maximized, the following equation defines $d_i(\hat{y}_i)$.

$$d_i(\hat{y}_i(x)) = \begin{bmatrix} 0 & \text{if} \\ \left(\dfrac{\hat{y}_i(x) - L_i}{T_i - L_i}\right)^S & \text{if} \quad L_i \le \hat{y}_i(x) \le U_i \\ 1 & \text{if} \quad \hat{y}_i(x) > U_i \end{bmatrix} \tag{31.2}$$

Table 31.5 reveals the predicted and experimental values for MRR with machining factors. In the present sets of experiments, the multiobjective desirability-based particle swarm optimization (PSO) program is arranged and ran in the MATLAB interface. The initial population for individual machining constraints was set at 50. The PSO factors are as follows: the archive size is 100, the weight of inertia is 0.4, and the cognitive and social parameters (C_1 and C_2) are considered as 2. As shown below, the iteration is achieved as shown. The considered problem of PSO can be obtained as:

Minimize Y

Subjected to: $X_{min}. \le X \le X_{max}$

Where, Y is the objective function, Y = 1/1 + DF and X stands for the input parameter as $X = \{P_{Cair}, Z, T_{ap}, GS_{ap}\}$ within the following ranges:

$4 \le P \le 7$, $20 \le T_{ap} \le 110$, $4 \le Z \le 10$, and $109 \le GS_{ap} \le 745$.

31.6 COST ANALYSIS AND SUSTAINABILITY ASSESSMENT OF HOT ABRASIVE JET CUTTING

31.6.1 COST ANALYSIS

Cost awareness in the process of machining is an important undertaking for an efficient industrial system. Vital concepts are defined based on component precision, form-converting, and tooling processes to measure the cost of output for the machining system. In order to work in ideal economic conditions, the machining techniques need to be reviewed due to the large expenditures involved. In view of the best possible working conditions suggested by the PSO technology, the cost analysis in terms of overall machining costs per component is used for implementing a detailed direct and indirect cost estimate in HAJMing as shown in Table 31.6.

TABLE 31.5 PSO of HAJMing Factors and Outcome.

Type of model	Process parameters				Response values
	P_{Cair} (kgf/cm^2)	Z (mm)	T_{ap} (°C)	GS_{ap} (μm)	MRR (g/s)
Predicted	7	10	20	525	0.051712
Experimental	7	10	20	525	0.05123
Error %					1

TABLE 31.6 Cost Evaluation of HAJMing Cutting.

Specific basic variables	In hours	Costs	In Indian rupees
Traverse rate			
Investment cost ($C_{INV.}$)	Rupees/h	$C_{INV} = I/DL_a = 58$	58
Electrical energy cost ($C_{ELEC.}$)	Rupees/h	$C_{ELEC} = c_e \times E = 62/h$	62
Abrasive cost ($C_{ABR.}$)	Rupees/h	$C_{ABR} = c_a \times m_a = 40/h$	40
Wearing part cost(C_{WP})	Rupees/h	25/h	25
Maintenance cost (C_M)	Rupees/h	30/h	30
Labor cost (C_L)	Rupees/h	50/h	50
Cost per meter of cut	Rupees/m	$C_{overall} = 1/v\,(C_{INV} + C_{ELEC} + C_{ABR} + C_{WP} + C_M + C_L)$	265

31.6.2 SUSTAINABILITY ASSESSMENT

Sustainability assessment (SA) is a dynamic form of estimation. It is carried out in a broad environmental, economic, and social sense to support decision-making and policy, and transcends a solely technical/scientific assessment. Sustainability is often described as meeting current needs without sacrificing future generations' ability to cope with their own needs. It has three main pillars: economic, ecological, and social. These three pillars are referred to informally as persons, the earth, and income. By measuring the success of social, environmental, and economic values, sustainability is calculated. Although an optimal target is a fair treatment of all three, it is not always feasible. A sustainability indicator, as described earlier, is a measurable component of environmental, economic, or social systems that is useful for tracking improvements in system features related to the continuation of the well-being of humans and the environment. The worst-to-best rule is followed when assigning the score for each feature. This implies that most outcomes are allocated to “2”, while “1” is allocated to the lowest. Finally, by summarizing the results obtained from all sustainable factors (refer to Table 31.7), a description of the overall results for both AJM and HAJM was obtained and this system guaranteed the highest overall score, which was considered to be the best machining method after analysis. After calculating (i.e., 12), the net score obtained with the Kiviat diagram (i.e., 9, as shown in Figure 31.6) is slightly lower than that of HAJM. From this sustainability assessment point of view, it can be inferred that the HAJMing process is cost-effective as well as socially beneficial. In order to make such a decision, however, further study is also needed.

TABLE 31.7 Pugh Matrix for Evaluation of Sustainability Using Weightage Allocation of Parameters.

Sustainability evaluation constraints	Weightage	
	AJM	HAJM
Abrasive consumption	1	2
Energy consumption	2	1
Level of noise	1	2
Surface finish	1	2
Material crosion ratc	1	2
Operator safety	2	1
Cost of abrasive	1	2
Total cost	9	12

31.7 CONCLUSION

This study scientifically executed experimental analysis, optimization, and cost evaluation in FB-HAJMing setup, paying specific attention to the application of heating-cum-mixing chamber as well as hot abrasives. Based on HAJMing of K-92 ceramic, several conclusions are drawn which are as follows:

- The surface effect plot analysis of ANOVA reported that the abrasive temperature parameter only has an incredible effect for improving the removal rate (MRR) of HAJM.
- Because of the higher R^2 value (1.000), model probability *P*-value below 0.05 and higher AD-test *P*-value (0.0014), the predictive model proposed for MRR outcome using response surface methodology is efficient in terms of statistically acceptable, significant, and probabilistically validated.
- The multiple use of PSO for multitarget optimization achievable under optimal production conditions of the machining constraints were P_{Cair} = 7 kgf/cm², Z = 10 mm, T_{ap} = 20°C and GS_{ap} = 525 µm. The optimum response value for MRR is 0.051712 g/s.
- The overall total cost of the abrasive jet cutting for each component of only Rs. 265 in Indian rupees was estimated under optimal conditions (as suggested by PSO technique), which guarantees economic benefits due to reduced machining time and downtime.

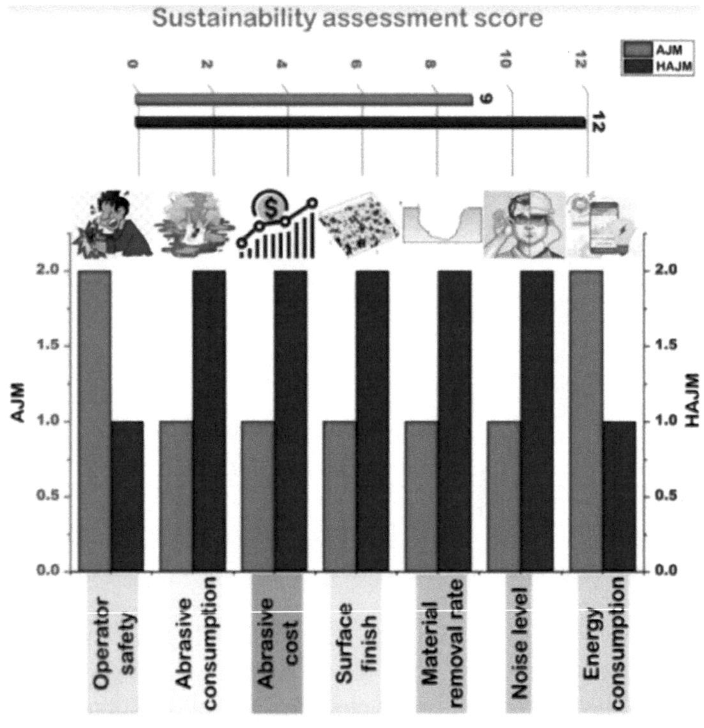

FIGURE 31.6 Kiviat diagram in between AJM and HAJM for comparative sustainability evaluation.

ACKNOWLEDGMENT

The authors would like to acknowledge the All India Council for Technical Education (AICTE) for their grant (Ref. No.: 8–32///RIFD///RPS–NDF/// Policy–1///2018–19) for their technical and financial support.

KEYWORDS

- **K-92 ceramic**
- **silicon carbide**
- **FB-HAJM**
- **material removal rate**
- **cost analysis**

REFERENCES

1. Jagannath, N.; Hiremath, S.; Sadashivappa, K. Analysis and Parametric Optimization of Abrasive Hot Air Jet Machining for Glass Using Taguchi Method and Utility Concept. *Int. J. Mech. Mater. Eng.* **2012,** *7* (1), 9–15.
2. Jagannatha, N.; Hiremath, S.; Sadashivappa, K.; Arun, K. V. Machining of Soda Lime Glass Using Abrasive Hot Air Jet. *Mach. Sci. Technol.* **2012,** *16*, 459–472.
3. Pradhan, S.; Behera R. K.; Nanda, B. K.; Das, S. R.; Dhupal, D. Drilling of K-60 Alumina Ceramic with Different Grades Abrasives at Various Temperatures using Fluidized BED-Hot Abrasive Jet Machining (FB-HAJM) Process. *IOP Conf. Series: Mater. Sci. Eng.* **2018,** *455*, 012076
4. Nanda, B. K.; Das, D.; Nayak R. K.; Mishra, S. B.; Routara, B. C.; Dhupal D. Fluidized Bed Hot Abrasive Jet Machining of Borosilicate Glass with Zircon Sand. *Mater. Today: Proc.* **2019,** *18*, 2993–3000.
5. Pradhan, S.; Tripathy, S. S.; Sahu, S.; Das, S. R.; Jena, P. C.; Dhupal, D. Investigation on MRR and DOC of the Micro-Holes Generated on Quartz Using Silicon Carbide by FB-HAJM. *Mater. Today: Proc.* **2020.** https://doi.org/10.1016/j.matpr.2020.02.437.
6. Tomy, A.; Hiremath, S. S. Machinability and Characterization of Machined Hole on Quartz Using Developed μ-AJM set-up. *Adv. Mater. Process. Technol.*, **2019**.
7. Nanda, B. K.; Dhupal, D.; Buda, D.; Das, S. R. Abrasive Jet Drilling of Alumina Ceramic with Pressurized-Fluidized Bed Set-Up. *Mater. Today: Proc.* **2018,** *5*, 12570–12578.
8. Nanda, B. K.; Dhupal, D. Effect of Machining Parameters in Drilling of Glass Fibre Reinforced Polymer Composite with Modified AJM Process. *Int. J. Manuf. Technol. Manage.*

9. Mogul, Y. I.; Nasir, I.; Myler, P. Investigation and Optimization for Depth of Cut and Surface Roughness for Control Depth Milling in Titanium Ti6AL4V with Abrasive Water Jet Cutting. *Mater. Today: Proc.* **2020,** 4–10.
10. El-Hofy, M.; Helmy, M. O.; Escobar-Palafox, G.; Kerrigan, K.; Scaife, R.; El-Hofy, H. Abrasive Water Jet Machining of Multidirectional CFRP Laminates. *Procedia CIRP* **2018,** *68* (April), 535–540.
11. Natarajan, Y.; Murugesan, P. K.; Mohan, M.; Khan, S. A. L. A. Abrasive Water Jet Machining Process: A State of Art of Review. *J. Manuf. Process.* **2020,** *49* (January 2019), 271–322.
12. Wang, Z.; Li, H. N.; Yu, T. B.; Chen, H.; Zhao, J. On the Predictive Modelling of Machined Surface Topography in Abrasive Air Jet Polishing of Quartz Glass. *Int. J. Mech. Sci.* **2019,** *152* (December 2018), 1–18.
13. Pradhan, S.; Dhupal, D.; Das, S. R.; Jena, P. C. Experimental Investigation and Optimization on Machined Surface of Si3N4 Ceramic Using Hot SiC Abrasive in HAJM. *Mater. Today Proc.* **2021.** https://doi.org/10.1016/j.matpr.2020.12.066.
14. Pradhan, S.; Das, S. R.; Nanda, B. K.; Jena, P. C.; Dhupal, D. Experimental Investigation on Machining of Hardstone Quartz with Modified AJM Using Hot Silicon Carbide Abrasives. *J. Braz. Soc. Mech. Sci. Eng.* **2020.** https://doi.org/10.1007/s40430-020-02644-4.
15. Pradhan, S.; Das, S. R.; Nanda, B. K.; Jena, P. C.; Dhupal, D. Machining of Hardstone Quartz with Modified AJM Process Using Hot SiC Abrasives: Analysis, Modelling, Optimization, and Cost Analysis. *Surf. Rev. Lett.* **2020.** https://doi.org/10.1142/S0218625X20500493.
16. Pradhan, S.; Das, S. R.; Dhupal, D. Performance Evaluation of Recently Developed New Process HAJM during Machining Hardstone Quartz Using Hot Silicon Carbide Abrasives: an Experimental Investigation and Sustainability Assessment. *Silicon* **2020.** https://doi.org/10.1007/s12633-020-00641-9.

CHAPTER 32

A REVIEW ON THE EFFECT OF PROCESS PARAMETERS IN ELECTROCHEMICAL MACHINING OF ADVANCED MATERIALS

MADHUSUDAN PAINULY, RAVI PRATAP SINGH*, and RAJEEV TREHAN

Department of Industrial and Production Engineering, Dr. B. R Ambedkar National Institute of Technology, Jalandhar, India

**Corresponding author. E-mail: singhrp@nitj.ac.in*

ABSTRACT

This study explores several types of research works carried out in the domain of electrochemical machining (ECM) of advanced engineering materials. It has been revealed that the SS304 poses a multidimensional application in various industries. Apart from its applications, its positive machining is still unanswered. Therefore, an attempt is made to critically review the numerous studies in this regard. A thorough literature review concludes that ECM and its allied methods can process this advanced material more effectively.

32.1 INTRODUCTION

Long ago in the nineteenth century, the electrochemical machining (ECM) theory (1791–1867) was discovered by Michael Faraday. In 1833, Faraday developed the electrolysis rules, which are the basis of both the better-known techniques of electrodeposition and electrodissolution. Russian researcher

Modern Manufacturing Systems: Trends and Developments. Rajiv Kumar Garg, Ravi Pratap Singh, Rajeev Trehan, & Ramesh Singh (Eds.)

W. Gussef first invented the process of anodic dissolution of metal and started metal machining anodically in 1929. The main advancement of ECM took place in the 1950s, when the shaping of high strength alloys was studied as a tool on ECM. In 1959, the commercial application of anodic metal machining was developed by Anocut Engg., Chicago, IL, USA.

Advancement of ECM took place day by day by the investigators, and then the typical task for the investigators was to machine hard to cut metals, alloys, and the machining of large components such as parts of aircraft, gas turbine, and aerospace industries. The operations requirement was finishing deburring, milling, and shaping.[1–4] Even though ECM was originally designed for the machining of hard to cut alloys and any metals, ECM was used in the application of aerospace, biomedical, and automotive firms, and which are progressively its user until now.

In the field of ECM, lots of research has been done on SS304. SS304 has wide features that make it valuable material for industrial applications. It has corrosion resistance properties, excellent weldability, high strength, and toughness in temperatures. Chemical composition of SS304 is manganese 2.00%, nickel 8.00–12.00%, silicon .75%, carbon 0.08%, phosphorus .045%, sulfur .030%, nitrogen .10%, chromium 18–20.00%, iron-balanced.

SS304 can be used in almost all industrial applications, but most commonly it can be found in heat exchangers, architectural paneling, kitchen equipment, springs, braces, marine, threaded fasteners, railing and trim, and food processing equipment[5,6] (Table 32.1).

TABLE 32.1 Mechanical and Electrical Properties of SS304.

Vickers hardness (HV)	129
Specific gravity (kg/m^3)	8000
Tensile strength (MPa)	515
Electrical resistivity (n Ω-m)	7220
Melting point (°C)	1450
Thermal conductivity (W/m–K)	16.20

32.2 WORKING PRINCIPLE

Electrochemical machining works on Faraday's law of electrolysis, Faraday's law states the amount of electrical charges transferred to the electrode is proportional to the mass altered by the ECM process. This is the inverse process of the electroplating process, in case of ECM altered material from the workpiece or anode flushed out before reaching to tool or cathode (Figure 32.1).

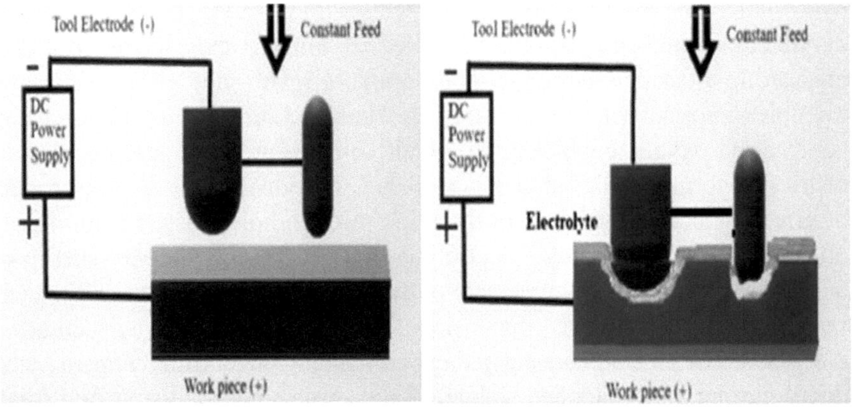

FIGURE 32.1 Electrochemical machining working.

32.3 LITERATURE REVIEW

Lilong et al.[7] worked on magnetic-assisted ECM and investigated the localization of ECM by applying the magnetic field in the machining zone, which is categorized by the drilling technique of ECM. For experiments, two models were designed based on the effect of magnetic field on the workpiece (anodic) dissolution. The first was built by the 304 SS ($FeCl_3$) system, where the magnetic field reduces the anodic (workpiece) dissolution and improves the ECM localization. It was found on the basis of experiments that due to the magnetic field, the stray dissolution reduced in this system of ECM drilling, which is the main cause of improved ECM localization. While the second was based on the Cu (NaCl) system and concluded that the stray dissolution promoted by the magnetic field and also increased the anodic dissolution by which has no beneficial influence on it.[8–9] They also found the elemental (in first case the main elements of 304 SS ($FeCl_3$) which are Fe, Cr, and Ni) behavior related to the magnetic field and highlighted a conclusion that in the case of only chromium (Cr), pore's inlet diameter reduced (Cr anode). The inlet pore diameter of Cr anode was reduced, which is the main cause of the reduction in the anodic dissolution and increase in ECM localization. In the second case of Cu | NaCl system (The main elements of Cu (NaCl) systems are Cu, O, and Cl) at magnetic field intensity of 450 mT, the content of O on Cu anode surface reduced by 4.95–1.95%, and Cu was reduced 0.91–0.21% which is not sufficient to improve the machining localization.

Pan et al.[10] suggested through-mask electrochemical micromachining (TMEMM) for the fabrication of microdimple arrays by using high-pressure hydrostatic electrolytes. On the surface of the workpiece electrical field distribution's uniformity should be increased. For that they suggested a high-pressure hydrostatic electrolyte. High-pressure hydrostatic electrolyte played multiple important roles in this research. High-pressure hydrostatic electrolyte compressed oxygen bubbles to stop their conversion into bigger bubbles and restrict the formation of bubbles as well as nonconductive unsolvable products of electrolytic absconding from the thick through-hole mask. Uniform accumulation of the bubbles and electrolytic products on the dimples' surfaces is needed to improve the homogeneity of the electrical field. The uniform electric field provides the improved dimensional precision and machining accuracy of the arrays. The process investigated experimentally under different parameters, direct current (DC), applied voltages, and during high-pressure hydrostatic TMEMM to know the impact on dimensional accuracy and machining accuracy.

Two unique phenomena were observed: at low voltages, the dimensions of the microdimple arrays were superior than at high voltages microdimple arrays, and no effect on machining accuracy at both high and low voltages; it is high in both cases but in the case of intermediate voltages low accuracy found. At 32 V DC, they successfully fabricated a 14000 micro-dimple's array with a depth and diameter of 9.79 and 105.95 μm, respectively, and with depth deviation of 0.21 and diameter deviations of 0.59 μm.

Han et al.[11] proposed a new theory which violates the old theory. It highlighted a new theory that increment in the frequency has a null effect on improving the machining accuracy of ECM. The new finding was that precision of ECM cannot be improved by the frequency of pulsed power. The experimental results show that the machining precision under the pulsed power supply is determined by the service cycle, the effective average voltage value, irrespective of the frequency of the pulse peak and the pulsed waveform, relative to the DC power supply. According to past researchers' beliefs if DC power supply was replaced with the pulsed power supply and higher frequency pulsed power was used, it will provide improved precision and accuracy directly. But according to Wei Chen et al.'s experimental results, extremely short power supply line form tool electrode to workpiece have produced certain capacitance and inductance effect in the circuit; thus, when the frequency of pulsed power supply is very high, the distortion of impulse waveform will be rather obvious.[12–14] This is the reason for massive reduction in the ECM phase of the effective average voltage value. By its very nature, due to the reduction of the effective average voltage value at the ultra-high frequency, the machining precision is increased.

Liu et al.[15] proposed a pulse electrochemical sawing micromachining for an annular grove in a 304 SS tube by using a $NaClO_3$ electrolyte (pH-neutral).

For enhancement in electrolytic product removal and distribution of electrolyte, they used a rotating dentate cathode. The motive after use of rotating dentate cathode was to minimize the effects of concentration polarization and enhance the removal of machining debris on an electrochemical reaction. The bubble layer structure was kept in contact with the surface of the rotating dentate cathode and photographed by using a high-speed camera for studying the effects of rotating dentate cathode on the electrochemical sawing micromachining process. To study pulse electrochemical sawing machining (PESM) process micro-inner annular groves on a 304 SS tube, they mainly considered four process parameters: a rotational speed of rotating dentate cathode, initial gap, a radius of the dentate cathode, and feed rate at different levels.

Experiments showed that PESM has a great potential for machining micro-inner annular grove on an SS tube and by using the rotating dentate cathode, effect of concentration polarization reduced in the electrochemical reaction. They fabricated a micro-inner annular grove of 263 μm width and 340 μm in-depth on a 304 SS tube.

Liu et al.[15] proposed a study on the Electrochemical micromachining of SS 304 with an optimized electrolyte. The objectives of the study were to improve the process stability and machining efficiency. They focused on the selection and optimization of the electrolytes. To generate soluble elements with metal ions, they utilized the complexing agents, and microelectrochemical experiment was performed according to the properties of the complexing agents. For the selection and optimization of electrolytes, they created a valuation criteria in which they included material removal rate (MRR), product deposition on the electrode surface, ratio of removal rate to current (MRC), and Inlet diameter. According to the optimization result, an electrolyte that contained 0.01 mol/L sodium citrate and 1.0 mol/L sodium nitrate was chosen as an optimum constituent, and with these constituent, microholes of diameter, 160~190 μm with 0.5 mm thickness was achieved on SS tube 304. With the optimized parameters MRR increased by 35%, and by using these optimized compositions of electrolyte, good machining localization was achieved and no deposition collected on the workpiece and tool electrode was found.

Wang et al.[16] proposed the machining of a small helical hole (diameter 1.5 mm with an aspect ratio >10) with a low-frequency vibration of the tool in ECM. They developed a 3D flow field model. They used electrode shape, according to the required shape and size with an electromagnetic coil for low-frequency vibrations. The objective of their work was to obtain increased uniformity and stability of internal ribbed holes along the flow direction. To analyze the flow property in the machining gap and find the tools (low-frequency vibration tool) effectiveness, they started the study with the development of physical models based on the flow field in the interelectrode gap. Second, for the validation of the model, they performed an experiment with the variation in acceleration, frequency, and vibration amplitude. They concluded according to the simulation result that, during the machining process due to the fluctuation in fluid velocity periodically, electrolyte flushing in the IEG was enhanced. According to the results, grove depth improved with 15% over the nonvibrating control tests and it also enhanced uniformity.

Hung et al.[17] explored a new technique for enhancing the efficiency and accuracy of the electrochemical jet machining process (ECJM). They tried to mix gas with electrolytes for spraying on the workpiece and analyzed the result with and without the mixture of gas in the electrolyte. The objective of the study was to improve the machining efficiency and analyze the influence of various parameters. They also investigated the influence of current on the hole diameters and nozzle feed rate. They used the SUS304 as a work material with the 25 × 25 (Length × Width) and 1 mm height and tool electrode material was composed of nickel and stainless steel. The results revealed that due to the mixed gas with electrolyte in the microscale blind hole, the conductive area decreased between the workpiece and electrode, by which the current density increased, and it increased the material removal rate. If the current density is reduced then the tendency of the stray material corrosion may occur, which affected the electrolyte flow field and is the cause of poor surface topography.

In the case of microgroove machining, due to the increment in feed rate, machining time (per unit area) shortened thereby reducing the machining depth, side width, and overcut. Experimental results concluded that machining efficiency could be improved by mixing the gas with the electrolyte and surface accuracy could be affected due to the blocking of excessive surrounding stray currents generation.

Chen et al.[18] tried to enhance machining localization and minimize the undercutting of microdimple with the help of conductive mass jet electrochemical machining. In their method, an insulated patterned mask was replaced with a conductive patterned mask and during the machining, it was directly covered on the workpiece. They worked on the concept of reducing the electric field intensity, which could reduce the electric field intensity at the edge of a microdimple. For stable columnar jet flow and regeneration of electrolyte in the machining area, they included a metallic nozzle with a 20mm ID that also works as a cathode in case of electrochemical jet machining. They also investigated the electrical field distribution in the machine area with a different mask, for which they built the mathematical model and it also helped in the prediction of the evolution process of the microdimple. According to the simulation results, due to the use of a conductive mask, the electrical field intensity reduced at the edge of microdimple and it also reduced the undercutting of a microdimple. When 55 μm deep microdimples were machined with both insulated mask and conductive mask, the microdimples' etch factor (EF) increased from 1.47 to 6.11. It happened when the conductive mask was used in place of an insulating mask, and machining localization was also improved.[19,20] Direct current was more suitable in the case of generating deep microdimple in conductive mask jet electromicromachining when compared with pulse current. In the case of conductive mask jet electrochemical machining, low undercutting and high machining localization were found they concluded.

Aixi et al.[27] proposed formation of a metal microhole with the help of laser machining and electromicromachining. They established a modeling of metal microhole ECM and modeling of laser machining by which they found the optimum process parameters for LM-ECM (Laser machining – ECM). Experimental results reveal the effects of process parameters and optimum data. In their experiments, they compared single machining results with combined LM-ECM results. According to these experimental results, the cleaners' surface with no molten slag and no recast layer was found in the case of the LM-ECM process when they compared it with laser machining. When they compared LM-ECM with only electrochemical machining, they found a 51.35% improved efficiency and a 30.43% improved accuracy.

Fang et al.[21] proposed a method of wire electrochemical micromachining (WECMM) with a rotary helical electrode. They wanted to achieve a high machining accuracy and faster electrode feed rate with decreasing machine gap by increasing the electrolyte refreshment. They used the simulation of an electric field for the rotary helical electrode and the localized current density. They compared the rotary helical electrode with a cylindrical electrode and found that the rotary helical electrode reduced the machining gap from 94 to 19 μm. They also investigated the effect of process parameters on the interelectrode gap width.

They revealed that by using the rotary helical electrode, a narrower slit could be fabricated, which is typical in the case of a cylindrical electrode of the same diameter. For obtaining a precision slit with clean-cut edges, the combination of low duty cycle, high-frequency pulsed voltage, low electrolyte concentration, and low amplitude is effective. They concluded that complex microstructures with good accuracy, fabricated by using optimal parameters.

Kawanaka et al.[22] proposed a study to find the solution of low surface finish around the jet impinging zone in electrolyte jet machining, which occurred due to low current density. They suggested three methods to obtain a mirror finish on the surface of arbitrary shapes. In the first method, they used a high jet translation speed which required an X-Y table with high accuracy. In the second method, they used a monopolar current with long interval and short period of time in which the machining rate was low, and in the third method, they used a bipolar pulse for the increment in the machining rate.

Finally, they concluded that a higher surface finish can be attained with a 100 mm/s translating speed. At a very low translation speed (1 mm/s), they achieved a mirror-like finish which was smoother than Rz 0.2 mm within a short duration (100 ms) monopolar pulse current and duty factor of lower than 10% was used with low material removal rate. Therefore, they suggested a bipolar current method with a duty factor higher than 50%.

Chen et al.[23] suggested a technique to minimize the undercutting in ECM of microdimple arrays. They suggested that if during the machining, oxygen bubbles formed on the workpiece surface, it might reduce the undercutting. For covering the oxygen bubbles, thick mask was used, which prevented the oxygen bubble from escaping. Growth of oxygen bubbles created a shield at the edges of microdimples and decrease the etching on it, which might reduce undercutting and show improvement in the machining accuracy.

They revealed that by generating oxygen bubbles on a 250-μm thick polydimethylsiloxane (PDMS) mask, only 3 μm undercutting were produced, and there was an increment in machining time.Voltage applied does not affect undercutting, but an advantage of little increment in depth was found. Finally, they concluded that for generating a microdimple array of different depths with low undercutting, direct and pulse currents can be used with controlling machining time irrespective of the applied voltage.

Fang et al.[28] proposed a method to improve the surface profile, material removal rate, and heat transfer in electrochemical machining by pulsating electrolyte flow, via a servo-valve in the electrolytic supply pipe. In this study, ion concentration, electrolytic temperature along the flow direction near the workpiece surface, and the variations of electrolytic velocity were used.

They developed a model of heat coupling which was based on multiphysics, transport of diluted species electric, and the feasibility of the proposed process verified by the experiment results. The results of the simulation showed that pulsating flow has a major effect on gas fraction distributions, the temperature along the flow path near the workpiece surface, and velocity. The experimental results verified the simulation results. In experiments with optimum pulsating parameters, the surface profile and material removal rate are enhanced (Table 32.2).

TABLE 32.2 Literature Review on Past Investigations.

Sr. no.	Author(s)	Tool material/ electrolyte	Process parameters considered	Machining characteristics investigated	Results/conclusions
1	Lilong et al. [7]	H62 brass/3Fe Cl_3 and NaCl	Voltage feed IEG Spindle speed	Effect of magnetic field on machining localization and anodic dissolution	ECM localization is not always increased by the magnetic field. The localization in ECM is increases when the magnetic field suppresses the anodic dissolution.
2	Pan et al. [29]	Through-mask/ $NaNO_3$	Applied DC voltage Electrolyte pressure, IEG	The machining accuracy and surface roughness	In high-pressure hydrostatic TMEMM, applied voltage should be higher than 24 V for higher surface accuracy.
3	Han et al. [11]	Copper pipe/ sodium nitrate, NaCl	Pulse Frequency 40, 60, 80, and 100 kHz Duty cycle 50% Average voltage 12 V	Machining accuracy	Due to the distortion in the high-frequency pulse transmitted from the power supply to the distance between the electrode and the work-piece in the processing area, high-frequency pulsed power will increase ECM accuracy.
4	Liu et al. [30]	stainless steel dentate cathode/ pH-neutral $NaClO_3$	Applied voltage (U) Pulsed frequency (F) Duty ratio (D) Feed rate (*vf*)	Machining accuracy and stability	Proper controlling of the rotational speed of rotating dentate cathode is crucial to balance machining efficiency and accuracy
5	Guodong et al. [24]	$NaNO_3$ and sodium citrate	Voltage pulse width interpulse width, inflow pressure	MRR, machining efficiency and stability	MRR of microholes increased by 35% with improved machining efficiency and stability
6	Wang et al. [31]	$NaNO_3$ electrolyte	Applied voltage, Machining time (min)	MRR, Machining Stability, Grove Depth	The uniformity of the groove depth improves and its depth increases by more than 15% with the increase of the vibration amplitude

TABLE 32.2 *(Continued)*

Sr. no.	Author(s)	Tool material/ electrolyte	Process parameters considered	Machining characteristics investigated	Results/conclusions
7	Hung et. al. [32]	$NaNo_3$	Feed rate, electrolyte with or without gas, machining time, Applied voltage	MRR, Machining surface topography	MRR increased with uniformity of the machining surface topography.
8	Chen et al. [18]	$NaNO_3$	Voltage, Pulse duty cycle, Frequency	Machining localization, undercutting, MRR	The appropriate type of current to produce deep micro dimple in conductive mask JEM was direct current and pulse current
9	Sun et al. [25]	$NaNO_3$	Applied voltage, Pump current, Laser pulse width,	Machining accuracy and efficiency	The accuracy and efficiency of LMECM are enhanced by 30.43 and 51.35%, respectively with no recast layer
10	Xiaolong et al. [26]	Electrolyte concentration 20 g/L $NaNO_3$	Applied voltage, Feedrate, Spindle speed	Machining efficiency, maximum machining depth	Low amplitude, duty cycle, low electrolyte conc. and high frequency pulsed voltage is effective in obtaining a precision slit with clean cut edges.
11	Kawanaka et al. [22]	Aqueous solution of $NaNO_3$ with 20 weight %	current density, duty factor and translating speed	surface roughness	To obtain mirror like surface of Rz (0.2 mm), duty factors should be five times greater than the monopolar pulse method at the translating speed of 1 mm/s.
12	Chen et al. [33]	$NaNO_3$ Electrolyte concentration 100 g/L	Applied voltage, Thickness of PDMS mask,	Undercutting, Machining accuracy	The undercutting of the microdimples was only approximately 3 μm with no change in undercutting
13	Fang et al. [34]	$NaNO_3$, 20 g/L	Pulse frequency Pulsating amplitude (MPa), Machining time	MRR and quality and profile of Surface.	MRR, quality and profile of surface, heat transfer capacity was improved.

32.4 CONCLUSION

The following conclusion can be drawn from this study:

1. Electrochemical machining has a wide capability to effectively and productively process the advanced engineering materials, that is, SS304, etc.
2. It is observed that the allied methods of the ECM process have been used to machine a variety of smart materials.
3. The applied voltage and electrolyte concentration have been considered as the most dominating process variables in ECM of SS304.

KEYWORDS

- **SS304**
- **ECM**
- **PMECM**
- **feed**
- **hybrid**
- **review**

REFERENCES

1. Singh, R. P.; Singhal, S. Rotary Ultrasonic Machining: A Review. *Mater. Manuf. Process.* **2016,** 31, 1795–1824.
2. Singh, R. P.; Singhal, S. Investigation of Machining Characteristics in Rotary Ultrasonic Machining of Alumina Ceramic. *Mater. Manuf. Process.* **2017a,** 32, 309–326.
3. Singh, R. P.; Singhal, S. Experimental Investigation of Machining Characteristics in Rotary Ultrasonic Machining of Quartz Ceramic. *J. Mater. Des. Appl.* **2018a,** *232,* 870–889.
4. Singh, R. P.; Singhal, S. Rotary Ultrasonic Machining of Alumina Ceramic: Experimental Study and Optimization of Machining Responses. *J. Eng. Res.* **2018c,** 6, 01–24.
5. Singh, R. P.; Singhal, S. Rotary Ultrasonic Machining of Macor Ceramic: An Experimental Investigation and Microstructure Analysis. *Mater. Manuf. Process.* **2017b,** *32*, 927–939.
6. Singh, R. P.; Singhal, S. An Experimental Study on Rotary Ultrasonic Machining of Macor Ceramic. *J. Eng. Manuf.* **2018b,** *232,* 1221–1234.
7. Lilong, B.; Cheng, P.; Yun, K.; Yin, P. Effect of Magnetic Field on the Electrochemical Machining Localization. *Int. J. Adv. Manuf. Technol.* **2019,** *102* (1–4), 949–956. doi: 10.1007/s00170-018-3185-1.

8. Singh, R. P.; Tyagi, M.; Kataria, R. Selection of the Optimum Hole Quality Conditions in Manufacturing Environment Using MCDM Approach: A Case Study. In *Operations Management and Systems Engineering*; Springer: Singapore, 2019a; pp 133–152.
9. Singh, R. P.; Kataria, R.; Singhal, S. Decision-Making in Real-Life Industrial Environment through Graph Theory Approach. In *Computer Architecture in Industrial, Biomechanical and Biomedical Engineering*; IntechOpen, 2019b. DOI: 10.5772/intechopen.82011
10. Pan, Y.; Hou, Z.; Qu, N. Improvement in Accuracy of Micro-Dimple Arrays Prepared by Micro-Electrochemical Machining with High-Pressure Hydrostatic Electrolyte. *Int. J. Adv. Manuf. Technol.* **2019,** *100* (5–8), 1767–1777. doi: 10.1007/s00170-018-2822-z.
11. Han, F.; Chen, W.; Ying, W.; Zhang, J. Effects of Polarization on Machining Accuracy in Pulse Electrochemical Machining. *Procedia CIRP* **2018,** *68*, 493–498. doi: 10.1016/j.procir.2017.12.080.
12. Singh, R.; Singh, R. P.; Tyagi, M.; Kataria, R. Investigation of Dimensional Deviation in Wire EDM of M42 HSS Using Cryogenically Treated Brass Wire. *Mater. Today: Proc.* **2019c**.
13. Singh, R.; Singh, R. P.; Trehan, R. State of the Art in Processing of Shape Memory Alloys with Electrical Discharge Machining: A Review. In *Proceedings of the Institution of Mechanical Engineers, Part B: Journal of Engineering Manufacture*, 2020. 09544054 20958771.
14. Tyagi, M.; Panchal, D.; Singh, R. P.; Sachdeva, A. Modeling and Analysis of Critical Success Factors for Implementing the IT-Based Supply-Chain Performance System. In *Operations Management and Systems Engineering*; Springer: Singapore, 2019; pp 51–67.
15. Liu, G. X.; Zhang, Y. J.; Jiang, S. Z.; Liu, J. W.; Gyimah, G. K.; Luo, H. P. Investigation of Pulse Electrochemical Sawing Machining of Micro-Inner Annular Groove on Metallic Tube. *Int. J. Mach. Tools Manuf.* **2016,** *102*, 22–34. doi: 10.1016/j.ijmachtools.2015.12.001.
16. Wang, M.; Zhang, Y.; Xu, X.; Chen, G.; Clare, A. T.; Ahmed, N. Effects of Tool Intermittent Vibration on Helical Internal Hole Processing in Electrochemical Machining. *Proc. Inst. Mech. Eng. Part C J. Mech. Eng. Sci.* **2019,** *233* (12), 4102–4111. doi: 10.1177/0954406218792591.
17. Hung, J. C.; Liu, J. H.; Fan, Z. W. Fabrication of Microscale Concave and Grooves through Mixed-Gas Electrochemical Jet Machining. *Precis. Eng.* **2019,** *55* (December 2017), 310–321. doi: 10.1016/j.precisioneng.2018.09.020.
18. Chen, X. L.; Dong, B. Y.; Zhang, C. Y.; Wu, M.; Guo, Z. N. Jet Electrochemical Machining of Micro Dimples with Conductive Mask. *J. Mater. Process. Technol.* **2018,** *257*, 101–111. doi: 10.1016/j.jmatprotec.2018.02.035.
19. Singh, R. P.; Kumar, J.; Kataria, R.; Singhal, S. Investigation of the Machinability of Commercially Pure Titanium in Ultrasonic Machining Using Graph Theory and Matrix Method. *J. Eng. Res.* **2015,** 3, 75–94.
20. Singh, R. P.; Kataria, R.; Kumar, J.; Verma, J. Multi-response Optimization of Machining Characteristics in Ultrasonic Machining of WC-Co Composite through Taguchi Method and Grey-fuzzy Logic. *AIMS Mater. Sci.* **2018,** 5, 75–92.
21. Fang, X.; Qu, N.; Zhang, Y.; Xu, Z.; Zhu, D. Effects of Pulsating Electrolyte Flow in Electrochemical Machining. *J. Mater. Process. Technol.* **2014,** *214* (1), 36–43. doi: 10.1016/j.jmatprotec.2013.07.012.

22. Kawanaka, T.; Kunieda, M. Mirror-like Finishing by Electrolyte Jet Machining. *CIRP Ann.—Manuf. Technol.* **2015,** *64* (1), 237–240. doi: 10.1016/j.cirp.2015.04.029.
23. Chen, X.; Qu, N.; Fang, X.; Zhu, D. Reduction of Undercutting in Electrochemical Micro-Machining of Micro-Dimple Arrays by Utilizing Oxygen Produced at the Anode. *Surf. Coatings Technol.* **2015,** *277*, 44–51. doi: 10.1016/j.surfcoat.2015.07.028.
24. Guodong, L.; Yong, L.; Quancun, K.; Hao, T. Selection and Optimization of Electrolyte for Micro Electrochemical Machining on Stainless Steel 304. *Procedia CIRP* **2016,** *42* (Isem Xviii), 412–417. doi: 10.1016/j.procir.2016.02.223.
25. Sun, A.; Chang, Y.; Liu, H. Metal Micro-Hole Formation without Recast Layer by Laser Machining and Electrochemical Machining. *Optik (Stuttg)* **2018,** *171* (June), 694–705. doi: 10.1016/j.ijleo.2018.06.099.
26. Xiaolong, F.; Pengfei, Z.; Yongbin, Z.; Ningsong, Q.; Di, Z. Enhancement of Performance of Wire Electrochemical Micromachining Using a Rotary Helical Electrode. *J. Mater. Process. Technol.* **2016,** *227*, 129–137. doi: 10.1016/j.jmatprotec.2015.08.014.

INDEX

G

N

P

R

S

T

U

V